Metal Organic Frameworks as Heterogeneous Catalysts

RSC Catalysis Series

Series Editor:
Professor James J Spivey, *Louisiana State University, Baton Rouge, USA*

Advisory Board:
Krijn P de Jong, *University of Utrecht, The Netherlands*, James A Dumesic, *University of Wisconsin-Madison, USA*, Chris Hardacre, *Queen's University Belfast, Northern Ireland*, Enrique Iglesia, *University of California at Berkeley, USA*, Zinfer Ismagilov, *Boreskov Institute of Catalysis, Novosibirsk, Russia*, Johannes Lercher, *TU München, Germany*, Umit Ozkan, *Ohio State University, USA*, Chunshan Song, *Penn State University, USA*

Titles in the Series:
 1: Carbons and Carbon Supported Catalysts in Hydroprocessing
 2: Chiral Sulfur Ligands: Asymmetric Catalysis
 3: Recent Developments in Asymmetric Organocatalysis
 4: Catalysis in the Refining of Fischer–Tropsch Syncrude
 5: Organocatalytic Enantioselective Conjugate Addition Reactions:
    A Powerful Tool for the Stereocontrolled Synthesis of Complex Molecules
 6: *N*-Heterocyclic Carbenes: From Laboratory Curiosities to Efficient
    Synthetic Tools
 7: *P*-Stereogenic Ligands in Enantioselective Catalysis
 8: Chemistry of the Morita–Baylis–Hillman Reaction
 9: Proton-Coupled Electron Transfer: A Carrefour of Chemical Reactivity
    Traditions
10: Asymmetric Domino Reactions
11: C–H and C–X Bond Functionalization: Transition Metal Mediation
12: Metal Organic Frameworks as Heterogeneous Catalysts

How to obtain future titles on publication:
A standing order plan is available for this series. A standing order will bring delivery of each new volume immediately on publication.

For further information please contact:
Book Sales Department, Royal Society of Chemistry, Thomas Graham House, Science Park, Milton Road, Cambridge, CB4 0WF, UK
Telephone: +44 (0)1223 420066, Fax: +44 (0)1223 420247
Email: booksales@rsc.org
Visit our website at www.rsc.org/books

Metal Organic Frameworks as Heterogeneous Catalysts

Edited by

Francesc X. Llabrés i Xamena
Consejo Superior de Investigaciones Científicas and Universidad Politécnica de Valencia, Spain
Email: fllabres@itq.upv.es

Jorge Gascon
Delft University of Technology, The Netherlands
Email: j.gascon@tudelft.nl

RSCPublishing

RSC Catalysis Series No. 12

ISBN: 978-1-84973-572-8
ISSN: 1757-6725

A catalogue record for this book is available from the British Library

Published by The Royal Society of Chemistry,
Thomas Graham House, Science Park, Milton Road,
Cambridge CB4 0WF, UK

Registered Charity Number 207890

For further information see our web site at www.rsc.org

Printed in the United Kingdom by CPI Group (UK) Ltd, Croydon, CR0 4YY

Preface

Catalysis *Quo Vadis?*

Catalysis is of all times. In ancient times mankind used fermentation for production of alcoholic beverages and conservation purposes. Catalysis was also at the origin of the agricultural revolution through the synthesis of ammonia. In the industrialization era the increasing demand for basic chemicals focused the catalyst development for large scale production of a single specific bulk chemical. In the past century these developments continued, and, under strong societal drive for sustainability, focused on improved and intensified production routes with higher feedstock and energy efficiency.

Further, catalytic attention increasingly focused on the production of fine chemicals, pharmaceutics and food additives. Here, however, relatively large fractions of undesired by-products are formed in the multistep production routes, while intermediate separation treatments can be pretty energy intensive.

Ideally these chemical production routes should take place in a single reaction environment with different immobilized specific catalytic centres acting in concert, without intermediate separation of product mixtures and removal of catalyst, evolving towards the functioning of a cell, the wet dream of many catalytic scientists.

Such an integration of reaction sequences requires combined efforts in catalysis and engineering. In the last decades a strong evolution towards structuring of catalytic systems can be observed, in heterogeneous catalysis (zeolites, nanocrystallites), homogeneous catalysis (well-defined local environment of the active centre) and engineering (reactor internals, structured catalyst bodies, microreactors), covering the whole relevant range characteristic time and length scales of processes.

Combining various reactions in one process unit requires the presence of different catalytic centres, either in different catalysts, or in one particle where

RSC Catalysis Series No. 12
Metal Organic Frameworks as Heterogeneous Catalysts
Edited by Francesc X. Llabrés i Xamena and Jorge Gascon
© The Royal Society of Chemistry 2013
Published by the Royal Society of Chemistry, www.rsc.org

sometimes close proximity is required, e.g. like in the industrial bifunctional hydroisomerization catalyst.

In the last decade new types of hybrid materials drew increasing attention of the scientific community, the metal organic frameworks, MOFs, or porous coordination polymers, PCPs. Like zeolites these are crystalline porous materials, but built up regularly from organic and inorganic building blocks, having a vast variability in composition, porosity and functionality, much larger than the classical inorganic porous materials.

Because of this huge variability, MOFs have been attributed a large application potential in various fields, including adsorption, separation, storage, sensing, optoelectronics, magnetism….and catalysis. The latter will be obvious to those skilled in the art. MOFs can be functionalized at the organic or inorganic linker, or catalytic units can be accommodated in their pore space. In principle each linker or node can be or transformed into an active site, resulting in combinations of high dispersion and high loading, while several different functionalities can be combined in one system.

There seems to be no limit to which system is incorporated in a MOF, inorganic centres, metallo-organic complexes or organo-catalytic centres, and even enzymes can be immobilized. In this sense MOFs hold promises as the link between homogeneous and heterogeneous catalysis, realising the wet dream of many catalytic scientists, provided that turnover numbers can be achieved that are large enough to be competitive.

MOF have been discovered by the catalysis community and rapid developments are taking place. This book is the first of its kind completely devoted to this topic. With contributions of major players in the field it is not just a literature review with the developments until 2012, but also the strategies behind these developments are discussed. As such it has a lasting didactic and reference value and is a must for both experts and novices.

Freek Kapteijn
Avelino Corma

Contents

RSC Catalysis Series No. 12
Metal Organic Frameworks as Heterogeneous Catalysts
Edited by Francesc X. Llabrés i Xamena and Jorge Gascon
© The Royal Society of Chemistry 2013
Published by the Royal Society of Chemistry, www.rsc.org

CHAPTER 1

Introduction

FRANCESC X. LLABRÉS I XAMENA[a] AND
JORGE GASCON*[b]

[a] Instituto de Tecnología Química UPV-CSIC, Universidad Politécnica de
Valencia, Consejo Superior de Investigaciones Científicas, Avda. de los
Naranjos, s/n, 46022 Valencia, Spain; [b] Catalysis Engineering, Technical
University of Delft, Julianalaan 136, 2628 BL Delft, The Netherlands
*Email: j.gascon@tudelf.nl

1.1 Introduction

The last few decades have witnessed the unprecedented explosion of a new
research field built around metal organic frameworks (MOFs). The first reports
on metal organic frameworks (MOFs) or, more widely speaking, on coordination polymers date from the late 1950s[1] and early 1960s,[2–6] although it was
not until the end of the last century when Robson and co-workers[7,8] followed
by Kitagawa et al.,[9,10] Yaghi and coworkers,[11] and Ferey et al.[12] rediscovered
and boosted the field. Metal organic frameworks are crystalline compounds
consisting of metal ions or clusters coordinated to often rigid organic molecules
to form one- two-, or three-dimensional pore structures. The combination of
organic and inorganic building blocks into highly ordered, crystalline structures
offers an almost infinite number of combinations, enormous flexibility in pore
size, shape and structure, and plenty opportunities for functionalization,
grafting and encapsulation. These materials hold very high adsorption
capacities, specific surface areas and pore volumes. Their porosity is much
higher than that of their inorganic counterpart zeolites (up to 90%). In contrast
to other nano-structured materials, many MOFs display a remarkable flexibility and respond to the presence of guests and external stimuli. Their

RSC Catalysis Series No. 12
Metal Organic Frameworks as Heterogeneous Catalysts
Edited by Francesc X. Llabrés i Xamena and Jorge Gascon
© The Royal Society of Chemistry 2013
Published by the Royal Society of Chemistry, www.rsc.org

thermostability is sometimes unexpectedly high, reaching temperatures above 400 °C and their chemo-stability is acceptable in many cases.

Indeed MOFs are fascinating porous solids. The assembly of organic and inorganic struts allows, in theory, the facile tuning of properties, either by the chemical functionalization of the organic building units or by selection of the inorganic constituents. Even within such a relatively short time span, the field has rapidly evolved from an early stage, in which the main scope was the discovery of new structures, to a more mature stage in which dozens of applications are currently being explored. High adsorption capacities and easy tunability have crystallized in perspective applications in gas storage, separation and molecular sensing.[13–17] The possibility of synthesizing bio-compatible scaffolds infers a very promising future for medical applications.[18,19] Magnetic, semi-conductor and proton conducting MOFs will certainly find their way towards advanced applications in several research fields.[20] The easy compatibilization of MOFs with either organic or inorganic materials opens the door to advanced composites with applications varying from (opto)electronic devices to food packaging materials and membrane separation.[21,22] Last but not least, their tunable adsorption properties and pore size and topology, along with their intrinsic hybrid nature, all point at MOFs as very promising heterogeneous catalysts,[23,24] the topic of this book (see Figure 1.1 for a general picture).

According to the classical definition, a catalyst is a substance that increases the rate of a reaction towards equilibrium without being appreciably consumed. The word "catalysis" stems from Greek: "κατα" means "down" and "λνσισ" means "loosening". The eastern approach to catalysis is different. The Chinese characters for catalyst refer to a marriage broker, emphasizing the fact that a catalyst brings together two different "species" resulting in a mechanism of production.[25] By using a satisfactory catalyst the desired reactions proceed with a higher rate and selectivity at relatively mild conditions.

It is convenient to distinguish between heterogeneous and homogeneous catalysis. In the former case the catalyst and reactants are present in different phases, whereas in the latter case we are dealing with a single-phase system, usually a solution. Strictly speaking heterogeneous catalysis is not limited to solid catalysts. For instance, a system consisting of a liquid phase catalyst dispersed in a continuous liquid phase is heterogeneous. However, in practice only in the case of solid catalysts the term "heterogeneous catalysis" is used. How important is catalysis in practice? In the production of bulk chemicals catalysis is visibly present in nearly all plants. In the same lines, the role of catalysis is crucial in environmental protection, especially in emission control. In contrast, in the production of fine chemicals and pharmaceuticals, catalysis is developing at a slower pace, mostly due to the lack of efficient catalysts and to the high added value of the products.

With the discovery and explosion of MOFs, it was only a matter of time until the first catalytic applications were explored.[23] First reports mostly consisted of demonstrating that a certain MOF contained the necessary catalytic centers to catalyze a given reaction. In many cases, the performance of the material was poor and many concerns existed regarding the stability of the materials under

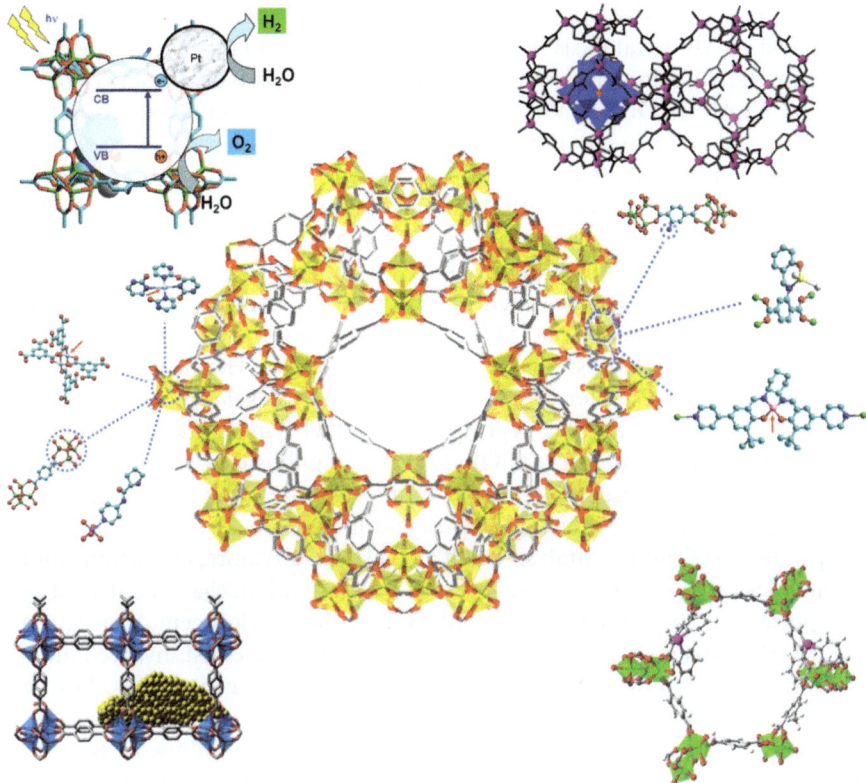

Figure 1.1 Metal organic frameworks offer a great number of possibilities for catalysis engineers: from semiconductor based photo-catalysis, to the encapsulation of different moieities and nanoparticles and from single site metal catalysis to the fine tuning of the organic moieties, either following pre or post-synthetic modifications.

reaction conditions. The current challenge is to develop truly efficient and selective catalytic processes using MOFs, ideally exploiting the versatility of these materials. In this sense, catalysis by MOFs is at this moment a *hot topic* in research, with new catalytic applications being continuously described, including new materials and new reactions. Indeed, it would not be overly controversial to state that we have already passed from poor "proof-of-concept" solids to highly active catalysts, in some cases with performances comparable to (or even surpassing) state-of-the-art catalysts.

Because the field is reaching now a stage of maturity, we strongly believe that this is the perfect timing to publish, to the best of our knowledge, the first book fully devoted to MOF catalysis. We would like to stress that this book does not intend to be just a literature review of the main advances in MOF catalysis until 2012 but a lasting reference book with a didactic spirit, where results and synthetic strategies are thoroughly discussed rather than simply highlighted.

The book contains outstanding contributions from some of the main players in the field of MOF catalysis. In its second part, after Llabrés i Xamena *et al.* introduce the different strategies for the inclusion of catalytically active sites in MOFs (Chapter 7), Dirk de Vos and colleagues explain in detail the possibilities of MOF metal nodes as catalytic sites along with synthetic strategies to enhance activity and selectivity (Chapter 8). In Chapter 9 Joseph T. Hupp and co-workers challenge the reader with the almost infinite possibilities of catalysis at the organic linker. Gascon and co-workers explore in Chapter 10 the utilization of the MOF porosity to host slightly bigger catalytic species. Finally, Wenbin Lin *et al.*, in Chapter 11, thoroughly investigate the limits of MOFs in asymmetric catalysis, probably one of the most promising catalytic applications together with photocatalysis, as rationally explained by Hermenegildo García and Belén Ferrer in Chapter 12. Since this is a rapidly developing research field, already outstanding catalytic reports on "brother" materials, the so-called Covalent Organic Frameworks (COFs) have been published during the last few years. In Chapter 13, Regina Palkovits, pioneer in the catalytic application of these materials, discusses the advantages and limitations of COFs.

As in the 21st Century catalysis is not a black-box anymore, characterization, rational design by synthesis, adsorptive properties and mechanistic insight are as important as the catalytic cycle itself. Indeed the development of structure–activity relationships in catalysis is the dream of every scientist involved in this field and the key towards rational design of new catalyst generations. For this reason, in the first part of the book, MOF synthesis and post-synthesis strategies are thoroughly discussed by Norbert Stock and Andrew D. Barrows, in Chapters 2 and 3 respectively. Carlo Lamberti, Silvia Bordiga and co-workers teach the reader on the most advanced spectroscopic and diffraction techniques for the characterization and structural determination of MOFs in Chapters 4 and 5. Last but not least, Evgeny Pidko and Emiel J. M. Hensen gather computation chemistry and MOF catalysis in Chapter 6.

The book finishes with a last Chapter where we not only speculate about future directions but also emphasize some of the main barriers that MOFs need to overcome to finally reach industrial catalytic applications.

We want to warmly acknowledge all the authors for their excellent contributions and the editorial team at RSC for their efforts on behalf of this book. We hope that this book will be valuable to the catalysis community both in industry and academia and especially to undergraduate students. We can only wish the reader as much joy as we had when editing this book.

References

1. Y. Kinoshita, I. Matsubara, T. Higuchi and Y. Saito, *Bull. Chem. Soc. Jpn.*, 1959, **32**, 1221–1226.
2. A. A. Berlin and N. G. Matveeva, *Russ. Chem. Rev.*, 1960, **29**, 119–128.
3. B. P. Block, E. S. Roth, C. W. Schaumann, J. Simkin and S. H. Rose, *J. Am. Chem. Soc.*, 1962, **84**, 3200.
4. F. W. Knobloch and W. H. Rauscher, *J. Polym. Sci.*, 1959, **38**, 261–262.

5. M. Kubo, M. Kishita and Y. Kuroda, *J. Polym. Sci.*, 1960, **48**, 467–471.
6. E. A. Tomic, *J. Appl. Polym. Sci.*, 1965, **9**, 3745.
7. S. R. Batten, B. F. Hoskins and R. Robson, *J. Am. Chem. Soc.*, 1995, **117**, 5385–5386.
8. B. F. Hoskins and R. Robson, *J. Am. Chem. Soc.*, 1990, **112**, 1546–1554.
9. S. Kitagawa, S. Kawata, Y. Nozaka and M. Munakata, *J. Chem. Soc., Dalton Trans.*, 1993, 1399–1404.
10. S. Kitagawa, S. Matsuyama, M. Munakata and T. Emori, *J. Chem. Soc., Dalton Trans.*, 1991, 2869–2874.
11. O. M. Yaghi and H. L. Li, *J. Am. Chem. Soc.*, 1995, **117**, 10401–10402.
12. D. Riou and G. Ferey, *J. Mater. Chem.*, 1998, **8**, 2733–2735.
13. G. Ferey, C. Serre, T. Devic, G. Maurin, H. Jobic, P. L. Llewellyn, G. De Weireld, A. Vimont, M. Daturi and J.-S. Chang, *Chem. Soc. Rev.*, 2011, 550–562.
14. L. E. Kreno, K. Leong, O. K. Farha, M. Allendorf, R. P. Van Duyne and J. T. Hupp, *Chem. Rev.*, 2012, **112**, 1105–1125.
15. J. R. Li, R. J. Kuppler and H. C. Zhou, *Chem. Soc. Rev.*, 2009, **38**, 1477–1504.
16. L. J. Murray, M. Dinca and J. R. Long, *Chem. Soc. Rev.*, 2009, **38**, 1294–1314.
17. K. Sumida, D. L. Rogow, J. A. Mason, T. M. McDonald, E. D. Bloch, Z. R. Herm, T.-H. Bae and J. R. Long, *Chem. Rev.*, 2012, **112**, 724–781.
18. P. Horcajada, T. Chalati, C. Serre, B. Gillet, C. Sebrie, T. Baati, J. F. Eubank, D. Heurtaux, P. Clayette, C. Kreuz, J. S. Chang, Y. K. Hwang, V. Marsaud, P. N. Bories, L. Cynober, S. Gil, G. Ferey, P. Couvreur and R. Gref, *Nature Mater.*, 2010, **9**, 172–178.
19. P. Horcajada, R. Gref, T. Baati, P. K. Allan, G. Maurin, P. Couvreur, G. Ferey, R. E. Morris and C. Serre, *Chem. Rev.*, 2012, **112**, 1232–1268.
20. M. Kurmoo, *Chem. Soc. Rev.*, 2009, **38**, 1353–1379.
21. J. Gascon, F. Kapteijn, B. Zornoza, V. Sebastián, C. Casado and J. Coronas, *Chem. Mater.*, 2012, **24**, 2829–2844.
22. B. Zornoza, C. Tellez, J. Coronas, J. Gascon and F. Kapteijn, *Microporous Mesoporous Mater.*, 2013, **166**, 67–78.
23. A. Corma, H. Garcia and F. X. Llabrés i Xamena, *Chem. Rev.*, 2010, **110**, 4606–4655.
24. D. Farrusseng, S. Aguado and C. Pinel, *Angew. Chem., Int. Ed.*, 2009, **48**, 7502–7513.
25. J. A. Moulijn, M. Makkee and A. van Diepen, Chemical Process Technology, John Wiley & Sons, Chichester, England, 2001.

Part A
Synthesis and Characterization of MOFs

CHAPTER 2

Synthesis of MOFs

NORBERT STOCK, HELGE REINSCH AND
LARS-HENDRIK SCHILLING

Christian-Albrechts-Universität, Max-Eyth-Straße 2, D-24118 Kiel,
Germany
*Email: stock@ac.uni-kiel.de

2.1 Introduction

Metal organic frameworks are a highly diverse class of compounds, which are based on the assembly of defined organic and inorganic building units. One reason for the tremendous interest may lie in the structural beauty and variability of the framework compounds, which attracts chemists with a background in solid-state chemistry as well as in coordination chemistry. The potential applications of MOFs, related to their porous nature, also arouse interest among engineers and material scientists. This overlap in scientific background can be considered as an explanation for the manifold developments in this specific field of research. However, no matter which specific property of a MOF is of interest, the synthesis of the respective compound is always the beginning of the experiment. Various methods and approaches towards the understanding of MOF formation have been reported[1] which cover a large variety of analytical methods and chemical parameters as well as reaction conditions. Nevertheless, up to now the designed synthesis of a new material must be considered as nearly impossible, especially since the variety of possible inorganic building units and topologies prohibits the prediction of the structure of a reaction product.

Understanding the principles of crystallization may give scientists a route towards reaction conditions and chemical parameters, which allow for the synthesis of new MOFs (section 2.2). The methods and common strategies for

RSC Catalysis Series No. 12
Metal Organic Frameworks as Heterogeneous Catalysts
Edited by Francesc X. Llabrés i Xamena and Jorge Gascon
© The Royal Society of Chemistry 2013
Published by the Royal Society of Chemistry, www.rsc.org

the discovery of new MOFs and their synthesis optimization are summarized in section 2.3. The concepts described can also give an insight into the complexity of the chemical systems and the diversity of possible structures that can be obtained. Attempts to achieve control over the morphology of the crystals and strategies for creating hierarchically porous materials are especially of interest for the application of MOFs in catalysis (section 2.4). Both aspects are important for the accessibility of catalytically active sites. Proper purification and complete activation is essential for the use of MOFs in catalysis (section 2.5), as the existence of impurities makes an understanding of the catalytic performance at least difficult, incomplete activation leads to a lower degree of porosity and thus to materials with inferior properties.

2.2   Mechanisms and Methods of Crystallisation

Generally, MOFs are crystallised from solution. Water and especially organic solvents have been shown to lead to highly porous materials in which the pores are filled with guest molecules such as solvent, structure directing agent or unreacted linker molecules. From an energetic point of view this is astounding, since dense structures are thermodynamically favoured, *i.e.* "Nature abhors open space in solid-state materials".[2] Thus, the incorporation of guest molecules and especially the kinetics of the formation of the inorganic building units play a crucial role in the formation of MOF structures.

In general crystallisation can be regarded as an equilibrium reaction between the dissolved precursors and the solid compound (*i.e.* the MOF). The thermodynamics of this reaction at constant pressure is described with the Gibbs–Helmholtz equation (eqn (1)).

$$\Delta G = \Delta H - T\Delta S \tag{1}$$

Due to the smaller number of microstates the entropy of a solid body is far lower than the entropy of a liquid or solution. It directly follows that higher temperature will cause the equilibrium to shift towards the dissolved compound, as depicted in Figure 2.1 (centre, left). Also, an increase in concentration will lead to precipitation, since the solubility of the reactants is finite. Thus, cryst-allisation can be induced by influencing the concentration and temperature of the solution. A simple example is recrystallisation for the purification of a substance, in which an increase in temperature and/or amount of solvent leads to dissolution of the substance and the recrystallisation can be caused by evaporating the solvent and/or decreasing the temperature.

The formation of a crystal can be seen as a two-step process in which nucleation is followed by crystal growth. Nucleation is the assembly of ions or molecules to form a cluster. Below a certain size the cluster is not stable and re-dissolves. Once the cluster attains a minimum size, the so called critical size r_c, which is in the nm-range, it is thermodynamically stable and is called a nucleus. The Gibbs free energy of crystallisation (ΔG_N) is composed of two terms, the surface term (ΔG_S) and the volume term, which scale with r^2 and r^3,

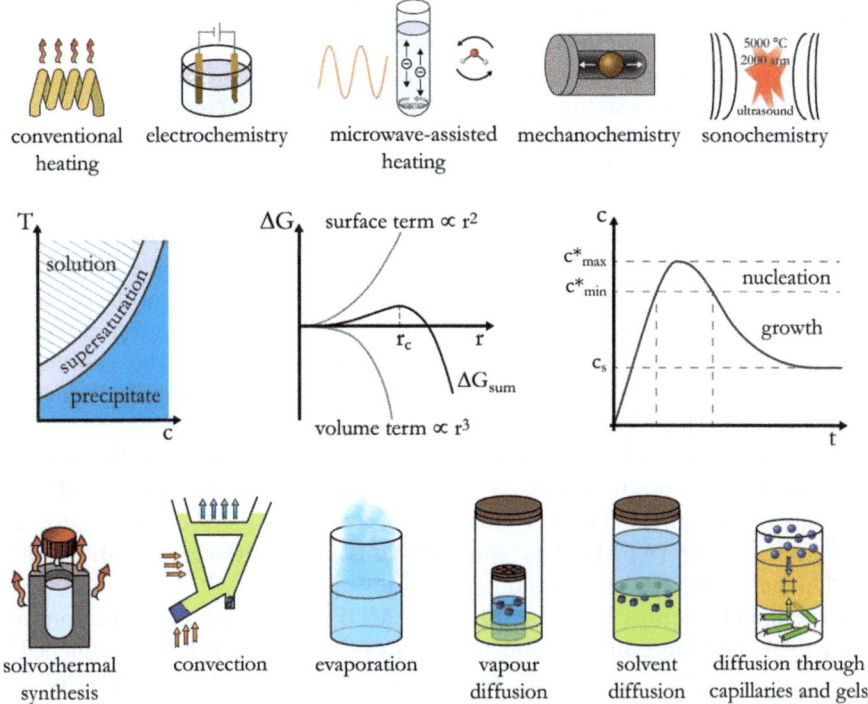

Figure 2.1 Aspects of crystallisation in synthesis of solid compounds.

respectively (Figure 2.1, centre). For a spherical body the following equation applies (eqn (2)):

$$\Delta G_N = \tfrac{4}{3}\pi r^3 \Delta G_V + 4\pi r^2 \gamma \tag{2}$$

Since ΔG_V is a negative and the surface energy a positive term, the change in Gibbs free energy is positive up to the critical size r_c. Once this point has been surpassed, ΔG_N rapidly decreases and the growth of the crystal becomes an exergonic process ($\Delta G_N < 0$).

The crystallisation process depends not only on thermodynamic but also on kinetic factors. The time-dependent growth of a crystal from a solution can be described by the La Mer-diagram (Figure 2.1 centre, right). At $t = 0$ the reactants are combined. The reaction leads to the formation of precursors, *i.e.* inorganic building units and/or deprotonated organic linker molecules, and the concentration c increases. The concentration surpasses the thermodynamical solubility c_s (described by the solubility product) and a supersaturated solution is formed. In this concentration regime heterogeneous nucleation, *i.e.* nucleation on surfaces (*e.g.* on a glass wall, an impurity, a seed crystal, bubbles *etc.*) can occur. Homogeneous nucleation takes place without preferential nucleation sites above the critical nucleation concentration c^*_{min}. After the period of nucleation, these seeds grow to form larger crystals until the concentration is lowered to c_S. In addition, Ostwald ripening can occur by which larger crystals are formed at the expense of smaller ones.

The crystallisation process can be investigated by *in situ* characterisation methods (*i.e.* in the reactor while the reaction occurs).[3] Nucleation is much harder to probe, since particles are small and their atomic arrangement cannot be easily determined. The mere size of particles can for example be measured by dynamic or static light scattering (DLS and SLS). *In situ* energy dispersive X-ray diffraction (EDXRD) allows the phase identification during the crystallisation and crystal growth can be monitored. In some cases crystalline intermediate products can also be detected and mathematical analysis of these data allows for the calculation of rate constants and the activation energies. Based on theoretical models, reaction mechanisms can also be proposed. Various studies on the formation of MOFs have been recently reported, such as CAU-1-NH$_2$, HKUST-1, Al-MIL-101-NH$_2$.

An imaging method was used to study layer by layer the crystallisation of HKUST-1. The product was synthesised at room temperature and the subsequent growth of the {111} facet was measured *in situ* using an atomic force microscope (AFM). The study demonstrated surface nucleation and attachment of 1.5 nm large building units during the growth.

Experimental setups for the synthesis of MOFs are shown at the bottom of Figure 2.1. The concentration and solubility of the precursors can be influenced by variation of the temperature, evaporation of the solvent or addition of an anti-solvent. Historically two different approaches to the field of MOFs emerged somewhat independently from one another from two different disciplines: while coordination chemists, specialising in coordination polymers, applied mild conditions to assemble porous compounds, chemists with a background in zeolites and zeolite-related materials started to look at the use of organic molecules not only as structure-directing agents but also as reactants to be incorporated as a part of the framework structure. The latter employed solvothermal synthesis for the formation of MOFs, a method in which the reactants are sealed in an autoclave and heated to a temperature above the boiling point of the solvent at standard pressure. While conventionally electric heating has been used, recently other ways of energy input have been employed.

Microwave irradiation can be used to minimise reaction times. In contrast to conventional heating where the heat is conducted through the walls of the vessel, the interaction of the reaction mixture with microwave radiation rapidly heats up the entire reaction volume. Often an increase in the nucleation rate is observed that leads to smaller product particles. In sonochemistry, ultrasonic oscillation causes the solvent to move leading to areas of rarefaction and compression on a microscopic scale. Bubbles are formed, which collapse under their own instability causing extremely high temperatures and pressures (up to 5000 °C and 2000 atm). This so called cavitation leads to the introduction of highly localised mechanical or thermal energy, which can be used to drastically accelerate nucleation and crystallisation in some cases by increasing the dissolution rate of hard to dissolve reactants. In mechanochemistry, mechanical energy is introduced using a ball mill or pestle and mortar. This method has only sparsely been utilised in the synthesis of MOFs, but has some advantages: it is a simple process, no or only small amounts of solvents are necessary, and the amount of waste is reduced. Very selectively, electrochemistry has been used for the synthesis of Cu- and Zn-based MOFs. The metal ions are formed *in situ* by redox processes on electrodes. Since the precursor solution is formed in the immediate vicinity of the electrodes even thin films of the MOF can be grown on the electrode.

In contrast to solvothermal reactions, reactions under mild conditions can be carried out in less elaborate reactor systems, such as simple beakers or specially constructed glass vessels (Figure 2.1, bottom). Concentration gradients can be induced by temperature differences along the reactor, which leads to convection and crystallisation at a colder spot of the reactor. Other means to form concentration gradients are the use of gels or capillaries, through which the reactants have to diffuse from opposite directions, the slow evaporation of the solvent or the use of an anti-solvent. The latter can be introduced through the gas phase *via* vapour diffusion into a clear solution or by liquid/liquid diffusion by carefully layering a low density anti-solvent on top of the reaction solution.

2.3 Strategies for the Synthesis of MOFs

In principle, the discovery of a new MOF structure is purely explorative. With a given organic linker molecule, the large number of inorganic building units, in combination with possible topologies, makes prediction of the structure of a new reaction product almost impossible, even if both building units are known. An excellent example is the structure of Cr–MIL-100 (MIL = Materiaux d'Institute Lavoisier).[4] Based on two well defined building units, namely a trimeric chromium cluster and the trimesate moiety (1,3,5-benzene tricar-boxylate) which were expected to form a large tetrahedral subunit, at least three different zeotypic topologies were determined to be energetically favorable, but only one of them was observed. Nevertheless, there exist several strategies and methods that are helpful tools for the discovery of new or isoreticular compounds and for the optimization of synthesis procedures. These methods are summarized on the following pages.

2.3.1 High-throughput Methods

The conventional method for the synthesis of a new MOF is very often based on the use of Teflon-lined steel autoclaves. Reactor volumes larger than 20 mL are commonly used and solvothermal synthesis conditions are often chosen. For these reactions and reactions at lower temperatures, single batch reactors are usually employed. Thus, only few reactions are conducted simultaneously, which limits the amount of extractable information and impedes the discovery of new MOFs and their optimal synthesis conditions.

High-throughput-methods (HT methods) are based on the miniaturization, parallelization and automation of various steps in the workflow.[5] In most cases, the reactors are miniaturized to a volume of 2 mL or even less, which reduces the amount and costs of the reactants, and allows the use of chemicals which are available only in small amounts. The latter point is of particular importance for MOF discovery where large linker molecules are tested, which can only be obtained from multi-step syntheses in small quantities. The parallelized synthesis in so-called multiclaves or well-plates allows the simultaneous reaction of 24 to 96 reaction mixtures, in which the molar ratios and concentrations of the reactants can be varied systematically. Automation of the dosing of the reactants and especially the characterization – for example by automated X-ray powder diffraction – accelerates the discovery of crystalline products and eases the recognition of reaction trends. The combination of these approaches leads to a powerful tool which speeds up the discovery of new compounds and allows the rapid optimization of the composition of the reaction mixture.[6] However, it must be mentioned that especially the size of the reactors can be a disadvantage, since only small amounts of MOF are synthesized (Table 2.1).

An impressive example for the discovery of new compounds using HT methods was given for the synthesis of zeolitic imidazolate frameworks (ZIFs) under subsolvothermal reaction conditions.[7] These MOFs are built up by the connection of tetrahedrally coordinated metal ions by imidazolate ions. In this study 9600 reactions were carried out in 96-well-plates at temperatures between 65 °C and 100 °C. Dimethylformamide (DMF) and diethylformamide (DEF) were employed as solvents, zinc or cobalt nitrate was used as inorganic reactant and nine different imidazole derivatives were used as organic building units. The coordination geometry of the building units did not change throughout this study and therefore the main influence on the product formation was the functionalisation of the imidazole molecules. This rather extensive study led to the discovery of 16 new compounds, of which five exhibited unprecedented zeotypic topologies. However, this article showed also a drawback of HT-methods: sometimes reaction procedures cannot be scaled up, with only 7 out of 16 new ZIFs obtained in amounts that allowed bulk characterisation.

A similarly comprehensive example for the synthesis of carboxylate based MOFs was given for the system Fe^{3+}/aminoterephthalic acid/solvent/base.[8] Instead of varying the funtionalisation of the organic building units, the influence of the solvent, the metal source, pH-value, concentration and the

Table 2.1 Advantages and Disadvantages of the discussed HT-methods.

Pros	*Cons*
Systematic screening for discovery	High investment for equipment
Rapid synthesis optimization	Small amounts of products
Low consumption of chemicals	Up-scaling sometimes problematic
High reproducibility	Risk of generating large amounts of meaningless data
Identification of reaction trends	
Time-saving	

molar ratios on product formation was thoroughly investigated. Furthermore, the reactions were carried out at different temperatures. It was possible to synthesize NH_2-functionalized Fe-MOFs exhibiting the well known MIL-53, MIL-88 and MIL-101 topology, respectively. This study also revealed that the solvent has the most profound influence on the topology of the obtained products and that the synthesis conditions for topologically identical framework compounds can differ remarkably. Besides the solvent, the reaction temperature and the concentration of the reactants play major roles in the formation of these MOFs.

The limited amount of data on Ni-based MOFs inspired another HT-investigation.[9] In this study, the parameters solvent (*N*,*N*-dimethyl-formamide) and the metal ion (Ni^{2+}) were kept constant. The influence on the metal source, the temperature, the reaction time and the addition of different bases was investigated, employing aromatic polycarboxylic acids varying in size and geometry. The use of Ni^{2+} and aromatic tricarboxylic acids resulted in the formation of the Ni-based analogues of HKUST-1 and MOF-14, which are based on the well know paddle-wheel units as the inorganic building block. Changing to a linear linker geometry by using terephthalic acid, and increasing the complexity of the solvothermal system by the addition of a diamine, resulted in two polymorphs being obtained. While the one is based on a square-grid of Ni-paddlewheel units and terephthalate ions, the other is based on a kagome-grid. The layers are interconnected by the diamine to form highly porous framework compounds. The use of microwave-assisted heating in a HT-investigation on MOFs was also reported for the first time in this study.

2.3.2 Isoreticular Synthesis

Perhaps the most striking concept in the chemistry of MOFs is isoreticular synthesis. This principle deals with the replacement of inorganic or organic building units by topologically similar or identical building blocks. Thus, the pore size can be extended by employing larger linker molecules and the surface chemistry can be modulated by using functionalized organic moieties or by replacing the inorganic cations (Figure 2.2).[10] This approach also allows tuning of the properties of a material.

Although on paper this seems to be trivial, the replacement may require substantial effort in finding and optimizing the necessary reaction conditions. This is exemplified by the synthesis of Cr-MIL-101-NDC, the naph-thalenedicarboxylate based analogue of the terephthalate based Cr-MIL-101,[11] which was intentionally synthesized using HT-methods. More than 600 reactions were performed until the use of acetic acid as an appropriate additive led to the formation of a highly crystalline product, exhibiting giant pores with diameters up to 48 Å.[12] Nevertheless, there exist few examples where similar reaction conditions lead to a family of isoreticular compounds.

The best known example is the IRMOF-series, which is based on the primitive cubic net as observed for MOF-5.[10] In one study, sixteen isoreticular MOFs were synthesized, all based on linear dicarboxylate molecules varying in

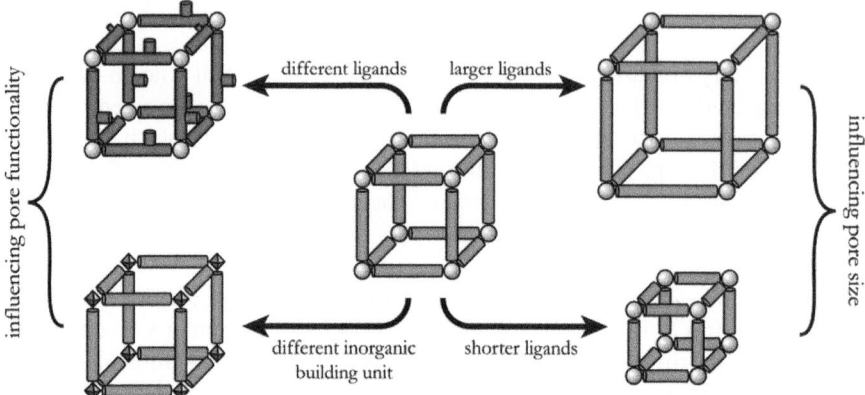

Figure 2.2 Schematic representation of the concept of isoreticular synthesis.

size and bearing different functional groups. By this approach, the sorption properties were drastically altered. This study also highlighted a limit of the isoreticular synthesis concept: compounds containing larger linker molecules often exhibit interpenetrated structures, which means that one or more identical frameworks crystallise/grow within each others' pore networks. Thus the sorption capacity is substantially reduced in comparison with non-interpenetrated structures, but this effect may also enhance the thermal and chemical stability and improve selectivity.[13]

Another highly diverse example for the isoreticular synthesis of MOFs is based on the MIL-53 topology. In this case, the organic as well as the inorganic building units can be replaced. These compounds with the general formula [M(OH)(dicarboxylate)] are based on trivalent metal ions that are forming *trans*-connected chains of MO_6-octahedra.[14] These chains are interconnected by linear dicarboxylate molecules, which results in the formation of rhombic channels. The smallest linker molecule that was used for the synthesis of a MIL-53-topology is fumaric acid,[15] while the largest one is 4,4'-biphenyl dicarboxylic acid.[16] Furthermore, a still increasing number of differently functionalized terephthalic acids have been used as organic moiety,[17,18] and even the use of a mixture of linker molecules was shown to be possible.[19] In addition, the metal ion can be broadly varied, for example Al^{3+},[20] Sc^{3+},[21] Fe^{3+},[17] Cr^{3+},[14] In^{3+},[22] and Ga^{3+},[23] were reported to form the MIL-53 framework. Thus, the pore size and functionalization can be altered using different linker molecules. Since the framework opens and shrinks reversibly depending on the occluded guest molecules, the structures are denoted as "breathing" MOFs. This breathing is strongly dependent on the incorporated metal ion, and a still increasing number of framework conformations could be identified, depending strongly on the inorganic brick. Moreover, by extending the size of the linker molecule this breathing effect can be suppessed.[16,24]

A series of isoreticular MOFs that is yet unparalleled regarding stability is based on the topology of UiO-66. All of these compounds are based on a

$Zr_6O_4(OH)_4{}^{12+}$-cluster, which is twelvefold bridged to adjacent inorganic building units.[25] The use of the tetravalent Zr^{4+} results in a remarkable stability, which is also exhibited by the functionalized members of the UiO-66-series.[26] The smallest molecules used up to now for the synthesis of a MOF based on this topology is the fumaric acid,[27] while the largest reported molecule is based on terphenyldicarboxylic acid.[28] If the linker size is further extended, interpenetrated frameworks with UiO-66-topology are formed, which can also be synthesized with various functional groups.[29]

2.3.3   The Precursor Approach

The use of precursors has been shown to be a useful approach towards the synthesis of new materials. Organic as well as inorganic units have been used as precursors. The organic building unit can be formed *in situ* and subsequently incorporated into the framework structure. Most often, chemical deprotection of the organic linker leads to the generation of coordinating groups.[30] Protection chemistry can alter the coordination geometry of the linker molecules,[34] while non-coordinating functional groups can be modified during the synthesis.[35] Moreover, the complete linker molecules can be formed in some cases during the synthesis.[36] Ligand replacement reactions on pre-synthesized inorganic clusters can lead to interconnection and a framework structure.[37]

The deprotection strategy has been used for the synthesis of phase-pure Al-MIL-100, which has only been obtained up to now starting from the trimethylester of 1,3,5-benzentricarboxylic acid (trimesic acid).[30] The acid-catalysed slow solvolysis of this precursor leads to the formation of a highly crystalline MOF, which could not be obtained by the direct use of trimesic acid. This precursor was also successfully employed for the synthesis of other trimesate-based MOFs, such as Al^{3+}-, In^{3+}- and Ga^{3+}-based MOFs exhibiting MIL-96 topology.[31–33] The *in situ* protection of one functional group of 1,3,5-benzenetricarboxylic is observed during the synthesis of the Cu-based MOF STAM-1.[34] The addition of methanol to the reaction mixture leads to the partial esterification of trimesic acid during the synthesis.[34] In comparison to the framework of HKUST-1, which is also based on trimesate molecules and can be obtained under similar conditions, a completely different framework structure was observed, due to the conversion of the tritopic ligand into a V-shaped ditopic ligand. A further example of *in situ* conversion of func-tionalized organic building units was reported for the Al-based MOF CAU-1-NH_2. The reaction is conducted in methanol, and depending on the reaction progress, the NH_2-group at the linker molecule is methylated by the solvent in varying degree.[35] Thus, only very short reaction times lead to the formation of a non-methylated framework compound, which can be used for further post-synthetic modification. *In situ* synthesis of the organic linker has been demonstrated in the conversion of acetonitrile and sodium azide into a tetrazole linker molecule during MOF-synthesis. The reaction leads to the formation of 5-methyl-1H-tetrazole, and this heterocyclic molecule is further deprotonated and incorporated into the framework of a Cd-based MOF.[36]

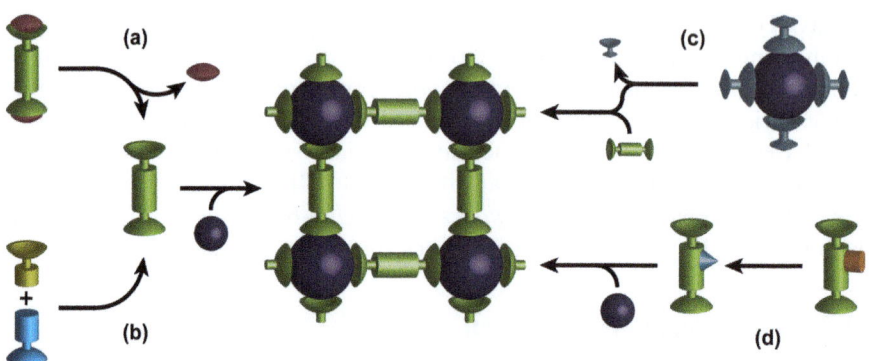

Figure 2.3 Schematic representation of the different strategies leading to MOFs: (a) deprotection of the organic linker, (b) *in situ* formation of the organic linker molecule, (c) use of an inorganic precursor and ligand exchange, (d) *in situ* protection/conversion of the organic linker.

The use of an inorganic precursor has been less intensively studied, most probably due to the elaborate synthesis procedures of inorganic cluster compounds. The examples that can be found in literature are rather convincing. Thus, for the formation of isoreticular compounds with MOF-5-topology, the inorganic units were pre-prepared as acetates based on Zn^{2+} and Be^{2+} with the formula $[M_4O(O_2CCH_3)_6]$. These molecular species were subsequently successfully incorporated into the framework by replacing the mono-carboxylate ions by the ditopic terephthalate ions.[37] The inorganic unit of UiO-66 could be obtained as a methacrylate complex and was successfully used for the synthesis of a MOF with UiO-66 topology based on *trans,trans*-muconate (Figure 2.3).[38]

To confirm the structural integrity of the inorganic precursor during synthesis, EXAFS studies were carried out. Iron acetate $[Fe_3O(O_2CCH_3)_6(H_2O)_3]ClO_4$, consisting of the trimeric inorganic building units, was used as the precursor for the synthesis of MIL-89.[39] The EXAFS spectra of the initial reaction mixture showed the presence of an X-ray amorphous intermediate with a very similar spectrum to the final product, therefore indicating the structural integrity of the trimeric unit during the synthesis.

2.3.4 Structure Directing Agents

In principle, most chemicals employed in the synthesis of MOFs can be considered as structure directing agents (SDAs). A chemical may be classified as an SDA if it is involved in adjusting the pH-value and/or serving as a counter-ion for charged frameworks, preventing the collapse of a framework structure (pore filler) or acting as a true templating molecule. The pH-value of a reaction may be influenced by counterions of metal salts employed as well as acids, bases or the solvent used. The pore filling by reactants or solvent molecules is nearly always observed in MOF synthesis. However, only few examples of the "true" templating of structures are known which are summarized in this section. In these structures the additives act as a negative of the pore structure and cannot be replaced in the synthesis mixture. For the synthesis of porous compounds, such an approach can be very challenging, since the templates must usually be removed in an activation step to make the pores accessible.

The use of amines as an SDA was reported for the synthesis of In^{3+}-based MOFs. Using 4,5-imidazoledicarboxylic acid, three different structures were synthesized using three different amines.[40] Surprisingly, all of them exhibit anionic zeolitic frameworks, in which the protonated amines act as charge-balancing counterions. The protonated amine in one of these MOFs could also be exchanged for Na^+ ions, which resulted in a porous framework. Similar results were reported for the templated synthesis of two Cd^{2+}-based MOFs.[41] *cis,cis*-Cyclohexanetricarboxylate was employed as linker molecule and piperazine or triethanloamine were used as SDAs. In the two MOFs obtained, the protonated amines were exchanged by K^+ ions. Thus, in these two examples the cationic organic templates could be removed by ion-exchange. In contrast to this, it is also possible to use inorganic, anionic templates.[42] Phosphotungstic acid $H_3PW_{12}O_{40}$ was employed as an inorganic SDA for the formation of the well-known HKUST-1 at room temperature. The MOF only formed if the reactants were mixed in the sequence $H_3PW_{12}O_{40}$, Cu^{2+} source, linker molecule. Therefore, the proposed reaction mechanism comprises the coordination of Cu^{2+} ions by the acid, and the subsequent coordination of linker molecules, which is followed by the crystallization. The driving force for the templating was postulated to be the perfect matching of the template to one of the pores. The addition of the acid during the synthesis was also proven to accelerate the formation of the product.[43]

2.3.5 Rational Approach Towards Topology and Function

While all the aforementioned methods for the synthesis of new MOFs are mainly explorative, sometimes the deduction of possible structures based on synthetic experience and topological requirements can be promising. This is especially the case if synthesis conditions have been established under which the inorganic building units are frequently observed. Thus in some cases, the topology of a framework compound can be anticipated based on the geometry of the inorganic and the organic building units.

An example for the successful application of a topological approach towards the synthesis of new MOFs was given in the case of some Cu^{2+}-based MOFs.[44] Since the occurrence of Cu-based paddle-wheel units is fairly common, the formation of a MOF based on these inorganic units can be considered as very plausible, if the linker molecules allow for a suitable topology. Combining the rather large 5,5′,5″,5‴-1,2,4,5-phenyltetramethyleneoxytetraisophthalic acid as an octatopic linker with the paddle-wheel units led to the formation of a MOF. Even the use of an expanded ligand was demonstrated. In both cases the synthesis yielded highly porous MOFs with the anticipated **tbo** topology.

The possibility to synthesize isoreticular structures can be well exemplified by two polymorphs with the composition [V(OH)BDC], MIL-47 and MIL-68. MIL-47 exhibits a square-grid structure[45] while MIL-68 possesses a kagome structure.[46] The analogues of MIL-47 based on several trivalent cations, such as Al^{3+},[20] In^{3+},[22] Ga^{3+}[23] and Fe^{3+}[20] were reported shortly afterwards. Although the MIL-68 analogues should be synthetically accessible, it took several years until the isoreticular frameworks based on Al^{3+},[47] Fe^{3+},[48] as well as Ga^{3+} and In^{3+},[49] were reported.

Specific functionalities, such as catalytically active centers, can be incorporated into MOF structures by using functionalized organic linker molecules. If the MOF can be successfully synthesized, this approach can be used to introduce well-defined properties of a molecular catalyst into a solid framework. This was shown for the aminofunctionalized MOF UiO-66-NH_2. The basic character of the amino-group makes this MOF an efficient catalyst for aldol-reactions.[50] In porphyrrin-based MOFs, the complexation of metal ions into the porphyrrin core results in the formation of compounds which can be used in heterogeneous catalysis.[51] Thus, the properties of the building units predefine the potential properties of MOFs.

Another promising way to introduce catalytically active sites is the modification of a suitable MOF compound, after its synthesis. This so-called post-synthetic approach will be discussed in detail in the next chapter of the book.

2.4 Crystal Morphology and Hierarchical Porosity

For the use of MOFs as catalysts, the tailoring of the crystal morphology and the generation of hierarchical porosity is one of the most important aspects since the crystal structure itself determines the functionality, but the accessibility of reactive sites is also an important factor. Diffusion of reactants and products to and from the catalytically active sites is often the rate limiting step. Effective diffusion might be achieved through the use of nano-sized crystallites or the introduction of meso- or macropores. Some promising approaches for the tailoring of crystal size and shape and the generation of hierarchical porosity are described in the following paragraphs.

2.4.1 Influencing Crystal Habitus

Various approaches are known to tune the crystal size of a material. The variation of process parameters such as temperature, heating time or rate can strongly affect the size of the crystallites. Furthermore, the particle size can also depend on the method of energy application. The use of microwave heating or sonochemistry is often found to yield nano-sized products.[1] In contrast to these methods, mechanochemical synthesis routes using ball milling or grinding limit the particle size due to the application of strong shear forces. Moreover, the chemical composition of the reaction mixture has probably the most profound influence on the particle size. This includes the solvent, the metal salt employed and especially the use of additives. Since the influence of the synthesis parameters on the crystal growth is usually not predictable, the systematic investigation of these parameters is often necessary for the formation of crystallites of the desired size. This was, for example, demonstrated employing the aforementioned HT methods for the synthesis of HKUST-1.[52]

Different heating rates and methods were shown to have a drastic influence on the nucleation rate and therefore on the size of the particles. An interesting example was given for the microwave-assisted synthesis of the Al-based MOF CAU-1-(OH)$_2$.[53] Variation of the heating method and the reaction time resulted in altered particle sizes of the obtained product: the higher the reaction temperature and the shorter the reaction time, the smaller the particles that were isolated. This was attributed to the uniform microwave-assisted heating, which results in fast and homogeneous nucleation. Subsequent crystal growth is limited by the remaining concentration of reactants or can be stopped by quenching.

The use of ultrasound for the synthesis of monodisperse nanoparticles was reported for the iron fumarate MIL-88A.[54] The use of ultrasound is commonly seen as a fast, efficient and environmental-friendly method for synthesis. MIL-88A could be obtained by various synthetic routes, employing conventional heating, microwave-assisted heating and ultrasonication. However, only employing sonochemical methods or microwave-assisted heating resulted in monodisperse products.[54] All other synthetic routes resulted in the formation of polydisperse materials, although a broad range of conditions were tested.

An example of the influence of chemical parameters was reported for the Al^{3+}-based MOF [Al(OH)(1,4-NDC)] which exhibits the MIL-53-topology.[55] This compound is based on 1,4-naphtalenedicarboxylic acid, and was originally obtained as a microcrystalline powder which is non-porous towards nitrogen, although the diameter of the channels (7 Å) should allow the diffusion of nitrogen molecules into the pores.[55] Changing the synthesis temperature, the reaction time and especially the solvent led to the formation of nanosized crystals (diameter of ~ 50 nm). Surprisingly the micropores of this compound are readily accessible for the adsorption of nitrogen.[56] Thus, the crystal size can also have an impact on the sorption properties.

A very systematic approach to manipulate the crystal size is the use of additives, which are not incorporated into the framework. Monocarboxylic acids were shown to be ideal candidates for this so called modulation approach. Modulators were first employed in the synthesis of HKUST-1 (Figure 2.4).[57] The use of an aliphatic carboxylic acid in varying concentrations to influence crystal growth, resulted in the formation of single crystals with different morphologies.

Moreover it was also possible to combine this approach with different synthesis methods, which allowed further control over the size of the resulting crystals.[58] The modulating effect was explained by the formation of inorganic precursors which contain the additive coordinated to the metal ions. Therefore, the nucleation rate for the crystallization can be altered, and thus the growth of the particles can be controlled and slowed down. This has also proven to be a valuable method for the synthesis of zirconium-based MOFs.[28]

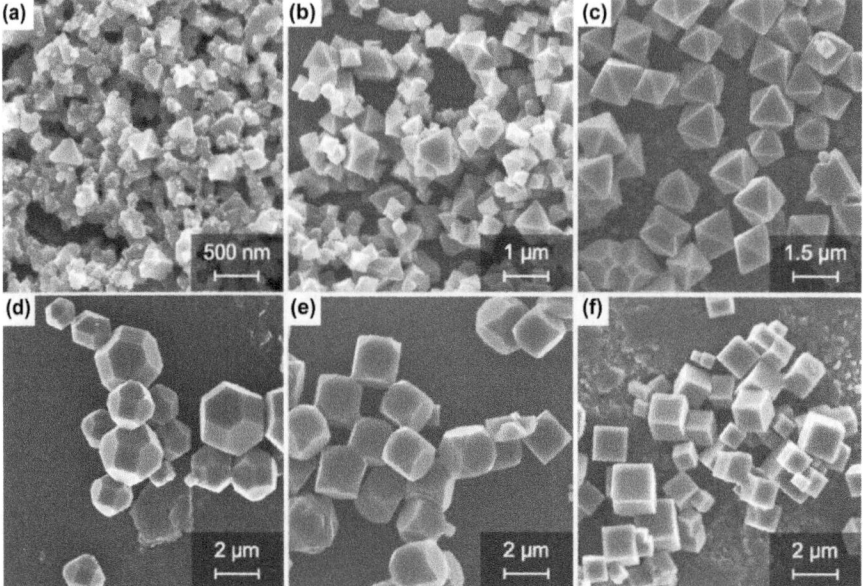

Figure 2.4   The different crystal sizes and shapes that could be obtained using a modulator.
Reprinted with permission from ref. 57. Copyright 2011 American Chemical Society.

2.4.2   Hierarchical Porosity and Mesopores

Hierarchically structured porous materials have been intensively investigated, especially for catalytic applications of zeolites.[59] Various methods, established for zeolites and metal oxides have been transferred to MOF science. Thus, hierarchical porosity is often achieved by the use of surfactants, which have proven to be appropriate additives in the synthesis of ordered mesoporous materials.[60] Furthermore the generation of mesopores can be induced by the modulation of the crystal growth.[62] In addition to micellar templates, small droplets as observed in water in oil emulsions for example, can be employed as well.[63] However, sometimes the generation of additional large pores can result in a decreased crystallinity, which means that the crystalline structure itself may be disturbed. Therefore, the synthesis of MOFs which exhibit intrinsic mesopores must also be mentioned.

An example for the use of block-copolymers that are often employed for the synthesis of mesoporous oxides was given for a hierarchically porous material based on the MIL-53-topology. The addition of the surfactant in the synthesis of Al-MIL-53 resulted in the formation of the crystalline MOF, which possesses additional mesopores due to the aggregation of crystals to meso-porous particles.[60] Surprisingly the addition of ethanol to the reaction mixture was shown to lead to the formation of mesopores to the otherwise microporous material. The mesopores possess diameters between 4 and 8 nm, while the structural micropores exhibit a diameter of ~ 1 nm. This observation indicates that the templating effect does not necessarily follow the same mechanism as in the synthesis of mesoporous oxides, in which the polymers can not be replaced by alcohols.

Using the cationic surfactant cethyltrimethylammonium bromide (CTAB), it was possible to synthesize HKUST-1 possessing additional well-ordered mesopores with diameters between 3.8 and 5.6 nm or even up to 31 nm.[61] These mesoporous channels are hexagonally arranged and their pore walls are made of crystalline microporous HKUST-1 with pore diameters up to 1 nm, thus the templating effect seems to be similar to the known micellar templating in mesoporous materials.

An approach to yield hierarchical pore systems by modulation of crystal growth was pursued by the incorporation of monocarboxylate ions into single crystals of MOF-5.[62] The implementation of *para*-functionalized benzoate bearing alkoxy chains results in the formation of a crystalline material with MOF-5 topology, but the aliphatic chains at the aromatic ring interrupt the crystal growth, which results in the formation of meso- and macropores within the crystal. This pore system with pore diameters between 10 and 100 nm is formed of microporous crystalline walls, with structural pores of 1.2 nm in diameter, and this altered structure results in drastically changed sorption properties.

One unique example for the generation of hierarchical porosity in the macropore regime was given by the synthesis of hollow capsules of HKUST-1.[63] Dropwise addition of a clear reactant solution into a non-miscible solvent yields macroporous capsules (diameter of several μm) with microporous walls. This approach towards the formation of crystals from a clear solution seems to be generally applicable for HKUST-1, since it also allows the localized growth of crystals in a predefined shape.[64]

One big advantage of MOFs is the fact that, in some systems, intrinsically mesoporous structures are formed without using additives or templates. An extraordinary example is the MIL-101-framework. Although the rather small linker molecule terephthalic acid is employed, the hierarchical, zeotypic MTN framework structure contains both micro- and mesopores. Usually larger linker molecules are used for the synthesis of large-pore MOFs.[65] 1,3,5-Benzene-tribenzoic acid is one of the most frequently used reagents, and there exist examples for MOFs based on this molecule, including UMCM-1[66] or NU-100.[67] Such MOFs show remarkable gas sorption capacities, but they also have the potential to adsorb large dye molecules or even enzymes.[68] Furthermore, the easily accessible pore space allows chemical functional-ization, if there exist modifiable groups inside the cavities.[69,70]

2.4.3 Films and Membranes

The control of crystal growth is also important in order to form MOF films and membranes on a surface. Thick membranes are interesting for molecular separation processes while thin films are often investigated for sensor applications. The formation of MOF films is often achieved by the use of dispersions containing very small crystallites of the MOF to be deposited (forming layers by either coating or chemical solution deposition).[71] The synthesis of MOF films can be induced by dipping the substrate alternately into solutions of the building units (layer-by-layer growth).[72] Furthermore, the crystallization can be induced on the surface directly from solution (direct growth) or after seeding of few nano-crystals onto the surface (seeded growth).[73] One of the most elegant methods for the synthesis of membranes is the electrochemical pathway, which induces the nucleation of crystals on the surface of an electrode.[74]

A versatile approach towards the synthesis of MOFs as a thin film is the use of SAMs (self assembling monolayers). α-ω-Bifunctionalized long chain alkanes are attached to a gold-surface by thiol-groups and the second functional group can be used to act as nucleation site or to anchor the crystals onto the surface. Carboxylic acid or amine groups are especially versatile for this method, but hydroxyl groups can be used as well. Growth of crystals by this route can be achieved by alternately dipping the SAMs into solutions of the inorganic and the organic reactants, similar to the layer-by-layer approach. This was demonstrated for various MOFs, including HKUST-1.[75] Another method uses seeding solutions, for example the filtrate of a bulk synthesis. Placing SAMs in such a seeding solution is a versatile method to yield surface bound crystals. A variety of MOFs could be prepared as a film, for example a functionalized mesoporous Zr-based MOF, which could be post-synthetically modified by covalent bonding of dye molecules.[76] In the synthesis of a CAU-1-film it was shown that the orientation of the crystal growth and thus the orientation of the pores not only depends on the terminal functional group on the SAM, but also on the synthesis conditions.[77]

It is also possible to use MOF dispersions, which are deposited on a surface by spin-coating or dip-coating.[78] If dispersions of different materials are used, this approach can be successfully employed to grow photonic crystals incorporating MOFs.[71]

The direct growth of membranes with a remarkable thickness can be achieved using a seed-covered or bare substrate, employing a secondary or direct growth approach, respectively. Such membranes could be important for use in separation processes. For example, membranes of ZIF-8 could be grown on a TiO_2-support by microwave-assisted heating in excellent quality, using the secondary growth approach.[73]

The use of seeded substrates also allowed the fabrication of membranes of other ZIFs like ZIF-90, which can be further chemically modified to increase separation properties.[79] In an alternative electrochemical approach, thick layers of MOFs were grown utilizing metallic anodes as the substrates and reactant.[74] Thus, HKUST-1 is deposited electrochemically on the surface of a Cu-electrode. This approach was also used to construct patterned HKUST-1-films.[80] In another example, the electrochemical synthesis of porous films of MOF-5 was demonstrated.[81] At the cathode, the reduction of nitrate ions to nitrite and hydroxide ions takes place. The latter induced the deprotonation of linker molecules near the electrode surface.

2.5 Purification and Activation

All steps necessary to purify and activate MOFs strongly depend on the chemical and thermal stability of the compound of interest. The purity of a MOF is one of the most important aspects for catalysis, since the existence of by-products can mimic catalytic activity and lower the sorption capacity of the material.[82] The absence of crystalline by-products is normally proven by X-ray powder diffraction. X-Ray amorphous by-products can be detected using thermogravimetric or elemental analyses. The purity of the material can be further proven by sorption experiments, which also give information about the degree of activation. Deviations from theoretical values indicate that the MOF was not appropriately purified or activated or that the structural information is incomplete.

The removal of impurities – which means by-products that are not occluded inside the pores – is a challenging task (Figure 2.5). Metal oxides are likely to form as a by-product under various conditions and must be removed from the product.[83] Other frequently observed impurities are recrystallised linker molecules.[11,20] In some cases, dense hybrid compounds are observed as by-products.[84] Purification can be accomplished by sieving or based on the densities of the synthesized products. Most often impurities are removed by solvent treatment, often at elevated temperatures.

Activation of a MOF, *i.e.* the removal of guest molecules from the pores (Figure 2.5), can be difficult, since the thermal removal of guest molecules can lead to structural collapse, especially if strong host–guest interactions are present or if the guest molecules necessitate high activation temperatures.

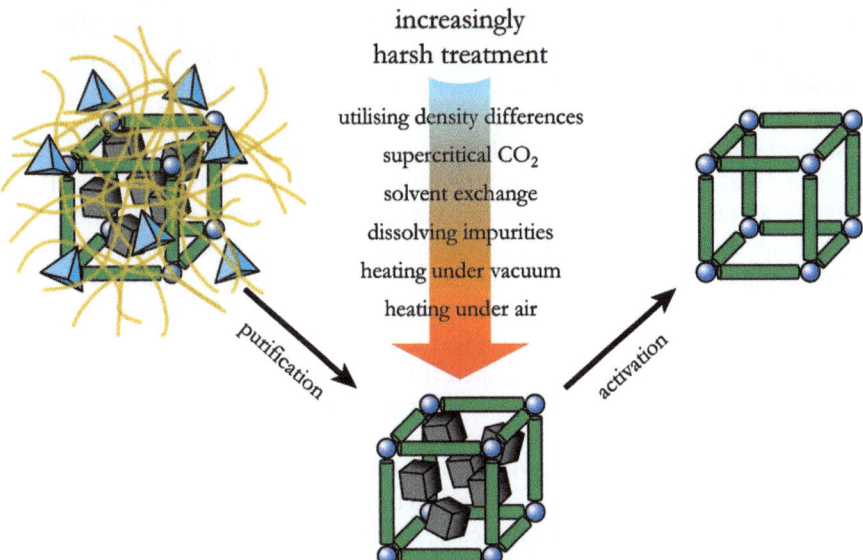

increasingly
harsh treatment

utilising density differences
supercritical CO_2
solvent exchange
dissolving impurities
heating under vacuum
heating under air

purification

activation

Figure 2.5 Schematic representation of purification and activation of as-synthesized MOF products using different approaches. Tetrahedra/strings symbolize impurities and cube guest species in the pores.

Various techniques can be used to extract occluded linker molecules and to exchange solvent molecules with more volatile reagents.

Common inorganic impurities such as oxides, hydroxides or halides can occur as X-ray amorphous[83] or crystalline[14] by-phases. Usually they are removed using a suitable solvent at elevated temperatures. The MOF must be stable in this solvent, while the by-phase must be soluble. For example, during the synthesis of the Al-based MOF CAU-3-BDC, an amorphous aluminium containing by-product could be removed by treatment of the as synthesised product in DMF at 150 °C.[83]

Organic impurities, such as recrystallized linker molecules, are often observed, especially if they are sparingly soluble in the synthesis solvent. This has been observed, for example, in the synthesis of Cr-MIL-101.[11] In principle the organic linker molecules can be dissolved in an organic solvent. In few cases, large crystals of terephthalic acid, which is hardly soluble in water under ambient conditions, can be removed from the as-prepared product by filtration, as was reported for Cr-MIL-101.

The separation of two phases of different density was reported for $Cu_3(TATB)_2(H_2O)_3$.[85] Employing a solvent mixture with a density between the less dense, porous compound and the more dense, non-porous by-product, the porous MOF could be skimmed from the upper region of the mixture, while the dense phase settled at the bottom of the flask.

Activation, *i.e.* the removal of linker or solvent molecules from the pores, is crucial to achieve the highest possible porosity. A way to remove linker molecules from the pores is to dissolve them in an appropriate solvent, which can be assisted by thermal treatment. This is also an appropriate method to remove terephthalic acid, as was demonstrated for the activation of Al-MIL-53, for example.[20] The chromium-based analogue Cr-MIL-53 is highly stable at elevated temperatures, therefore it was possible to remove unreacted terephthalic acid by heating in air at 300 °C.[14]

Solvent molecules are most often removed from the pores by thermal treatment under reduced pressure. Sometimes the thermal stability of the MOF does not allow the application of this standard procedure. In such cases, the solvent molecules are exchanged with more volatile compounds like di- or trichloromethane.[29] These solvents can be often removed without thermal treatment or at least at rather low temperatures. A very mild approach used in aerogel formation has been also applied for MOF activation, which is solvent exchange followed by the use of supercritical CO_2. In this case, full activation can be achieved, even if the framework of the MOF is quite fragile.[86]

During all steps that are necessary for the purification and activation, the integrity of the framework must be reinvestigated. High temperatures during the activation can induce intra-framework reactions, as was shown for the COOH-functionalized Al-MIL-53.[87] In this case, the functional carboxylic acid groups are partially removed and condense to form anhydrides at temperatures above 300 °C in vacuum. Furthermore, the activation atmosphere must be also taken into account. The thermal activation of CAU-1-NH_2 in air leads to an exchange of methanolate ions by hydroxide ions in the inorganic building block.[88]

References

1. N. Stock and S. Biswas, *Chem. Rev.*, 2012, **112**, 933.
2. G. A. Ozin and A. C. Arsenault, *Nanochemistry – A Chemical Approach to Nanomaterials*, RSC Publishing, 2005, p. 379.
3. N. Pienack and W. Bensch, *Angew. Chem., Int. Ed.*, 2011, **50**, 2014.
4. G. Ferey, C. Serre, C. Mellot-Draznieks, F. Millange, S. Surble, J. Dutour and I. Margiolaki, *Angew. Chem., Int. Ed.*, 2004, **43**, 6296.
5. N. Stock, *Micropor. Mesopor. Mater.*, 2010, **129**, 287.
6. K. Sumida, S. Horike, S. S. Kaye, Z. R. Herm, W. L. Queen, C. M. Brown, F. Grandjean, G. J. Long, A. Dailly and J. R. Long, *Chem. Sci.*, 2010, **1**, 184.
7. R. Banerjee, A. Phan, B. Wang, C. Knobler, H. Furukawa, M. O'Keeffe and O. M. Yaghi, *Science*, 2008, **319**, 939.
8. S. Bauer, C. Serre, T. Devic, P. Horcajada, J. Marrot, G. Ferey and N. Stock, *Inorg. Chem.*, 2008, **47**, 7568.
9. P. Maniam and N. Stock, *Inorg. Chem.*, 2011, **50**, 5085.
10. M. Eddaoudi, J. Kim, N. L. Rosi, D. T. Vodak, J. Wachter, M. O'Keeffe and O. M. Yaghi, *Science*, 2002, **295**, 469.
11. G. Férey, C. Mellot-Draznieks, C. Serre, F. Millange, J. Dutour, S. Surblé and I. Margiolaki, *Science*, 2005, **23**, 2040.
12. A. Sonnauer, F. Hoffmann, M. Fröba, L. Kienle, V. Duppel, M. Thommes, C. Serre, G. Ferey and N. Stock, *Angew. Chem., Int. Ed.*, 2009, **48**, 3791.
13. B. Chen, M. Eddaoudi, S. T. Hyde, M. O'Keeffe and O. M. Yaghi, *Science*, 2001, **291**, 1021.
14. C. Serre, F. Millange, C. Thouvenot, M. Noguès, G. Marsolier, D. Louër and G. Férey, *J. Am. Chem. Soc.*, 2002, **124**, 13519.
15. M. Gaab, N. Trukhan, S. Maurer, R. Gummaraju and U. Müller, *Micropor. Mesopor. Mater.*, 2012, **157**, 131.
16. I. Senkovska, F. Hoffmann, M. Fröba, J. Getzschmann, W. Böhlmann and S. Kaskel, *Micropor. Mesopor. Mater.*, 2009, **122**, 93.
17. T. Devic, P. Horcajada, C. Serre, F. Salles, G. Maurin, B. Moulin, D. Heurtaux, G. Clet, A. Vimont, J. Grenèche, B. Le Ouay, F. Moreau, E. Magnier, Y. Filinchuk, J. Marrot, J. Lavalley, M. Daturi and G. Férey, *J. Am. Chem. Soc.*, 2010, **132**, 1127.
18. S. Biswas, T. Ahnfeldt and N. Stock, *Inorg. Chem.*, 2011, **50**, 9518.
19. T. Lescouet, E. Kockrick, G. Bergeret, M. Pera-Titus and D. Farrusseng, *Dalton Trans.*, 2011, **40**, 11359.
20. T. Loiseau, C. Serre, C. Huguenard, G. Fink, F. Taulelle, M. Henry, T. Bataille and G. Férey, *Chem.–Eur. J.*, 2004, **10**, 1373.
21. J. P. S. Mowat, V. R. Seymour, J. M. Griffin, S. P. Thompson, A. M. Z. Slawin, D. Fairen-Jimenez, T. Düren, S. E. Ashbrook and P. A. Wright, *Dalton Trans.*, 2012, **41**, 3937.
22. E. V. Anokhina, M. Vougo-Zanda, X. Wang and A. J. Jacobson, *J. Am. Chem. Soc.*, 2005, **127**, 15000.

23. C. Volkringer, T. Loiseau, N. Guillou, G. Férey, E. Elkaïm and A. Vimont, *Dalton Trans.*, 2009, **12**, 2241.
24. T. Loiseau, C. Mellot-Draznieks, H. Muguerra, G. Férey, M. Haouas and F. Taulelle, *C. R. Chimie*, 2005, **8**, 765.
25. J. H. Cavka, S. Jakobsen, U. Olsbye, N. Guillou, C. Lamberti, S. Bordiga and K. P. Lillerud, *J. Am. Chem. Soc.*, 2008, **130**, 13850.
26. M. Kandiah, M. H. Nilsen, S. Usseglio, S. Jakobsen, U. Olsbye, M. Tilset, C. Larabi, E. A. Quadrelli, F. Bonino and K. P. Lillerud, *Chem. Mater.*, 2010, **22**, 6632.
27. G. Wißmann, A. Schaate, S. Lilienthal, I. Bremer, A. M. Schneider and P. Behrens, *Micropor. Mesopor. Mater.*, 2012, **152**, 64.
28. A. Schaate, P. Roy, A. Godt, J. Lippke, F. Waltz, M. Wiebcke and P. Behrens, *Chem.–Eur. J.*, 2011, **17**, 6643.
29. A. Schaate, P. Roy, T. Preuße, S. J. Lohmeier, A. Godt and P. Behrens, *Chem.– Eur. J.*, 2011, **17**, 9320.
30. C. Volkringer, D. Popov, T. Loiseau, G. Férey, M. Burghammer, C. Riekel, M. Haouas and F. Taulelle, *Chem. Mater.*, 2009, **21**, 5695.
31. T. Loiseau, L. Lecroq, C. Volkringer, J. Marrot, G. Férey, M. Haouas, F. Taulelle, S. Bourrelly, P. L. Llewellyn and M. Latroche, *J. Am. Chem. Soc.*, 2006, **128**, 10223.
32. C. Volkringer and T. Loiseau, *Mater. Res. Bull.*, 2006, **41**, 948.
33. C. Volkringer, T. Loiseau, G. Férey, C. M. Moraisc, F. Taulelle, V. Montouillout and D. Massiot, *Micropor. Mesopor. Mater.*, 2007, **105**, 111.
34. M. I. H. Mohideen, B. Xiao, P. S. Wheatley, A. C. McKinlay, Y. Li, A. M. Z. Slawin, D. W. Aldous, N. F. Cessford, T. Düren, X. Zhao, R. Gill, K. M. Thomas, J. M. Griffin, S. E. Ashbrook and R. E. Morris, *Nature Chem.*, 2011, **3**, 304.
35. T. Ahnfeldt and N. Stock, *CrystEngComm.*, 2012, **14**, 505.
36. H. Zhao, Z. R. Qu, H. Y. Ye and R. G. Xiong, *Chem. Soc. Rev.*, 2008, **37**, 84.
37. S. Hausdorf, F. Baitalow, T. Böhle, D. Rafaja and F. O. R. L. Mertens, *J. Am. Chem. Soc.*, 2010, **132**, 10978.
38. V. Guillerm, S. Gross, C. Serre, T. Devic, M. Bauer and G. Ferey, *Chem. Commun.*, 2010, **46**, 767.
39. S. Surble, F. Millange, C. Serre, G. Ferey and R. I. Walton, *Chem. Commun.*, 2006, 1518.
40. Y. L. Liu, V. Kravtsov, R. Larsen and M. Eddaoudi, *Chem. Commun.*, 2006, 1488.
41. Q.-R. Fang, G.-S. Zhu, M. Xue, J.-Y. Sun and S.-L. Qiu, *Dalton Trans.*, 2006, 2399.
42. S. R. Bajpe, C. E. A. Kirschhock, A. Aerts, E. Breynaert, G. Absillis, T. N. Parac-Vogt, L. Giebeler and J. A. Martens, *Chem.–Eur. J.*, 2010, **16**, 3926.
43. S. R. Bajpe, E. Breynaert, D. Mustafa, M. Jobbagy, A. Maes, J. A. Martens and C. E. A. Kirschhock, *J. Mater. Chem.*, 2011, **21**, 9768.

44. J. F. Eubank, H. Mouttaki, A. J. Cairns, Y. Belmabkhout, L. Wojtas, R. Luebke, M. Alkordi and M. Eddaoudi, *J. Am. Chem. Soc.*, 2011, **133**, 14204.
45. K. Barthelet, J. Marrot, D. Riou and G. Férey, *Angew. Chem., Int. Ed.*, 2002, **41**, 281.
46. K. Barthelet, J. Marrot, G. Férey and D. Riou, *Chem. Commun.*, 2004, 520.
47. M. Schubert, U. Muller and S. Marx, *WO Pat.* 2008/129051, 2008.
48. A. Fateeva, P. Horcajada, T. Devic, C. Serre, J. Marrot, J. Grenèche, M. Morcrette, J. Tarascon, G. Maurin and G. Férey, *Eur. J. Inorg. Chem.*, 2010, **24**, 3789.
49. C. Volkringer, M. Meddouri, T. Loiseau, N. Guillou, J. Marrot, G. Férey, M. Haouas, F. Taulelle, N. Audebrand and M. Latroche, *Inorg. Chem.*, 2008, **47**, 11892.
50. F. Vermoortele, R. Ameloot, A. Vimont, C. Serre and D. De Vos, *Chem. Commun.*, 2011, **47**, 1521.
51. O. K. Farha, A. M. Shultz, A. A. Sarjeant, S. T. Nguyen and J. T. Hupp, *J. Am. Chem. Soc.*, 2011, **133**, 5652.
52. E. Biemmi, S. Christian, N. Stock and T. Bein, *Micropor. Mesopor. Mater.*, 2009, **117**, 111.
53. T. Ahnfeldt, J. Moellmer, V. Guillerm, R. Staudt, C. Serre and N. Stock, *Chem.–Eur. J.*, 2011, **23**, 6462.
54. T. Chalati, P. Horcajada, R. Gref, P. Couvreur and C. Serre, *J. Mater. Chem.*, 2011, **21**, 2220.
55. A. Comotti, S. Bracco, P. Sozzani, S. Horike, R. Matsuda, J. Chen, M. Takata, Y. Kubota and S. Kitagawa, *J. Am. Chem. Soc.*, 2008, **130**, 13664.
56. J. Zhang, L. Sun, F. Xu, H.-Y. Zhou, Y.-L. Liu, Z. Gabelica and C. Schick, *Chem. Commun.*, 2012, **48**, 759.
57. A. Umemura, S. Diring, S. Furukawa, H. Uehara, T. Tsuruoka and S. Kitagawa, *J. Am. Chem. Soc.*, 2011, **133**, 15506.
58. S. Diring, S. Furukawa, Y. Takashima, T. Tsuruoka and S. Kitagawa, *Chem. Mater.*, 2010, **22**, 4531.
59. S. Lopez-Orozco, A. Inayat, A. Schwab, T. Selvam and W. Schwieger, *Adv. Mater.*, 2011, **23**, 2602.
60. D. Xuan-Dong, H. Vinh-Thang and S. Kaliaguine, *Micropor. Mesopor. Mater.*, 2011, **141**, 135.
61. L. Qiu, T. Xu, Z. Li, Y. Wu, X.-Y. Jiang and L.-D. Zhang, *Angew. Chem., Int. Ed.*, 2008, **47**, 9487.
62. K. Choi, H. Jeon, J. Kang and O. M. Yaghi, *J. Am. Chem. Soc.*, 2011, **133**, 11920.
63. R. Ameloot, F. Vermoortele, W. Vanhove, M. B. J. Roeffaers, B. F. Sels and D. E. De Vos, *Nature Chem.*, 2011, **3**, 382.
64. D. Witters, N. Vergauwe, R. Ameloot, S. Vermeir, D. De Vos, R. Puers, B. Sels and J. Lammertyn, *Adv. Mater.*, 2012, **24**, 1316.
65. P. Thanasekaran, T. Luo, J. Wu and K. Lu, *Dalton Trans.*, 2012, **41**, 5437.

66. K. Koh, A. G. Wong-Foy and A. J. Matzger, *Angew. Chem., Int. Ed.*, 2008, **47**, 677.
67. O. K. Farha, A. Ö. Yazaydın, I. Eryazici, C. D. Malliakas, B. G. Hauser, M. G. Kanatzidis, S. T. Nguyen, R. Q. Snurr and J. T. Hupp, *Nature Chem.*, 2010, **2**, 944.
68. V. Lykourinou, Y. Chen, X. Wang, L. Meng, T. Hoang, L. Ming, R. L. Musselman and S. Ma, *J. Am. Chem. Soc.*, 2011, **133**, 10382.
69. K. K. Tanabe, C. Allen and S. M. Cohen, *Angew. Chem., Int. Ed.*, 2010, **49**, 9730.
70. S. Bernt, V. Guillerm, C. Serre and N. Stock, *Chem. Commun.*, 2011, **47**, 2838.
71. F. M. Hinterholzinger, A. Ranft, J. M. Feckl, B. Rühle, T. Bein and B. V. Lotsch, *J. Mater. Chem.*, 2012, **22**, 10356.
72. O. Shekhah, H. Wang, D. Zacher, R. A. Fischer and C. Wöll, *Angew. Chem., Int. Ed.*, 2009, **48**, 5038.
73. H. Bux, F. Liang, Y. Li, J. Cravillon, M. Wiebcke and J. Caro, *J. Am. Chem. Soc.*, 2009, **131**, 16000.
74. U. Müller, H. Puetter, M. Hesse and H. Wessel, *World Pat.* WO 2005/049892, 2005.
75. O. Shekhah, J. Liu, R. A. Fischer and C. Wöll, *Chem. Soc. Rev.*, 2011, **40**, 1081.
76. F. M. Hinterholzinger, S. Wuttke, P. Roy, T. Preuße, A. Schaate, P. Behrens, A. Godt and T. Bein, *Dalton Trans.*, 2012, **41**, 3899.
77. F. Hinterholzinger, C. Scherb, T. Ahnfeldt, N. Stock and T. Bein, *Phys. Chem. Chem. Phys.*, 2010, **12**, 4515.
78. E. A. Flügel, A. Ranft, F. Haaseab and B. V. Lotsch, *J. Mater. Chem.*, 2012, **22**, 10119.
79. A. Huang and J. Caro, *Angw. Chem., Int. Ed.*, 2011, **51**, 4979.
80. R. Ameloot, L. Stappers, J. Fransaer, L. Alaerts, B. F. Sels and D. E. De Vos, *Chem. Mater.*, 2009, **21**, 2580.
81. M. Li and M. Dincă, *J. Am. Chem. Soc.*, 2011, **133**, 12926.
82. J. Hafizovic, M. Bjørgen, U. Olsbye, P. D. C. Dietzel, S. Bordiga, C. Prestipino, C. Lamberti and K. P. Lillerud, *J. Am. Chem. Soc.*, 2007, **129**, 3612.
83. H. Reinsch, M. Feyand, T. Ahnfeldt and N. Stock, *Dalton Trans.*, 2012, **41**, 4164.
84. T. D. Keene, D. J. Price and C. J. Kepert, *Dalton Trans.*, 2011, **40**, 7122.
85. O. K. Farha, K. L. Mulfort, A. M. Thorsness and J. T. Hupp, *J. Am. Chem. Soc.*, 2008, **130**, 8598.
86. A. P. Nelson, O. K. Farha, K. L. Mulfort and J. T. Hupp, *J. Am. Chem. Soc.*, 2009, **131**, 458.
87. N. Reimer, B. Gil, B. Marszalek and N. Stock, *CrystEngComm*, 2012, **14**, 4119.
88. T. Ahnfeldt, D. Gunzelmann, J. Wack, J. Senker and N. Stock, *CrystEngComm*, 2012, **14**, 4126.

CHAPTER 3

Post-synthetic Modification of MOFs

ANDREW D. BURROWS

Department of Chemistry, University of Bath, Claverton Down,
Bath BA2 7AY, UK
Email: a.d.burrows@bath.ac.uk

3.1 Introduction

Over the past fifteen years or so, the study of materials in which metal centres or aggregates are linked into extended network structures by bridging organic ligands has become a major area of chemistry.[1] These materials are referred to by a number of different names including coordination polymers and metal organic frameworks (MOFs) and, although there is debate about how inter-changeable these are,[2,3] this review uses the term 'metal organic framework' for any type of coordination network structure, regardless of the nature of the ligands.

The current level of interest in MOFs has arisen largely as a consequence of their porosity. Many MOFs, especially those based on polycarboxylates and imidazolates, show permanent porosity on removal of the solvent molecules from the pores and channels of the 'as-synthesised' materials, and this has been exploited for a wide range of potential applications including hydrogen storage,[4] carbon sequestration,[5] separations,[6] catalysis[7] and drug delivery.[8]

For many of these applications, the ability to chemically functionalise the pores in a particular manner allows for tuning of the MOF properties. While it is possible to include relatively simple functional groups into a MOF through the use of a functionalised linker in the synthesis, this approach is limited as

RSC Catalysis Series No. 12
Metal Organic Frameworks as Heterogeneous Catalysts
Edited by Francesc X. Llabrés i Xamena and Jorge Gascon
© The Royal Society of Chemistry 2013
Published by the Royal Society of Chemistry, www.rsc.org

many desired functionalities are either not tolerant to the conditions used for MOF synthesis (typically solvothermal) or non-innocent, and hence lead to changes in the nature of the network formed. Post-synthetic (or postsynthetic) modification (PSM) provides an alternative strategy to forming functionalised MOFs. In this approach, the nature of the pores is altered *after* synthesis of the MOF, thus allowing sensitive and/or coordinating groups to be introduced into the pores of a pre-formed network structure.

3.1.1   The Scope of Post-Synthetic Modification

The concept of modifying the pores within an extended network was first suggested by Hoskins and Robson in 1990,[9] and the first reports were published independently by the groups of Lee,[10] Kim[11] and Williams[12] in 1999 and 2000. After this initial flurry, the field lay dormant for several years before being revived by Wang and Cohen in 2007,[13] who also introduced the term 'postsynthetic modification' by analogy to the posttranslational modification of proteins. The PSM protocol is not unique to MOFs, and has also been employed for other materials including mesoporous silicas,[14] hydrogels[15] coordination oligomers,[16] organic network structures[17] and biomolecules.[18]

As a general concept in MOF chemistry, post-synthetic modification can refer to any change of a MOF structure following its synthesis. For the purposes of this review, PSM is divided into four main categories, which are shown schematically in Figure 3.1. These are:

(a) *Covalent PSM* – this is the most well-developed type of PSM in MOF chemistry, and involves covalent modification of a linker ligand.
(b) *Dative PSM* – this involves coordination of a metal centre to a linker.
(c) *Inorganic PSM* – this involves a modification of the secondary building units,[2] which form the nodes of the MOF network.
(d) *Ionic PSM* – this involves exchange of a counter-ion in a cationic or anionic MOF.

For covalent or dative PSM, the MOF normally needs to have a reactive group present on a linker. This group is often referred to as a 'tag', which is defined as a group or functionality that is stable and innocent (that is, non-structure-defining) during MOF formation, but that can be transformed by a post-synthetic modification.[19] Sometimes, sequential PSM reactions are carried out. These are referred to as *tandem PSM reactions*, and they typically involve two or more consecutive covalent modifications or a covalent modification followed by a dative modification, though other combinations of PSM reactions are also possible.

There are other possible MOF transformations that can be viewed as post-synthetic modifications, though space constraints do not allow these to be dealt with in detail here. These include the formation of nanoparticles within the pores,[20] and the growth of a MOF on the surface of another to form a

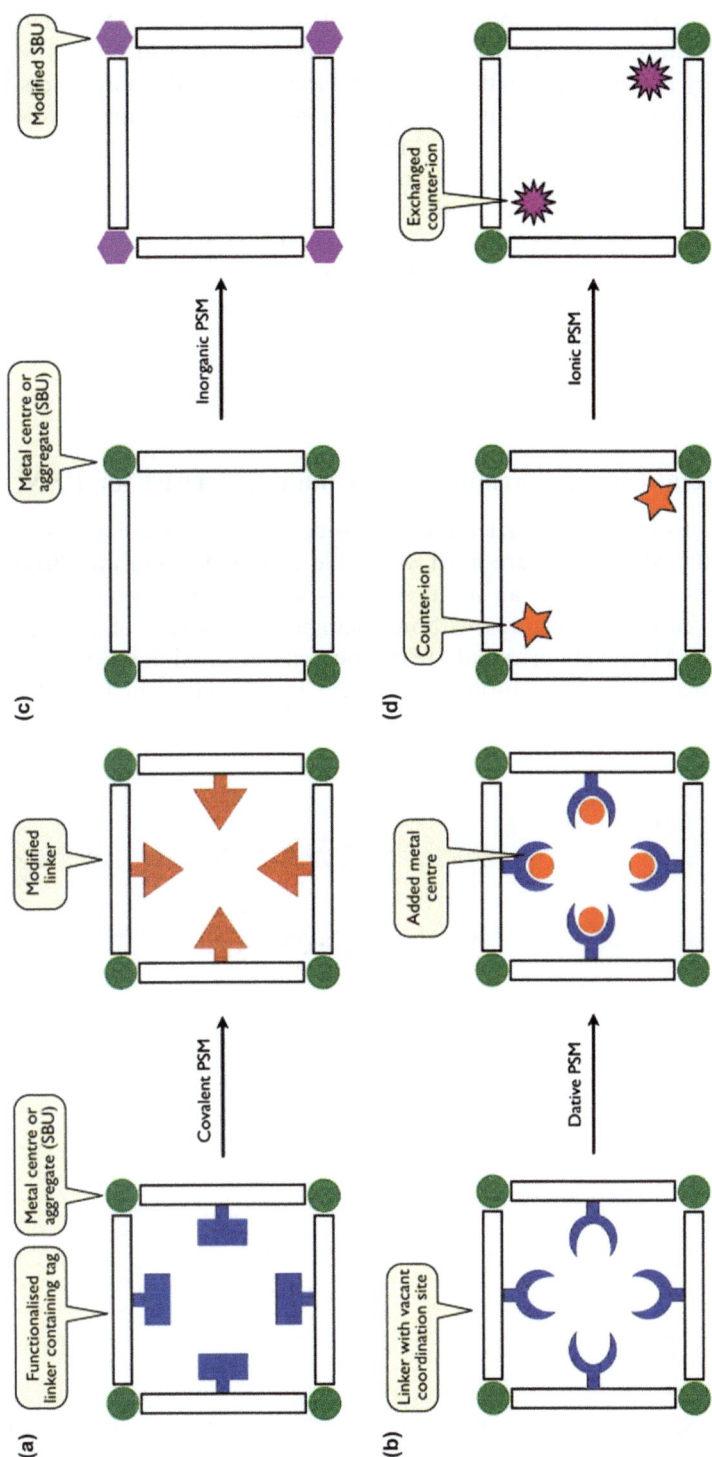

Figure 3.1 Schematic representation of the four categories of post-synthetic modification covered in this review.

'core-shell' MOF.[21] One important and well-studied area is the exchange or removal of guest molecules from the pores. Indeed, for most porous MOFs, activation by removal of included solvent molecules from the pores is essential for activity. Guest removal can occur with changes in pore shape, as observed in flexible MOFs such as [M(OH)(bdc)] (MIL-53, M = Al, Cr; bdc = 1,4-benzenedicarboxylate),[22] and occasionally with more dramatic changes in the network.[23] In addition, reactions of guest molecules within the pores can lead to the isolation and characterisation of very reactive species which are stabilised by confinement within the network.[24] Li, Stoddart, Yaghi *et al.* have shown that supramolecular interactions between electron-rich and electron-poor aromatics can lead to MOFs that are capable of including specific guests as pseudorotaxanes.[25]

3.1.2   Metal Organic Frameworks Commonly Used for PSM

PSM reactions have been carried out on a wide variety of MOF architectures. However, the MOFs that are most amenable to modification are those that form isoreticular series,[26] as in these cases a tag group can typically be introduced into the structures without changing the nature of the network adopted. The MOF networks that are most commonly used in PSM studies are shown in their non-tagged forms in Figure 3.2 alongside their formulae and

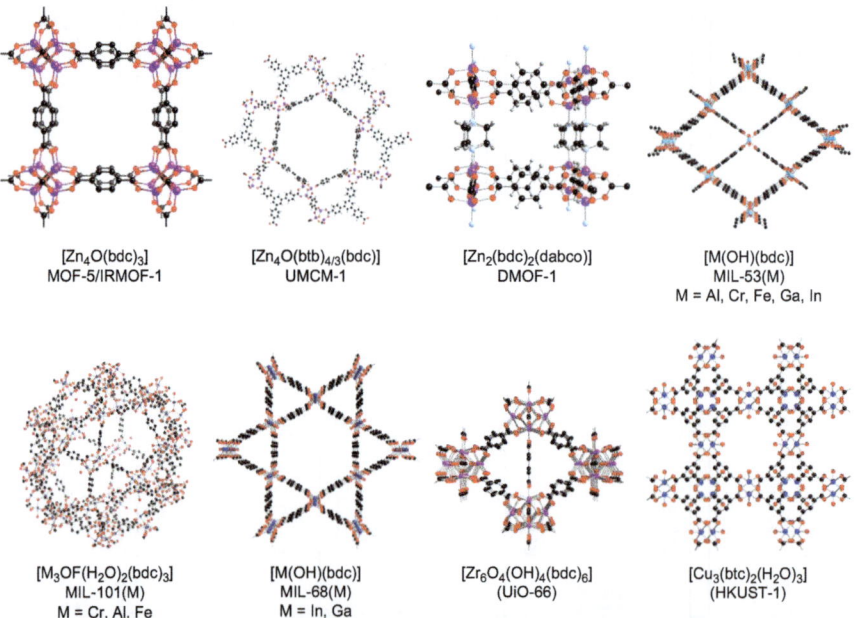

Figure 3.2   Some of the key MOF networks that have been employed in PSM studies.[12,26–32] For ligand abbreviations, see Figure 3.3.

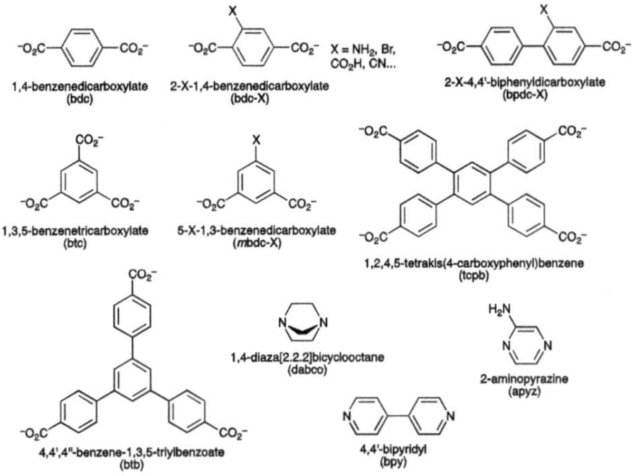

Figure 3.3 Examples of framework ligands and tagged ligands used in post-synthetic modification studies.

acronyms, whereas commonly employed ligands in PSM studies are shown in Figure 3.3.

In this review, solvent molecules incorporated in the pores and unco-ordinated to a metal centre are not included in the formulae presented. The acronym for a MOF structure that can be formed by more than one metal is denoted with the framework metal in brackets, for example MIL-53(Al), and a ligand functionalised with a tag group has the functionality shown after the normal ligand abbreviation, for example 2-amino-1,4-benzenedicarboxylate is abbreviated as bdc-NH_2. For a functionalised MOF that is isostructural to an unfunctionalised one, the functional group is shown in the acronym for the MOF. For example, MIL-53(Al)-NH_2 has the formula [Al(OH)(bdc-NH_2)] – it has the MIL-53(Al) structure, but contains bdc-NH_2 instead of bdc.

3.1.3 Evidence for Post-Synthetic Modification

In an ideal world, PSM processes would occur in single-crystal-to-single-crystal (SCSC) transformations and the modified group would be unambiguously identified through single crystal X-ray crystallography. In reality, SCSC transformations are unusual, and when they do occur the modified group is often not observable due to disorder. There are a few cases in which ordered, modified groups have been observed crystallographically,[19,33,34] but these are exceptions. In the absence of crystallographic data, a range of complementary techniques are normally employed together to provide evidence for the occurrence and extent of PSM.

Powder X-ray diffraction (PXRD) is important for verifying that the network structure remains unaltered during the PSM process. Normally for a PSM reaction to be useful, crystallinity needs to be maintained. For flexible

MOFs such as those based on MIL-53(Al), changes to the PXRD patterns are often observed on PSM, as the reactions can alter the shape of the pores and hence the unit cell parameters.

PSM reactions often do not go to completion, and when that is the case it is difficult to work out the degree of conversion from X-ray methods, as even with a crystal structure, disorder and low resolution may limit the information available. For covalent and dative PSM, NMR spectroscopy generally provides the best evidence for the degree of conversion. Typically, the modified MOFs are digested and the resultant solutions analysed by ^{1}H NMR spectroscopy.[13] Comparison of the integrals for the signals of the tagged and modified linkers allows the degree of conversion to be calculated. Zinc MOFs, such as those based on MOF-5, UMCM-1 and DMOF-1, are readily digested in DCl/DMSO-d_6,[35] however aluminium, chromium and zirconium MOFs do not generally decompose under these conditions. For MOFs based on MIL-53(Al) and UiO-66, digestion in HF is commonly used,[36,37] whereas those based on MIL-101(Cr) can be digested in base.[38] In all cases, care needs to be taken to ensure that digestion does not alter the modified tag. Other nuclei can give useful data, and solid state NMR is also informative.[39–41]

Electrospray ionisation mass spectrometry (ESI-MS) is useful for determining that conversion has occurred, and like most NMR methods it is carried out on digested MOFs. The high sensitivity of the technique means that modified linkers can be observed even when the degree of conversion is low. This makes it difficult to use this technique quantitatively. Aerosol time-of-flight mass spectrometry (ATOF-MS) has recently been used to show the occurrence of ligand exchange reactions in zirconium MOFs.[42]

Infrared spectroscopy provides evidence of changes in functional groups, and has been used to estimate the degree of conversion, especially in cases in which the modified tag is destroyed by digestion.[36,43] Hydrogen/deuterium exchange has been used to unambiguously assign NH(D) peaks.

Gas adsorption measurements are useful for showing that a MOF remains porous following PSM, though typically BET surface areas decrease with increasing size of the modified group due to the increase in the framework mass.

For dative and inorganic PSM processes that involve introduction or exchange of metal centres, a range of spectroscopic techniques have been employed to provide evidence of the metal centres and their environments. These include atomic absorption spectroscopy (AAS),[44] inductively coupled plasma atomic emission spectroscopy (ICP-AES),[45,46] energy-dispersive X-ray spectroscopy (EDX),[47] X-ray photoelectron spectroscopy[48] and extended X-ray fine structure (EXAFS) spectroscopy.[49]

3.2 Covalent Post-Synthetic Modification

The most common type of PSM in MOF chemistry involves a covalent transformation of the linking ligands. In this section, the different covalent modifications that have been reported are described.

3.2.1   Transformations of Amines

Post-synthetic modifications of amine groups have been well studied, with the vast majority of reports employing MOFs with 2-amino-1,4-benzenedicarboxylate (bdc-NH$_2$) linkers (Figure 3.3). Typically, the MOFs used are isoreticular to those formed with 1,4-benzenedicarboxylate (bdc). The first reactions were reported using [Zn$_4$O(bdc-NH$_2$)$_3$] (IRMOF-3), but subsequently efforts have expanded the range of substrates to include other zinc-based MOFs such as [Zn$_4$O(btb)$_{4/3}$(bdc-NH$_2$)] (UMCM-1-NH$_2$) and [Zn$_2$(bdc-NH$_2$)$_2$(dabco)] (DMOF-1-NH$_2$) as well as MOFs based on gadolinium, aluminium, iron, zirconium, indium and chromium.

3.2.1.1   Conversion to Amides

The conversion of an amine to an amide is the most studied PSM reaction in MOF chemistry. The reaction is typically carried out by reacting the MOF with an acid anhydride in chloroform or dichloromethane at room temperature (reaction (3.1)), though conversions are often improved by heating.

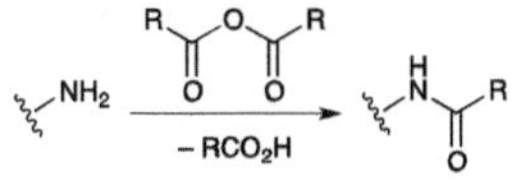

$$\text{(3.1)}$$

The first example of the amine to amide transformation was reported by Cohen and Wang in 2007. They showed that IRMOF-3 reacts with acetic anhydride in chloroform with the reaction going to completion in 2–3 days at room temperature.[13] This reaction has subsequently been developed to form part of an undergraduate experiment.[50]

The Cohen group subsequently demonstrated that this transformation could be undertaken with a wide range of acid anhydrides, and that the degree of conversion decreases with increasing chain length. For example, complete conversion to the amide occurs for (Me(CH$_2$)$_n$CO)$_2$O with $0 \leq n \leq 4$, but this decreases to 46% with $n = 8$ and to 20% with $n = 15$.[51] As the chain size increases, the MOFs become less sensitive to moisture and more hydrophobic.[52] IRMOF-3 membranes have been post-synthetically modified with heptanoic anhydride, thereby changing their surface properties. IRMOF-3 membranes preferentially adsorb CO$_2$ over n-propane, but this is reversed following modification.[53] The reaction with an anhydride has also been used to introduce chiral centres and to generate free carboxylic acid groups using cyclic reagents such as maleic anhydride (reaction (3.2)).[54] The introduced carboxylic acid groups have been further modified by coordination to a metal centre (see Section 3.3.2).

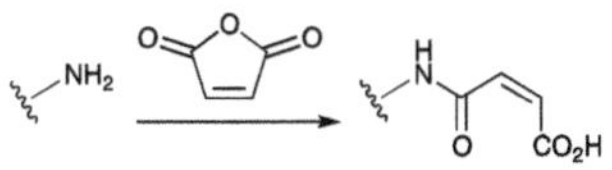

$$\text{(3.2)}$$

DMOF-1-NH$_2$ has smaller pores than IRMOF-3, and Cohen *et al.* demonstrated that reasonable conversions were only observed with short chain anhydrides. The degree of conversion is often improved by heating, and it was postulated that thermal energy allows the benzene rings to rotate and place the amino groups in sterically more favourable positions to react.[35] In the DMOF-1-NH$_2$ system, PSM has been used to modify breathing of the MOF. The unmodified MOF has open, large pores, as does the modified MOF to which a chain of 4–6 carbons has been added. However, on introduction of a smaller alkyl chain, the pores are narrower, which leads to much reduced BET surface areas. Reaction with (PrCO)$_2$O led to a product with a stepwise adsorption profile, consistent with transitions between large pore and narrow pore forms.[55]

When the reactions with acid anhydrides were carried out on UMCM-1-NH$_2$, slightly higher conversions than for IRMOF-3 were typically observed due to the larger pores present. However, the substituents generated by PSM are likely to occupy the smaller pores in this framework since the larger hexagonal pores are not lined by bdc-NH$_2$ ligands. This ensures that modified UMCM-1-NH$_2$ materials retain large BET surface areas.[35,56] The reactions of UMCM-1-NH$_2$ with 3-hydroxyphthalic anhydride and 2,3-pyrazinedicar-boxylic anhydride (reactions (3.3, 3.4)) gave MOFs containing metal-binding sites that can be further functionalised with catalytically active metal centres (see Section 3.3.2).[57,58]

(3.3, 3.4)

Cohen *et al.* showed that inclusion of phenyl groups into the substituents increases hydrogen adsorption, both in terms of capacity and heat of adsorption.[59] Electron-withdrawing aromatic substituents have been shown to enhance CO$_2$ adsorption to a greater extent than alkyl substituents.[60]

Although the conversion of an amine to an amide usually involves reaction with an acid anhydride, this is not always the case. Gamez *et al.* showed that [Gd$_2$(bdc-NH$_2$)$_3$(DMF)$_4$] reacts with acetic acid to give the methyl amide. In this compound only one of the three independent bdc-NH$_2$ groups in the structure can be modified due to steric constraints.[33]

In recent years, there has been an increased focus on functionalising MOFs that are more robust and less moisture sensitive than those formed by zinc. Stock *et al.* demonstrated that MIL-53(Al)-NH$_2$ reacts with formic acid to give [Al(OH)(bdc-NHC(O)H)].[61] Cohen *et al.* showed that MIL-53(Al)-NH$_2$ reacts with acid anhydrides, and as with IRMOF-3, increasing the size of the alkyl chain makes the MOF more hydrophobic. Indeed, using pentanoic anhydride leads to materials that are superhydrophobic, with water contact angles of greater than 150°.[52] Burrows, Marken *et al.* showed that the reaction of MIL-53(Al)-NH$_2$ with ferrocenecarboxylic anhydride allowed the introduction of redox-active ferrocene groups into the pores. They subsequently studied these electrochemically.[62]

Cohen *et al.* reported that the reaction of MIL-53(Al)-NH$_2$ with the cyclic reagents succinic and maleic anhydride (reaction (3.2)) goes with reasonable conversion (40–43%), and the product derived from maleic acid acts as a Brønsted acid catalyst for the methanolysis of small epoxides. Recyclability of the catalyst is poor due to methanolysis of the carboxylic acid, but activity was restored by treatment with HCl(aq) to regenerate the carboxylic acid group.[63]

Furukawa and Kitagawa *et al.* showed that core-shell MOFs containing DMOF-1-NH$_2$ on the outside could be modified on reaction with succinic anhydride, and that the modified MOF selectively accumulates *N,N*-dimethylaniline.[64] Fröba *et al.* recently prepared [Cu$_3$(L^1)$_2$], an analogue of HKUST-1, and showed that the amine groups could be modified with acetic anhydride.[65]

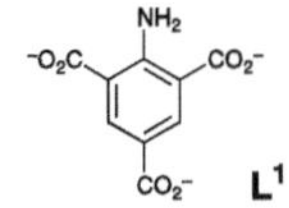

L^1

[M$_3$OF(H$_2$O)$_2$(bdc)$_3$] (MIL-101; M = Cr, Fe) is a stable, large pore MOF which is attractive for post-synthetic modification processes. Unfortunately, in the iron system, carrying out the MOF synthesis using H$_2$bdc-NH$_2$ instead of H$_2$bdc led to a product with a different, less porous structure. However, by using of a mixture of H$_2$bdc and H$_2$bdc-NH$_2$, Lin *et al.* showed that up to 17% bdc-NH$_2$ can be incorporated into the MIL-101(Fe) structure. Nanoparticulate MOFs were reacted with the ethoxysuccinato-cisplatin prodrug with 37% of the amino groups functionalised (reaction (3.5)). The nanoparticulate MOFs can be further modified by adding a silica shell in order to control release of the prodrug.[66]

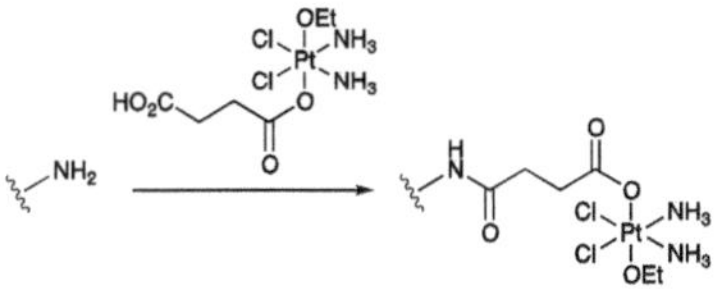

(3.5)

MIL-101(Cr)-NH$_2$ reacts with acid chlorides to form amides. This reaction was used by Stock *et al.* to prepare compounds such as [Cr$_3$OF(H$_2$O)$_2$(bdc-NH$_2$)$_{3-x}$(bdc-NHCOC$_6$H$_4$N = NPh)$_x$]. They then investigated photoisomerism reactions of the azo groups.[67]

The groups of Cohen and Tilset independently showed that the zirconium MOF UiO-66-NH$_2$ reacts with acid anhydrides in either chloroform or dichloromethane.[37,43] As with the more well-studied zinc systems, the degree of conversion decreases with increasing chain length. Vapour-phase post-synthetic modification was also reported on this system, and shown to increase the extent of the reaction.[68]

Farrusseng *et al.* demonstrated that IRMOF-3 and the diaminotriazine-based MOF [ZnF(L^2)] both react with nicotinyl chloride to give pyridyl amides (reaction (3.6)) that are catalytically active for aza-Michael condensation reactions and transesterifications.[69]

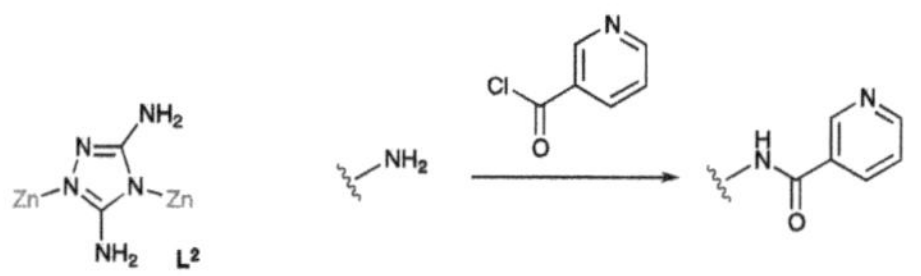

$$(3.6)$$

They also treated MIL-68(In)-NH$_2$ with L-Fmoc-Pro-OH together with bromo-tris-pyrrolidinophosphonium hexafluorophosphate and 4-(dimethyl-amino)pyridine (DMAP) to give the Fmoc-protected proline-modified MOF. Removal of the protecting group with piperidine gave MIL-68(In)-NHPro (reaction (3.7)),[70] which was studied by NMR spectroscopy.[41] A similar process gave the alanine-modified MOF, and in both cases the degree of conversion was 10%.

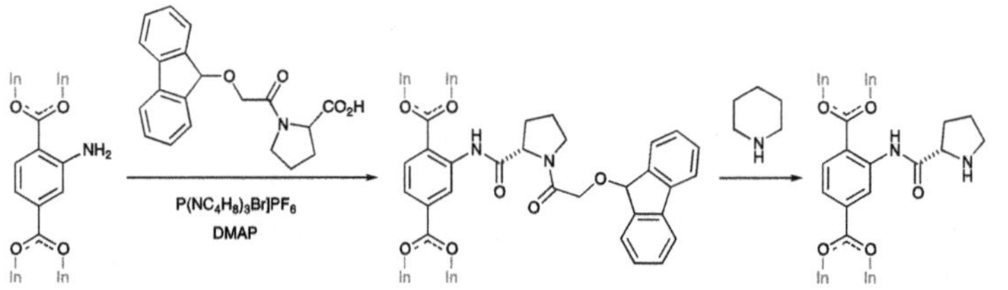

$$(3.7)$$

3.2.1.2 Conversion to (Thio)Ureas and (Thio)Cyanates

The reaction of an amino group with an isocyanate is expected to yield a urea, whereas the analogous reaction with an isothiocyanate is expected to give a thiourea (reactions (3.8,3.9)).

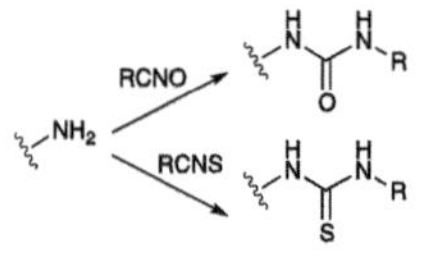

$$(3.8)$$

$$(3.9)$$

The first reported example of a reaction between an amino-functionalised MOF and an isocyanate was by Gamez *et al.*, though their reaction of [Gd$_2$(bdc-NH$_2$)$_3$(DMF)$_4$] with EtNCO did not lead to the expected urea. Instead, hydrolysis of the urea to a urethane (bdc-NHCO$_2$H) was observed. As with the reaction of [Gd$_2$(bdc-NH$_2$)$_3$(DMF)$_4$] with acetic acid (Section 3.2.1.1), modification occurs only on one of the three crystallographically independent bdc-NH$_2$ ligands due to steric constraints.[33]

Cohen *et al.* demonstrated that IRMOF-3 reacts with isocyanates to give urea-functionalised MOFs. Straight chain alkyl isocyanates gave decreasing conversion with increasing chain length, in a similar manner to that observed with acid anhydrides. Although *tert*-butyl isocyanate gave essentially no conversion, the isosteric trimethylsilyl isocyanate led to quantitative conversion, though in this case the product was the simple primary urea, formed following hydrolysis.[71] Phenyl isocyanate gave a higher degree of conversion than cyclohexyl isocyanate, and the product showed enhanced H_2 adsorption in comparison with IRMOF-3.[59]

Metzler-Nolte and Fischer *et al.* investigated the reaction of IRMOF-3 with fluorescein isothiocyanate (ArNCS) (reaction (3.10)). They used confocal laser scanning microscopy to show that fluorescence originated only from a very thin layer at the outer surfaces of the crystals. This showed that PSM is restricted to the crystal surfaces, with the steric bulk of the fluorescein group preventing the isocyanate from penetrating into the bulk.[72]

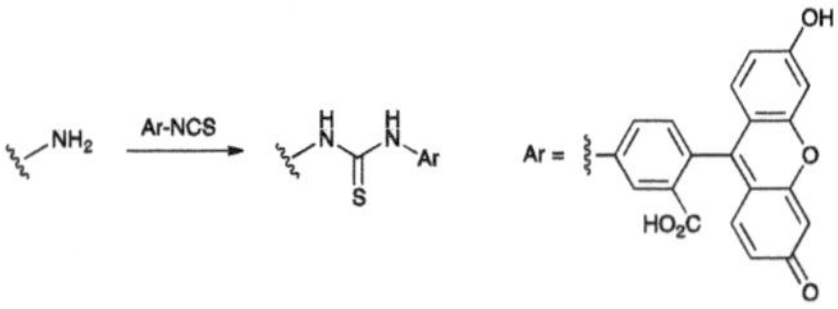

(3.10)

Bein *et al.* used a similar approach to anchor the dye Rhodamine B isothiocyanate onto thin films of the zirconium UiO-66 derivative $[Zr_6O_4(OH)_4(L^3)_6]$.[73]

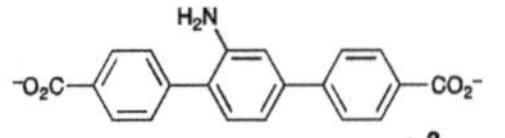

L^3

Stock *et al.* demonstrated that MIL-101(Cr)-NH$_2$ reacts with isocyanates and used this to introduce azo groups into the MOF. They found the *cis/trans* switching behaviour to be different to that in the amide-bound analogues.[67]

Shekhah *et al.* investigated the PSM of [Cu$_2$(bdc-NH$_2$)$_2$(dabco)] grown on Au(111) surfaces covered with an OH-terminated self-assembled monolayer fabricated from 11-mercaptoundecanol. They reacted these so-called SURMOFs with isocyanates, and used X-ray photoelectron spectroscopy to determine the extent of the conversion.[74]

Amino groups react with diphosgene or thiophosgene to form cyanates and thiocyanates, respectively. This reaction liberates HCl, so cannot be employed on zinc MOFs such as IRMOF-3 and UMCM-1-NH$_2$ as modification is accompanied by MOF disintegration. However, Volkringer and Cohen showed that MIL-53(Al)-NH$_2$ reacts with COCl(OCCl$_3$) and CSCl$_2$ to give the cyanate and thiocyanate, respectively, with 90% conversion (reactions (3.11–3.14). The products were further modified to ureas or thioureas on reaction with amines, and to carbamates or thiocarbamates on reaction with alcohols.[36]

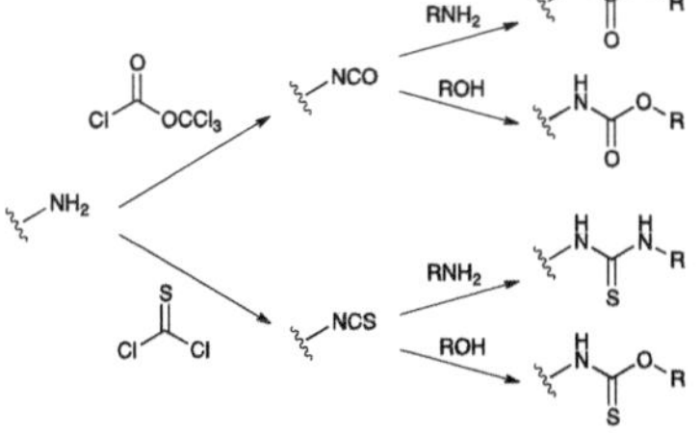

(3.11)

(3.12)

(3.13)

(3.14)

3.2.1.3   Conversion to Imines and Further Functionalisation

This section describes the reactions of amine-tagged MOFs with aldehydes, whereas the complementary reactions of aldehyde-tagged MOFs with amines are detailed in Section 3.2.5.1. Amines generally react with aldehydes to give imines (reaction (3.15)).

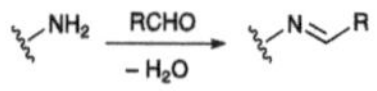

$$\text{\scriptsize NH}_2 \xrightarrow[-\,H_2O]{\text{RCHO}} \text{\scriptsize N}=\text{R} \qquad\qquad (3.15)$$

Reactions of amine-tagged MOFs with simple aldehydes have attracted only limited attention, in part due to the susceptibility of the products towards hydrolysis. Henderson *et al.* showed that $[Zn_2(cam)_2(apyr)]$ (cam $=(+)$-camphorate, apyr $=$ 2-aminopyrazine) reacts with acetaldehyde (MeCHO) to give functionalisation of the neutral apyr linker.[75] Yaghi *et al.* studied the reaction of UiO-66-NH$_2$ with acetaldehyde by solid state NMR spectroscopy and discovered that that imine was not formed. Instead, they concluded that the reaction gave a mixture of the hemiacetal and aziridine (reaction (3.16)), with heating under reduced pressure converting the former into the latter.[39]

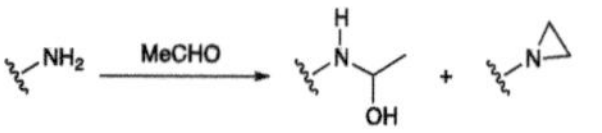

$$\qquad\qquad (3.16)$$

Rosseinsky *et al.* demonstrated that IRMOF-3 reacts with salicaldehyde, with a maximum degree of conversion of 13% (reaction (3.17)). The product was reacted with $[VO(acac)_2]$ to yield a catalytically active MOF (see Section 3.3.2).[76]

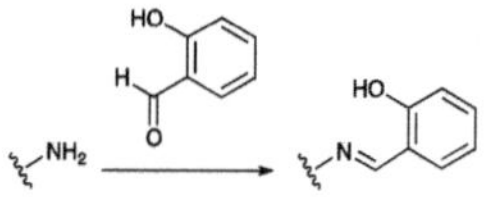

$$\qquad\qquad (3.17)$$

A similar approach was taken by Corma *et al.*, though under their conditions only 3% conversion could be achieved without framework collapse.[77]

Yaghi *et al.* treated UMCM-1-NH$_2$ with 2-pyridinecarboxaldehyde to produce the iminopyridine with quantitative conversion (reaction (3.18)), and then reacted the product with $[PdCl_2(MeCN)_2]$ (Section 3.3.1).[49]

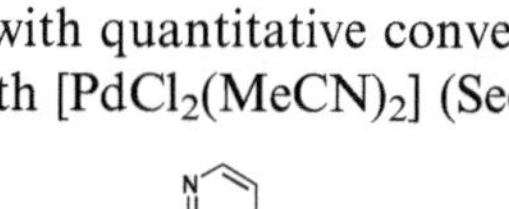

$$\qquad\qquad (3.18)$$

Ranocchiari and Van Bokhoven *et al.* reported that vapour-phase PSM could be used to enhance the degree of modification while decreasing reaction time. In this process, the reagent is in contact with the MOF only as a vapour. They reported that the reaction between IRMOF-3 and salicaldehyde vapour went to completion in 16 h, and that the analogous reaction with UiO-66-NH$_2$ also proceeded with a greater degree of conversion than under the normal solution conditions.[68]

Burrows and Keenan investigated tandem reactions on IRMOF-3, seeking to improve the scope of the reaction with an aldehyde by reacting the imine to give a product that is stable to hydrolysis. They showed that the imines formed on reaction with alkyl aldehydes could be reduced to secondary amines with sodium cyanoborohydride (reaction (3.19).[78]

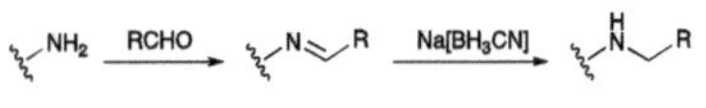

$$\text{\textasciitilde NH}_2 \xrightarrow{\text{RCHO}} \text{\textasciitilde N=R} \xrightarrow{\text{Na[BH}_3\text{CN]}} \text{\textasciitilde N\textasciitilde R} \qquad (3.19)$$

3.2.1.4 Other Transformations of Amines

Britt and Yaghi *et al.* demonstrated that IRMOF-3 reacts with 1,3-propane-sultone and 2-methylaziridine leading to ring-opened products (reactions (3.20,3.21)).

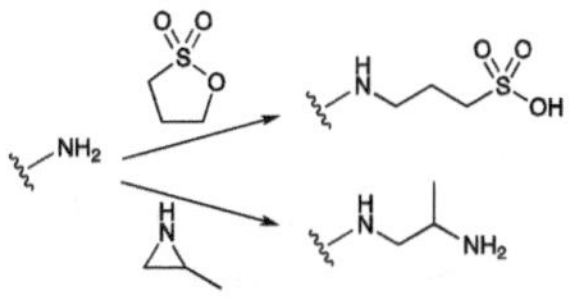

$$(3.20)$$

$$(3.21)$$

The reaction with 1,3-propanesultone led to 57% conversion to the alkylsulfonate. The reaction with 2-methylaziridine is more complicated; the product is a primary amine so more than one addition per linker is possible. Elemental analysis of the product suggested a mixture with zero, one, and two or more additions.[79]

Cohen *et al.* demonstrated that IRMOF-3 and UMCM-1-NH$_2$ react with NO gas to form *N*-diazenium diolates (reaction (3.22)). NO was released on digestion of the MOFs in a phosphate buffer.[80]

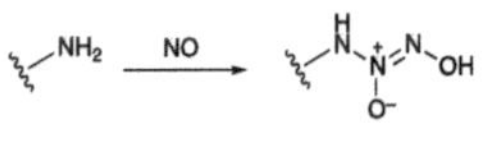

$$\text{\textasciitilde NH}_2 \xrightarrow{\text{NO}} \qquad (3.22)$$

Farrusseng *et al.* showed that DMOF-1-NH$_2$ and MIL-68(In)-NH$_2$ react with *tert*-butyl nitrite and trimethylsilyl azide with conversion of the amine groups to azides (reaction (3.23)).[81] They subsequently extended their study to [Al$_4$(OH)$_2$(OMe)$_4$(bdc-NH$_2$)$_3$] (CAU-1), MIL-53(Al)-NH$_2$ and MIL-101(Fe)-NH$_2$, in all cases observing complete conversion to the azide, though reactions were notably slower for MIL-68(In)-NH$_2$ and MIL-53(Al)-NH$_2$ perhaps reflecting the one-dimensional channels present in these structures.[82]

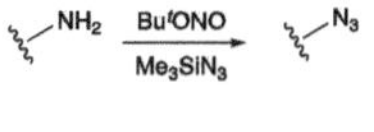

$$\text{\textasciitilde NH}_2 \xrightarrow[\text{Me}_3\text{SiN}_3]{\text{Bu}^t\text{ONO}} \text{\textasciitilde N}_3 \qquad (3.23)$$

MIL-68(In)-N$_3$ was subsequently converted into the isocyanate on reaction with CO at 120 °C, this providing an alternative pathway to the use of diphosgene (reaction (3.11)) to form these derivatives.[83]

Lin *et al.* showed that MIL-101(Fe) containing bdc-NH$_2$ reacts with the optical imaging contrast agent Br-BODIPY, in a process that converts the primary amine into a secondary amine (reaction (3.24)).[66]

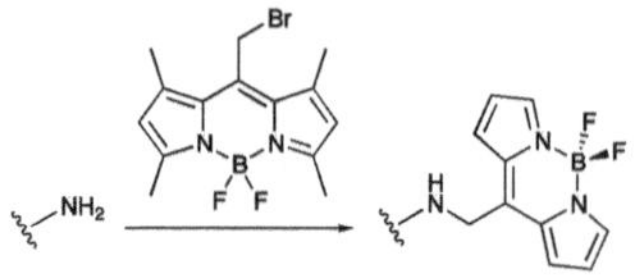

$$(3.24)$$

Yoo and Jeong demonstrated that IRMOF-3 reacts with cyanuric chloride to form an aryl amine (reaction (3.25)). The reaction generates HCl as a by-product, and this was observed to form macroscopic trenches in the crystals in a self-limited etching process.[84]

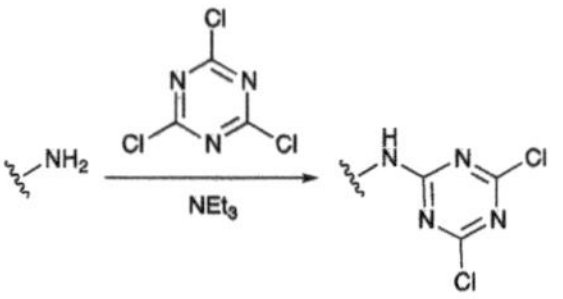

$$(3.25)$$

Alkylation has also been observed for $[Al_4(OH)_2(OMe)_4(bdc\text{-}NH_2)_3]$ (CAU-1). In a detailed study of the synthesis of this MOF, Senker, Stock *et al.* showed that $[Al_4(OH)_2(OMe)_4(bdc\text{-}NHMe)_3]$ was generally also formed, and was derived from post-synthetic methylation of CAU-1. Long reaction times also led to the formation of $[Al_4(OH)_2(OMe)_4(bdc\text{-}NMe_2)_3]$.[85]

Burrows *et al.* treated MIL-101(Cr)-NH_2 with nitrite and an acid to form the diazonium salts, and isolated the tetrafluoroborate. The chloride salt reacts with aqueous iodide to give MIL-101(Cr)-I (reaction (3.26)), and with phenol and sodium carbonate to give an azo-coupled product (reaction (3.27)). Heating the tetrafluoroborate salt led to MIL-101(Cr)-F (reaction 3.28).[86]

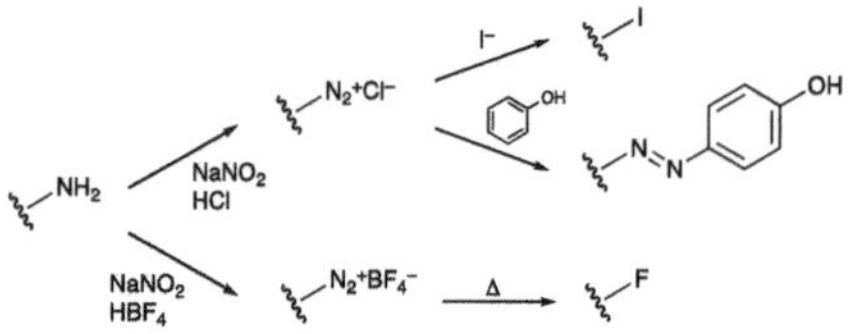

$$(3.26)$$

$$(3.27)$$

$$(3.28)$$

3.2.2 Transformations of Azides

Azide groups undergo [3 + 2] cycloaddition reactions in the presence of a copper(I) catalyst, the so-called 'click reaction', to form 1,2,3-triazoles. Azides introduced into MOFs both pre- and post-synthetically have been shown to undergo this reaction (3.29).

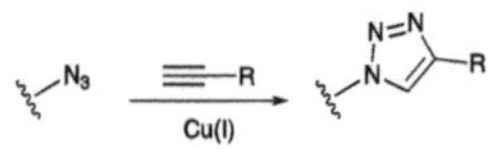

$$(3.29)$$

The first reported examples were from Sada *et al.*, who demonstrated that a zinc MOF isoreticular to MOF-5 and including the functionalised ter-phenylenedicarboxylate L^4 reacts with the terminal alkynes $CH{\equiv}CCO_2Me$, $CH{\equiv}CCH_2OH$ and $CH{\equiv}CBu$ in the presence of CuBr to give 1,2,3-triazoles.

In contrast, use of $CH\equiv CCH_2NH_2$ and $CH\equiv CCO_2H$ led to framework dissolution.[87]

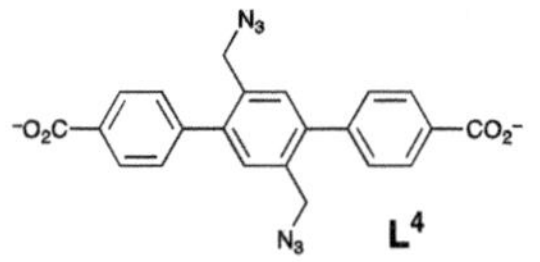

Farrusseng *et al.* prepared a range of azide-tagged MOFs from the PSM of amines (Section 3.2.1.4) and have used these to prepare 1,2,3-triazoles. They carried out reactions with $CH\equiv CPh$, $CH\equiv CCH_2NEt_2$, $CH\equiv CBu$, $CH\equiv CCH_2OH$, $CH\equiv CCO_2H$ and $CH\equiv CCH_2NH_2$ in the presence of $[Cu(MeCN)_4]PF_6$, and demonstrated that conversions can be quantitative.[81,82] For the MIL-68(In) system, the triazole derived from $CH\equiv CCH_2NH_2$ was shown to retain copper, presumably coordinated to the bidentate ligand, and on heating under vacuum this catalysed partial reduction of unfunctionalised azide groups to primary amines.[88] Triazole-containing MOFs derived from DMOF-1-NH$_2$ were shown to be transesterification catalysts, with a combination of introduced tertiary amino groups and phenyl groups used to enhance the activity.[89]

Rosi *et al.* used a strain-promoted 'click reaction' to react azides with cyclooctyne derivatives, thus eliminating the need for a catalyst. This approach (reaction (3.30)) was used to introduce reactive groups onto the $[Zn_8O_2(adeninate)_4(bpdc)_6]^{4-}$ (bio-MOF-100) framework. By using an appropriate R group, the products were able to be further modified on reaction with peptides.[90]

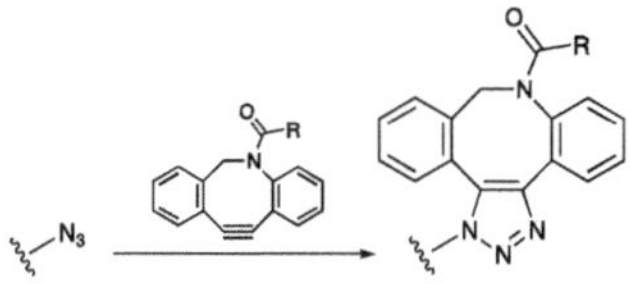

(3.30)

Azides can also be photoactivated to form nitrenes. Kitagawa *et al.* used 5-azido-1,3-benzenedicarboxylate (*m*bdc-N$_3$) to form $[Zn_2(m\text{bdc-N}_3)_2(bpy)_2]$ (CID-N$_3$) and then irradiated this with ultraviolet radiation to convert the azide groups into nitrenes. Formation of triplet nitrenes was confirmed by ESR spectroscopy, and single crystal X-ray diffraction measurements show the transformation occurs with a change in space group. The nitrene groups react with oxygen to give nitro- and nitroso- functionalities, and with CO to give isocyanate groups.[34]

Sada *et al.* demonstrated that azides can be converted to primary amines by a Staudinger reduction using triphenylphosphine and water. The bulky phosphine reagent limits the reaction to the crystal surfaces. The amines were detected by further reaction with the *N*-hydroxysuccinic ester of carboxy-fluorescein, to convert them into fluorescent amides.[91]

3.2.3   Transformations of Alcohols

Post-synthetic modification of alcohol groups has received less attention than amine transformations, in part because of the difficulties of including unreacted hydroxyl tags into MOFs. However, there have been a few examples of conversions of alcohols into esters, including the first example of covalent PSM. Lee *et al* prepared a range of silver networks with functionalised trinitrile ligands such as L^5. They demonstrated that $[Ag(L^5)]OTf$ reacts with $(CF_3CO)_2O$ vapour to convert the pendant alcohol into an ester (reaction (3.31)).[10]

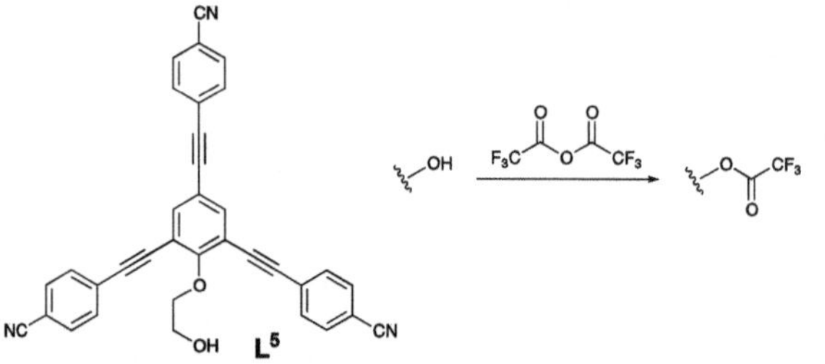

$$\tag{3.31}$$

Hupp, Nguyen *et al.* prepared $[Zn(tcpb)(L^6)]$ (tcpb $= 1,2,4,5$-tetrakis(4-carboxyphenyl)benzene, Figure 3.3) which has a three-dimensional structure with OH groups on the neutral pillaring ligands. Reaction with succinic anhydride converts almost all of the alcohol groups into esters, generating a MOF containing free carboxylic acid functionalities (reaction (3.32)).[92]

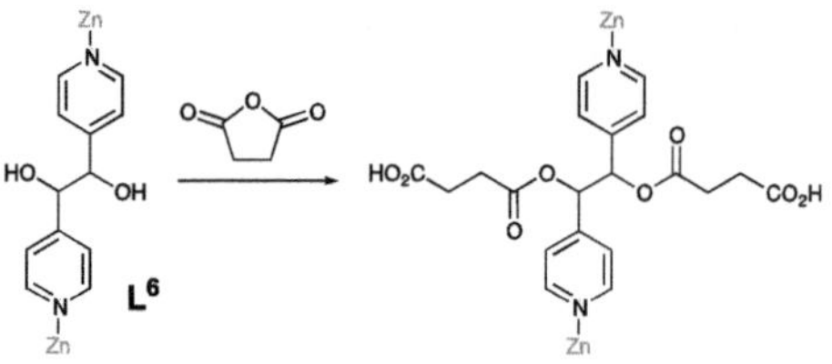

$$\tag{3.32}$$

3.2.4   Transformations of C–C Multiple Bonds

Jones and Bauer reported the synthesis of the stilbene-based MOF $[Zn_4O(sdc)_3]$ (sdc $= trans$-4,4'-stilbenedicarboxylate), and showed that post-synthetic modification with Br_2 in chloroform gave 60% conversion after 48 h at room temperature and full conversion after 24 h at 100 °C in 1,1,2-trichloroethane (reaction (3.33)).

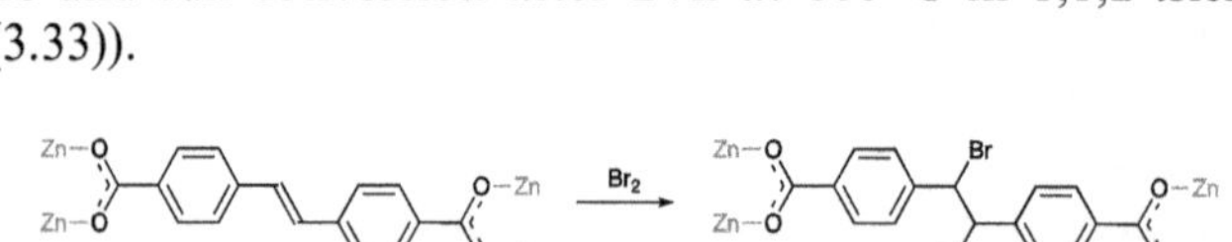

$$\tag{3.33}$$

Only the *meso* isomer of the brominated linker was observed. Formation of this dicarboxylate inside the MOF prevents C–C rotation and, as a consequence, the *rac* isomer was not produced.[93]

Rieger *et al.* prepared [Zn$_2$(tdc)$_2$(**L^7**)] (tdc = 9,10-triptycenedicarboxylate) in which **L^7** is a modified 4,4′-bipyridyl linker containing an alkene group. They treated this MOF with dimethyldioxirane (DMDO) to quantitatively form the epoxide. They then reacted the epoxide-functionalised MOF with ethanethiol (EtSH) to ring open the product giving a sulfide with 50% conversion (reaction (3.34)).[94]

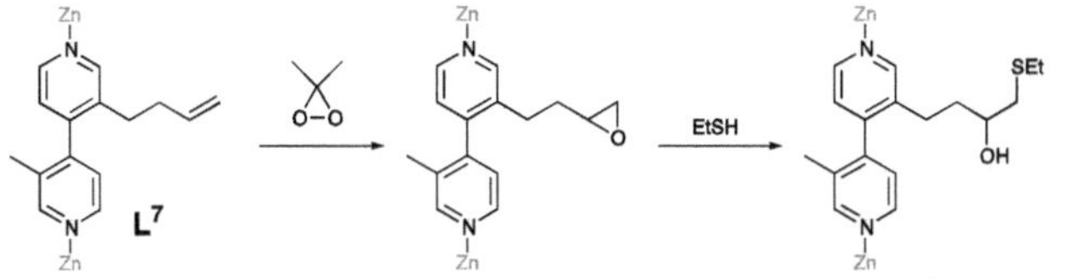

(3.34)

Hupp and Nguyen *et al.* investigated tandem reactions starting with the (trimethylsilyl)ethynyl functionalised MOF [Zn$_2$(2,6-ndc)$_2$(**L^8**)] (ndc = 2,6-naphthalenedicarboxylate). They showed that the trimethylsilyl groups on the crystal surfaces could be removed by reacting with a fluoride, and that the primary alkyne products underwent copper-catalysed [3 + 2] cycloaddition reactions with ethidium bromide monoazide (reaction (3.35)). This reagent was chosen because of its fluorescence, which was used to demonstrate that the reaction had occurred.[95]

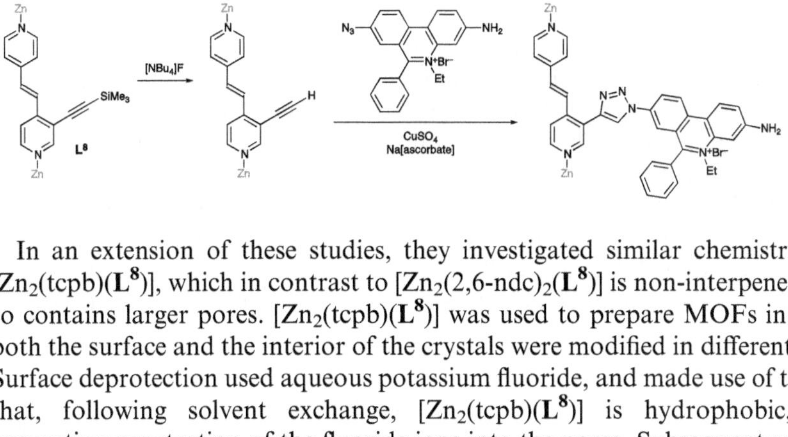

(3.35)

In an extension of these studies, they investigated similar chemistry with [Zn$_2$(tcpb)(**L^8**)], which in contrast to [Zn$_2$(2,6-ndc)$_2$(**L^8**)] is non-interpenetrated, so contains larger pores. [Zn$_2$(tcpb)(**L^8**)] was used to prepare MOFs in which both the surface and the interior of the crystals were modified in different ways. Surface deprotection used aqueous potassium fluoride, and made use of the fact that, following solvent exchange, [Zn$_2$(tcpb)(**L^8**)] is hydrophobic, thus preventing penetration of the fluoride ions into the pores. Subsequent reaction with ethidium bromide monoazide converted the terminal alkyne groups into triazoles. This surface-modified material was subsequently treated with NEt$_4$F to deprotect the interior alkynes, followed by benzyl azide, which converted the interior terminal alkynes into triazoles. In this way, all of the silyl-protected alkyne groups were converted into triazoles, but with different substituents on the surfaces and in the interior of the crystals.[96]

Duan *et al.* prepared [Zn$_2$(*m*bdc-OCH$_2$C≡CH)$_2$(bpy)$_2$], and demonstrated that the alkyne group reacts with chiral azides based on L-AMP or D-AMP under copper-catalysed conditions (reaction (3.36)) with 80% conversion.[97]

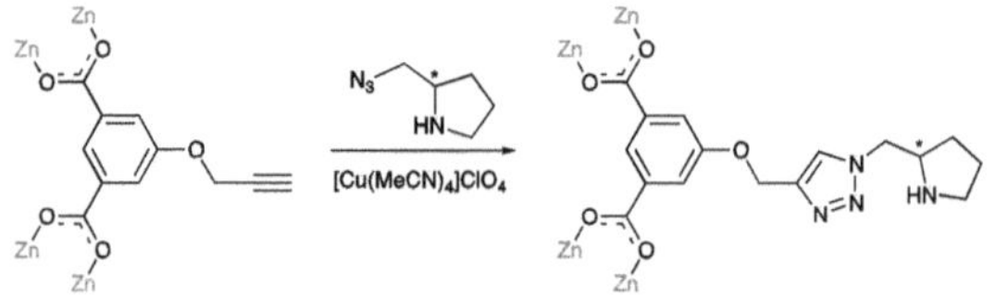

(3.36)

The resulting chiral MOFs are active catalysts for asymmetric aldol reactions between aromatic aldehydes and cyclohexanone. The MOFs showed better enantioselectivity than L-AMP under homogeneous conditions. Godt *et al.* also demonstrated the use of 'click reactions', in this case between alkyne-functionalised zirconium MOFs and azides. In addition, they showed that furan-functionalised analogues react with maleimides (reaction (3.37)) or maleic anhydride in Diels–Alder reactions.[98]

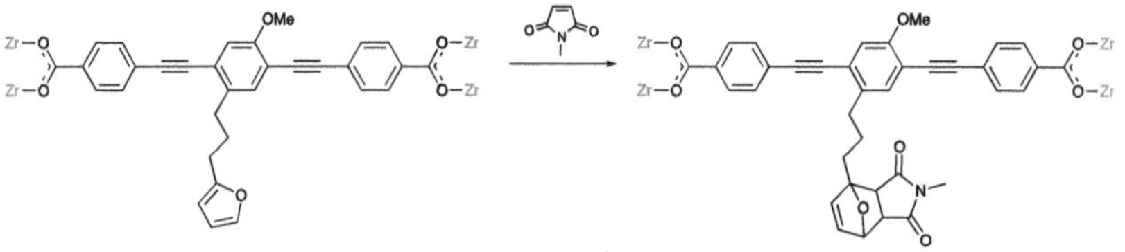

(3.37)

It is well established that alkenes can undergo $[2+2]$ photodimerisation reactions in the solid state if the double bonds are close enough together, and crystal engineering approaches have used hydrogen bonds to facilitate this.[99] Recent reports have demonstrated that this cycloaddition reaction can occur in a coordination network. Lang *et al.* showed that a silver coordination polymer containing *trans*-1,2-bis(4-pyridyl)ethylene (bpe) could undergo photodimerisation.[100] They later showed that the cadmium coordination polymer $[Cd(bpe)(O_2CC_6H_4Cl-4)_2]$ underwent the same reaction in a SCSC transformation.[101] Vittal *et al.* showed that compounds such as $[Zn(bpe)(fumarate)]$ formed interpenetrated 3D networks with the double bonds from the two networks less than 4 Å apart. Following UV irradiation, the compounds were converted into the cyclobutane derivatives in SCSC transformations (reaction (3.38)).[102]

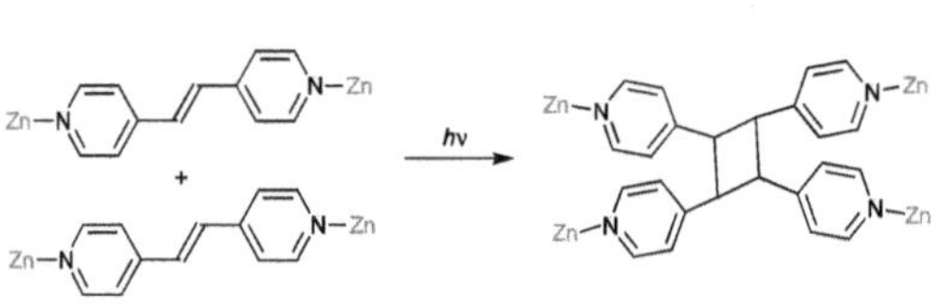

(3.38)

They subsequently showed that the cyclobutane rings in related photo-dimerised products could be cleaved by heating.[103]

Lang *et al.* showed that $[2+2]$ cycloaddition reactions could occur with coordination polymers containing 1,4-bis(2-(4-pyridyl)ethenyl)benzene (bpeb), which contains two double bonds. With this ligand there are two possible dimerisation products, depending on the relative positions of the ligands in the solid state. For $[Zn_4(OH)_2(mbdc-SO_3)_2(bpeb)_2]$, dimerisation occurs out-of-phase (reaction (3.39)), whereas for $[Cd_2(pda)_2(bpeb)_2]$ (pda = 1,3-phenylenediacetate) it occurs in-phase, with both double bonds undergoing the reaction (reaction (3.40)).[104]

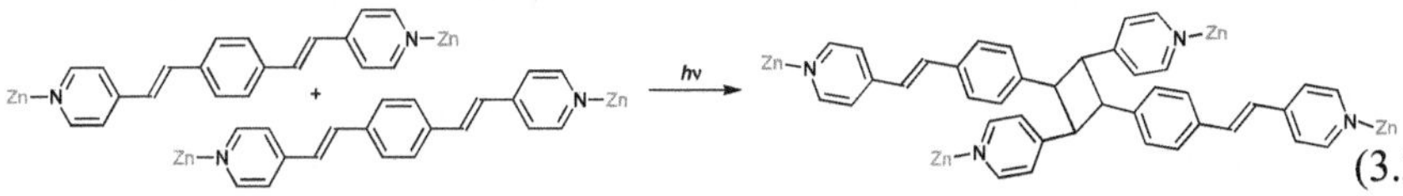

$$(3.39)$$

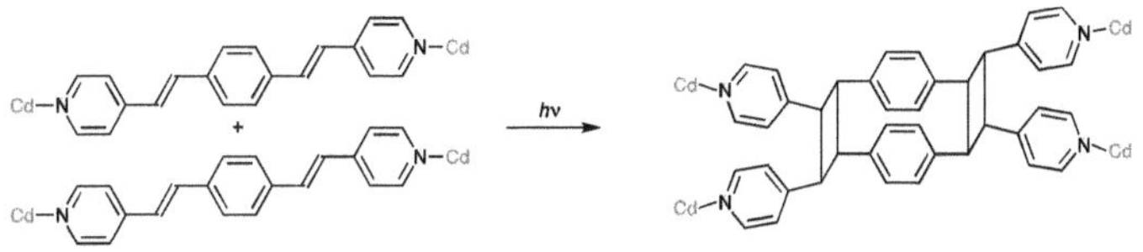

$$(3.40)$$

3.2.5 Other Transformations at Carbon

3.2.5.1 Reactions of Aldehydes and Carboxylic Acids

The reaction of amine-tagged MOFs with aldehydes to form imines was described in Section 3.2.1.3. Imines can also be formed from the reaction of an aldehyde-tagged MOF with an amine. Burrows *et al.* used an aldehyde-tagged 4,4'-biphenyldicarboxylate (bpdc-CHO) linker, and showed that $[Zn_4O(bpdc-CHO)_3(H_2O)_2]$ reacts with 2,4-dinitrophenylhydrazine ($ArNHNH_2$) with conversion of 60% of the aldehyde groups into hydrazones (reaction (3.41)). Use of the mixed-ligand MOF $[Zn_4O(bpdc-OMe)_{2.7}(bpdc-CHO)_{0.3}]$, containing fewer reactive tag groups, enabled the PSM reaction to proceed to completion.[19]

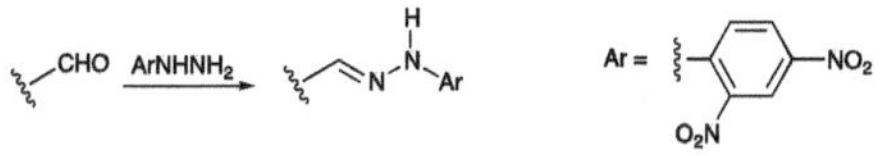

$$(3.41)$$

Yaghi *et al.* reported that the aldehyde-tagged imidazolate framework $[Zn(L^9)_2]$ (ZIF-90) reacts with ethanolamine with quantitative conversion to the imine after 3 h at 60 °C (reaction (3.42)).[105] Huang and Caro used this reaction to improve the H_2/CO_2 selectivity of ZIF-90 membranes.[106]

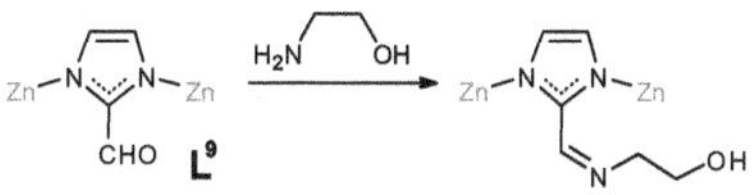

$$(3.42)$$

Yaghi *et al.* also demonstrated that ZIF-90 was reduced by sodium borohydride in methanol at 60 °C to give the primary alcohol (reaction 3.43). Conversion was approximately 80% after 24 h.[105]

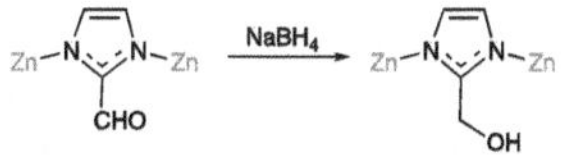

$$(3.43)$$

Aguado, Farrusseng *et al.* carried out studies with the ZIF material $[Zn(mica)_2]$ (SIM-1, mica = 4-methyl-5-imidazolecarboxaldehyde). SIM-1 has the sodalite structure, and the aldehyde groups were modified by reaction with amines, with the dodecylamine product shown to be hydrophobic. The modified product exhibits enhanced catalytic activity in Knoevenagel condensations.[107] The same modification reaction has been shown to take place when SIM-1 is coated onto alumina beads or tubes, with the latter efficient for CO_2/N_2 separation under humid conditions.[108,109]

MOFs with carboxylic acid tags are uncommon, but Stock *et al.* showed that $[Al(OH)(bdc-CO_2H)]$ (MIL-53(Al)-CO_2H) can be prepared directly from aluminium chloride and 1,2,4-benzenetricarboxylic acid using hydrothermal conditions. On heating the product under reduced pressure, they observed condensation of the carboxylic acid tags to form acid anhydrides (reaction (3.44)). In addition, partial decarbonylation was observed at high temperatures.[110]

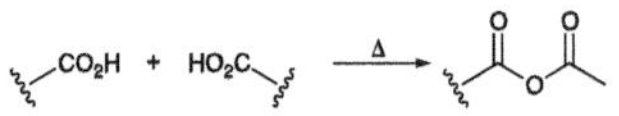

$$(3.44)$$

3.2.5.2 Reactions of CH Groups

There are a few examples of covalent PSM that do not require the presence of a tag group. Stock *et al.* showed that the framework of MIL-101(Cr) is sufficiently stable to acid that the aromatic rings can be nitrated with a mixture of concentrated sulfuric and nitric acids at $0\,^{\circ}C$, the reaction going to completion within 5 h. The resultant nitro groups were reduced to amines using tin(II) chloride, and these were further functionalised by reaction with ethyl isocyanate (reaction (3.45)).[38]

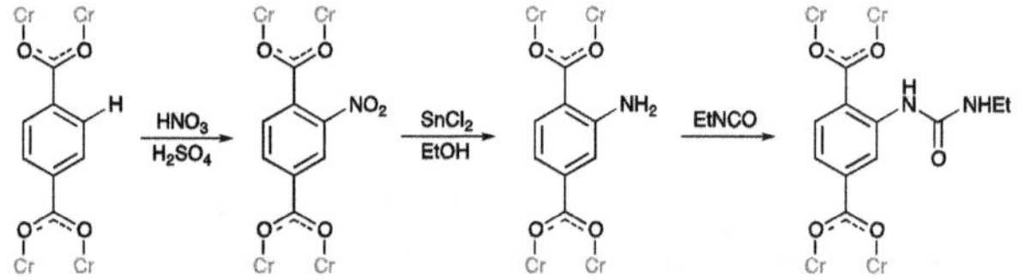

$$(3.45)$$

Glorius *et al.* showed that a CH group on an indole-dicarboxylate ligand could be substituted with a phenyl group. The UMCM-1 analogue $[Zn_4O(btb)_{4/3}(\mathbf{L^{10}})]$ reacts regioselectively with $[Ph_2I]BF_4$ in DMF at room temperature in the presence of palladium(II) acetate (reaction 3.46).[111]

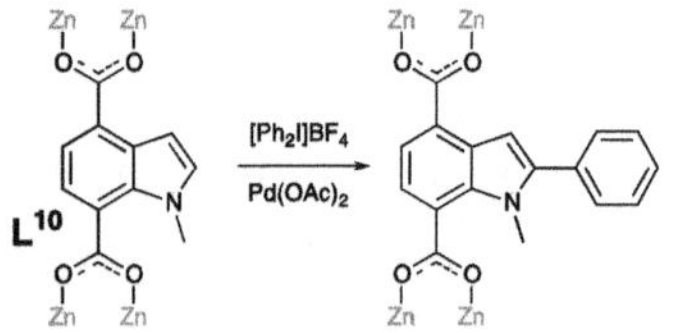

$$(3.46)$$

The conversion was quantitative after 5 days at room temperature, though attempted reactions with the MOF-5 analogue $[Zn_4O(\mathbf{L^{10}})_3]$ gave no conversion due to the smaller windows into the pores.

Goesten, Gascon *et al.* developed a route to add chloromethylene groups into MOFs. They treated MIL-53(Al)-NH_2 and MIL-101(Cr) with methoxyacetyl chloride and aluminium chloride, and demonstrated the potential of the approach by further reacting the modified MIL-101(Cr) material with LiPPh$_2$ to give a MOF in which 8% of the bdc linkers were modified with phosphine groups (reaction (3.47)).[112]

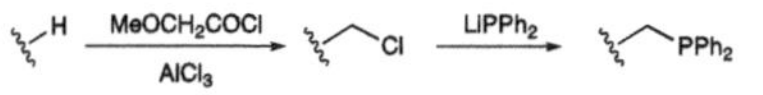

$$\text{\begin{array}{ccc} \text{\small H} & \xrightarrow[\text{AlCl}_3]{\text{MeOCH}_2\text{COCl}} & \text{\small Cl} & \xrightarrow{\text{LiPPh}_2} & \text{\small PPh}_2 \end{array}} \qquad (3.47)$$

Gascon *et al.* also demonstrated that MIL-101(Cr) and MIL-53(Al) could be sulfonated using trific anhydride and sulfuric acid.[113]

Richardson *et al.* showed that the allyloxy-functionalised MOF [Zn$_4$O(L^{11})$_3$] undergoes a solventless Claisen rearrangement on heating to 260 °C (reaction (3.48)). The progress of the reaction could be followed by simultaneous thermogravimetric-differential thermal analysis.[114]

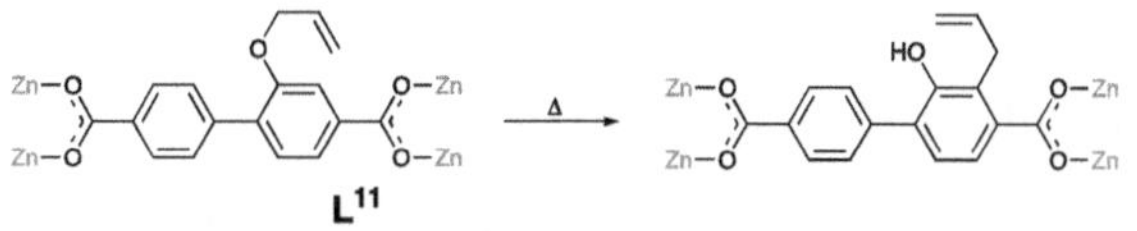

$$\qquad (3.48)$$

$$\mathbf{L^{11}}$$

3.2.5.3 Substitution of Halides

Cohen *et al.* prepared the zirconium MOF [Zr$_6$O$_4$(OH)$_4$(bdc-Br)$_6$] (UiO-66-Br), and showed that this can undergo post-synthetic cyanation. The reaction with CuCN in DMF at 140 °C gave 43% conversion after 24 h, though use of microwave irradiation enabled 90% conversion after 10 min in *N*-methyl-2-pyrrolidone.[115] Mixed-ligand zirconium MOFs of the general formula [Zr$_6$O$_4$(OH)$_4$(bdc-NH$_2$)$_{6-x}$(bdc-Br)$_x$] were also prepared, and the two tags modified sequentially, with the amine and bromide converted to an amide and nitrile respectively (reactions (3.49, 3.50)).

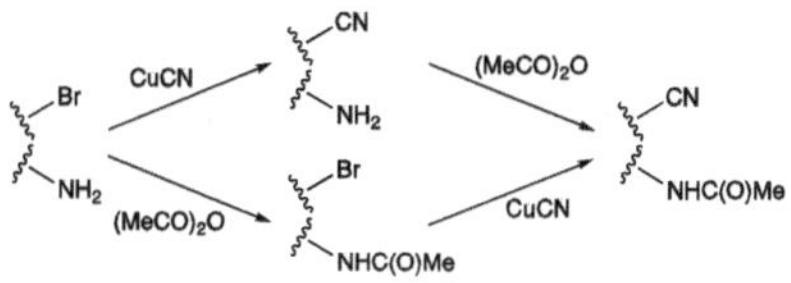

$$\qquad (3.49)$$

$$\qquad (3.50)$$

3.2.6 Reactions at Heteroatoms

One of the earliest PSM transformations in MOF chemistry was the quaternation of pyridine groups. Kim *et al.* showed that [Zn$_3$O(L^{12})$_6$] (D-POST-1) has a structure in which three of the pyridine groups connect to neighbouring Zn$_3$O units, whereas the other three are uncoordinated. Treatment of D-POST-1 with alkyl iodides led to the conversion of the free pyridine rings to *N*-alkylpyridinium ions, thus making the framework cationic. The pore volume shrinks by 14% and 60% upon alkylation with iodomethane and 1-iodohexane respectively.[11]

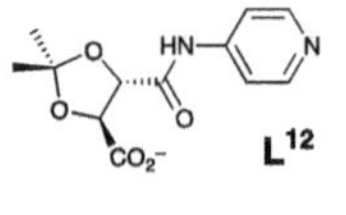

L^{12}

Burrows *et al.* looked at the oxidation of sulfide tags, using dimethyl-dioxirane (DMDO) as a small molecular oxidant (reaction (3.51)). They showed that when $[Zn_4O(bpdc\text{-}CH_2SMe)_3]$, which has a doubly-interpenetrated structure containing a linear sulfide, was treated with DMDO, conversion to the sulfone was incomplete. However, in an analogous system containing a cyclic sulfide, 100% modification was observed.[116]

$$(3.51)$$

3.2.7 Deprotection Reactions

The majority of PSM reactions involve increasing the size and complexity of the functional groups projecting into the pores. However, the role of a recently developed type of PSM reaction is to make this group smaller, simultaneously unmasking a protected functional group. This class of reactions has been termed post-synthetic deprotection (PSD) by Cohen,[117] and PSD has been achieved by thermolysis and photolysis.

3.2.7.1 Thermolysis Reactions

Telfer *et al.* showed that a *tert*-butylcarbamate-functionalised 4,4′-biphe-nyldicarboxylate (bpdc-NHBoc) could be incorporated into an IRMOF structure, giving $[Zn_4O(bpdc\text{-}NHBoc)_3]$. This forms a non-interpenetrated network due to the steric demands of the NHBoc group. On heating to 150 °C, the NHBoc group decomposed to a primary amine (reaction (3.52)), generating carbon dioxide and 2-methylpropene which were readily lost due to their volatility.[118]

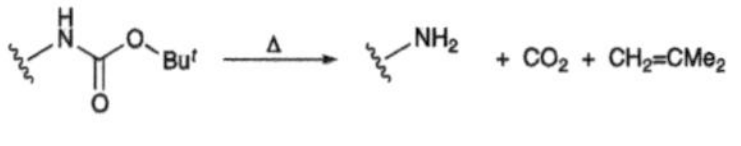

$$(3.52)$$

The group used a similar methodology (reaction (3.53)) to form a proline-functionalised IRMOF, $[Zn_4O(L^{13})_3]$ which was shown to be an active catalyst for the asymmetric aldol reaction.[119]

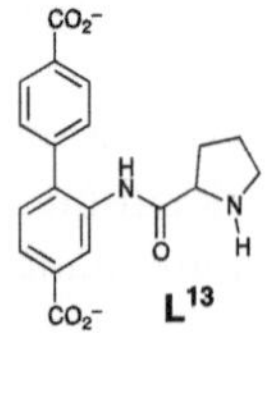

L^{13}

$$(3.53)$$

3.2.7.2 Photolysis Reactions

Nitrobenzyl groups are well established as photocleavable protecting groups for alcohols and amines. This is important on MOF chemistry, as MOFs containing bdc-OH and bdc(OH)$_2$ are difficult to prepare directly. Cohen *et al.* showed that one or two nitrobenzyl ether groups could be appended to 1,4-benzenedicarboxylate, and that the resulting ligands could be combined with zinc(II) and H$_3$btb (Figure 3.3) to form analogues of UMCM-1 with the formulae [Zn$_4$O(btb)$_{4/3}$(bdc-OCH$_2$C$_6$H$_4$NO$_2$-2)] and [Zn$_4$O(btb)$_{4/3}${bdc-(OCH$_2$C$_6$H$_4$NO$_2$)$_2$-2,3}]. On irradiation with laser light for 24 h, [Zn$_4$O(btb)$_{4/3}$(bdc-OCH$_2$C$_6$H$_4$NO$_2$-2)] was converted quantitatively to [Zn$_4$O(btb)$_{4/3}$(bdc-OH)] (reaction (3.54)), whereas for the photolysis of [Zn$_4$O(btb)$_{4/3}$(bdc-{OCH$_2$C$_6$H$_4$NO$_2$)$_2$-2,3}], 75% conversion to [Zn$_4$O(btb)$_{4/3}${bdc-(OH)$_2$-2,3}] was observed.[120]

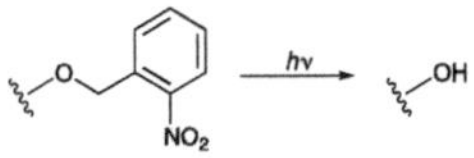

$$\tag{3.54}$$

Nitrobenzyl-ether-functionalised ligands have also been incorporated into IRMOF structures such as [Zn$_4$O(bdc-OCH$_2$C$_6$H$_4$NO$_2$-2)$_3$], and photolysis used to prepare [Zn$_4$O(bdc-OH)$_3$].[121]

Telfer *et al.* showed that the same nitrobenzyl groups could be incorporated into bpdc linkers, and these were included into [Zn$_4$O(bpdc-OCH$_2$C$_6$H$_4$NO$_2$-2)$_3$] which has an IRMOF structure. Photolysis of this 2-nitrobenzyl ether was achieved by irradiation of this compound with laser light giving [Zn$_4$O(bpdc-OH)$_3$]. On powdered samples, the photolysis reaction went to completion, though with large single crystals it stopped at about 50% conversion.[122]

3.2.8 Reduction and Oxidation of Bridging Ligands

It has been established both theoretically and experimentally that inclusion of alkali metal centres into MOFs can enhance their H$_2$ adsorption properties.[123] With this in mind, Hupp *et al.* prepared [Zn$_2$(2,6-ndc)$_2$(L^{14})] (2,6-ndc = 2,6-naphthalenedicarboxylate)[124,125] and [Zn$_2$(2,6-ndc)$_2$(L^{15})],[126] both of which have three-dimensional structures with reducible pillaring ligands.

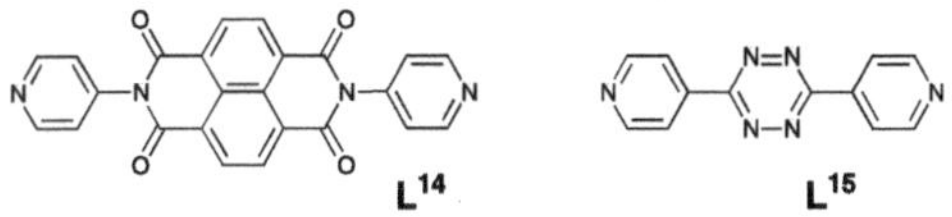

On treatment with lithium, sodium or potassium naphthalenide, the L^{14} or L^{15} linkers were partially reduced, giving MOFs that formally contain both the neutral linker and its monoanion, with the charges balanced by the included Group 1 cations. The most promising materials for H$_2$ adsorption were those in which the cation incorporation was low. For example, the H$_2$ uptake in

$Na_{0.24}[Zn_2(2,6\text{-}ndc)_2(\mathbf{L^{15}})]$ was enhanced by 43% over that for $[Zn_2(2,6\text{-}ndc)_2(\mathbf{L^{15}})]$.

Cheon and Suh showed that the bridging ligand in $[Zn_3(ntb)_2(EtOH)_2]$ (ntb = 4,4′,4″-nitrilotrisbenzoate) was oxidized to a nitrogen radical by palladium(II), which in turn was reduced to palladium nanoparticles. The resultant MOF showed enhanced uptake of H_2.[127]

Lin *et al.* reported that the ruthenium complex $\mathbf{L^{16}}$ could be incorporated into IRMOF-type structures, and both doubly-interpenetrated and non-interpenetrated supramolecular isomers of $[Zn_4O(\mathbf{L^{16}})_3]$ were isolated. These could be reduced with $LiBEt_3H$, converting ruthenium(III) to ruthenium(II), and re-oxidised in air in SCSC processes. The non-interpenetrated Ru(II)-containing MOF was shown to be an asymmetric cyclopropanation catalyst.[128]

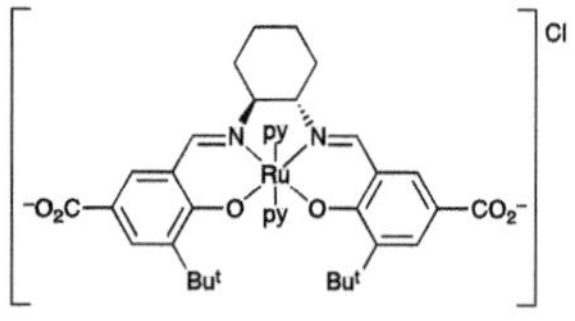

3.3 Dative Post-synthetic Modification

This section describes post-synthetic coordination of a metal centre to one of the linker ligands. In many cases, the ligands are assembled by post-synthetic modification reactions of the type described in Section 3.2, so the dative PSM reaction forms the second part of a tandem process. In this review, dative PSM reactions are divided into two types – those in which the metal coordinates to a neutral site on the ligand, and those in which coordination occurs together with deprotonation.

3.3.1 Coordination to a Neutral Ligand

The first example of post-synthetic coordination of a metal centre to a ligand in a MOF was reported by Kaye and Long in 2007. They described how $[Zn_4O(bdc)_3]$ (MOF-5) reacts with $[Cr(CO)_6]$ at 140 °C to form $[Zn_4O\{bdc\text{-}Cr(CO)_3\}_3]$, in which $Cr(CO)_3$ fragments coordinate to the aryl rings of the bdc ligands (reaction (3.55)).[129]

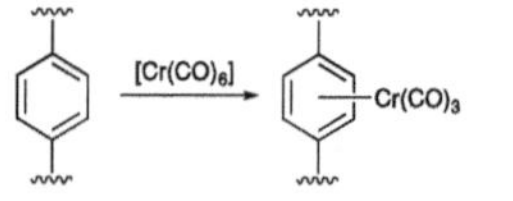

(3.55)

Photolysis of this product in the presence of N_2 or H_2 led to substitution of one of the carbonyl groups and formation of $[Zn_4O\{bdc\text{-}Cr(CO)_2(N_2)\}_3]$ and $[Zn_4O\{bdc\text{-}Cr(CO)_2(H_2)\}_3]$, respectively. Studies have been carried out on UiO-66, which reacts with $[Cr(CO)_6]$ in an analogous manner. Groppo *et al.* analysed the products by a range of spectroscopic techniques and demonstrated that, like the MOF-5 analogue, $[Zr_6O_4(OH)_4\{bdc-Cr(CO)_3\}_6]$ undergoes photolysis, with replacement of a carbonyl by N_2.[130]

Yaghi *et al.* showed how an iminopyridine functionalised UMCM-1 derivative, obtained by covalent PSM (reaction (3.18)), reacts with $[PdCl_2(MeCN)_2]$ to give complete metallation of the bidentate coordination sites (reaction (3.56)). Extended X-ray absorption fine structure (EXAFS) spectroscopy was used to confirm the palladium coordination environment.[49]

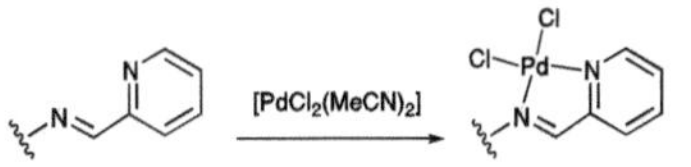

(3.56)

Wang *et al.* modified IRMOF-3 by reaction with glyoxal (OHC-CHO), then treated the product with CuI (reaction (3.57)). Although the degree of modification was relatively low (4%), the copper-containing MOF was shown to catalyse the A^3 coupling of aldehydes, alkynes and amines.[131]

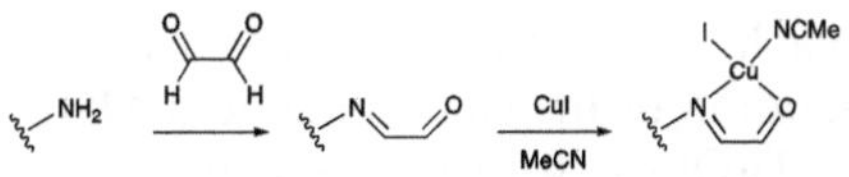

(3.57)

Long, Yaghi *et al.* reported the first example of a MOF containing free 2,2′-bipyridyl units. They prepared $[Al(OH)(bpydc)]$ (MOF-253, bpydc = 2,2′-bipyridine-5,5′-dicarboxylate) and demonstrated that it reacts with $PdCl_2$ (reaction (3.58)) and $Cu(BF_4)_2$, filling 83% and 97% of the 2,2′-bipyridyl sites respectively.[132]

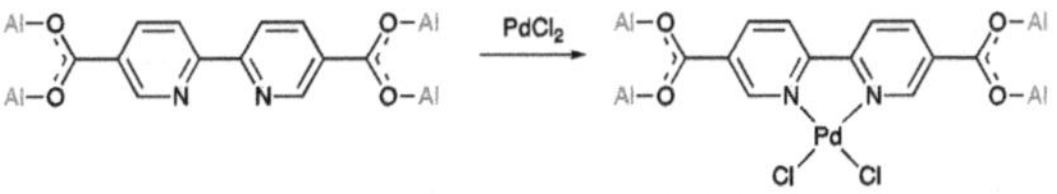

(3.58)

$[Al(OH)(bpydc)]$ also reacts with $[H(DMSO)_2][RuCl_4(DMSO)_2]$, with $RuCl_3(DMSO)$ fragments filling up to 13% of the 2,2′-bipyridyl sites. The ruthenium-containing MOFs were employed as catalysts in alcohol oxidation reactions.[133]

Cohen *et al.* have prepared zinc and zirconium MOFs containing 2-phenylpyridine-5,4′-dicarboxylate (dcppy), and shown they undergo cyclo-metalation reactions to introduce iridium and rhodium (reaction (3.59)) fragments. The reactions with $[Zn_2(dcppy)_2(bpy)]$ are notable as cyclometa-lation only occurs on ligands along one of the crystallographic axes due to steric interactions between the interpenetrated frameworks.[134]

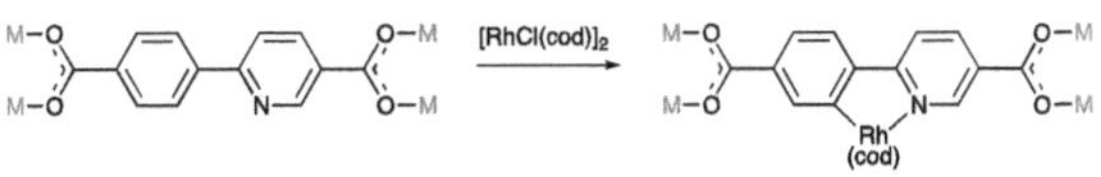

(3.59)

Hardie *et al.* used the ligand L^{17} to prepare $[Zn(mbdc)(L^{17})]$, and demonstrated that the product was able to take up copper(II) ions from DMF solution. The crystal structure of the product shows coordination of copper(I) to the 2,2′-bipyridyl sites of two ligands.[135]

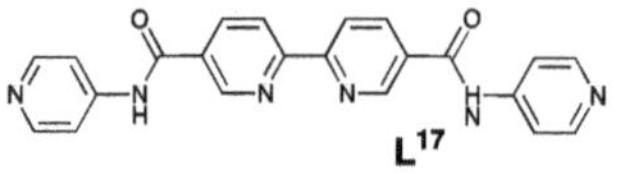

$\mathbf{L^{17}}$

Humphrey *et al.* used phosphine-carboxylate ligands to prepare coordination networks. They showed that Ca(OH)$_2$ reacts with P(C$_6$H$_4$CO$_2$H-4)$_3$ to generate a material which when desolvated has the formula [Ca$_3${P(C$_6$H$_4$CO$_2$-4)$_3$}$_2$] (PCM-10) and contains uncoordinated phosphine groups. PCM-10 reacts with [AuCl(SMe$_2$)], and a combination of analytical techniques suggested the product contains AuCl groups appended to the phosphines. The aurated product selectively binds 1-hexene over *n*-hexane.[136]

3.3.2 Coordination Accompanied by Deprotonation

Several groups have employed the conversion of hydroxyl groups to alkoxides as a means of incorporating metal centres. The Lin group provided the first example of this approach. They used dipyridyl-substituted binaphthol ligands as pillars in the preparation of the chiral cadmium MOF [Cd$_3$Cl$_6$(L^{18})$_3$]. This product reacted with Ti(OPr^i)$_4$ with deprotonation of the hydroxyl groups (reaction (3.60)) to form a material that catalyses the addition of diethylzinc to aromatic aldehydes.[137]

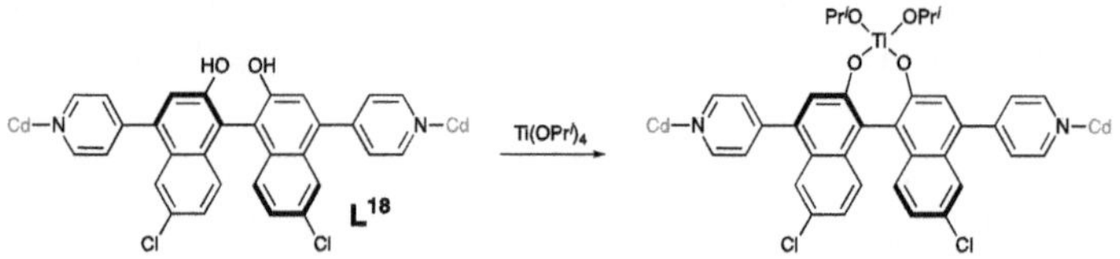

(3.60)

On changing the cadmium source from CdCl$_2$ to Cd(NO$_3$)$_2$ or Cd(ClO$_4$)$_2$, the nature of the framework structure was altered. For steric reasons, the perchlorate derivative was unable to react with Ti(OPr^i)$_4$ whereas the nitrate did. The product derived from [Cd$_3$(L^{18})$_4$(NO$_3$)$_6$] is also a good catalyst for the enantioselective addition of ZnEt$_2$ to aromatic aldehydes.[138]

Hupp *et al.* treated [Zn(tcpb)(L^6)] with lithium *tert*-butoxide and Mg(OMe)$_2$ to form lithium and magnesium alkoxides respectively, with the degree of conversion depending on the stirring rate and reaction time. Small amounts of lithiation led to enhanced H$_2$ uptake.[139]

Jeong *et al.* used the chiral ligand 2,2′-dihydroxy-6,6′-dimethyl(1,1′-biphenyl)-4,4′-dicarboxylate (L^{19}) to prepare the copper MOF [Cu$_2$(L^{19})$_2$(H$_2$O)$_2$]. The hydroxyl groups were modified by reaction with Ti(OPr^i)$_4$, and the product was shown to catalyse hetero-Diels–Alder reactions. [Cu$_2$(L^{19})$_2$(H$_2$O)$_2$] also reacts with ZnMe$_2$, and the product was used to catalyse the carbonyl-ene reaction in the cyclisation of 3-methyl-geranial.[140]

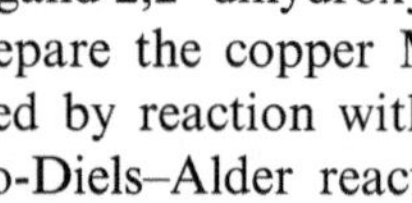

$\mathbf{L^{19}}$

Kim *et al.* prepared $[Zn_4O(\mathbf{L}^{20})_3]$ and showed that this could by modified by reaction with isopropoxytitanatrane.[141]

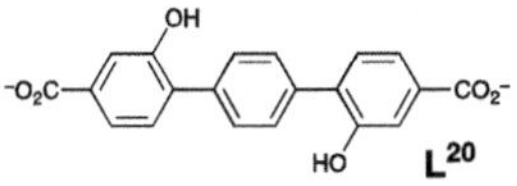

$\mathbf{L}^{20}$

Tandem reactions that use a covalent PSM process to form a bidentate coordination site, followed by reaction with a metal source have attracted considerable attention, especially for the inclusion of Lewis acidic metal centres. Rosseinsky *et al.* demonstrated that IRMOF-3 reacts with salicaldehyde (Section 3.2.3) and that the product reacts with $[VO(acac)_2]$ to append a vanadyl group to the MOF (reaction 3.61).[76]

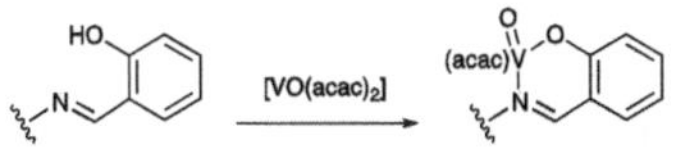

(3.61)

The product catalysed the oxidation of cyclohexene with Bu^tOOH. Similar salicaldehyde-tagged MOFs were prepared by Corma *et al.* and metallated by reaction with $[AuCl_4]^-$. The gold-containing product was used as a catalyst for domino coupling and cyclisation reactions.[77] Ahn *et al.* recently showed it was possible to combine the post-synthetic modification of an amine to form an imine-based ligand together with coordination into a single step. They reported that IRMOF-3 reacts with $[Mn(acac)_2]$ to form a manganese-functionalised MOF (reaction (3.62)).[142]

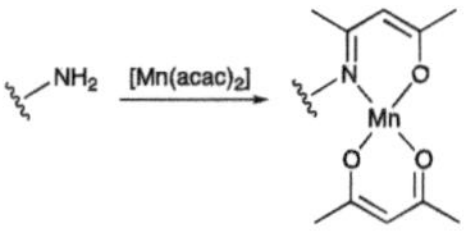

(3.62)

The product catalysed the epoxidation of alkenes using oxygen and an aldehyde oxidant precursor. Both conversion and selectivity were enhanced over use of $[Mn(acac)_2]$ as the catalyst.

Tanabe and Cohen showed that functionalised MOFs formed by the reactions of UMCM-1-NH_2 with 3-hydroxyphthalic anhydride and 2,3-pyrazinedicarboxylic anhydride (Section 3.2.1.1) were capable of further reaction with $[Fe(acac)_3]$ or $[Cu(acac)_2]$ (reaction (3.63, 3.64)). The iron-containing product was shown to act as a catalyst in the Mukaiyama-aldol reaction.[57]

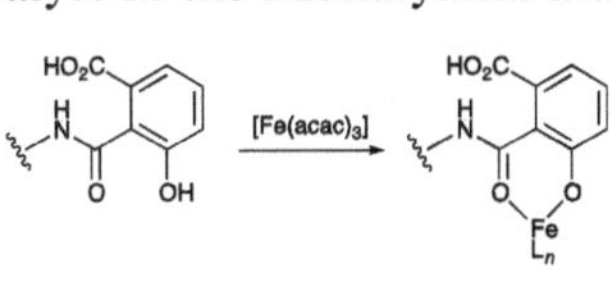

(3.63)

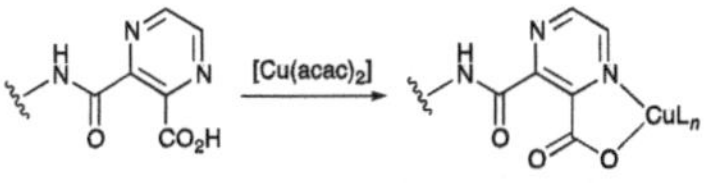

(3.64)

These functionalised amides also reacted with $[In(acac)_3]$, with more than 70% of the coordination sites metallated. The indium MOFs were shown to be good catalysts for the ring opening of epoxides with trimethylsilyl azide to give β-azido alcohols and β-amino alcohols.[58]

Wu *et al.* used the dicarboxylate **L²¹** to construct MOFs containing imidazolium moieties. The imidazolium groups can be converted post-synthetically to *N*-heterocyclic carbenes on reaction with Pd(OAc)₂, with the palladium centres coordinated in the product. The resultant MOFs were shown to catalyse the Suzuki–Miyaura coupling reaction of phenylhalides with phenylboronic acids[48] and alkene hydrogenation reactions.[143]

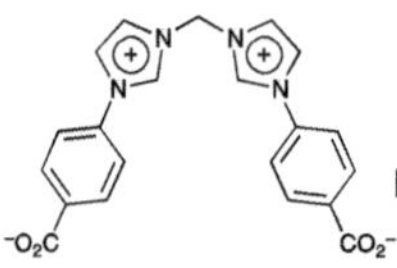

Post-synthetically modified MOFs have been used to sequester metal ions from solution. [Zn(tcpb)(**L⁶**)] was used to prepare carboxylic acid decorated MOFs following reaction with succinic anhydride (reaction (3.32)). This MOF was shown to sequester copper(II) from aqueous solution, reducing the concentration from 75 ppm to 20 ppm after 1 hour.[92] Xu *et al.* used the sulfide-appended 1,4-benzenedicarboxylate **L²²** to form [Zn₄O(**L²²**)₃], which adopts an IRMOF structure. They demonstrated that this product could remove mercury(II) from an ethanolic solution, with the concentration reduced from 84 ppm to 5 ppm after 6 days.[144]

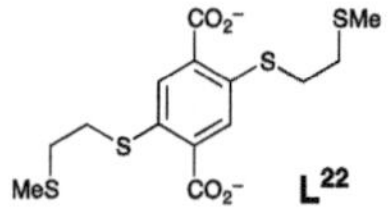

3.4 Inorganic PSM – Modification of the Metal Coordination Sphere

3.4.1 Substitution of Labile Terminal Ligands

Many MOFs contain labile ligands, often coordinated solvent molecules, attached to the secondary building units. These may be removed together with the solvent molecules included in the pores in the activation process to generate a porous material with open coordination sites, and although this is a post-synthetic modification, this form of activation is beyond the scope of the current review. However, a number of reactions have been described in which these labile ligands are substituted for another ligand, and this has often led to enhanced adsorption or catalytic properties.

The first example of this type of substitution was reported by Williams *et al.* in 1999 in their seminal report of [Cu₃(btc)₂(H₂O)₃] (HKUST-1). Each copper(II) centre in the as-synthesised MOF contains a terminal water ligand that can be removed by heating, and the activated material reacts with pyridine to form [Cu$_3$(btc)$_2$(py)$_3$].[12] Rosseinsky *et al.* looked at the analogous reaction of HKUST-1 with 4-(methylamino)pyridine (map), and showed it too could substitute for the water ligands to give [Cu$_3$(btc)$_2$(map)$_x$(H$_2$O)$_{3-x}$]. They showed that the 60%-functionalised material adsorbs NO, converting the secondary amine into an *N*-diazenium diolate.[145]

Qiu *et al.* demonstrated that, once activated, HKUST-1 reacts with dithioglycol (HSCH$_2$CH$_2$SH) to give a material in which thiol groups are projected into the pores. They showed that the modified MOF is effective at sequestrating mercury(II) ions from solution, removing over 90% of ions even when the initial concentration was as low as 81 ppb.[146] Barea, Navarro *et al.* used diamines in place of dithioglycol, and showed that the resultant MOFs showed enhanced CO$_2$ adsorption.[147]

Hupp *et al.* showed that [Zn$_2$(tcpb)(DMF)$_2$] forms a three-dimensional network with DMF molecules projecting into the pores. These were removed by heating under vacuum at 150 °C, and replaced by a range of pyridine derivatives including 4-vinylpyridine.[148]

MIL-101(Cr) also contains coordinated water ligands that can be substituted for another terminal ligand. Chang, Férey *et al.* showed that MIL-101(Cr) reacts with amines such as 1,2-diaminoethane (en) to form MOFs in which primary amine groups project into the pores. The modified MOFs showed enhanced catalytic activity for the Knoevenagel condensation of benzaldehyde and ethyl cyanoacetate.[149] Hwang and Lee *et al.* increased the range of diamines incorporated in this process and concluded that the catalytic activity relates to the basicities of the amine,[150] whereas Ahn *et al.* introduced 1,4,7-triazaheptane (dien) and showed that palladium ions could be coordinated to the pendant diamine.[151]

Kim *et al.* treated MIL-101(Cr) with pyridyl-functionalised L-proline derivatives such as **L**[23], and demonstrated that the resultant chiral MOFs were catalysts for asymmetric aldol reactions, displaying much higher enantio-selectivity than the ligands themselves.[152]

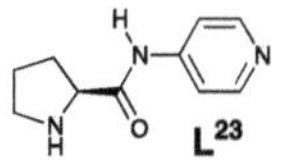

Hupp, Farha, Nguyen *et al.* treated MIL-101(Cr) with dopamine (**L**[24]), coordinating the amine to the chromium centres, then metallated the catechol group by reaction with [VO(acac)$_2$]. They showed that the loading was relatively low due to the size of dopamine, and that the remaining co-ordinatively unsaturated metal centres could be functionalised on reaction with 2-(methylthio)ethylamine. The vanadium derivative catalyses the oxidation of thioanisole with *tert*-butyl hydroperoxide to give the sulfoxide and sulfone.[153]

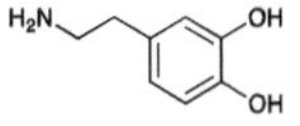

Long *et al.* prepared the porous material $H_3[(Cu_4Cl)_3(L^{25})_8]$ which contains coordinatively unsaturated copper(II) centres. This MOF reacts with 1,2-diaminoethane in a similar manner to MIL-101(Cr), and the modified material showed a greater uptake of CO_2 at low pressures than $H_3[(Cu_4Cl)_3(L^{25})_8]$.[154]

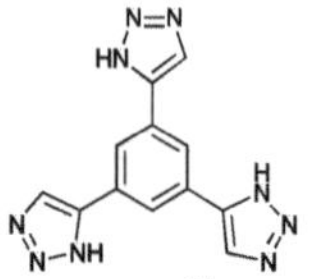

H_3L^{25}

Suh *et al.* reported the structure of $[Zn_2(L^{26})(H_2O)_2]$ (SNU-30), a three-dimensional network in which each zinc(II) centre is coordinated to a water molecule. They demonstrated that SNU-30 reacts with L^{15} in a SCSC transformation, in which L^{15} molecules substitute for the water molecules and bridge between two zinc centres. The L^{15} molecules were removed by placing the crystals in DMF, also in a SCSC transformation.[155]

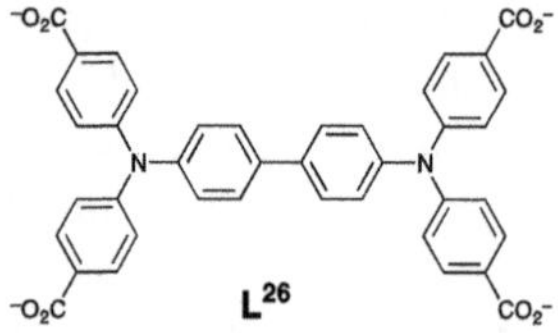

L^{26}

Shi *et al.* used a combination of inorganic and organic PSM to prepare a bifunctional MIL-101(Cr) derivative. They first modified the unsaturated metal sites with L^{27} before sulfonating the aromatic ring and then heating to remove the Boc protecting group on L^{27} and yield a MOF with the average formula $[Cr_3O(bdc)_{1.92}(bdc-SO_3)_{1.08}(en)_{0.85}]$. This product was employed as a catalyst in a tandem reaction involving acetal hydrolysis and a Henry reaction.[156]

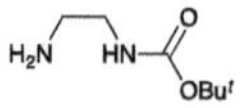

L^{27}

Long *et al.* demonstrated that $[Mg_2(dobdc)]$ (dobdc = 1,4-dioxido-2,5-benzenedicarboxylate, MOF-74(Mg)) was modified on reaction with $LiOPr^i$ to give a material containing *iso*-propoxide groups grafted to the magnesium centres. The Li^+ ions are free to move within the channels, leading to enhanced conductivity.[157] Horike and Kitagawa *et al.* modified the zinc analogue, $[Zn_2(dobdc)]$ with histamine (L^{28}), and used solid state NMR techniques to inform on the positions of the histamine molecules and the proton conductivity.[158]

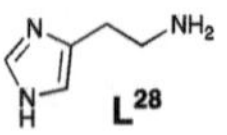

L^{28}

3.4.2 Reaction at the Anionic Part of the SBU

MIL-53(Al) contains infinite secondary building units (SBUs) in which hydroxide ions bridge between the Al^{3+} centres. Fischer *et al.* showed the hydroxide groups could be modified on reaction with 1,1′-ferrocenediyldimethylsilane in the gas phase (reaction (3.65)), with 25% of the hydroxyl groups converted into siloxy groups.[159]

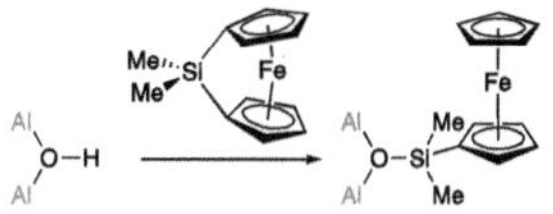

$$(3.65)$$

The product is a redox catalyst for the oxidation of benzene with aqueous hydrogen peroxide to form phenol.

Stock *et al.* showed that the methoxy groups in [Al$_4$(OH)$_2$(OMe)$_4$(bdc-NH$_2$)$_3$] (CAU-1) were replaced by hydroxyl groups on heating in air at 190 °C.[85] Larabi and Quadrelli showed that the hydroxyl groups in [Zr$_6$O$_4$(OH)$_4$(bpdc)$_6$] (UiO-67) could be deuterated, and that reaction with AuMe(PMe$_3$) occurred to give [Zr$_6$O$_4$(OH)$_3$(OAuPMe$_3$)(bpdc)$_6$].[160]

3.4.3 Oxidation and Reduction of the Framework Metal Centres

Post-synthetic oxidation and reduction of ligands was described in Section 3.2.8. Framework metal centres can also undergo post-synthetic changes in oxidation state. For example, Tarascon *et al.* reduced MIL-53(Fe) electrochemically in a cell with a lithium negative electrode. The process involves the uptake and removal of Li$^+$ ions together with the reversible reduction of Fe^{3+} into Fe^{2+}. Confirmation of the presence of both Fe^{2+} and Fe^{3+} in the product was obtained from Mössbauer spectroscopy.[161]

Vimont *et al.* showed that [Fe$_3$OF$_{0.81}$(OH)$_{0.19}$(H$_2$O)$_2$(btc)$_2$] (MIL-100(Fe)) undergoes a partial reduction from iron(III) to iron(II) on thermal treatment. The presence of iron(II) enhances interactions with π-acceptor guests such as CO and alkenes.[162]

Fischer *et al.* showed that [VO(bdc)] (MIL-47(V)) reacts with cobaltocene, [Co(η^5-C$_5$H$_5$)$_2$], to form [Co(η^5-C$_5$H$_5$)$_2$][VO(bdc)]$_2$. The included cobaltocene has been oxidised to cobaltocenium, and the framework contains a 1 : 1 mixture of vanadium(III) and vanadium(VI).[163] Clet *et al.* demonstrated that MIL-47(V) can be formed with different proportions of vanadium(III) and vanadium(IV), depending on the activation protocol. They showed that the vanadium(III) MOF is isostructural with the MIL-53 series and has a flexible structure, in contrast to the rigid structure of MIL-47 with vanadium(IV).[164]

3.4.4 Substitution of Metal Centres

There has been a growing number of reports involving post-synthetic exchange of one metal centre for another within a MOF. Hou *et al.* reported porous polymeric structures containing ferrocene-1,1′-disulfonate (**L^{29}**) such as

[Cd(4,4′-bpy)$_2$($\mathbf{L^{29}}$)], and showed that they underwent post-synthetic metal exchange with copper(II) in a SCSC transformation.[165] They later demonstrated that the related zinc compound [Zn(4,4′-bpy)$_2$(O$_3$SC$_6$H$_4$Fc-3)$_2$] (Fc = ferrocenyl) exchanges zinc for cadmium(II), lead(II) and copper(II).[166]

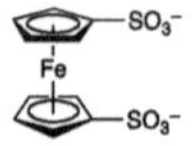

Kim *et al.* used this approach with Cd$_{1.5}$(H$_3$O)$_3$[(Cd$_4$O)$_3$($\mathbf{L^{30}}$)$_8$], and showed that this compound reversibly exchanges framework Cd^{2+} for Pb^{2+} while maintaining crystallinity. The substitution went to completion after 2 days in water, as witnessed by inductively coupled plasma atomic emission spectroscopy (ICP-AES), but mixed Cd-Pb MOFs were observed using shorter reaction times. The substitution is reversible, though the exchange of lead by cadmium was slower, with compete substitution taking 3 weeks.[46]

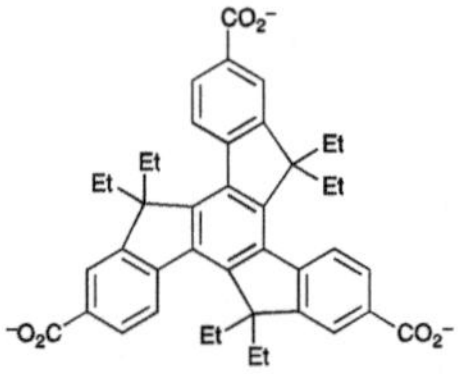

They also studied a related manganese MOF, and showed that Mn^{2+} ions could be replaced completely by Co^{2+}, Ni^{2+} and Fe^{2+} but that reaction with Cu^{2+} stopped after 66% of the metal ions had been exchanged. None of the new compounds could be prepared directly.[167]

Volkmer *et al.* showed that four of the five zinc centres in [Zn$_5$Cl$_4$($\mathbf{L^{31}}$)$_3$] were exchanged for cobalt centres by treatment with cobalt(II) chloride in DMF at 140 °C. The gas phase redox activity of [ZnCo$_4$Cl$_4$($\mathbf{L^{31}}$)$_3$] was investigated by cyclic temperature-programmed oxidation, and showed reversible oxidation with molecular oxygen at 80 °C.[47]

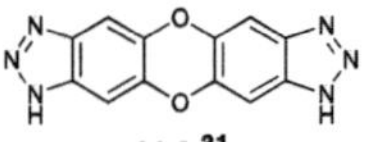

Suh *et al.* prepared [Zn$_2$($\mathbf{L^{32}}$)(DMF)$_3$] (SNU-51) and demonstrated that the zinc(II) centres were exchanged for copper(II) centres without change of the gross structure. This is noteworthy, as direct reaction of copper(II) with H$_4$$\mathbf{L^{32}}$ gave a network with a different topology, illustrating the power of PSM to form products that are not accessible by other means.[45]

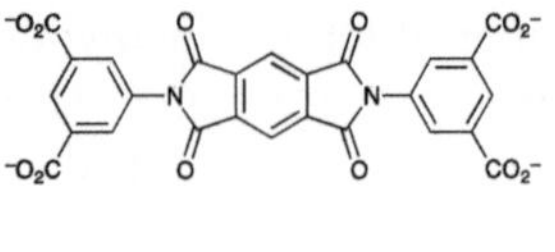

Brozek and Dincă treated MOF-5 with solutions of nickel(II) nitrate and showed that up to 25% of the tetrahedral zinc centres could be substituted by octahedral Ni^{2+} ions. Loss of the nickel-bound solvent molecules to form 4-coordinate nickel(II) and subsequent addition of ligands to form 5- and 6-coordinated nickel(II) were observed.[168]

Hupp, Nguyen *et al.* reacted zinc(II) nitrate, H_4tcpb and the dipyridyl-salen complex L^{33} to form $[Zn_2(tcpb)(L^{33})]$. They showed that the manganese centre from the salen complex within the MOF was removed by treatment with hydrogen peroxide, with control of conditions enabling this reaction to occur only on the crystal surfaces.[169] The 'de-manganated' MOF was modified by remetallating the salen struts with a range of other first row *d*-block ions giving a series of isoreticular MOFs. Most of these materials cannot be prepared directly due to competitive self-association.[170]

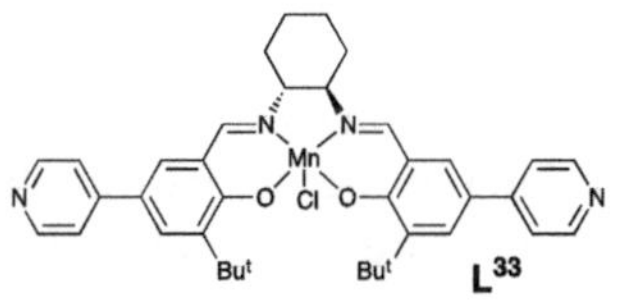

Zou *et al.* prepared $[M_6(btb)_4(bpy)_3]$ (M = Zn, Co) and showed that the zinc MOF reacts with copper(II) nitrate in DMF to substitute 95% of the metal centres within 3 days, with the reverse reaction occurring more slowly. The copper(II) MOF was not able to be prepared by direct combination of metal salt and ligands.[171]

Mukherjee and Biradha reported the synthesis of a series of cationic two-dimensional network structures of the form $[M(L^{34})_2(H_2O)_2]X_2$ (X = PF_6, SbF_6, ClO_4) in the presence of templating aromatic guests. They showed that these compounds could undergo post-synthetic exchange of the metal centres in a SCSC manner, with Zn^{2+} and Cd^{2+} exchanged for Cu^{2+}, and Cu^{2+} exchanged for Cd^{2+}.[44]

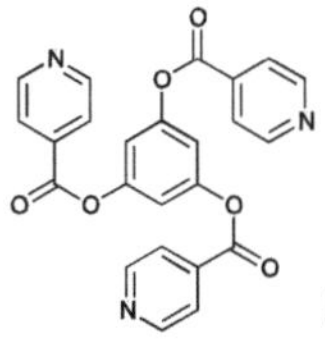

Cohen *et al.* demonstrated that metal exchange could be readily carried out on MOFs previously believed to be robust. They showed that Al^{3+} and Fe^{3+} ions were exchanged between MIL-53(Al)-Br and MIL-53(Fe)-Br, and furthermore demonstrated that titanium(IV) and hafnium(IV) centres could be incorporated into the UiO-66 structure, although MIL-101(Cr) remained inert to metal exchange reactions.[172]

3.4.5 Substitution of Framework Ligands

Choe *et al.* provided the first examples of linker-replacement reactions in MOF chemistry. They showed that the porphyrin-based MOF [$Zn_2(L^{35})(L^{14})$] reacts with 4,4′-bipyridyl to form [$Zn_2(L^{35})(bpy)$]. The transformation involves a decrease between layers from 21.2 Å to 12.8 Å, consistent with the shorter length of the bpy linker.[173]

Cohen and Prather *et al.* used aerosol time-of-flight mass spectrometry (ATOF-MS) to show that UiO-66 derivatives could undergo post-synthetic ligand exchange. When a mixture of UiO-66-Br and UiO-66-NH_2 was suspended in water, the majority of particles underwent ligand exchange to give a MOF containing both bdc-Br and bdc-NH_2 without loss of crystallinity. Post-synthetic exchange also occurred between a MOF and the ligand in solution so, for example, UiO-66-Br undergoes ligand exchange with salts of bdc-NH_2. This approach was used to introduce functionalities into a MOF and, for example, UiO-66 was treated with an aqueous solution of 2-azido-1,4-benzenedicarboxylate, to give azido-functionalised MOFs. The degree of post-synthetic exchange is time dependent, with 50% incorporation achieved after 5 days at ambient conditions.[42] In subsequent work they demonstrated ligand exchange with MIL-53(Al) and MIL-68(In) analogues and with [$Zn(dcim)_2$] (dcim = 4,5-dichloroimidazolate, ZIF-71).[172]

In contemporary studies, Hupp, Farha *et al.* also showed that ligand exchange could occur in imidazolate networks, a process they termed solvent-assisted linker exchange (SALE). They demonstrated that the linkers in [$Cd(eim)_2$] (eim = 2-ethylimidazolate) could be exchanged for 2-nitro-imidazolate and 2-methylimidazolate (mim), the latter reaction generating a new ZIF in a crystal-to-crystal transformation.[174] They subsequently showed that [$Zn(mim)_2$] (ZIF-8) undergoes linker exchange with imidazolate, with up to 85% completion. On treatment with *n*-BuLi, the product exhibits Brønsted base catalysis in contrast to ZIF-8.[175]

Kitagawa, Furukawa *et al.* recognised that the crystal surfaces of MOFs contain coordinatively unsaturated ligands, such as dicarboxylates coordinated through only one carboxylate group. They studied substitution reactions of these surface ligands with monocarboxylates, selecting a carboxylate containing a fluorescent BODIPY dye as a means of following the reaction. They showed that only the carboxylate-terminated surfaces of [$Zn_2(1,4$-ndc$)_2$(dabco)] (1,4-ndc = 1,4-naphthalenedicarboxylate) were fluorescent after modification (reaction (3.66)), whereas all surfaces of HKUST-1 became fluorescent.[176]

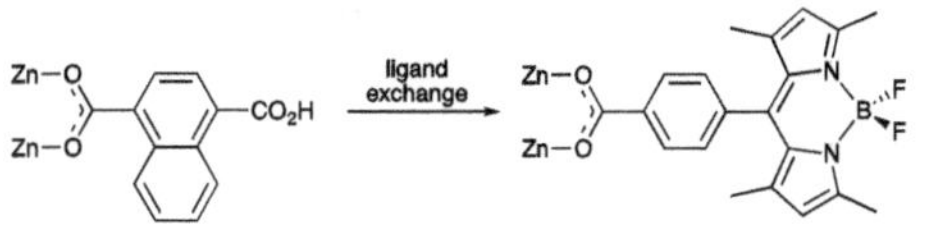

$$(3.66)$$

3.5 Ionic PSM

Ionic MOFs are those in which the framework has a cationic or anionic charge, with counter-ions present in the pores to balance these charges. Many ionic MOFs can undergo post-synthetic ion-exchange processes during which the cations or anions initially present in the pores are substituted for others. Space does not allow for a comprehensive account of this area, but a few representative examples are included.

Many MOFs have been isolated with dimethylammonium cations, derived from hydrolysis of the DMF solvent molecules, to balance the framework charge.[177] These cations can often be exchanged: for example, the $NMe_2H_2^+$ cations present with $[M_3(L^{36})_2]^{2-}$ (M = Co, Mn) and $[Ga_6(btc)_8]^{6-}$ were substituted by Group 1 cations,[178,179] as were those in $[Zn_3(L^{37})_2(HCO_2)]^-$.[180]

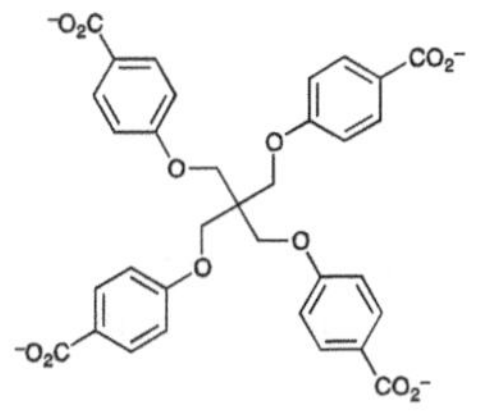

Rosi *et al.* prepared $(Me_2NH_2)_2[Zn_8O(adeninate)_4(bpdc)_6]$ (bio-MOF-1), and showed that the dimethylammonium ions can be exchanged for other organic cations including the biologically active procainamide.[181] Furthermore, exchange for other substituted ammonium ions provides a means of tuning the porosity and leads to enhanced CO_2 adsorption.[182]

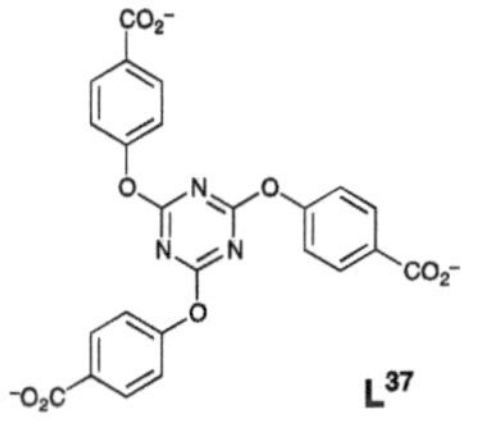

Anionic indium tetracarboxylate networks with dimethylammonium or piperazinium cations were prepared by Champness, Schröder *et al.* The organic cations block channels and act as pore gates. These ions can be substituted by Li^+ to give networks in which the pores are unblocked. The Li^+-containing frameworks showed enhanced H_2 adsorption.[40] Eddaoudi *et al.* have prepared a range of anionic zeolite-like MOFs such as $[In(Hida)_2]^-$ (Hida = 4,5-imidazoledicarboxylate) and $[In(pmdc)_2]^-$ (pmdc = 4,6-pyrimidinedicarboxylate) that undergo ion-exchange reactions.[183,184]

Zaworotko *et al.* showed that the reaction between cadmium(II) chloride and biphenyl-3,4′,5-tricarboxylic acid (H₃bpt) in the presence of a methylpyridinium-functionalised porphyrin gave a MOF in which Cd-porphyrin cations are encapsulated in an anionic cadmium carboxylate framework, of formula $[Cd_6(bpt)_4Cl_4(H_2O)_4]^{4-}$. They showed that the cadmium in the porphyrin could be post-synthetically exchanged for manganese(II) or copper(II).[185] Looking at a related system, they showed that both cation and anion binding is possible, with metal exchange accompanied by chloride binding to the porphyrin. The modified materials had enhanced selectivity for CO_2 adsorption.[186]

3.6 Conclusions

The examples presented in this Chapter illustrate that post-synthetic modification is an area of growing importance and diversity in MOF chemistry. The increase in the number of reactions possible within the pores of a MOF observed over the past few years, together with the development of tandem processes which allow sequential modifications, demonstrates the increasing versatility of the approach. PSM has already been exploited to prepare MOFs that cannot be prepared by direct combination, and materials that have greater catalytic activity,[142,149,152] gas adsorption properties,[59,127] and sequestration ability[144,146] than their unmodified analogues.

Many of the early examples of PSM involved covalent organic transformations, and this remains an important area, with new reactions being demonstrated to extend the utility of the approach. Increasingly, however, transformations involving the 'inorganic' part of the MOF are being developed, with exchange of both terminal and linker ligands ways of increasing complexity and functionality, as too is exchange of the metal ions themselves.

PSM reactions do not always go to completion, and this raises the question of how the modified groups are distributed within the MOF crystals. Are they evenly distributed, or are they concentrated in the outer layers and crystal surfaces? This is likely to vary from case to case, with the size of the introduced group and its potential for blocking access to channels a major factor. The rates of diffusion of the reactants into the crystals (and of any by-products out again) are also important considerations, as is the actual rate of reaction.[51] Incomplete PSM can be problematic in applications like catalysis and gas adsorption if access to the functionalised pores is restricted. One way around this problem is to dope particular functionalities into a network using a mixed-component MOF approach, where use of a combination of isosteric ligands in the synthesis ensures that only some of the linkers contain tag groups and are available for modification.[19,187]

Control of the distribution of modified groups within a MOF is likely to attract increasing interest as a way of further tuning the properties. Yaghi et al. have shown that mixed-component MOFs (multivariate, or MTV-MOFs) can have enhanced properties over the analogous single-component MOFs, and have postulated that this is a result of sequences of functionalities within the MOF.[188] Post-synthetic modification of such species might allow for even greater control of the pore architecture. It is already possible to limit PSM to surfaces through the use of bulky reagents or those with a different hydrophilic/hydrophobic nature to the pores, and this has allowed the formation of

MOFs with different functional groups on the crystal surfaces and in the core. The PSM of 'core-shell' MOFs[21,64] is likely to provide further control. The steric constraints imposed by interpenetration can also impart some selectivity to PSM processes.[120]

From the studies described in this review, it is clear that PSM has already become an essential tool for the preparation of functionalised MOFs. With anticipated developments further increasing the scope of the approach, PSM is likely to grow in importance and be key to the development of the next generation of functionalised MOF materials.

References

1. Recent issues of *Chem. Soc. Rev.* (2009, **38**(5)) and *Chem. Rev.* (2012, **112**(2)) are dedicated to this type of compound.
2. D. J. Tranchemontagne, J. L. Mendoza-Cortés, M. O'Keeffe and O. M. Yaghi, *Chem. Soc. Rev.*, 2009, **38**, 1257–1283.
3. R. Robson, *Dalton Trans.*, 2008, 5113–5131.
4. M. P. Suh, H. J. Park, T. K. Prasad and D.-W. Lim, *Chem. Rev.*, 2012, **112**, 782–835.
5. K. Sumida, D. L. Rogow, J. A. Mason, T. M. McDonald, E. D. Bloch, Z. R. Herm, T.-H. Bae and J. R. Long, *Chem. Rev.*, 2012, **112**, 724–781.
6. J.-R. Li, J. Sculley and H.-C. Zhou, *Chem. Rev.*, 2012, **112**, 869–932.
7. A. Corma, H. García and F. X. Llabrés i Xamena, *Chem. Rev.*, 2010, **110**, 4606–4655.
8. P. Horcajada, R. Gref, T. Baati, P. K. Allan, G. Maurin, P. Couvreur, G. Férey, R. E. Morris and C. Serre, *Chem. Rev.*, 2012, **112**, 1232–1268.
9. B. F. Hoskins and R. Robson, *J. Am. Chem. Soc.*, 1990, **112**, 1546–1554.
10. Y.-H. Kiang, G. B. Gardner, S. Lee, Z. Xu and E. B. Lobkovsky, *J. Am. Chem. Soc.*, 1999, **121**, 8204–8215.
11. J. S. Seo, D. Whang, H. Lee, S. I. Jun, J. Oh, Y. J. Jeon and K. Kim, *Nature*, 2000, **404**, 982–986.
12. S. S.-Y. Chui, S. M.-F. Lo, J. P. H. Charmant, A. G. Orpen and I. D. Williams, *Science*, 1999, **283**, 1148–1150.
13. Z. Wang and S. M. Cohen, *J. Am. Chem. Soc.*, 2007, **129**, 12368–12369.
14. A. Mehdi, C. Reye and R. Corriu, *Chem. Soc. Rev.*, 2011, **40**, 563–574.
15. R. T. Chen, S. Marchesan, R. A. Evans, K. E. Styan, G. K. Such, A. Postma, K. M. McLean, B. W. Muir and F. Caruso, *Biomacromolecules*, 2012, **13**, 889–895.
16. M. Wang, W.-J. Lan, Y.-R. Zheng, T. R. Cook, H. S. White and P. J. Stang, *J. Am. Chem. Soc.*, 2011, **133**, 10752–10755.
17. P. A. Kerneghan, S. D. Halperin, D. L. Bryce and K. E. Maly, *Can. J. Chem.*, 2011, **89**, 577–582.
18. J. Schoch, M. Wiessler and A. Jäschke, *J. Am. Chem. Soc.*, 2010, **132**, 8846–8847.
19. A. D. Burrows, C. G. Frost, M. F. Mahon and C. Richardson, *Angew. Chem. Int. Ed.*, 2008, **47**, 8482–8486.

20. M. Müller, O. I. Lebedev and R. A. Fischer, *J. Mater. Chem.*, 2008, **18**, 5274–5281.
21. S. Furukawa, K. Hirai, K. Nakagawa, Y. Takashima, R. Matsuda, T. Tsuruoka, M. Kondo, R. Haruki, D. Tanaka, H. Sakamoto, S. Shimomura, O. Sakata and S. Kitagawa, *Angew. Chem., Int. Ed.*, 2009, **48**, 1766–1770.
22. S. Bourrelly, P. L. Llewellyn, C. Serre, F. Millange, T. Loiseau and G. Férey, *J. Am. Chem. Soc.*, 2005, **127**, 13519–13521.
23. A. D. Burrows, C. G. Frost, M. F. Mahon, P. R. Raithby, C. Richardson and A. J. Stevenson, *Chem. Commun.*, 2010, **46**, 5064–5066.
24. T. Kawamichi, T. Haneda, M. Kawano and M. Fujita, *Nature*, 2009, **461**, 633–635.
25. Q. Li, W. Zhang, O. S. Miljanić, C.-H. Sue, Y.-L. Zhao, L. Liu, C. B. Knobler, J. F. Stoddart and O. M. Yaghi, *Science*, 2009, **325**, 855–859.
26. M. Eddaoudi, J. Kim, N. Rosi, D. Vodak, J. Wachter, M. O'Keeffe and O. M. Yaghi, *Science*, 2002, **295**, 469–472.
27. K. Koh, A. G. Wong-Foy and A. J. Matzger, *Angew. Chem., Int. Ed.*, 2008, **47**, 677–680.
28. H. Chun, D. N. Dybtsev, H. Kim and K. Kim, *Chem.–Eur. J.*, 2005, **11**, 3521–3529.
29. F. Millange, C. Serre and G. Férey, *Chem. Commun.*, 2002, 822–823.
30. G. Férey, C. Mellot-Draznieks, C. Serre, F. Millange, J. Dutour, S. Surblé and I. Margiolaki, *Science*, 2005, **309**, 2040–2042.
31. C. Volkringer, M. Meddouri, T. Loiseau, N. Guillou, J. Marrot, G. Férey, M. Haouas, F. Taulelle, N. Audebrand and M. Latroche, *Inorg. Chem.*, 2008, **47**, 11892–11901.
32. J. H. Cavka, S. Jakobsen, U. Olsbye, N. Guillou, C. Lamberti, S. Bordiga and K. P. Lillerud, *J. Am. Chem. Soc.*, 2008, **130**, 13850–13851.
33. J. S. Costa, P. Gamez, C. A. Black, O. Roubeau, S. J. Teat and J. Reedijk, *Eur. J. Inorg. Chem.*, 2008, 1551–1554.
34. H. Sato, R. Matsuda, K. Sugimoto, M. Takata and S. Kitagawa, *Nature Mater.*, 2010, **9**, 661–666.
35. Z. Wang, K. K. Tanabe and S. M. Cohen, *Inorg. Chem.*, 2009, **48**, 296–306.
36. C. Volkringer and S. M. Cohen, *Angew. Chem., Int. Ed.*, 2010, **49**, 4644–4648.
37. S. J. Garibay and S. M. Cohen, *Chem. Commun.*, 2010, **46**, 7700–7702.
38. S. Bernt, V. Guillerm, C. Serre and N. Stock, *Chem. Commun.*, 2011, **47**, 2838–2840.
39. W. Morris, C. J. Doonan and O. M. Yaghi, *Inorg. Chem.*, 2011, **50**, 6853–6855.
40. S. Yang, G. S. B. Martin, J. J. Titman, A. J. Blake, D. R. Allan, N. R. Champness and M. Schröder, *Inorg. Chem.*, 2011, **50**, 9374–9384.
41. A. J. Rossini, A. Zagdoun, M. Lelli, J. Canivet, S. Aguado, O. Ouari, P. Tordo, M. Rosay, W. E. Maas, C. Copéret, D. Farrusseng, L. Emsley and A. Lesage, *Angew. Chem., Int. Ed.*, 2012, **51**, 123–127.

42. M. Kim, J. F. Cahill, Y. Su, K. A. Prather and S. M. Cohen, *Chem. Sci.*, 2012, **3**, 126–130.
43. M. Kandiah, S. Usseglio, S. Svelle, U. Olsbye, K. P. Lillerud and M. Tilset, *J. Mater. Chem.*, 2010, **20**, 9848–9851.
44. G. Mukherjee and K. Biradha, *Chem. Commun.*, 2012, **48**, 4293–4295.
45. T. K. Prasad, D. H. Hong and M. P. Suh, *Chem.–Eur. J.*, 2010, **16**, 14043–14050.
46. S. Das, H. Kim and K. Kim, *J. Am. Chem. Soc.*, 2009, **131**, 3814–3815.
47. D. Denysenko, T. Werner, M. Grzywa, A. Puls, V. Hagen, G. Eickerling, J. Jelic, K. Reuter and D. Volkmer, *Chem. Commun.*, 2012, **48**, 1236–1238.
48. G.-Q. Kong, X. Xu, C. Zou and C.-D. Wu, *Chem. Commun.*, 2011, **47**, 11005–11007.
49. C. J. Doonan, W. Morris, H. Furukawa and O. M. Yaghi, *J. Am. Chem. Soc.*, 2009, **131**, 9492–9493.
50. K. Sumida and J. Arnold, *J. Chem. Educ.*, 2011, **88**, 92–94.
51. K. K. Tanabe, Z. Wang and S. M. Cohen, *J. Am. Chem. Soc.*, 2008, **130**, 8508–8517.
52. J. G. Nguyen and S. M. Cohen, *J. Am. Chem. Soc.*, 2010, **132**, 4560–4561.
53. Y. Yoo, V. Varela-Guerrero and H.-K. Jeong, *Langmuir*, 2011, **27**, 2652–2657.
54. S. J. Garibay, Z. Wang, K. K. Tanabe and S. M. Cohen, *Inorg. Chem.*, 2009, **48**, 7341–7349.
55. Z. Wang and S. M. Cohen, *J. Am. Chem. Soc.*, 2009, **131**, 16675–16677.
56. M. Kim, J. A. Boissonnault, C. A. Allen, P. V. Dau and S. M. Cohen, *Dalton Trans.*, 2012, **41**, 6277–6282.
57. K. K. Tanabe and S. M. Cohen, *Angew. Chem., Int. Ed.*, 2009, **48**, 7424–7427.
58. K. K. Tanabe and S. M. Cohen, *Inorg. Chem.*, 2010, **49**, 6766–6774.
59. Z. Wang, K. K. Tanabe and S. M. Cohen, *Chem.–Eur. J.*, 2010, **16**, 212–217.
60. N. Ko and J. Kim, *Bull. Korean Chem. Soc.*, 2011, **32**, 2705–2710.
61. T. Ahnfeldt, D. Gunzelmann, T. Loiseau, D. Hirsemann, J. Senker, G. Férey and N. Stock, *Inorg. Chem.*, 2009, **48**, 3057–3064.
62. J. E. Halls, A. Hernán-Gómez, A. D. Burrows and F. Marken, *Dalton Trans.*, 2012, **41**, 1475–1480.
63. S. J. Garibay, Z. Wang and S. M. Cohen, *Inorg. Chem.*, 2010, **49**, 8086–8091.
64. K. Hirai, S. Furukawa, M. Kondo, M. Meilikhov, Y. Sakata, O. Sakata and S. Kitagawa, *Chem. Commun.*, 2012, **48**, 6472–6474.
65. K. Peikert, F. Hoffmann and M. Fröba, *Chem. Commun.*, 2012, **48**, 11196–11198.
66. K. M. L. Taylor-Pashow, J. Della Rocca, Z. Xie, S. Tran and W. Lin, *J. Am. Chem. Soc.*, 2009, **131**, 14261–14263.
67. A. Modrow, D. Zargarani, R. Herges and N. Stock, *Dalton Trans.*, 2012, **41**, 8690–8696.

68. M. Servalli, M. Ranocchiari and J. A. Van Bokhoven, *Chem. Commun.*, 2012, **48**, 1904–1906.
69. M. Savonnet, S. Aguado, U. Ravon, D. Bazer-Bachi, V. Lecocq, N. Bats, C. Pinel and D. Farrusseng, *Green Chem.*, 2009, **11**, 1729–1732.
70. J. Canivet, S. Aguado, G. Bergeret and D. Farrusseng, *Chem. Commun.*, 2011, **47**, 11650–11652.
71. E. Dugan, Z. Wang, M. Okamura, A. Medina and S. M. Cohen, *Chem. Commun.*, 2008, 3366–3368.
72. M. Ma, A. Gross, D. Zacher, A. Pinto, H. Noei, Y. Wang, R. A. Fischer and N. Metzler-Nolte, *CrystEngComm*, 2011, **13**, 2828–2832.
73. F. M. Hinterholzinger, S. Wuttke, P. Roy, T. Preuße, A. Schaate, P. Behrens, A. Godt and T. Bein, *Dalton Trans.*, 2012, **41**, 3899–3901.
74. O. Shekhah, H. K. Arslan, K. Chen, M. Schmittel, R. Maul, W. Wenzel and C. Wöll, *Chem. Commun.*, 2011, **47**, 11210–11212.
75. J. A. Rood, B. C. Noll and K. W. Henderson, *Main Group Chem.*, 2009, **8**, 237–250.
76. M. J. Ingleson, J. Perez Barrio, J.-B. Guilbaud, Y. Z. Khimyak and M. J. Rosseinsky, *Chem. Commun.*, 2008, 2680–2682.
77. X. Zhang, F. X. Llabrés i Xamena and A. Corma, *J. Catal.*, 2009, **265**, 155–160.
78. A. D. Burrows and L. L. Keenan, *CrystEngComm*, 2012, **14**, 4112–4114.
79. D. Britt, C. Lee, F. J. Uribe-Romo, H. Furukawa and O. M. Yaghi, *Inorg. Chem.*, 2010, **49**, 6387–6389.
80. J. G. Nguyen, K. K. Tanabe and S. M. Cohen, *CrystEngComm*, 2010, **12**, 2335–2338.
81. M. Savonnet, D. Bazer-Bachi, N. Bats, J. Perez-Pellitero, E. Jeanneau, V. Lecocq, C. Pinel and D. Farrusseng, *J. Am. Chem. Soc.*, 2010, **132**, 4518–4519.
82. M. Savonnet, E. Kockrick, A. Camarata, D. Bazer-Bachi, N. Bats, V. Lecocq, C. Pinel and D. Farrusseng, *New J. Chem.*, 2011, **35**, 1892–1897.
83. J. G. Vitillo, T. Lecscouet, M. Savonnet, D. Farrusseng and S. Bordiga, *Dalton Trans.*, 2012, **41**, 14236–14238.
84. Y. Yoo and H.-K. Jeong, *Chem. Eng. J.*, 2012, **181–182**, 740–745.
85. T. Ahnfeldt, D. Gunzelmann, J. Wack, J. Senker and N. Stock, *CrystEngComm*, 2012, **14**, 4126–4136.
86. D. Jiang, L. L. Keenan, A. D. Burrows and K. J. Edler, *Chem. Commun.*, 2012, **48**, 12053–12055.
87. Y. Goto, H. Sato, S. Shinkai and K. Sada, *J. Am. Chem. Soc.*, 2008, **130**, 14354–14355.
88. M. Savonnet, J. Canivet, S. Gambarelli, L. Dubois, D. Bazer-Bachi, V. Lecocq, N. Bats and D. Farrusseng, *CrystEngComm*, 2012, **14**, 4105–4108.
89. M. Savonnet, A. Camarata, J. Canivet, D. Bazer-Bachi, N. Bats, V. Lecocq, C. Pinel and D. Farrusseng, *Dalton Trans.*, 2012, **41**, 3945–3948.
90. C. Liu, T. Li and N. L. Rosi, *J. Am. Chem. Soc.*, 2012, **134**, 18886–18888.

91. S. Nagata, H. Sato, K. Sugikawa, K. Kokado and K. Sada, *CrystEngComm*, 2012, **14**, 4137–4141.
92. T. Gadzikwa, O. K. Farha, K. L. Mulfort, J. T. Hupp and S. T. Nguyen, *Chem. Commun.*, 2009, 3720–3722.
93. S. C. Jones and C. A. Bauer, *J. Am. Chem. Soc.*, 2009, **131**, 12516–12517.
94. K. Hindelang, S. I. Vagin, C. Anger and B. Rieger, *Chem. Commun.*, 2012, **48**, 2888–2890.
95. T. Gadzikwa, G. Lu, C. L. Stern, S. R. Wilson, J. T. Hupp and S. T. Nguyen, *Chem. Commun.*, 2008, 5493–5495.
96. T. Gadzikwa, O. K. Farha, C. D. Malliakas, M. G. Kanatzidis, J. T. Hupp and S. T. Nguyen, *J. Am. Chem. Soc.*, 2009, **131**, 13613–13615.
97. W. Zhu, C. He, P. Wu, X. Wu and C. Duan, *Dalton Trans.*, 2012, **41**, 3072–3077.
98. P. Roy, A. Schaate, P. Behrens and A. Godt, *Chem.–Eur. J.*, 2012, **18**, 6979–6985.
99. L. R. MacGillivray, G. S. Papaefstathiou, T. Friščić, T. D. Hamilton, D.-K. Bučar, Q. Chu, D. B. Varshney and I. G. Georgiev, *Acc. Chem. Res.*, 2008, **41**, 280–291.
100. D. Liu, H.-X. Li, Z.-G. Ren, Y. Chen, Y. Zhang and J.-P. Lang, *Cryst. Growth Des.*, 2009, **9**, 4562–4566.
101. D. Liu, N.-Y. Li and J.-P. Lang, *Dalton Trans.*, 2011, **40**, 2170–2172.
102. M. H. Mir, L. L. Koh, G. K. Tan and J. J. Vittal, *Angew. Chem., Int. Ed.*, 2010, **49**, 390–393.
103. A. Chanthapally, G. K. Kole, K. Qian, G. K. Tan, S. Gao and J. J. Vittal, *Chem.–Eur. J.*, 2012, **18**, 7869–7877.
104. D. Liu, Z.-G. Ren, H.-X. Li, J.-P. Lang, N.-Y. Li and B. F. Abrahams, *Angew. Chem., Int. Ed.*, 2010, **49**, 4767–4770.
105. W. Morris, C. J. Doonan, H. Furukawa, R. Banerjee and O. M. Yaghi, *J. Am. Chem. Soc.*, 2008, **130**, 12626–12627.
106. A. Huang and J. Caro, *Angew. Chem., Int. Ed.*, 2011, **50**, 4979–4982.
107. J. Canivet, S. Aguado, C. Daniel and D. Farrusseng, *ChemCatChem*, 2011, **3**, 675–678.
108. S. Aguado, J. Canivet and D. Farrusseng, *J. Mater. Chem.*, 2011, **21**, 7582–7588.
109. S. Aguado, J. Canivet, Y. Schuurman and D. Farrusseng, *J. Catal.*, 2011, **284**, 207–214.
110. N. Reimer, B. Gil, B. Marszalek and N. Stock, *CrystEngComm*, 2012, **14**, 4119–4125.
111. T. Dröge, A. Notzon, R. Fröhlich and F. Glorius, *Chem.–Eur. J.*, 2011, **17**, 11974–11977.
112. M. G. Goesten, K. B. Sai Sankar Gupta, E. V. Ramos-Fernandez, H. Khajavi, J. Gascon and F. Kapteijn, *CrystEngComm*, 2012, **14**, 4109–4111.
113. M. G. Goesten, J. Juan-Alcañiz, E. V. Ramos-Fernandez, K. B. S. Sankar Gupta, E. Stavitski, H. van Bekkum, J. Gascon and F. Kapteijn, *J. Catal.*, 2011, **281**, 177–187.

114. A. D. Burrows, S. O. Hunter, M. F. Mahon and C. Richardson, *Chem. Commun.*, 2013, **49**, 990–992.
115. M. Kim, S. J. Garibay and S. M. Cohen, *Inorg. Chem.*, 2011, **50**, 729–731.
116. A. D. Burrows, C. G. Frost, M. F. Mahon and C. Richardson, *Chem. Commun.*, 2009, 4218–4220.
117. S. M. Cohen, *Chem. Rev.*, 2012, **112**, 970–1000.
118. R. K. Deshpande, J. L. Minnaar and S. G. Telfer, *Angew. Chem., Int. Ed.*, 2010, **49**, 4598–4602.
119. D. J. Lun, G. I. N. Waterhouse and S. G. Telfer, *J. Am. Chem. Soc.*, 2011, **133**, 5806–5809.
120. K. K. Tanabe, C. A. Allen and S. M. Cohen, *Angew. Chem., Int. Ed.*, 2010, **49**, 9730–9733.
121. C. A. Allen and S. M. Cohen, *J. Mater. Chem.*, 2012, **22**, 10188–10194.
122. R. K. Deshpande, G. I. N. Waterhouse, G. B. Jameson and S. G. Telfer, *Chem. Commun.*, 2012, **48**, 1574–1576.
123. S. S. Han, J. L. Mendoza-Cortés and W. A. Goddard III, *Chem. Soc. Rev.*, 2009, **38**, 1460–1476.
124. K. L. Mulfort and J. T. Hupp, *J. Am. Chem. Soc.*, 2007, **129**, 9604–9605.
125. K. L. Mulfort and J. T. Hupp, *Inorg. Chem.*, 2008, **47**, 7936–7938.
126. K. L. Mulfort, T. M. Wilson, M. R. Wasielewski and J. T. Hupp, *Langmuir*, 2009, **25**, 503–508.
127. Y. E. Cheon and M. P. Suh, *Angew. Chem., Int. Ed.*, 2009, **48**, 2899–2903.
128. J. M. Falkowski, C. Wang, S. Liu and W. Lin, *Angew. Chem., Int. Ed.*, 2011, **50**, 8674–8678.
129. S. S. Kaye and J. R. Long, *J. Am. Chem. Soc.*, 2008, **130**, 806–807.
130. S. Chavan, J. G. Vitillo, M. J. Uddin, F. Bonino, C. Lamberti, E. Groppo, K.-P. Lillerud and S. Bordiga, *Chem. Mater.*, 2010, **22**, 4602–4611.
131. J. Yang, P. Li and L. Wang, *Catal. Commun.*, 2012, **27**, 58–62.
132. E. D. Bloch, D. Britt, C. Lee, C. J. Doonan, F. J. Uribe-Romo, H. Furukawa, J. R. Long and O. M. Yaghi, *J. Am. Chem. Soc.*, 2010, **132**, 14382–14384.
133. F. Carson, S. Agrawal, M. Gustafsson, A. Bartoszewicz, F. Moraga, X. Zou and B. Martín-Matute, *Chem.–Eur. J.*, 2012, **18**, 15337–15344.
134. P. V. Dau, M. Kim and S. M. Cohen, *Chem. Sci.*, 2013, **4**, 601–605.
135. T. Jacobs, R. Clowes, A. I. Cooper and M. J. Hardie, *Angew. Chem., Int. Ed.*, 2012, **51**, 5192–5195.
136. A. J. Nuñez, L. N. Shear, N. Dahal, I. A. Ibarra, J. Yoon, Y. K. Hwang, J.-S. Chang and S. M. Humphrey, *Chem. Commun.*, 2011, **47**, 11855–11857.
137. C.-D. Wu, A. Hu, L. Zhang and W. Lin, *J. Am. Chem. Soc.*, 2005, **127**, 8940–8941.
138. C.-D. Wu and W. Lin, *Angew. Chem., Int. Ed.*, 2007, **46**, 1075–1078.
139. K. L. Mulfort, O. K. Farha, C. L. Stern, A. A. Sarjeant and J. T. Hupp, *J. Am. Chem. Soc.*, 2009, **131**, 3866–3868.

140. K. S. Jeong, Y. B. Go, S. M. Shin, S. J. Lee, J. Kim, O. M. Yaghi and N. Jeong, *Chem. Sci.*, 2011, **2**, 877–882.

141. J. Kim, D. O. Kim, D. W. Kim, J. Park and M. S. Jung, *Inorg. Chim. Acta*, 2012, **390**, 22–25.

142. S. Bhattacharjee, D.-A. Yang and W.-S. Ahn, *Chem. Commun.*, 2011, **47**, 3637–3639.

143. G.-Q. Kong, S. Ou, C. Zou and C.-D. Wu, *J. Am. Chem. Soc.*, 2012, **134**, 19851–19857.

144. J. He, K.-K. Yee, Z. Xu, M. Zeller, A. D. Hunter, S. S.-Y. Chui and C.-M. Che, *Chem. Mater.*, 2011, **23**, 2940–2947.

145. M. J. Ingleson, R. Heck, J. A. Gould and M. J. Rosseinsky, *Inorg. Chem.*, 2009, **48**, 9986–9988.

146. F. Ke, L.-G. Qiu, Y.-P. Yuan, F.-M. Peng, X. Jiang, A.-J. Xie, Y.-H. Shen and J.-F. Zhu, *J. Hazard. Mater.*, 2011, **196**, 36–43.

147. C. Montoro, E. García, S. Calero, M. A. Pérez-Fernández, A. L. López, E. Barea and J. A. R. Navarro, *J. Mater. Chem.*, 2012, **22**, 10155–10158.

148. O. K. Farha, K. L. Mulfort and J. T. Hupp, *Inorg. Chem.*, 2008, **47**, 10223–10225.

149. Y. K. Hwang, D.-Y. Hong, J.-S. Chang, S. H. Jhung, Y.-K. Seo, J. Kim, A. Vimont, M. Daturi, C. Serre and G. Férey, *Angew. Chem., Int. Ed.*, 2008, **47**, 4144–4148.

150. P. Kasinathan, Y.-K. Seo, K.-E. Shim, Y. K. Hwang, U.-H. Lee, D. W. Hwang, D.-Y. Hong, S. B. Halligudi and J.-S. Chang, *Bull. Korean Chem. Soc.*, 2011, **32**, 2073–2075.

151. S.-N. Kim, S.-T. Yang, J. Kim, J.-E. Park and W.-S. Ahn, *CrystEngComm*, 2012, **14**, 4142–4147.

152. M. Banerjee, S. Das, M. Yoon, H. J. Choi, M. H. Hyun, S. M. Park, G. Seo and K. Kim, *J. Am. Chem. Soc.*, 2009, **131**, 7524–7525.

153. H. G. T. Nguyen, M. H. Weston, O. K. Farha, J. T. Hupp and S. T. Nguyen, *CrystEngComm*, 2012, **14**, 4115–4118.

154. A. Demessence, D. M. D'Alessandro, M. L. Foo and J. R. Long, *J. Am. Chem. Soc.*, 2009, **131**, 8784–8786.

155. H. J. Park, Y. E. Cheon and M. P. Suh, *Chem.–Eur. J.*, 2010, **16**, 11662–11669.

156. B. Li, Y. Zhang, D. Ma, L. Li, G. Li, G. Li, Z. Shi and S. Feng, *Chem. Commun.*, 2012, **48**, 6151–6153.

157. B. M. Wiers, M.-L. Foo, N. P. Balsara and J. R. Long, *J. Am. Chem. Soc.*, 2011, **133**, 14522–14525.

158. M. Inukai, S. Horike, D. Umeyama, Y. Hijikata and S. Kitagawa, *Dalton Trans.*, 2012, **41**, 13261–13263.

159. M. Meilikhov, K. Yusenko and R. A. Fischer, *J. Am. Chem. Soc.*, 2009, **131**, 9644–9645.

160. C. Larabi and E. A. Quadrelli, *Eur. J. Inorg. Chem.*, 2012, 3014–3022.

161. G. Férey, F. Millange, M. Morcrette, C. Serre, M.-L. Doublet, J.-M. Grenèche and J.-M. Tarascon, *Angew. Chem., Int. Ed.*, 2007, **46**, 3259–3263.

162. H. Leclerc, A. Vimont, J.-C. Lavalley, M. Daturi, A. D. Wiersum, P. L. Llwellyn, P. Horcajada, G. Férey and C. Serre, *Phys. Chem. Chem. Phys.*, 2011, **13**, 11748–11756.

163. M. Meilikhov, K. Yusenko, A. Torrisi, B. Jee, C. Mellot-Draznieks, A. Pöppl and R. A. Fischer, *Angew. Chem., Int. Ed.*, 2010, **49**, 6212–6215.

164. H. Leclerc, T. Devic, S. Devautour-Vinot, P. Bazin, N. Audebrand, G. Férey, M. Daturi, A. Vimont and G. Clet, *J. Phys. Chem. C*, 2011, **115**, 19828–19840.

165. L. Mi, H. Hou, Z. Song, H. Han, H. Xu, Y. Fan and S.-W. Ng, *Cryst. Growth Des.*, 2007, **7**, 2553–2561.

166. L. Mi, H. Hou, Z. Song, H. Han and Y. Fan, *Chem.–Eur. J.*, 2008, **14**, 1814–1821.

167. Y. Kim, S. Das, S. Bhattacharya, S. Hong, M. G. Kim, M. Yoon, S. Natarajan and K. Kim, *Chem.–Eur. J.*, 2012, **18**, 16642–16648.

168. C. K. Brozek and M. Dincă, *Chem. Sci.*, 2012, **3**, 2110.

169. A. M. Shultz, O. K. Farha, D. Adhikari, A. A. Sarjeant, J. T. Hupp and S. T. Nguyen, *Inorg. Chem.*, 2011, **50**, 3174–3176.

170. A. M. Shultz, A. A. Sarjeant, O. K. Farha, J. T. Hupp and S. T. Nguyen, *J. Am. Chem. Soc.*, 2011, **133**, 13252–13255.

171. Q. Yao, J. Sun, K. Li, J. Su, M. V. Peskov and X. Zou, *Dalton Trans.*, 2012, **41**, 3953–3955.

172. M. Kim, J. F. Cahill, H. Fei, K. A. Prather and S. M. Cohen, *J. Am. Chem. Soc.*, 2012, **134**, 18082–18088.

173. B. J. Burnett, P. M. Barron, C. Hu and W. Choe, *J. Am. Chem. Soc.*, 2011, **133**, 9984–9987.

174. O. Karagiaridi, W. Bury, A. A. Sarjeant, C. L. Stern, O. K. Farha and J. T. Hupp, *Chem. Sci.*, 2012, **3**, 3256–3260.

175. O. Karagiaridi, M. B. Lalonde, W. Bury, A. A. Sarjeant, O. K. Farha and J. T. Hupp, *J. Am. Chem. Soc.*, 2012, **134**, 18790–18796.

176. M. Kondo, S. Furukawa, K. Hirai and S. Kitagawa, *Angew. Chem., Int. Ed.*, 2010, **49**, 5327–5330.

177. A. D. Burrows, K. Cassar, R. M. W. Friend, M. F. Mahon, S. P. Rigby and J. E. Warren, *CrystEngComm*, 2005, **7**, 548–550.

178. T.-F. Liu, J. Lü, C. Tian, M. Cao, Z. Lin and R. Cao, *Inorg. Chem.*, 2011, **50**, 2264–2271.

179. D. Banerjee, S. J. Kim, H. Wu, W. Xu, L. A. Borkowski, J. Li and J. B. Parise, *Inorg. Chem.*, 2011, **50**, 208–212.

180. H. J. Park and M. P. Suh, *Chem. Sci.*, 2013, **4**, 685–690.

181. J. An, S. J. Geib and N. L. Rosi, *J. Am. Chem. Soc.*, 2009, **131**, 8376–8377.

182. J. An and N. L. Rosi, *J. Am. Chem. Soc.*, 2010, **132**, 5578–5579.

183. F. Nouar, J. Eckert, J. F. Eubank, P. Forster and M. Eddaoudi, *J. Am. Chem. Soc.*, 2009, **131**, 2864–2870.

184. D. F. Sava, V. C. Kravtsov, F. Nouar, L. Wojtas, J. F. Eubank and M. Eddaoudi, *J. Am. Chem. Soc.*, 2008, **130**, 3768–3770.

185. Z. Zhang, L. Zhang, L. Wojtas, P. Nugent, M. Eddaoudi and M. J. Zaworotko, *J. Am. Chem. Soc.*, 2012, **134**, 924–927.
186. Z. Zhang, W.-Y. Gao, L. Wojtas, S. Ma, M. Eddaoudi and M. J. Zaworotko, *Angew. Chem., Int. Ed.*, 2012, **51**, 9330–9334.
187. A. D. Burrows, *CrystEngComm*, 2011, **13**, 3623–3642.
188. H. Deng, C. J. Doonan, H. Furukawa, R. B. Ferreira, J. Towne, C. B. Knobler, B. Wang and O. M. Yaghi, *Science*, 2010, **327**, 846–850.

CHAPTER 4

Characterization of MOFs. 1. Combined Vibrational and Electronic Spectroscopies

FRANCESCA BONINO,[a] CARLO LAMBERTI,[a] SACHIN CHAVAN,[a,b] JENNY G. VITILLO[a] AND SILVIA BORDIGA*[a]

[a] Department of Chemistry, NIS Centre of Excellence and INSTM Reference Center, Via Quarello 15, Università di Torino, 10135 Torino, Italy; [b] inGAP Centre of Research-based Innovation Department of Chemistry, University of Oslo, P.O. Box 1033, N-0315 Oslo, Norway
*Email: silvia.bordiga@unito.it

4.1 Introduction

The present chapter aims to provide the elementary background and few selected examples on the uses of vibrational and electronic spectroscopies in the characterization of metallorganic materials, such as metal organic frameworks (MOFs). It shows how spectroscopies allow fundamental steps such as MOF desolvation and successive adsorption of gases to be followed.

For didactic reasons we have divided vibrational (treated first in Section 4.2) from electronic spectroscopies, reported in Section 4.3 of the chapter. Each of these sections starts with a concise but essential introduction to the used techniques written with a didactic approach, then examples follow.

This chapter is not aimed at reporting an exhaustive review on the most relevant spectroscopic papers that have significantly contributed to

RSC Catalysis Series No. 12
Metal Organic Frameworks as Heterogeneous Catalysts
Edited by Francesc X. Llabrés i Xamena and Jorge Gascon
© The Royal Society of Chemistry 2013
Published by the Royal Society of Chemistry, www.rsc.org

understanding MOF structures and reactivities, so the reported bibliography is by no way exhaustive and has not been chosen on the basis of the scientific impact of the papers. The chapter is instead aimed at providing general guidelines and explaining the information that can be obtained using the different vibrational and electronic spectroscopies, to stress their complementarity and the importance in performing a broad characterization study based on the use of different techniques. We believe that this point is crucial in achieving a full understanding of the structure and the properties of complex materials such as MOFs. We further believe that the most used techniques, such as X-Ray Powder Diffraction (XRPD), BET[1] and thermo-gravimetry (TG), should be supported by combined, *ad hoc* chosen, vibrational and electronic spectroscopies. Moreover, the characterization should not be limited to the use of laboratory techniques but, when possible, neutron and synchrotron radiation sources should also be exploited to complete the investigation.

Although not directly described in this chapter, theoretical calculations provide important support to the experimental techniques, allowing the checking of, for example, the stability of a structure inferred from Rietveld refinement. They also allow a precise assignment of the vibrational and electronic features observed with the different spectroscopies. This is the reason why the spectroscopic results reported hereafter will often be compared to the corresponding calculated spectra. For more details on density functional theory (DFT) calculations the reader is referred to Chapter 7 of this book.[2]

Due to space limitations (and also for lack of competencies) spin resonance spectroscopies are not described in this chapter. This lack is particularly relevant for NMR spectroscopy, which has played a relevant role in supporting standard XRD refinements in determining MOF structures and in proving the effective grafting of functionalizing groups.

Before entering into the description of the chosen examples for vibrational (Section 4.2) and electronic (Section 4.3) spectroscopies, in Section 4.1.1 we report a brief discussion on the units that are usually reported in the abscissa axis of the different spectra.

4.1.1 General Considerations and Some Useful Relationships Among Units Used in Spectroscopies

Light-based spectroscopies are performed investigating the interaction between light and matter in the whole range of the electromagnetic spectrum: microwave (NMR and EPR), mid-infrared (IR), near-infrared (NIR), visible (Vis) and ultraviolet (UV) (Raman, optical and luminescence spectroscopies), far-UV (UPS), soft X-rays (XPS), hard X-rays (XANES, EXAFS,[3] XES), and γ (Mössbauer). In this chapter we will discuss the use of IR, Raman, optical and luminescence spectroscopies, XPS, XANES, XES, in the understanding of MOF vibrational and electronic properties, while in the next chapter of this book EXAFS will be widely debated.[3]

In this regard it is worth recalling that light is an electromagnetic wave characterized by a space periodicity or wavelength (λ) and a time periodicity or period (T) bound together by the constraint that the wave has to propagate at speed $c = \lambda/T = 2.998 \times 10^8$ m s^{-1}. Often used are also the reciprocal of the reported quantities, defined as time frequency ($v = 1/T$) and space frequency or wavenumber ($\tilde{v} = 1/\lambda$). The definition of the propagation speed can consequently be given in different ways: $c = \lambda/T = \lambda v = v/\tilde{v} = 1/(T\tilde{v})$. The theories of Planck and Einstein taught us that the electromagnetic wave is quantized and is constituted by photons carrying energy ($E = hv$), momentum ($P = hv/c$) and angular momentum ($L = h/2\pi$), with h being the Planck constant ($h = 6.626068 \times 10^{-34}$ J s). The relations written above imply that a single photon can be unequivocally identified by reporting only one of the following quantities: E (in eV), λ (in nm), v (in THz) or $\tilde{v}$ (in cm^{-1}). This degree of freedom has generated some heterogeneity in the spectroscopic data reported in the literature: spectra reported in eV, nm, THz, or cm^{-1} are found, depending on the scientific community and on the time period. The relationships among the different units can be easily extracted from the definition reported above and spectra can be converted; as an example:

$$E = hc\tilde{v}, i.e. 1 \, \text{eV} = 8065.73 \, \text{cm}^{-1}. \tag{4.1}$$

Since the development of Fourier Transform infrared instruments,[4] the most widely used unit has become cm^{-1} and most of the spectra are reported in wavenumber. We will follow this convention for IR and Raman data reported here. Luminescence and optical spectroscopies are those where eV, nm and cm^{-1} units are still widely found in the literature. UPS, XPS, XANES and XES spectra are usually reported in eV (or keV).

Concerning non-optical spectroscopies, only inelastic neutron scattering (INS)[5] will be briefly mentioned in this chapter. INS is performed with thermal neutrons,[6] that have a speed (v) of few thousand of m s^{-1}, for which the classical (non-relativistic) law of mechanics holds. Thermal neutrons are massive particles carrying energy ($E = \frac{1}{2}m_N v_N^2$), momentum ($P = m_N v_N$) and angular momentum ($L = \frac{1}{2}\hbar = \frac{1}{2}h/2\pi$), being $m_N = 1.6749 \times 10^{-27}$ kg, the neutron rest mass. INS is different from Raman (inelastic photon scattering) by two main facts. Firstly, for a given exchanged energy ΔE ($m_N v_N \Delta v_N$ or $h\Delta v$) a much larger momentum exchange is possible with neutrons as $\Delta P = m_N \Delta v_N$, thus giving access to the whole Brillouin zone, while for photons $\Delta P = h\Delta v/c$ is almost zero, as c is a large constant. Students may remember that $m_N \Delta v_N \gg h\Delta v/c$, for the same ΔE. Secondly, INS enhances all vibration (and rotation) modes involving H atoms, because neutrons interact much more strongly with hydrogen than any other nucleus,[6] (H has the largest neutron incoherent scattering length). Consequently, INS is a powerful technique in the studies of MOF dehydroxylation and of H_2 or CH_4 storage in MOFs. Thirdly, the optical selection rules that define IR or Raman active modes, *vide infra* eqn (4.9), does not hold for INS, which is potentially able to detect all transitions, when sufficiently intense.

4.2 Vibrational and Rotational Spectroscopies (IR, Raman and INS): Basic Theory and Examples

4.2.1 Basic Consideration of the Techniques

4.2.1.1 *Vibrational Modes*

Atoms in molecules and solids are not immobile but vibrate at specific frequencies that are defined by the masses of the involved atoms and by the strength of the chemical bonds that bind the atoms together (interatomic potential V) according to quantum mechanical laws. This means that each chemical group (–O–H; –C–H; –C–H$_2$, –C–H$_3$; –C–C–; –C=C–; –C≡C–; –N=N, *etc.* ....) has its own vibrational frequency that represent its vibrational fingerprint (vibrational mode).[7–12] Small perturbations of such vibrational modes may reflect the nature of the environment where chemical group has been embedded. An accurate vibrational study will allow determination of the chemical groups present in the sample and molecules hosted in the cavities. This is the reason why vibrational spectroscopies are so powerful and widely employed in chemistry, materials science, biology and solid state and molecular physics, and why they have been also extensively used in the investigation of MOFs.

In the simplest case of an AB diatomic molecule, V is just a 1D function of the interatomic distance r, that exhibits a stable minimum at the interatomic equilibrium distance r_0 (Figure 4.1). $V(r)$ can thus be expanded in to its Taylor series around r_0 as follows:[13]

$$V(\Delta r) = V(r_0) + (1/2)[\mathrm{d}^2 V/\mathrm{d}r^2]_{r=r_0}(\Delta r)^2 + (1/3!)[\mathrm{d}^3 V/\mathrm{d}r^3]_{r=r_0}(\Delta r)^3 + \mathrm{O}[(\Delta r)^4] \tag{4.2}$$

where $\Delta r = r - r_0$ represents the displacement from the equilibrium position, and where the term $\mathrm{O}[(\Delta r)^4]$ indicates that the error done truncating the series at the cubic term is proportional to $(\Delta r)^4$, *i.e.* negligible for small displacements. Of course the linear term in (Δr) is missing in eqn (4.2) because $[\mathrm{d}V/\mathrm{d}r]_{r=r_0} = 0$ at the equilibrium point. The Schrödinger equation with potential given by eqn (4.2) cannot be solved analytically and a numerical approach is required. An analytical solution is, however, available for the harmonic approximation where the potential is simplified as:[13]

$$V(\Delta r) \approx V(r_0) + (1/2)k(\Delta r)^2 \tag{4.3}$$

where $k = [\mathrm{d}^2 V/\mathrm{d}r^2]_{r=r_0}$. In this approximation, the Hamiltonian operator becomes:[14,15]

$$H = T + V(\Delta r) = (h^2/8\pi^2 \mu)[\mathrm{d}^2 V/\mathrm{d}(\Delta r)^2] + V(r_0) + (1/2)k(\Delta r)^2 \tag{4.4}$$

where h is the Planck constant and μ the reduced mass of the two atoms A and B: $\mu = (m_A m_B)/(m_A + m_B)$. The solutions of the Schrödinger equation $H\Psi = E\Psi$ result in eigenfunctions $\Psi_n(r)$ that are Hermite polynomials

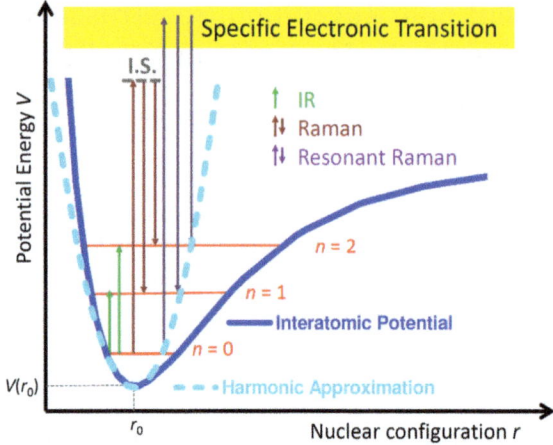

Figure 4.1 Schematic 1D-representation of the nuclear configuration of a general chemical group showing the typical Morse-like potential (full dark blue line), the ground vibrational state ($n=0$) and the first two excited states ($n=1, 2$). The harmonic approximated potential is also reported (dashed cyan line). r is the generalized coordinate allowing to define the nuclear positions: in the simplest case of a diatomic molecule, r coincides with the internuclear distance. The different colours of the vertical arrows allow the transitions that at the base of IR (green arrows), Raman (brown arrows) and resonant Raman (violet arrows) spectroscopies to be distinguished. In particular, photon absorption occurs in IR spectroscopy. Inelastic scattering is the base of Raman spectroscopy: it goes through an intermediate state (I.S.) that is usually a virtual state. In case the I.S. is not virtual but coincides with an empty electronic state, *i.e.* when the excitation line of the laser has the energy of an allowed electronic transition, we are dealing with the resonant Raman process that may enhance the intensity of some specific transitions by several orders of magnitude.

multiplied by Gaussian functions for the confinement, and where the eigenvalues are given by:[15]

$$E_n = V(r_0) + [n + 1/2](h/2\pi)\omega_0 \tag{4.5}$$

where $n=0$, 1, 2, 3... and $\omega_0 = (k/\mu)^{1/2}$. For $n=0,1$ the harmonic approximation is reliable in a large fraction of cases because the displacement Δr is relatively small, see Figure 4.1. In order to take into account anharmonicity eqn (4.5) has to be implemented as follows:[16]

$$E_n = V(r_0) + [n + 1/2](h/2\pi)\omega_0 + x_n[n + 1/2]^2(h/2\pi)\omega_0 \tag{4.6}$$

where x_n is the anharmonicity constant of the n-th vibrational level. It is a negative dimension-less number, typically ranging between –0.0001 and –0.002.[16]

Actually, experiments do not directly measure the eigenvalues E_n but measure the energy needed to allow the transition between the ground vibrational state ($n=0$) and the excited vibrational states. In fact, in general only the transitions from the $n=0$ state are observed at RT, because the excited vibrational states ($n=1$, 2, ...) are unoccupied, as $\Delta E_{0\to n} \gg k_B T = 25.852$ meV ($= 208.52$ cm^{-1}) at $T = 300$ K, where k_B is the Boltzmann constant ($k_B = 1.3806 \times 10^{-23}$ J K^{-1} = 8.6173×10^{-5} eV K^{-1}).

$$\Delta E_{0\to 1} = E_1 - E_0 = h/2\pi\omega_0 \qquad (4.7)$$

Experimentally, $\Delta E_{0\to 1}$ values are usually measured allowing a poly-chromatic light beam to interact with the sample. The $(n=0) \to (n=1)$ transition can be induced by an IR photon carrying exactly the energy needed for the transition $E = hv = \Delta E_{0\to 1} = h/2\pi\omega_0$. In such a case, the photon is absorbed and is no longer present in the final state, see green arrows in Figure 4.1. This phenomenon follows the standard Lambert–Beer law:[16]

$$I_1(\tilde{v}) = I_0(\tilde{v})\exp[-\mu(\tilde{v})x] \Rightarrow \mu(\tilde{v})x = \ln[I_0(\tilde{v})/I_1(\tilde{v})] \qquad (4.8)$$

where $I_0(\tilde{v})$ and $I_1(\tilde{v})$ are the incident and the transmitted (through the sample) IR beams (at the different wavenumbers $\tilde{v}$), respectively, x is the sample thickness and $\mu(\tilde{v})$ is the extinction coefficient. $\mu(\tilde{v})x$ is the physical quantity plotted in the ordinate of the IR spectra measured in transmission mode and reported in "absorbance units",[17] while usually wavenumber in cm^{-1} are used in the abscissa. Alternative ways of measuring an IR spectrum are diffuse reflectance (DRIFT)[18,19] and attenuated total reflection (ATR)[20] modes.

Alternatively, $\Delta E_{0\to 1}$ values can be determined using an intense and monochromatic photon beam (usually a laser) of energy $E_{ex} = hv_{ex}$ (significantly higher than the vibrational transition $\Delta E_{0\to 1}$) and measuring the energy of the inelastic scattered photons $E_{scattered} = hv_{scattered}$: $\Delta E_{0\to 1} = h/2\pi\omega_0 = hv_{ex} - hv_{scattered}$ (see up and down brown arrows in Figure 4.1). This is the Raman process.[21–22] In this case the spectrum is reported as counts (of scattered photons) as a function of the Raman shift $\Delta\tilde{v}$ (in cm^{-1}). The Raman shift is defined as $\Delta\tilde{v} = (hv_{ex} - hv_{scattered})/hc = \tilde{v}_{ex} - \tilde{v}_{scattered}$. Actually, the most probable decay channel of this process is the elastic scattering, where $\tilde{v}_{scatter} = \tilde{v}_{ex}$ and $\Delta\tilde{v} = 0$ cm^{-1}. This process carries no information and its efficiency can be up to 10^{+6} that of the inelastic process: this explains why laser sources are needed. Experimentally impressive improvements have been achieved in the last decade, either by analyzing the scattered photons with a Fourier interferometer (when E_{ex} falls in the NIR) or by collecting the scattered photon with strip detectors (when E_{ex} falls in the Vis-UV).

The low probability of the Raman process is related to the fact that the system has to be excited into an intermediate state (I. S. in Figure 4.1) that is, in most of the cases, a virtual state.[22] When the intermediate state coincides with a real electronic state of the system, we are dealing with a resonant Raman process (see up and down violet arrows in Figure 4.1).[21–22] In this cases, vibrational modes involved with the species that has been excited by the electronic transition and having the same symmetry will be enhanced by some order of magnitude.[23–25] In the case of INS, the neutron exchanges both energy and momentum with the system, promoting it into an excited vibrational state.[5]

So, in principle IR, Raman, Resonant Raman and INS allows access to the same information, see eqn (4.7), *i.e.* the energy needed to perform the $(n=0) \rightarrow (n=1)$ transition; the transition probability $(P_{0\rightarrow1})$ is, however, different in the different cases. In general, the transition probability $(P_{i\rightarrow f})$ from an initial state $|\Psi_i>$ to a final state $|\Psi_f>$ is given by the Fermi golden rule:[26–28]

$$P_{i\rightarrow f} = \rho_i(\text{occ})\rho_f(\text{unocc})|<\Psi_f|\mathbf{O}|\Psi_i>|^2 \qquad (4.9)$$

where $\rho_i(\text{occ})$ and $\rho_f(\text{unocc})$ are the densities of initial occupied and final unoccupied states, respectively, and where $<\Psi_f|\mathbf{O}|\Psi_i>$ is the matrix element that projects on the final state $|\Psi_f>$ the state obtained by the action of the operator $\mathbf{O}$ on the initial state $|\Psi_i>$. In IR spectroscopy (and generally in all cases where a photon is destroyed or created) the $\mathbf{O}$ operator is the electrical dipole moment $\mathbf{O}=q\mathbf{r}$, while in Raman spectroscopy it is the polarizability tensor $\mathbf{O}=\boldsymbol{\alpha}$.[22] Quantum mechanics allows the simulation of vibrational frequencies of molecules and solids,[29,30] that will be of great help in understanding the experimental results and attributing the different bands to the correct modes.

Several textbooks mention Laporte's law, stating that in all spectroscopies, where a photon is destroyed or created (such as IR), a given transition from the state $|\Psi_i>$ to a final state $|\Psi_f>$ can be observed only if the two states have a different symmetry. Laporte's law is a special case of the Fermi golden rule and refers to systems having an inversion center, because only in this case the related wave functions have a defined symmetry: either even (when $\Psi(-\mathbf{r})=\Psi(\mathbf{r})$) or odd (when $\Psi(-\mathbf{r})=-\Psi(\mathbf{r})$). In such cases, adopting the Schrödinger notation, the matrix element in eqn (4.9) can be rewritten as:

$$<\Psi_f|q\mathbf{r}|\Psi_i> = q \int \Psi_f^*(\mathbf{r})\Psi_i(\mathbf{r})\mathbf{r}d^3\mathbf{r} \qquad (4.10)$$

An integral over the whole space, as that reported in eqn (4.10) is equal to zero when the integrated function is an odd function. To understand this fact for one dimension functions, please simply consider the fact that $\int f(x)\, dx = 0$ for all odd functions because the integral from $-\infty$ to 0 exactly cancels the contribution from 0 to $+\infty$. Now, as the dipole operator $q\mathbf{r}$ is odd, then it is evident that if both $\Psi_f(\mathbf{r})$ and $\Psi_i(\mathbf{r})$ have the same symmetry then the integrated function, $\Psi_f^*(\mathbf{r})\,\Psi_i(\mathbf{r})\mathbf{r}$, is odd and $<\Psi_f|q\mathbf{r}|\Psi_i>=0$. This simple symmetry concept explains the Laporte rule.

From what stated, it emerges that there will be vibrational modes that cannot be observed by IR because $|<\Psi_f|q\mathbf{r}|\Psi_i>| =0$ but that would be observed by Raman, as $|<\Psi_f|\boldsymbol{\alpha}|\Psi_i>| \neq 0$, or *vice versa* and that there may be modes that can not be observed neither by IR nor by Raman because both matrix elements are zero. Students should refer to the character tables for the different space groups (reported in the appendices of most physical chemistry text books) to be able to discriminate IR and Raman active modes for systems having different local symmetries.[13,28] INS is in principle able to observe all vibration transitions because there are no selection rules.

The $(n=0) \rightarrow (n=2)$ transition, also drawn in Figure 4.1, is usually very weak because forbidden in a pure harmonic potential ($|<\Psi_2|q\mathbf{r}|\Psi_0>| =0$ as both

$|\Psi_0>$ and $\Psi_2>$ are even functions): however, it is measurable due to the anharmonicity of the potential. Indeed, the true interatomic potential is the Morse-like one, that has no center of inversion and so the true $|\Psi_0>$ and $\Psi_2>$ functions have also a small odd part that allows the matrix element to be not null. In such an anharmonic potential, $\Delta E_{0\rightarrow 2}<2\Delta E_{0\rightarrow 1}$ and then the difference $2h/2\pi\omega_0 - \Delta E_{0\rightarrow 2}$ provides an experimental measurement of the anharmonicity of the specific vibrational mode and a determination of the x_n constant, see eqn (4.6).

4.2.1.2   Diatomic Probe Molecules

A powerful method to get indirect information on a surface adsorption site S consists of dosing a probe molecule P and observing the modification induced on the vibrational modes of P upon formation of a S $\cdots$ P adduct.[31–42] In this regard, MOFs behave differently from standard medium/high surface area materials (*e.g.* oxides and zeolites), because their extremely high surface area and their high framework flexibility implies that the formation of the S $\cdots$ P adduct will also cause a significant perturbation of the MOF framework modes, making vibrational spectra even more informative for this class of materials,[43–45] *vide infra* the examples reported in Figures 4.8–4.10, and Figure 4.15.

For the sake of simplicity, we will limit the discussion to diatomic molecules, that are however the most used as surface probe. Diatomic molecules (AB) have only one vibrational mode, the $A - B$ stretching, designed as the $\nu(AB)$ mode. In these experiments, the most relevant observation is the frequency shift induced by the adsorption, defined as :

$$\Delta\tilde{\nu}(AB) = \tilde{\nu}(AB) - \tilde{\nu}_0(AB) \qquad (4.11)$$

Where $\tilde{\nu}_0(AB)$ is the wavenumber of the $(n=0)\rightarrow(n=1)$ transition for the unperturbed AB molecule in the gas phase.

In this simplest case of a diatomic molecule, the nuclear configuration, represented by the general r variable in the abscissa of Figure 4.1, coincides with the interatomic r_{AB} distance and the vibrational mode consists of a stretching of r_{AB} across the equilibrium distance (r_0 in Figure 4.1). For diatomic molecules the Fermi golden rule (eqn (4.10)) is also simplified and results in the selection rules affirming that homo-nuclear molecules (AA, *e.g.* H_2, N_2, O_2) are IR inactive but Raman active, while hetero-nuclear molecules (AB, *e.g.* CO, NO) are both IR and Raman active. The fact that $|<\Psi_1|q\mathbf{r}|\Psi_0>| = 0$ for homo-nuclear molecules is due to the fact that there is no variation in the dipole moment of the molecule in the transition as $|<\Psi_0|q\mathbf{r}|\Psi_0>| = |<\Psi_1|q\mathbf{r}$ $\Psi_1>| = 0$. This holds for unperturbed molecules in the gas phase, but this does not hold for an adsorbed molecule, because the adsorption causes a lowering of the symmetry of the electronic wavefunction of the molecule and the presence of a non-zero induced dipole moment: $S^+ \cdots A^{\delta-} - A^{\delta+}$ or $S^- \cdots A^{\delta+} - A^{\delta-}$. This means that IR spectroscopy can be used to monitor the adsorption of *e.g.* H_2, N_2 or O_2 molecules on MOFs, but that IR can observe only those that have a specific interaction with a surface site (represented by a metal center with coordination vacancies or by the linker), the remaining fraction fluctuating in the pores being IR inactive. In such cases, both the $\tilde{v}(AA)$ shift, $\Delta\tilde{v}(AA) = \tilde{v}(AA) - \tilde{v}_0(AA)$ and the band intensity will be informative on the nature of the S $\cdots$ AA adduct and thus, indirectly, on the S site.[46,47] The larger is the induced dipole moment, δR_{AA}, the larger will be the extinction coefficient of the $v(AA)$ mode of the $S^+ \cdots A^{\delta-} - A^{\delta+}$ or $S^- \cdots A^{\delta+} - A^{\delta-}$ complex.

4.2.1.3   INS Spectra of H_2 Adsorbed on MOFs: Insights on the Rotational Modes and More

We will not enter into the rules of quantum mechanics that describe the molecular rotation as it will be, by far, beyond the purposes of this chapter and the readers should refer to the classical physical chemistry text books for a proper description of the problem.[13] Here we will simply focus on the aspects that are relevant to investigate the adsorption of H_2 inside MOFs.

The total spin of the molecule is defined as J, that is the modulus of the vector sum on all electrons and nuclei spins. Neutrons, possessing spin $S = \hbar/2$, have the ability to change the total nuclear spin I of the hydrogen molecule, something that photons ($S = \hbar$) cannot do. This makes possible to observe the rotational $\Delta J = 1$ transitions, such as $(J=0) \rightarrow (J=1)$ of para-H_2 ($I=0$) and $(J=1) \rightarrow (J=2)$ of ortho-H_2 ($I=1$). Note that $\Delta J = 2n+1$ transitions are forbidden in optical, infrared (IR) and Raman spectroscopy of H_2, since they involve the ortho–para conversion. This peculiar property has been exploited by some authors who collected INS data to investigate the fundamental properties of molecular hydrogen adsorbed inside MOFs.[48–62]

As shown by Matanovic *et al.*,[62] the measured INS spectra contain a wealth of information about the excitations of H_2 center-of-mass translational motions as well as the rotational excitations of the hydrogen molecule at various binding sites, and through this, about the potential interactions of the guest molecule with the host material. Such information can be extracted and INS spectra can be fully understood once the Schrödinger equation can be solved with the following Hamiltonian:[62,63]

$$H = (h^2/8\pi^2 M)\nabla^2 + \mathbf{j}^2/(2I_\perp) + V(\mathbf{r}, \theta, \phi) \qquad (4.12)$$

where M is the mass of the H_2 molecule, $I_\perp = M/4r^2_{HH}$ is the corresponding momentum of inertia of the H_2 molecule with respect to one of the two equivalent axes perpendicular to the molecular axis, $\mathbf{j}$ is the angular momentum operator and $V(\mathbf{r},\theta,\phi)$ is the potential energy describing the intermolecular interaction between H_2 and the MOF framework, with $\mathbf{r}$ being the vector defining the position of the center of mass of the molecule inside the MOF unit cell and θ, ϕ the two polar angles defining the orientation of the molecule relative to the framework. Several textbooks report the molecule rotational constant defined as $B = (h/8\pi^2 cI_\perp)$, writing the eigenvalues of the angular part of eqn (4.12) as $Bj(j+1)$, being j the rotational quantum number. Care must be taken with this notation, as B is a spectroscopic wavenumber unit, measured in cm^{-1} and not an energy measured in eV. Actually such textbooks have adopted eqn (4.1) and the eigenvalues of the angular part of eqn (4.12) in eV should be given as: $hcBj(j+1)$.

It is finally worth recalling that such important information is obtained in the few meV range of neutron energy transfer (typically $5-50$ meV, corresponding to $40-400\,cm^{-1}$, *vide infra* Figure 4.14) *i.e.* between one and two orders of magnitude below the energy required in IR or Raman spectroscopies to excite the $(n=0)\rightarrow(n=1)$ vibrational transition of the H_2 molecule, typically about $4000\,cm^{-1}$, *vide infra* Figure 4.13.

4.2.2   Framework Modes: Linkers, Inorganic Cornerstones and Functional Groups

4.2.2.1   *Linkers and Inorganic Cornerstones*

It is well known that by keeping constant an inorganic unit and by choosing different organic ligands, different structures of MOFs with different properties can be synthesized. Figure 4.2 shows how vibrational spectroscopies (IR and Raman, $\lambda_{ex} = 514$ nm) can be informative in this sense in the case of three Ni-MOFs (CPO-27-Ni,[64] Ni-STA-12[65] and Ni-bpz[66]) synthesized starting from three different organic linkers: 2,5-dihydroxyterephthalic acid $(C_8H_6O_6)$, N,N'-piperazinebismethylenephosphonic acid $(C_8H_{16}N_2O_6P_2)$ and

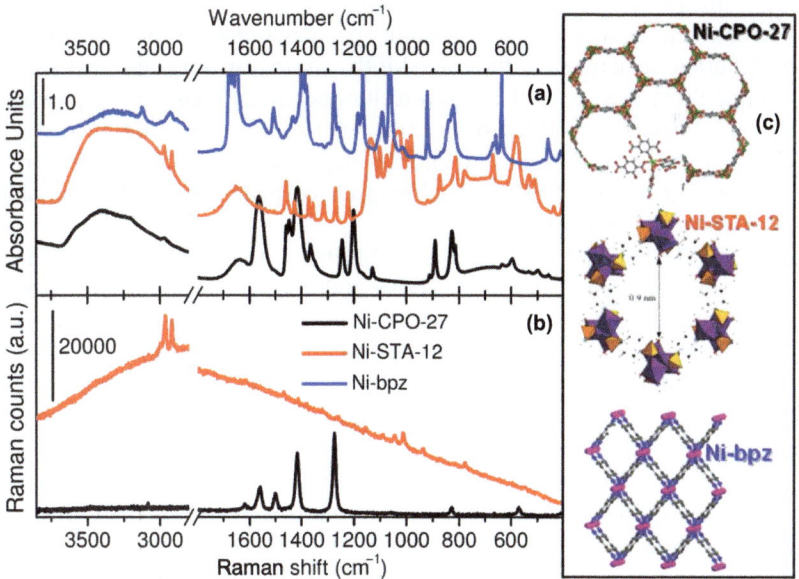

Figure 4.2   IR (part a) and Raman ($\lambda_{ex} = 514$ nm, part b) spectra of CPO-27-Ni (black curve), Ni-STA-12 (red curve) and Ni-bpz (blue curve) in air. Part (c): pictorial representations of the three considered materials. Previously unpublished figure reporting some spectra published in ref. 66, 69 and 71.

4,4'-bipyrazole ($C_6H_6N_4$), respectively. The three reported structures (see Figure 4.2) show clear examples on how a MOF can be built by means of oxygen and/or nitrogen donor ligands.

Very briefly, CPO-27-Ni (other common names adopted for this material in literature are MOF-74, Ni_2(dhtp), Ni_2-DOBDC) is a 3D honeycomb like structure (see Figure 4.2c, top panel), where all the oxygen atoms of the ligands are involved in the coordination of nickel atoms.[64] This accounts for five of ligands around each nickel centre, whereas the sixth coordinative bond is normally occupied by a water molecule that points toward the cavity. All the Ni(II) ions are identical with respect to the first coordination sphere, showing five oxygens that differ slightly from one another.

On the other hand, the Ni-STA-12 framework is based on inorganic columns of helical chains of edge-sharing NiO_5N octahedra (see Figure 4.2c, middle panel).[66] Each NiO_5N octahedron consists of nickel coordinated to four oxygen atoms belonging to two phosphonate tetrahedra (on the two shared edges) plus an oxygen atom from a coordinating water molecule and a nitrogen from the piperazine ring of the bis(phosphonic acid). Each chain is linked to three others *via* piperazine groups, so that each bis(phosphonate) ligand coordinates *via* two oxygen atoms and a nitrogen to each of two helical chains of nickel cations, leaving a phosphonate oxygen projecting towards the pore space. This is the first porous solid in which this ligand coordinates through both oxygen and nitrogen atoms. Finally, Ni-bpz possesses the same structural topology of the MIL-53 family,[67] built up by coupling the 4,4'-bipyrazole (H_2bpz) to square planar, coordinatively unsaturated Ni(II) ions.

As already anticipated, IR and Raman spectroscopies can give information on the different structures of these Ni-MOFs, based on very different linkers. The sample form used with these two techniques is different, being powdered samples used to collect Raman spectra, while IR spectra are recorded in transmission mode on thin film of samples deposited on silicon wafers. In the last case, in fact, the conventional self-supported pellets would have revealed too intense, out of scale, modes.[17]

In general, IR and Raman spectra of MOFs in the hydrated form can be divided in three regions: (i) signals above $3000\,cm^{-1}$, due to the physisorbed water or to framework –O–H groups,[43,45,68] the latter clearly visible only after desolvation. This region is characterized by an intense and broad band in the IR spectra (Figure 4.2a) while, as water is a very weak Raman scatterer, its presence is not directly detected in the Raman spectra (see Figure 4.2b where only the spectra for CPO-27-Ni and Ni-STA-12 are reported). (ii) Peaks around 3000 cm^{-1}, present in almost all the cases, associated with ν(C–H) vibrational modes deriving from the ligands. (iii) At lower frequency, typically $1800–400\,cm^{-1}$, the vibrational modes characteristic of the linker functional groups coordinating metal ions are present. Actually in this last region, not only linker functional groups are present, but also the linker itself and the framework modes (M–O).

For CPO-27-Ni the bands in the range $1500–1400\,cm^{-1}$ are due to ν(COO$^-$)$_{asym}$ and ν(COO$^-$)$_{sym}$ of the carboxylates. While Raman shows definite and sharp peaks, IR presents a very broad feature (for a more detailed assignment, please refer to section 4.2.3.4).[69,70] Conversely, Ni-STA-12 shows absorptions due to v(C–C) and v(C–N) combined with features associated with the v(P–O) of the phosphonate groups in the 1160–950 cm^{-1} range and bands ascribable to symmetric and antisymmetric bending modes of CPO$_3$ units in the 580–500 cm^{-1} range. In this case, the Raman spectrum is not so informative due to the fluorescence phenomenon.[71] Finally, Ni-bpz IR spectrum is of course very different from the previously discussed ones. In this case, it is more difficult to identify specific vibrational modes for the coordinating species (N atoms in the 4,4′-bipyrazolate) as they are part of rings and thus they undergo to collective ring modes and *ad hoc, ab initio* calculations are needed for a correct assignment.[66]

In summary, IR and Raman spectroscopies are able to detect the vibrational fingerprints of the linker molecules. In most of the cases the assignment of the linker bands is straightforward; in other cases the support of *ab initio* calculations is required.

4.2.2.2 Functionalization

The development in the field of MOF materials is moving from the discovery of new structures toward applications of the most promising materials. In most cases, specialized applications require incorporation of functional chemical groups. Here, a family of isoreticular MOFs, based on the UiO-66 structure[43,68] and obtained from three different linkers (H$_2$N-H$_2$BDC, O$_2$N-H$_2$BDC and Br-H$_2$BDC), is discussed.[72] UiO-66 parent material results from connecting together hexanuclear zirconium clusters with a simple, commercially available bridging ligand 1,4-benzenedicarboxylate (BDC), furnishing a robust, 3-dimensional porous structure[68] (see also Section 4.2.3.3). The physicochemical and chemical investigation of both UiO-66 and its functionalized derivatives demonstrates that this class of MOFs retains high thermal and chemical stabilities, even with functional groups present at the linker units. This high stability has been attributed to the combination of strong Zr–O bonds and on the ability of the inner Zr$_6$(O)$_4$(OH)$_4$ cluster to rearrange reversibly upon dehydroxylation or rehydratation of μ_3-OH groups, without detrimental effects on the stabilities of the connecting dicarboxylate bridges.[43,68]

The presence of the introduced functional groups on the linkers was evidenced by IR and Raman spectroscopies.[72] The IR spectra were collected on samples prepared as thin films and the measurements were performed after solvent removal at 100 °C in vacuum. The Raman spectra were obtained on powder samples in air. In Figure 4.3, the FTIR (left parts) and Raman (right parts) spectra of the tagged MOFs are compared with the spectra of the parent UiO-66 (bottom spectra); a pictorial representation of the considered materials is also given in the right panels.

The FTIR spectrum of the amino-tagged sample UiO-66-NH$_2$ is depicted in Figure 4.3a. Spectroscopic features due to the presence of the amino group can be discerned in the whole spectral region and in particular in the high frequency

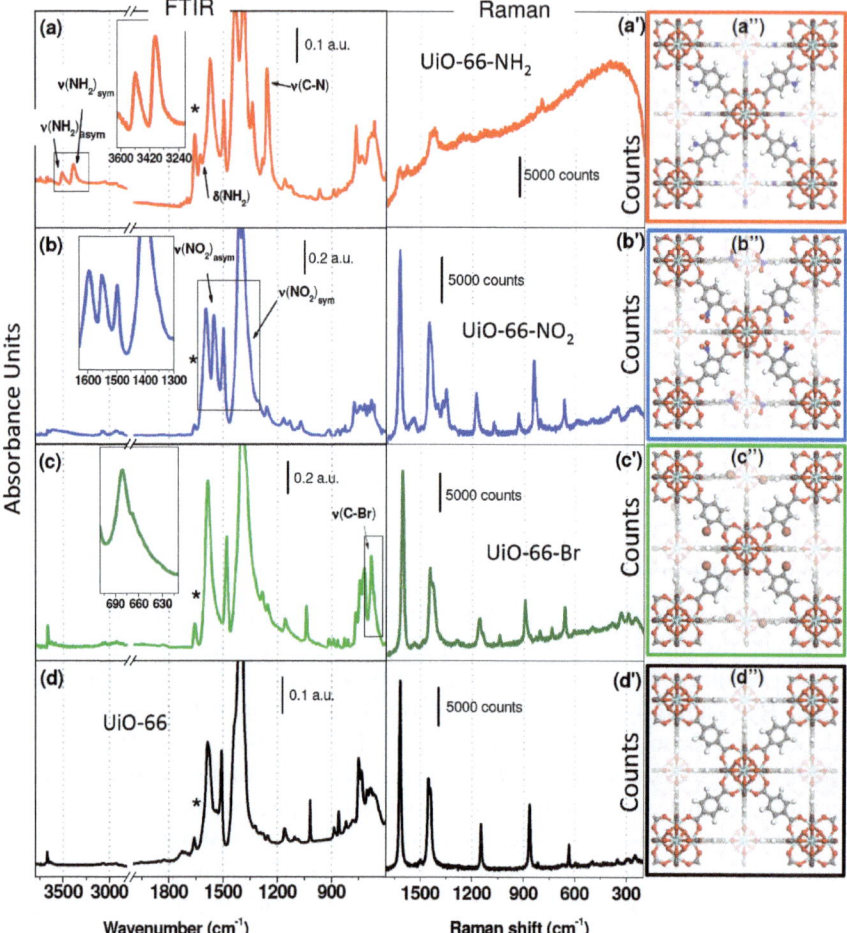

Figure 4.3   Left panels: FTIR spectra of tagged Zr-MOFs activated at 373 K in vacuum. (a) UiO-66-NH$_2$, (inset) magnification of the v(NH$_2$)$_{asym}$ and v(NH$_2$)$_{sym}$ region; (b) UiO-66-NO$_2$, (inset) magnification of the v(NO$_2$)$_{asym}$ and v(NO$_2$)$_{sym}$ region; (c) UiO-66-Br, (inset) magnification of the v(C–Br) region; (d) original, not-tagged, UiO-66. Vibrational modes due to the presence of functional groups are shown, while the symbol * denotes residual DMF. Middle panels: Raman spectra of tagged Zr-MOFs recorded in air: (a') UiO-66-NH$_2$; (b') UiO-66-NO$_2$; (c') UiO-66-Br; (d') original, not-tagged, UiO-66. Right panels: pictorial representation of the considered materials (a'')-(d'').
Adapted with permission from Kandiah *et al.*,[72] copyright American Chemical Society, 2010.

region where the primary aromatic amino group displays two medium absorptions, one at 3507 cm^{-1} and the other at 3384 cm^{-1}. These bands represent, respectively, the asymmetric and symmetric N–H stretching modes. Upon treatment with D$_2$O vapors, these bands underwent the expected isotopic shifts[73] with new bands appearing at 2618 and 2478 cm^{-1} which are assigned to asymmetric and symmetric N–D stretching modes, respectively (spectra not shown). Importantly, this H/D scrambling shows that the amino group is accessible as well as chemically reactive and therefore very interesting for further studies within the field of catalysis. An analysis of the lower frequency region of the IR spectra requires particular attention due to the presence of a great number of bands, largely assigned to skeletal vibration modes of the material (combination of modes due to the organic aromatic linker and the Zr cluster unit). Among these features, it is possible to distinguish two other characteristic bands of the amino group: the medium N–H bending (scissoring) vibration observed at 1626 cm^{-1} and the strong C–N stretching absorption distinctive of aromatic amines at 1356 cm^{-1}. Unfortunately, the Raman spectrum of UiO-66-NH$_2$ (Figure 4.3a′) is dominated by fluorescence and only the most intense spectroscopic features of the material could be detected but are characterized by a poor signal to noise ratio.

Part (b) in Figure 4.3 reports the FTIR spectrum of UiO-66-NO$_2$. The presence of the nitro group at the linker is confirmed by the presence of typical spectroscopic features arising from aromatic nitro compounds. In particular, the nitro group shows absorption because of asymmetric, $v(NO)_{asym}$, and symmetric, $v(NO)_{sym}$, stretching modes. Asymmetric modes typically result in a strong band in the 1550–1500 cm^{-1} region; symmetric modes absorb in the 1360–1290 cm^{-1} region. In case of UiO-66-NO$_2$, when compared with UiO-66 (Figure 4.3d), the two additional bands in the abovementioned regions are observed. The first is centered at 1554 cm^{-1} and the second, partially overshadowed by a strong band attributed to a carboxylate mode, appears as a shoulder at approximately 1355 cm^{-1}. These bands are ascribed to the NO$_2$ group. The Raman spectrum reported in Figure 4.3b′, is not disturbed by fluorescence and exhibits many features in the 1700–200 cm^{-1} range. These mostly resemble the typical Raman bands of parent UiO-66 shown in Figure 4.3d′, except for some minor shifts in some cases. Here, bands ascribed to benzene ring and carboxylate vibrational modes are readily distinguished, while the vibrational features involving the Zr-containing moiety are hardly detected. However, in UiO-66-NO$_2$ some differences can be appreciated. The most prominent is the new band centered at 1360 cm^{-1} ascribable to $v(NO_2)_{sym}$. Furthermore, other new bands are seen in the 1100–900 cm^{-1} range and at lower frequencies.

The FTIR spectrum recorded for UiO-66-Br shows (Figure 4.3c) a profile that is quite similar to the one obtained for UiO-66 (Figure 4.3d). Although there are no pure carbon–halogen stretching vibration modes for aromatic halogen compounds, the presence of peaks sensitive to the bromo functionality are evidenced (*e.g.* contribution at 680 cm^{-1}). The Raman spectrum of UiO-66-Br reported in Figure 4.3c′, presents again common features also seen for UiO-66 in Figure 4.3d′, except for some small band shifts. As previously observed, some new bands (1040, 736, 290 cm^{-1}) that are not straightforwardly assigned also appear. In this case, the assignment of vibrational modes involving Br is more difficult and support from *ab initio* calculations would be welcome.

As deeply discussed in another chapter of the present book,[74] post-synthetic MOF modification is a viable route for the introduction of surface sites with new chemical properties in this class of materials alternative to the functionalization during the synthesis, as described above. Chavan *et al.*[75] succeeded in grafting $Mo(CO)_3$ and $Cr(CO)_3$ groups on the benzene rings of the UiO-66 linkers, as discussed later (Sections 4.2.5.2 and 4.3.3). Kandiah *et al.*[76] demonstrated that it is possible to perform covalent post-synthetic modifications of the UiO-66-NH$_2$ MOF with four different acid anhydrides. In this view, FTIR is one of the techniques employed to monitor the occurrence of reactions and to quantify the extent of the reaction.[75]

The spectra of the compounds after 2 weeks of reaction are reported in Figure 4.4: they are dominated by the skeletal modes of the MOFs, mainly originating from the vibrations of the aromatic linkers (see region below $1600\ cm^{-1}$ in Figure 4.4b). In addition to these main peaks it is possible to observe key spectroscopic features due to the presence of functional groups created by the post-synthetic modification.

The main bands ascribed to the original amino group are summarized in the following (see curve 1 in Figure 4.4): $3515\ cm^{-1}$ $\nu_{sym}(NH_2)$, $3390\ cm^{-1}$ $\nu_{asym}(NH_2)$, $1629\ cm^{-1}$ $\delta(NH_2)$, 1340 and $1257\ cm^{-1}$ $\nu(C_{ar}-N)$. Post-synthetic reactions produce a complete or a partial erosion of the $-NH_2$ peaks, which can be considered a measure of the extent of the reaction. Thus, the approximate conversions were calculated based on the peak intensities. The conversions

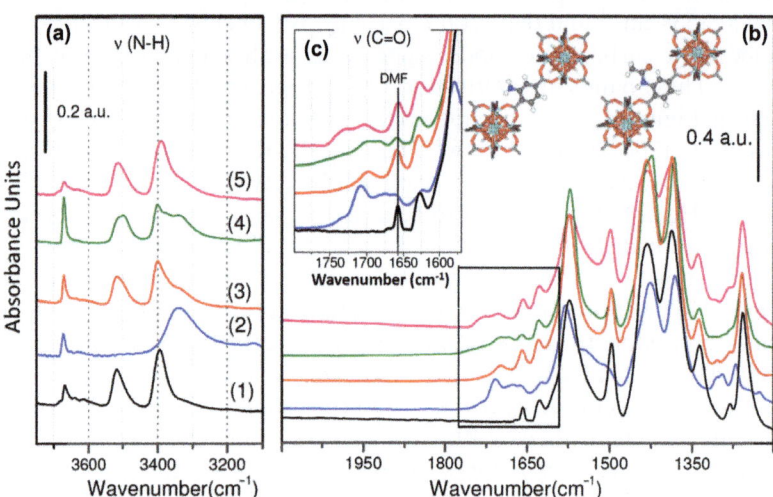

Figure 4.4   FTIR spectra of UiO-66–NH$_2$ (curve 1) and the post-synthetically modified MOFs, UiO-66-NH-COCH$_3$ (curve 2), UiO-66-NH-COCH$_2$CH$_3$ (curve 3), UiO-66-NH-COC(CH$_3$)=CH$_2$ (curve 4), and UiO-66-NH-CO–O–C(CH$_3$)$_3$ (curve 5) as obtained after two weeks of reaction. Part (a): N–H stretching region. Part (b): MOF skeletal modes region. Part (c): enlargement of the $\nu(C=O)$ region.
Adapted with permission from Kandiah *et al.*,[76] copyright Royal Society of Chemistry, 2010.

ranged from around 100% to around 7%, depending strongly on the bulkiness of the reagent applied. Notably, the reaction of UiO-66-NH$_2$ with the smallest reagent (acetic anhydride) gives full conversion after two weeks. The reduction of the IR amino features is accompanied by the appearance of new bands ascribed to the presence of secondary amides. In particular, for the reaction with acetic anhydride, the new v(C$=$O) band is seen at 1706 cm^{-1} (see Figure 4.4c, curve 2) with an accompanying v(N–H) single band at 3344 cm^{-1} (Figure 4.4a, curve 2). Secondary amides are characterized by the peak at 1544 cm^{-1} (observed as a shoulder) that corresponds to an interaction of the v(C–N) and the δ(CNH) vibrations. Likewise, a moderately strong band at 1305 cm^{-1} ascribed to δ(NH) and δ(OCN) is characteristic of secondary amides. Due to the partial conversion, the spectra recorded on the other samples (curves 3, 4 and 5 in Figure 4.4) are still dominated by bands assigned to the parent amino group. Nevertheless, the main spectroscopic features anticipated from the post-synthetic introduction of new functionalities can be clearly observed. The formation of secondary amino species is evidenced in all cases by the appearance of the characteristic v(N–H) band (observed as a shoulder at about 3340 cm^{-1}) together with the v(C$=$O) band at around 1700 cm^{-1} (see Figure 4.4c). It is worth noticing that reaction with the anhydrides modifies UiO-66-NH$_2$ without degradation of the framework, as documented by XRPD analysis even after extended reaction times and conversions.

Summarizing, IR and Raman spectroscopies are able to detect the vibrational fingerprints of the chemical groups that functionalize the MOF frameworks. Note that in case of partial functionalization, where the insertion (or grafting) occurs randomly, XRPD can fail in the detection of functionalized groups because of lack of long range order.[3] In such cases, vibrational spectroscopies result the most informative techniques.

4.2.3 Effect of Activation

4.2.3.1 Solvent Removal

Infrared spectroscopy performed in controlled atmosphere on thin self supported pellets allows for following of activation processes and solvent removal. Figure 4.5 compares IR spectra of the MOF-5 samples as synthesized (blue curve) and outgassed in vacuum at 523 K (red curve), respectively.[77] IR spectrum of DMF is reported for the sake of comparison (top curve). The two spectra show significant changes that can be associated with the removal of water and solvent (DMF). In particular, we observe almost total disappearance of the broad band centered at 3500 cm^{-1}, ascribed to v(OH) modes of H-bonded OH groups,[78,79] mainly due to adsorbed water. DMF elimination in thermally treated MOF-5 is confirmed by the disappearance of the bands at 2939, 2867, 1690, 1256, 1225, 1092, 1063, and 863 cm^{-1} (follow dashed lines up to the black spectrum). The bands ascribed to the v(CH) of the aromatic rings (band at 3072 cm^{-1}) are not affected by thermal treatments, confirming the stability of MOF-5 in the adopted activation range. The very weak broad

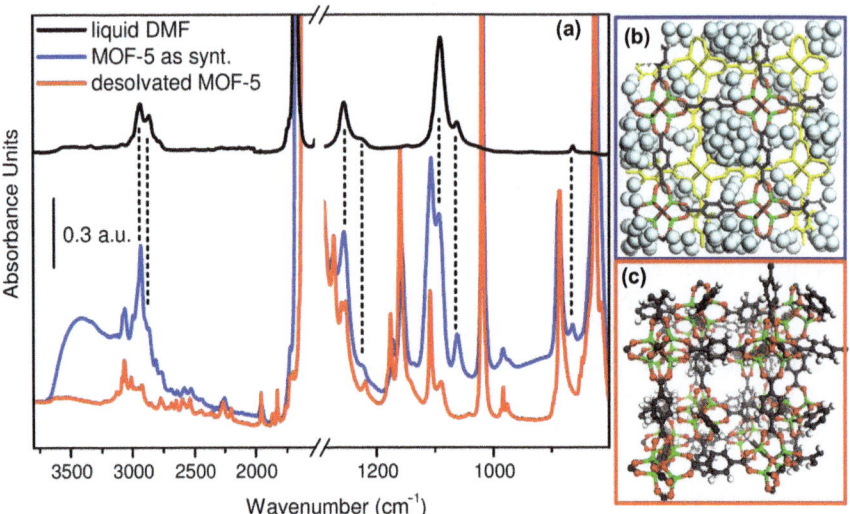

Figure 4.5 Part (a): comparison of MOF-5 IR spectra for the as prepared material
(blue curve) and for the heated and evacuated MOF-5 material (red
curve). Liquid DMF spectrum is reported for the sake of comparison
(top black curve). Vertical dashed lines evidenced the DMF modes. Parts
(b) and (c) report the structure of MOF-5 before and after desolvation,
respectively.
Unpublished figure reporting spectra published by J. Hafizovic *et al.*[77]

absorption, still present on the treated sample, can be explained in terms of
ν(OH) both associated to an extra phase and/or to hydroxyl groups located on
the external surfaces or at internal defects (for instance 1,4-benzenedicar-
boxylate vacancies). The frequency range that has been removed from the
picture is characterized by very strong bands associated to the carboxylates
groups. This spectral region is not substantially affected by solvent removal, so
it has not been reported.

4.2.3.2 Adsorbed Water in MOFs

In most of the cases, water is present either as a physisorbed or chemisorbed
adsorbate in MOFs and can be removed upon activation by performing a
thermal treatment in high vacuum. Nevertheless, in a few cases its presence has
been observed even in fully activated samples.[80,81] Figure 4.6a compares the IR
spectra of NiPBP (Ni$_8$(μ_4-OH)$_4$(μ_4-OH$_2$)$_2$(μ_4-PBP)$_6$, H$_2$PBP = 4,4'-bis(1H-
pyrazol-4-yl)biphenyl) obtained after degassing at room temperature and upon
activation at 373 K and 453 K, blue, cyan and red spectra, respectively. In
Figure 4.6b, the 3D representation of the MOF is reported. The progressive
elimination of water brings about a sharpening of the absorption band at
3590 cm^{-1}, attributed to the bridging OH groups capping the faces of the Ni$_8$
polyhedra (see inset in Figure 4.6a). Incidentally, this band lies at a significantly
lower frequency than that observed for the hydroxyl groups in Ni-based

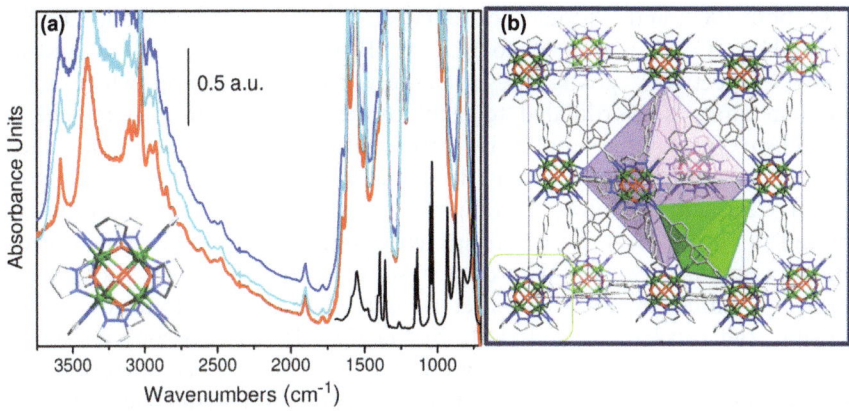

Figure 4.6 Part (a): effect of water removal on FTIR spectrum of NiPBP at room temperature (blue curve), at 373 K (cyan curve) and at 453 K (red curve) respectively. Black curve corresponds to IR spectrum of pyrazole. Part (b): 3D-representation of NiPBP MOF, highlighting one of the octahedral and one of the tetrahedral cavities: C, grey; H, light grey; Ni, green; N, blue; O, red. The inset in part (a) reports a magnification of a $Ni_8(OH)_4(OH_2)$ inorganic cornerstone.
Adapted with permission from Mino *et al.*,[81] copyright Royal Society of Chemistry, 2012.

phosphonates, reasonably because of the unusual μ_4-bridging mode adopted by the hydroxyl moieties in NiPBP. Notably, the band centered at $3390\,cm^{-1}$, persisting even after prolonged outgassing at 453 K, must be attributed to the structural water molecules capping the Ni_8 octanuclear clusters, whose removal would provoke the collapse of the framework. According to the crystal structure, the persistence of the water capping molecules in the activated material is not compatible with the formation of exposed metal sites. At lower frequency, framework modes of this material are substantially different from those observed in case of MOFs based on carboxylate units (such as MOF-5). Organic spacers, based on pyrazole linkers are characterized by sharp bands below $1600\,cm^{-1}$ (the spectrum of pyrazole is reported as black curve, at the bottom of Figure 4.6).

4.2.3.3 Vibrational Properties of Structural Hydroxyls

Among metallorganic frameworks, some structures are characterized by metal nodes that require the presence of hydroxyls groups to guarantee the electroneutrality (*e.g.* Al-MIL-53; In-MIL-68).[82] In these cases, even after prolonged activation, they preserve the presence of terminal hydroxyles groups that show weak Brønsted acidity (see next paragraph). A different behaviour is observed in the case of Zr based UiO metallorganic frameworks, where it is possible to have both hydroxylated and de-hydroxylated materials, taking advantages of the flexibility in the coordinative state of Zr(IV) species, that can show both 8- and 7-fold coordination.[43,45,68] The inorganic brick, $Zr_6O_4(OH)_4(CO_2)_{12}$, is bonded to BDC or BPDC linkers in UiO-66 and in UiO-67, respectively, giving rise to a very robust structure, where each zirconium atom is coordinated to eight oxygen atoms. One square face is formed by oxygen atoms supplied by carboxylates while the second square face is formed by oxygen atoms coming from the μ_3-O and μ_3-OH groups. Upon thermal treatment *in vacuo*, a complete de-hydroxylation of the cluster is achieved as two of the 4 μ_3-OH groups leave together with the two hydrogens from the remaining two μ_3-OH groups, resulting in a Zr_6O_6 inner cluster with seven-coordinated zirconium, see inset in Figure 4.7b. The phenomenon is fully reversible and can be followed easily by "*in situ*" IR spectroscopy collecting spectra at different thermal treatment stages.[43,83] This has been described by Valenzano *et al.*[43] on UiO-66 and by Chavan *et al.*[45] on UiO-67. Hereafter, we will focus on the data collected on the UiO-67, see Figure 4.7. As made UiO-67 (black curve in Figure 4.7) in the high frequency region (3800–1500 cm^{-1}) is dominated by a broad band centered at 3430 cm^{-1} and a doublet at 2930 and 2873 cm^{-1}. These are assigned to the hydrogen-bonded adsorbed water and v(C–H) stretching modes of the physisorbed DMF. Similarly, in the low frequency region (1800–700 cm^{-1}) bands at: (i) 1662 cm^{-1} (broad and intense absorption), (ii) 1256 cm^{-1} (medium intense band) and (iii) 1096 cm^{-1} (intense sharp band) with a shoulder at 1062 cm^{-1} are respectively assigned to the

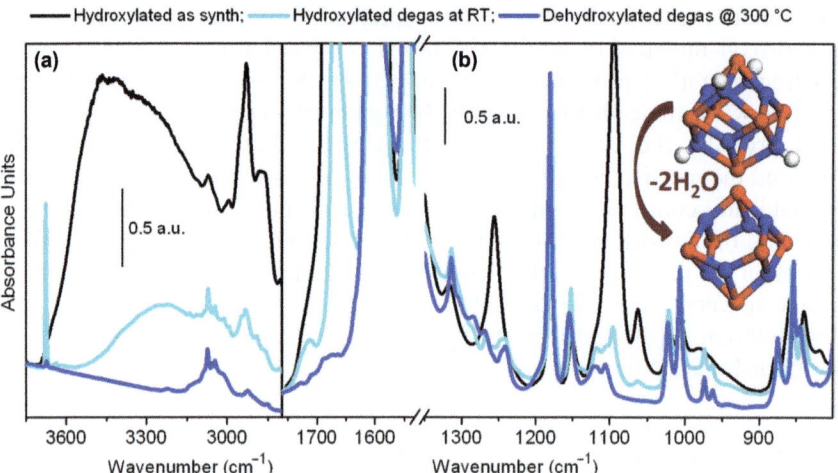

Figure 4.7   Effect of the desolvation on the FTIR spectra of UiO-67. Part (a): v(OH) stretching region. Part (b): skeletal modes region. As made sample (black curve); sample degassed at room temperature, representing the hydroxylated form (cyan curve); sample degassed at 573 K for 30 min, representing the dehydroxylated form (blue curve). Inset: evolution from of the inorganic cornerstone upon dehydroxylation: $Zr_6O_4(OH)_4$ to Zr_6O_6 (colour legend: white H, blue O, red Zr).
Adapted with permission from Chavan *et al.*,[45] copyright Royal Society of Chemistry, 2012.

$v(CO) + \delta(OH_2)$, $\delta(CH_3)$ and $v(C-N)$ vibrations of the physisorbed water and DMF. All these bands undergo substantial erosion upon degassing. The spectrum obtained upon degassing at room temperature is illustrated by cyan curve in Figure 4.7. In the high frequency region, the broad band at $3430\,cm^{-1}$ and the doublet at $2930-2870\,cm^{-1}$ show a significant decrease in intensity and a shift of maximum to $3240\,cm^{-1}$, associated to the removal of physisorbed solvent (H_2O and DMF). The growth of a sharp new component is observed at $3676\,cm^{-1}$ and it is assigned to the $v(O-H)$ stretching of μ_3-OH present at the Zr_6-octahedron.[84,85]

In the low frequency region, 1256, 1096 and $1062\,cm^{-1}$ bands associated to the solvent show a decrease in intensity. Conversely, the broad and intense absorption centered at $1662\,cm^{-1}$ is transformed into a narrow intense component centered at $1667\,cm^{-1}$ with a shoulder at $1712\,cm^{-1}$. This latter component can be assigned to the $v(C=O)$ mode of the residual DMF. Degassing at high temperature (573 K for 30 min) gives rise to the spectrum presented by blue curve in Figure 4.7. The spectrum shows the absence of all the bands associated to DMF and water and also the newly appeared $3676\,cm^{-1}$ component associated to the μ_3-OH present at the Zr_6-octahedron has almost disappeared. This indicates the complete de-solvation and also de-hydroxylation of the UiO-67 framework.

Thermal activation not only has the effect of removing adsorbates and surface hydroxyls. A closer inspection to the infrared spectrum allows verification that hydroxyl removal is accompanied by significant distortion of the Zr_6 metal cluster, while the XRPD pattern does not show any relevant change.[3] The system has been the object of deep characterization by combining spectroscopic results with an *ab initio* DFT (at the B3LYP level of theory) computational study carried out by using the CRYSTAL code[86,87] on the UiO-67 material in both its hydroxylated and dehydroxylated forms. The distortion of the Zr_6 metal cluster has also been independently demonstrated by Zr K-edge EXAFS spectroscopy,[43,45] see also Section 5.5.3 of this book.[3]

Figure 4.8 reports a direct comparison between experimental (part a) and theoretical (part b) framework modes of UiO-67 in its hydroxylated (cyan spectra) and dehydroxylated (blue spectra) forms, compared with those of the parent UiO-66 material both hydroxylated and dehydroxylated, orange and red spectra, respectively. A selection of the calculated vibrational frequencies (and the description of the corresponding normal modes) for UiO-66 and UiO-67 in both their hydroxylated and dehydroxylated forms is reported in Table 4.1. Mode animations can be visualized at the CRYSTAL web-page.[88] Experimental and computed spectra are in very good agreement, thus allowing an assignment of the main observed IR peaks.

4.2.3.4 Generation of Coordination Vacancies at Metal Sites Upon Solvent Removal

CPO-27-Ni is a 3D honeycomb like structure MOF with pores of $11\,\text{Å}$ in diameter[64] (see top right inset in Figure 4.9). All the oxygen atoms of the

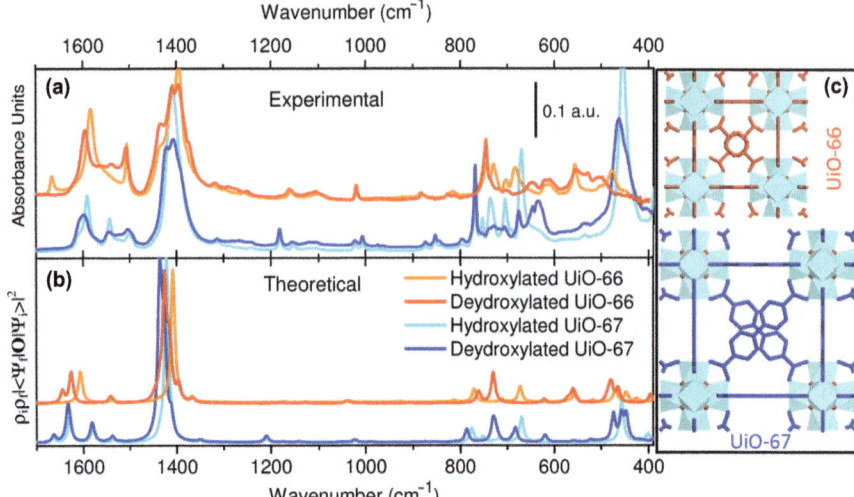

Figure 4.8   Vibrational framework modes of UiO-67 and UiO-66 in their hydroxylated (cyan and orange curves) and dehydroxylated (blue and red curves) forms. Part (a): experimental ATR IR spectra. Part (b): theoretical spectra, computed at the Γ point, based on the geometries optimized with periodic DFT calculation using CRYSTAL code.[86,87] A mass-weighed Hessian matrix obtained by numerical differentiation of the analytical first derivatives has been used.[45] No scaling factor has been applied to the theoretical curves. See Table 4.1 for the band assignment. Part (c): graphical representation of the UiO-66 (top) and UiO-67 (bottom) frameworks. The longer linker of UiO-67 guarantee a larger internal accessible volume.
Unpublished figure reporting spectra published in ref. 43,45 and structures published in ref. 45.

ligands are involved in the coordination of nickel atoms. This accounts for five of ligands around each nickel centre, whereas the sixth coordinative bond is normally occupied by a water molecule that points toward the cavity, see the "*" symbols in the central inset of Figure 4.9. All the Ni(II) ions are identical with respect to the first coordination sphere, showing five oxygen atoms that slightly differ one from the other.[69]

The IR spectrum of the hydrated material (curve 1 in Figure 4.9b) is dominated by an intense and broad signal above $3000\,cm^{-1}$ due to the physisorbed water while, as water is a very weak Raman scatterer, its presence is not directly detected in the Raman spectrum reported in Figure 4.9a (curve 2). Upon outgassing at 393 K, water is removed; in the dehydrated conditions the IR spectrum becomes dominated by the bands due to the organic part of the MOF (Figure 4.9b,c).[70]

As CPO-27-Ni is a weak Raman scatterer, the spectrum (Figure 4.9a) shows a modest number of bands that are not very intense (in comparison with IR, Figure 4.9b,c) mainly located at frequencies lower than $1650\,cm^{-1}$, mostly due to vibrations associated to the linker.[70] In particular, the bands at 1625 and

Table 4.1 Selection of calculated harmonic vibrational modes (in cm^{-1}) compared for the two structures. See the bottom part of Figure 4.8 for a graphical representation of the calculated spectra. No scaling factor has been adopted. Refer to the CRYSTAL web-page[88] for the full calculated spectra. Reproduced with permission from Chavan *et al.*,[45] copyright Royal Society of Chemistry, 2012.

UiO-67		UiO-66		
Hydroxylated Ref. 45	Dehydroxylated Ref. 45	Hydroxylated Ref. 43	Dehydroxylated Ref. 43	Description of the mode
1630, 1667	1634, 1665	1607	1568, 1627, 1648	OCO asymmetric stretching (in-phase/anti-phase)
1538	1540	1539	1541, 1545	CC ring
1422	1425, 1436	1408	1412, 1426	OCO symmetric stretching
1352	1350	1367	1364, 1367	CC ring
819	—	814	—	OH bending + CH bending (anti-phase)
776	787	771	762	OH bending + CH bending (in-phase)
717	—	711	—	OH bending + CC ring + OCO bending
670	730	673	730	μ_3-O stretching
—	620	—	622	Zr-(OC) symmetric stretching
552	559	556	560	Zr-(OC) asymmetric stretching
—	522	518	—	out-of-plane ring deformation
457	—	470	—	μ_3-OH stretching (in-phase)
—	474, 457, 448	—	478, 464	out-of-plane ring deformation
—	—	447	—	μ_3-OH stretching (anti-phase)
402	—	425	—	μ_3-OH stretching

$1561\ cm^{-1}$ can be ascribed to stretching modes of benzene ring, while the band at $586\ cm^{-1}$ is due to a ring deformation mode. Fingerprint of the linker is also the doublet at 1501 and $1427\ cm^{-1}$ due to $\nu(COO^-)_{asym}$ and $\nu(COO^-)_{sym}$ of the carboxylates, while the strong band at $1278\ cm^{-1}$ can be attributed to $\nu(C-O)$ vibration of the deprotonated species derived from the hydroxyl group. The minor feature at $833\ cm^{-1}$ is ascribed to ring C–H bending mode. At low frequency, bands associated with vibrational modes involving nickel–oxygen bonds appear at $430\ cm^{-1}$ and at $380\ cm^{-1}$. The Raman spectrum obtained on the dehydrated sample (curve 2 in Figure 4.9a) is characterized by an increase

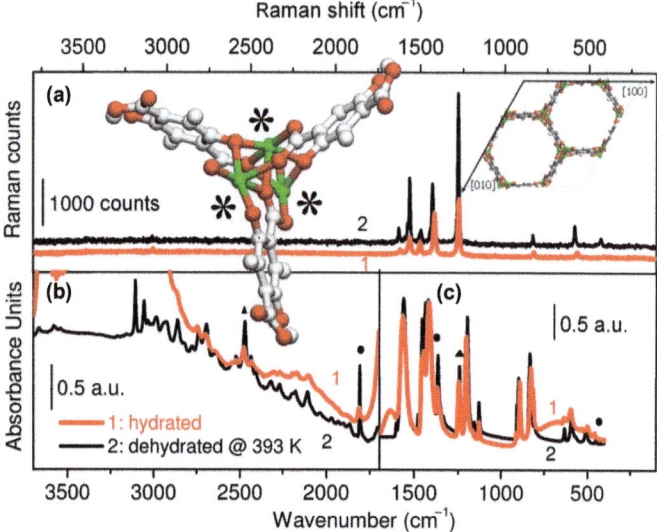

Figure 4.9 Vibrational spectra of hydrated CPO-27-Ni (curves 1) compared with the analogous spectra collected on the material outgassed at 393 K for 1 h (curves 2). Part (a): Raman spectra. Part (b): FTIR spectra collected on a self supported pellet. The circle (●) and triangle (▲) indicate the 1807 and 2472 cm^{-1} bands respectively. Part (c): FTIR spectra collected on a thin film deposited on a silicon wafer. Circles and triangles indicate the fundamental modes that originate the 1807 and 2472 cm^{-1} bands respectively. Top right inset: view along the [001] direction of the honeycomb structure of CPO-27. The C atoms are reported in gray, H atoms in white, O in red, and Ni in green. Central inset: detail of the Ni-spinal column of CPO-27-Ni. The three asterisks show the coordinative site on Ni atom for probe molecules.
Adapted with permission from Chavan *et al.*,[90] copyright Royal Society of Chemistry 2009.

of intensity and a systematic blue-shift of the bands. For this material, IR spectroscopy is potentially more informative (Figure 4.9b,c), but a single experiment is not able to investigate the whole mid-IR region covered in the reported Raman spectra. This is due to the high difference in the extinction coefficients of the modes in the low (below 1650 cm^{-1}) and in the high (above 1650 cm^{-1}) frequency regions. Using a standard self-supported pellet, like that used in the IR experiments, the modes in the 3200–1800 cm^{-1} range are correctly sampled (Figure 4.9b) while most of the bands below 1650 cm^{-1} (region not reported) are so intense that they go beyond the scale. In order to have access to the low frequency region, a thin film deposited on a silicon wafer was prepared and then treated in the same way (Figure 4.9c). In the high frequency region, Figure 4.9b, apart from the doublet above 3000 cm^{-1} (3100 cm^{-1} and 3050 cm^{-1}) associated to ν(CH) of the aromatic ring, the sharp bands in the 3000–1600 cm^{-1} range are mostly due to combination and overtones of fundamental modes that appear in the low frequency region

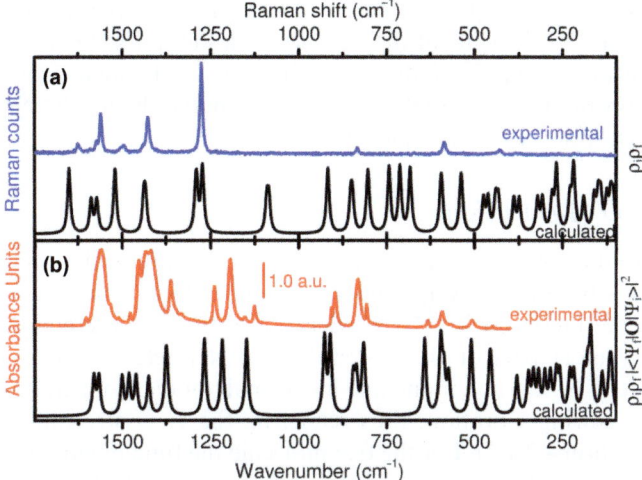

Figure 4.10   Comparison between computed and experimental vibrational spectra of dehydrated CPO-27-Ni: Raman (part a) and IR (part b) spectra (the simulated Raman spectrum represents the vibrational DOS, as no intensity calculations have been performed).
Adapted with permission from Valenzano *et al.*,[44] copyright Elsevier 2012.

(Figure 4.9c). In particular, two bands at 1807 and at 2472 cm^{-1} (marked by a triangle and a circle, respectively) strongly respond to solvent removal, becoming more intense and red shifted upon outgassing.[70] In the region of fundamental modes (Figure 4.9c), among all the peaks, the bands marked with the same symbols respond to the solvent removal process in the same way as the 1807 and 2472 cm^{-1} bands: intensity increase and red shift. The marked bands are thus good candidates to be the fundamental modes generating the high frequency doublet. In particular, the band at 2472 cm^{-1} can be ascribed to the first harmonic of the band at 1243 cm^{-1} ($2\tilde{v}$=2486 cm^{-1}, triangle), while the peak at 1807 cm^{-1} can be assigned to the sum of the 447 and 1360 cm^{-1} bands (circles). A complete understanding of the vibrational features of CPO-27-Ni material has been obtained from the periodic B3LYP-D* calculation performed with CRYSTAL code.[86,87] In Figure 4.10, both the IR and Raman spectra are reported and in both cases the main vibrational features are well reproduced by the calculations. These calculations indicate that none of these bands seem to be due to localized modes involving the carboxylates and/or the alcoholic groups, while they seem more reasonably due to collective modes that combine more than a single vibration.[44,89]

4.2.4   Probe Molecules Adsorption Followed by IR and Raman

Besides the vibrational spectrum of the materials, it is even more interesting from the characterization point of view to observe the perturbation of the vibrational modes of simple molecules (as CO and H_2) upon their adsorption on the materials.[42] The shift of their modes is in fact informative on the acid/basic nature and strength of the adsorption sites and then provides useful indications on the possible applications of the materials in different fields, as catalysis, adsorption and gas separation.[42]

4.2.4.1   CO Adsorption

In the IR studies of surface properties of a large variety of systems, carbon monoxide has been extensively used as a probe molecule and many data are available in the literature.[33–35,40,42,91] Because of the low interaction energy of CO with the most part of the materials, this characterization is in general performed at 77 K. Carbon monoxide has a small dipole moment ($\mu = 0.112$ D) which varies during vibration, thus rendering the C–O stretching mode IR-active (see Section 4.2.1.2). For the free molecule the fundamental $\tilde{v}_0(CO)$ value is 2143 cm^{-1}. Interaction with non-d charged sites is expected to produce a shift of the CO stretching frequency essentially based on the strength and type of interaction involved (as shown in Scheme 4.1).

Adsorption of carbon monoxide on metal organic frameworks can be used to probe the surface sites in terms of Lewis and Brønsted acidity. In particular, IR spectroscopy of CO adsorption at 77 K has been used to probe acidic Lewis sites generated by the presence of exposed metal cations and the Brønsted acidity of structural hydroxyl groups, respectively.[42,83,92,93] In both cases we are dealing with partially positively charged species and so with the formation of $M^+ \cdots CO$ and $M^+ \cdots OC$ adducts. Between these two possibilities, the carbon-end adducts are much more favorable, being the oxygen end adducts present only at high CO loadings. Extensive experimental evidence has proven that interaction of CO *via* the carbon-end with non transition metal cations is essentially of an electrostatic type.[32,42] This involves the interaction of the surface cation electric field with the CO charge distribution. This gives a hypsochromic shifts (positive shift) in the vibrational frequency of CO that depends on the cation size and charge. When transition metal ions are involved, the interaction of CO can lead to a more complex situation due to both

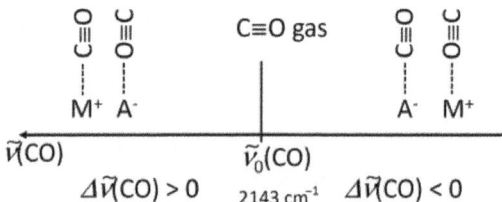

Scheme 4.1   Representation of the contribution of the electrostatic interaction to the overall frequency shift of carbon monoxide adsorbed on surface cationic (M^+) or anionic (A^-) sites. Both carbon-end and oxygen-end adducts have been considered. This simplified scheme holds for non-d sites. Previously unpublished scheme.

electrostatic and chemical type interaction. Chemical type interaction involves both ligand-to-metal σ donation (from the full 5σ orbital of CO to the empty d_z^2 orbital of the metal) with concomitant metal-to-ligand (from a full d_{xz} orbital of the metal to the empty $2\pi^*$ of CO) back donation. In these cases, adducts are also formed at room temperature and the sign of the $\Delta\tilde{v}$ (CO) shift, see eqn (4.11), will depend on the values obtained taking into account all the contributions.[42,94,95]

To illustrate the concept, two MOFs, CPO-27-M (M = Mg, Co, Ni) and MIL-53-Al are chosen as the representative of MOFs with structural hydroxyl groups and with exposed metal sites, respectively. Bordiga's group studied the CO adsorption on CPO-27-M by *in situ* IR spectroscopy. IR study was performed by varying the CO coverage at liquid nitrogen temperature. For the sake of simplicity, only low coverage spectra are reported in Figure 4.11. The main IR absorption bands 2177, 2160 and 2176 cm^{-1} were assigned to the stretching vibration of CO in the $M^{2+}\cdots$CO adducts (M = Mg, Co and Ni, respectively).[70,89] These main bands are all upward shifted with respect to the free CO molecule (dashed line in Figure 4.11). Upon progressive increase of P_{CO} (not shown), some relevant changes are observed. The main peak grows in intensity, broadens, and shifts to lower frequency, while on the low-frequency side a maximum at 2123 cm^{-1} (with a shoulder at 2132 cm^{-1}) is observed and

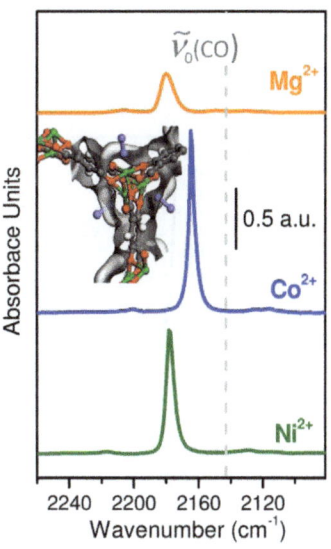

Figure 4.11 Background subtracted FTIR spectra of CO adsorbed at liquid nitrogen temperature on the M^{2+} site of CPO-27-M (M = Mg, Co, Ni, low CO equilibrium coverages only). Dashed vertical line marks the frequency of the unperturbed molecule: $\tilde{v}_0(CO) = 2143$ cm^{-1}. The inset shows a fraction of the framework.
Previously unpublished figure reporting new spectra reproducing the experiments published by Lamberti *et al.*[42]

on the high-frequency side two maxima at 2220 and 2210 cm^{-1} are detected. For the detailed assignments refer to ref. 70, 96. Following the reversibility of the main band by vacuum degassing, evidenced CO binding on CPO-27-Co > CPO-27-Ni > CPO-27-Mg. The frequency shift for the Ni^{2+}···CO and Co^{2+}···CO adducts with respect to free gas frequency indicates the presence of π-back donation responsible of the strong binding of CO on these two sites. These observations were supported by the enthalpy of adsorption for CO measured by calorimetric isotherm on CPO-27-Ni and by VTIR spectroscopy on CPO-27-Mg (55 ± 5 and 29 ± 5 kJ mol^{-1} for CPO-27-Ni and CPO27-Mg, respectively).[70,96] Moreover, CO adsorption revels the intermediate electrostatic field or polarizing power at the metal cations in CPO-27-M compared to the respective metal cations in zeolite and oxide.[97]

Ravon *et al.*[82] studied the acidity of structural hydroxyl group of MOFs, MIL-53-Al and IM-19 by low temperature CO adsorption followed by FTIR. CO forms the H-bonded complexes with acidic hydroxyl groups,[78,79] easily detectable by IR spectroscopy in ν(O–H) and ν(CO) regions (main and inset parts of Figure 4.12, respectively). In particular, the shift in the vibrational frequency induced by the H-bonding interaction is proportional to the charge present on the hydrogen and hence indirectly to its Brønsted acidity.

The larger and broader shift observed for IM-19 (Δν(OH) = − 50/ − 100 cm^{-1}) with respect to MIL-53(Al) (Δν(OH) = −30/ − 50 cm^{-1}) indicates a higher acidity for IM-19, although mild on absolute scale.[79] For example, ν(OH) in UiO-66 are shifted by –80 cm^{-1} ranking these hydroxyls more acidic than the ones present in MIL-53-Al and IM-19.[83]

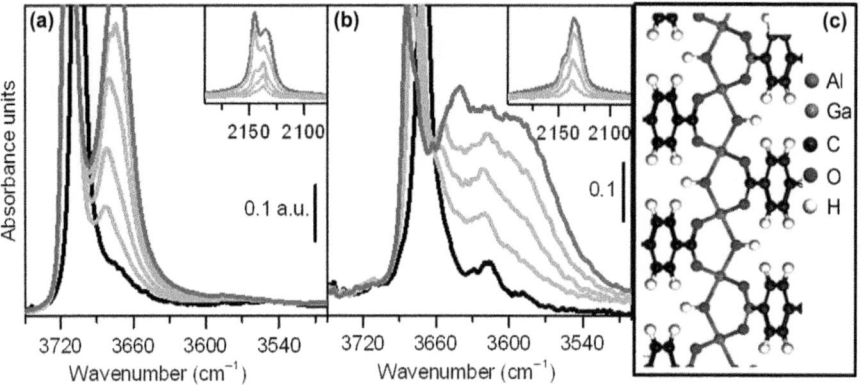

Figure 4.12   FTIR spectra of CO adsorbed at 100 K on MIL-53(Al) and on IM-19, parts (a) and (b), respectively. Black curves report spectra collected at RT *in vacuo*; dark gray curves are due to spectra collected at 100 K in the presence of 4×10^3 Pa of CO; light gray curves show spectra collected at decreasing coverage and increasing temperatures. Part (c): structure of MIL-53 as optimized by DFT calculations.
Reproduced with permission from Ravon *et al.*,[82] copyright Wiley VCH, 2010.

CO adsorption can be also adopted to provide evidence for the presence of defects in MOFs. The low temperature CO adsorption on HKUST-1 (Cu-Basolite C300 commercial sample, empirical formula: $C_{18}H_6Cu_3O_{12}$) investigated by Drenchev *et al.* showed the formation of multicarbonyl species upon the formation of copper oxide during the activation of the framework.[92,98]

4.2.4.2 Interaction with H_2: Vibrational Properties Investigated by IR

Over the past decade, hydrogen storage has been a consistent topic and controversial focus of research on a large variety of materials. The recent increased activity in the field is due to the growing recognition of the negative consequences of fossil fuel-based energy dependency. Given their potential for extraordinarily high surface areas as well as their tunable pore sizes and functionalities, MOFs are promising hydrogen storage media, among the porous materials. Several comparative studies on MOFs have shown a linear scaling of the hydrogen storage capacity with their specific surface area.[45,99,100] In fact, MOFs showing record surface areas surpassed the DOE targeted storage capacity but only at low temperature (77 K), due to weak MOF-H_2 interaction. Many efforts thus have been made to synthesize new MOFs or to post functionalize them to enhance their hydrogen affinity. A way to enhance this affinity is to introduce exposed metal sites in the structure.[101–103] It is thus essential to get a detailed understanding of the hydrogen interaction with the exposed metal sites in MOFs. In this regard, vibrational spectroscopy of hydrogen adsorption is a suitable choice. In fact, the $v(HH)$ stretching of H_2 is IR inactive; however, once adsorbed on polarizing centers, H_2 undergoes a perturbation with disruption of the local symmetry and activation of the IR mode. The perturbation in molecular hydrogen shifts the H-H stretching Raman-active mode to lower frequency (unperturbed $v(HH)$ modes at $\tilde{v}_0^{para}(HH) = 4161.1$ and $\tilde{v}_0^{ortho}(HH)$ 4155.2 cm^{-1} for para- and ortho-H_2, respectively). An evaluation of the number, distribution, and strength of specific interaction sites can be obtained by VTIR (Variable Temperature Infra-Red) spectroscopy of H_2 adsorption. Details on the VTIR method are given hereafter in section 4.2.4.5. Bordiga's group was the first to address on the role of exposed sites in the hydrogen adsorption followed by VTIR spectroscopy.[103] From the comparison of the IR spectra obtained on MOF-5, HKUST-1 and CPO-27-Ni, the conclusion was that MOFs with exposed metal sites show strong interaction with H_2, with the highest H_2-binding affinity found for the Ni^{2+} sites in CPO-27-Ni. The enthalpy of adsorption of -13 ± 1 kJ mol^{-1} obtained by VTIR was in agreement with enthalpy value obtained by other techniques. This is the highest reported value for H_2 in MOFs. These promising results motivated the investigation of isostructural CPO-27-M series.

IR spectra of hydrogen adsorbed at liquid helium temperature on CPO-27-M (Mg, Ni and Zn) are shown in Figure 4.13. For the sake of simplicity the picture reports two spectra for each sample, obtained at low and high coverages

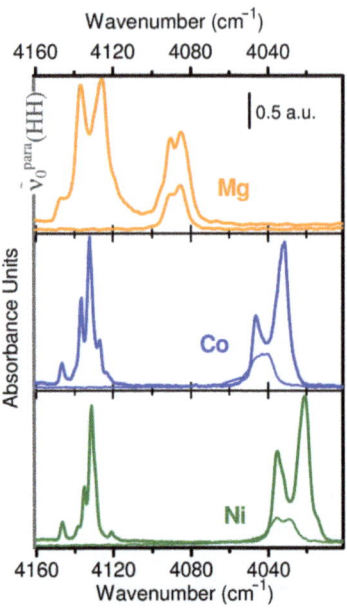

Figure 4.13 Background subtracted FTIR spectra of H_2 adsorbed at CPO-27-M ($M = Mg$, Co, Ni) isostructural materials at 20 K. The Raman –(HH) of the unperturbed *para*-hydrogen molecule coincides with the gray coloured, left ordinate axis.
Previously unpublished figure reporting new spectra reproducing the experiments published by Lamberti *et al.*[42]

respectively. For a complete description of the data we refer the reader to the original papers.[58,103,104] For all the CPO-27-Ms samples, hydrogen interaction with framework at low coverage is evident by the growth of doublet band in the IR spectrum. These doublets are red shifted by 140, 129 and 75 cm^{-1} with respect to $\tilde{v}_0^{para}$(HH). These doublets are assigned to *ortho*-H_2 and *para*-H_2 adsorbed at the metal sites, owing to their expected frequency difference (6 cm^{-1} for gas phase H_2), and by the observation of a spectral change at low temperature (*ca.* 20 K) and decreased equilibrium pressure. Based on the frequency shift (the larger the shift, the higher the perturbation of the H_2 molecule and then, in the first approximation, the interaction energy), the order of hydrogen perturbation is CPO-27-Ni > CPO-27-Co > CPO-27-Mg, in agreement with the results obtained by Zhou *et al.*[102] by means of a volumetric study conducted on the same systems. A linear correlation between temperature of adsorption, v(H–H) frequency shift and enthalpy of adsorption is seen for this isostructral CPO-27-M series. At higher hydrogen coverages, IR spectra evolve with the growth of a complex system of bands in the 4150–4110 cm^{-1} range, meanwhile the doublets seen at low coverages red shift by few wavenumbers. These changes are interpreted in terms of the gradual modification of the surrounding medium due to the filling of adsorption sites at the organic linkers, condensation of hydrogen, hydrogen–hydrogen interactions into the narrow channels of the CPO-27-M framework.[42]

4.2.4.3 Interaction with H_2: Rotational Properties Investigated by INS

As introduced in Section 4.2.1.3, INS spectra (in the few meV range of neutron energy transfer) of H_2 molecules adsorbed inside MOFs are potentially informative on the excitations of H_2 center-of-mass translational motions as well as the rotational excitations of the hydrogen molecule at various binding sites, and through this, about the potential interactions of the guest molecule with the host material.

Sumida *et al.*,[58] found two distinct adsorption sites in CPO-27-Mg for D_2 molecules by neutron diffraction methods, labeled as I and II in the inset of Figure 4.14a.[6,105] More precisely, at a loading of 0.6 D_2 molecules per Mg^{2+} site, the nuclear density map indicates the highest-affinity adsorption site to be just 2.45(4) Å from the metal center (site I in Figure 4.14a), in excellent agreement with the value of 2.50 Å found in the DFT study of Zhou *et al.*[102] At the same loading, a secondary binding site approximately 3.5 Å from the oxido group of the organic linker was identified,[58] (site II in Figure 4.14a).

The INS of evacuated spectrum recorded at 4 K for the evacuated CPO-27-Mg (not reported) is dominated by vibrational modes of the ligand, and bears close resemblance to the corresponding spectrum of $Zn_2(dobdc)$.[58] To highlight the contributions of the H_2 molecules, Figure 4.14b reports the subtraction between the INS spectra obtained after different H_2 dosages and the INS spectrum collected on the activated MOF and used as background. Upon dosage of 0.2, 0.4 and 0.6 *para*-H_2 per Mg^{2+} site (Figure 4.14b: blue, green and red full circles, respectively) the INS spectra are characterized by three new features at 6.8, 10.4, and 24.1 meV. These three new peaks are due to $(J=0) \rightarrow (J=1)$ transitions of *para*-H_2, while their multiplicity is to the splitting of the $J=1$ rotational level of H_2 into three non-degenerate m_j sublevels upon adsorption on the open Mg^{2+} site (site I in Figure 4.14a). In this regard, the computational study of Rowsell *et al.*[49] on H_2 adsorbed on the Zn^{2+} site of CPO-27-Zr predicts the $(J=0) \rightarrow (J=1)$ transitions of *para*-H_2 at 10.6, 14.0 and 21.6 meV, for the three non-degenerate m_j sublevels.

The analogous data for normal-H_2, which is a mixture of approximately 75% *ortho*-H_2 and 25% *para*-H_2 (open circles in Figure 4.14b), reveal three additional peaks at 14.1, 18.0, and 27.3 meV, respectively. These peaks have been attributed to the coupling of H_2 in the $J=1$ rotational state with the hydrogen phonons.[58] Upon increasing the loading to 1.2 *para*-H_2 per Mg^{2+} site (black full circles in Figure 4.14b), additional features between 15 and 20 meV are observed, and are attributable to H_2 adsorbed at site II (see inset of Figure 4.14a).[58]

Figure 4.14c shows, in the low neutron energy transfer region, the effect the cation nature (Mg^{2+}, Ni^{2+} or Co^{2+}) has on the first two $(J=0) \rightarrow (J=1)$ transitions of para-H_2.[57]

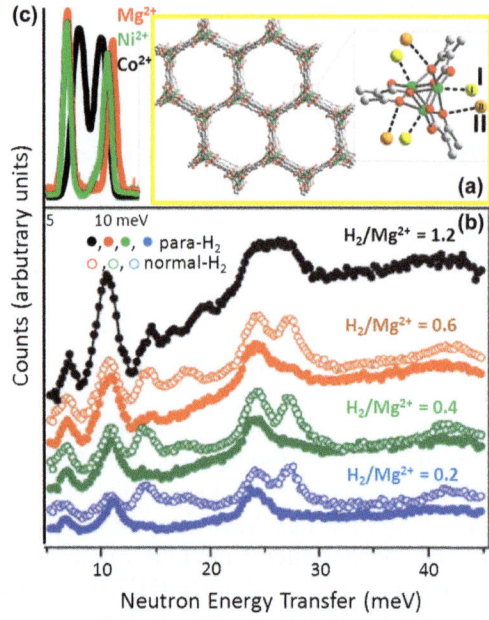

Figure 4.14 Part (a): a portion of the CPO-27-M as viewed down the [001] direction. Green, gray, and red spheres represent M, C, and O atoms, respectively (H atoms are omitted for clarity). Inset: D_2 binding sites (site I: yellow, site II: orange) as determined by neutron diffraction. Part (b): INS spectra recorded at 4 K following subtraction of the spectrum of evacuated CPO-27-Mg for loadings of 0.2 (blue), 0.4 (green), 0.6 (orange), and 1.2 (black) H_2 molecules per Mg^{2+} site. Filled and open symbols represent data for *para*-H_2 and normal-H_2 (i.e. $\sim 75\%$ *ortho*-H_2 + $\sim 25\%$ *para*-H_2), respectively. Part (c): INS spectra for 0.5 H_2 per metal site in the Ni (green), Mg (red) and Co-analogue of CPO-27. Spectra were collected at $T = 1.5$ K.
Parts (a) and (b) adapted with permission from Sumida *et al.*,[58] copyright Royal Society of Chemistry (2011); part (c) adapted with permission from Dietzel *et al.*,[57] copyright Royal Society of Chemistry (2010).

4.2.4.4 Adsorption of Different Gas Molecules at CPO-27-Ni

Valenzano *et al.*[44] have recently reviewed the structural, vibrational and energetic features of adsorption at the exposed Ni^{2+} site of CPO-27-Ni of industrially important gases (such as CO, H_2, N_2, H_2O, NO, CO_2 and C_2H_2). Combined use of experimental and theoretical methods improves the understanding of the effect of the adsorbate on the vibrational properties of the framework. Particularly, adsorption of NO significantly altered the CPO-27-Ni framework vibrations more than the adsorption of CO, indicating stronger interaction with NO than CO (see Figure 4.15a). This spectroscopic observation was confirmed by measured enthalpies of adsorption for NO and CO amounting to –92 and –59 kJ mol^{-1}, respectively (microcalorimetric measurements Figure 4.15b,c).

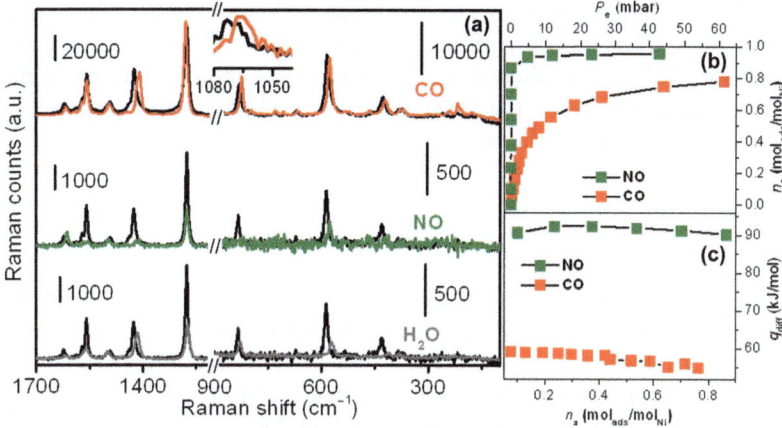

Figure 4.15 Part (a): Raman spectra of CPO-27-Ni upon interaction with probe molecules: CO, NO and H_2O, red, green and gray curves, respectively. In all cases the spectrum collected on the dehydrated material (activated 1 h at 120 °C for 1 h) is reported in black for comparison. Part (b): plot of the primary excess volumetric isotherms obtained for NO (green squares) and CO (red squares) adsorption on CPO-27-Ni at 303 K as a function of the equilibrium pressure P_e. The molar NO (and CO) amounts n_a have been normalized to the moles of Ni atoms present in the CPO-27-Ni sample, assuming all the Ni atoms are available to the interaction with the adsorbed molecule. Part (c): dependence of the differential molar adsorption heat q_{diff} on the molecular coverage n_a, obtained for NO (green squares) and CO (red squares).
Reproduced from Valenzano *et al.*,[44] copyright Elsevier, 2012.

The significant difference in the $-\Delta H_{ads}$ for H_2 (13 kJ mol^{-1} by VTIR method) and CO (59 kJ mol^{-1}, see Figure 4.15c) implies that CPO-27-Ni is an interesting material for the purification of H_2/CO mixtures to be used to feed fuel cells.

4.2.4.5 Energetic Aspects of the Molecular Adsorption Derived by VTIR: Comparison with Microcalorimetric and Quantum Mechanical Approaches

In general, a qualitative estimation of the energetic of the adsorption process can be obtained by comparing the shifts of the vibrational modes: in most of the cases, larger shifts are synonymous with higher interaction energies.[103,104,106,107] This can allow identification of the larger affinity of different probe molecules for the same material, or to compare different systems by adopting the same probe molecules. However, it is also possible to evaluate the site-specific adsorption enthalpy by performing the IR (or Raman) experiment at variable temperatures and analyzing the data by means of VTIR.[38,42,103,107–113] This experimental procedure has been already utilized for the weak adsorption of H_2 and other small molecules on a variety of materials such as polymers,[114] high surface area oxides,[39] zeolites,[109,112] and

MOFs.[44,103,107] Details on the VTIR approach and a sample calculation can be found in ref. 103. The adsorption enthalpy ΔH_{ads}^{VTIR} and the adsorption entropy ΔS_{ads}^{VTIR} are evaluated (in the Langmuir approximation) from the slope and the intercept of the linear plot of $\ln\{A/[A_M - A)(p/p^*]\}$ *vs.* $(1/T)$ respectively, according to the equation:[111]

$$\ln\left[\frac{A}{(A_M - A)(p/p^*)}\right] = -\Delta H_{ads}^{VTIR}\frac{1}{RT} + \Delta S_{ads}^{VTIR}\frac{1}{R}$$

where R is the gas constant $(8.314\,J\,K^{-1}\,mol^{-1})$, $A \equiv A(T)$ is the intensity of the selected IR manifestation of the adsorbed species as a function of the temperature T, A_M is the intensity of the same manifestation corresponding to the complete $(1:1)$ filling of the adsorbing sites, $p = p(T)$ is the equilibrium gas pressure and p^* is the reference pressure. Indeed ΔH_{ads}^{VTIR} and ΔS_{ads}^{VTIR} are the differences of the two thermodynamic functions (H and S) between two states: the adsorbed state where the molecule is adsorbed on the surface and the reference state where the molecule is in the gas phase (in the quoted references, usually p^* has been chosen to be 1 Torr). Note that H_{ads}^{VTIR} does not depend on the selected p^* value, while ΔS_{ads}^{VTIR} does, so when discussing ΔS_{ads}^{VTIR} values reported in the literature, care must be made to the adopted p^* reference value. Changing the reference pressure p^* from 1 Torr to 1 atm (760 Torr) will imply a change in the reported ΔS_{ads}^{VTIR} values of $R\,\ln(760) = 55.15\,J\,K^{-1}\,mol^{-1}$.

This approach has two main advantages with respect to calorimetry methods such as for example microcalorimetry: it is in fact site-specific (the value obtained is not averaged on different adsorption sites, as with most of the calorimetry methods) and also allows for determination of the enthalpy of adsorption for molecules with a very low interaction energy for which cryogenic temperatures are necessary to be used. For this reason, VTIR has been adopted in MOF characterization almost exclusively for H_2 adsorption where $\Delta H_{ads} < 15$ kJ mol^{-1} (see Table 4.2).[101,103] In general, the values obtained by VTIR are larger than the corresponding isosteric heats q_{iso}, being ΔH_{ads}^{VTIR} referring to a specific site, in general the most energetic sites, whereas q_{iso} represents in most of the cases an average value on different situations.

Being site-specific, between the experimental techniques available to quantify the energy of the adsorption, VTIR is particularly suitable for the comparison with results coming from computational techniques in order to validate the approximations adopted.[44] This is particularly evident for H_2 adsorption on MOF-5 for which, while the isosteric heat provides essentially an average between the binding enthalpies of the two sites, VTIR is able to distinguish them despite the tiny difference between the two values.[106] These values perfectly match the quantum mechanical ones[115] obtained at high computational level, the two methods validating each other. In all other cases, a larger difference is observed between the computational and experimental values. Concerning the interaction between H_2 and isoreticular CPO-27-M, the lower values obtained computationally are ascribable to the use of DFT methods that are known to not be able to take into account the dispersive part of the

Table 4.2 Comparison of experimental adsorption enthalpies obtained from VTIR ($-\Delta H_{ads}^{VTIR}$), isosteric heats of adsorption (q_{iso}) and adsorption enthalpies calculated by means of quantum mechanical methods ($-\Delta H_{ads}^{QM}$). Previously unpublished table reporting data from the quoted references.

MOF/adsorbate	$-\Delta H_{ads}^{VTIR}$ (kJ mol^{-1})	q_{iso} (kJ mol^{-1})	$-\Delta H_{ads}^{QM}$ (kJ mol^{-1})
CPO-27-Ni/H$_2$	13.5[103]	12.9[102]	7.33[102,a]
CPO-27-Mg/H$_2$	9.4[107]	10.1[102]	5.99[102,a]
CPO-27-Co/H$_2$	11.2[107]	10.7[102]	6.28[102,a]
HKUST-1/H$_2$	10.1[106]	6.6[116]	
MOF-5/H$_2$	7.4[106]	4.7[116]	7.1[115,b]
	3.5[106]		4.1–4.6[115,b]
CPO-27-Ni/N$_2$	17[44]		27.3[44,c]

[a]Binding energies, calculated at the PBE/PW level, cutoff energy of 544 eV and a 2×2×2 k-point mesh.
[b]Binding enthalpies calculated on cluster models at the MP2/def2-TZVP level.
[c]Binding energies, calculated at the B3LYP-D*/TZVP level.

interaction that accounts for the most important contribution to the process in these systems. Nevertheless, these methods are able to predict correctly the relative activity toward H$_2$ of this class of materials. For N$_2$/CPO-27-Ni, a dispersion-corrected DFT method was adopted. In this case, the energetic value obtained computationally is overestimated by about 38%. This difference can be only partially ascribed to the level of approximation adopted: part of the discrepancy is in fact due to the different nature of the experimental and computational values, being a binding enthalpy and energy respectively. By taking into account the zero-point and thermal contributions (by using the corresponding values calculated for CO/CPO-27-Ni at 77 K), the value reported in the Table 4.2 is decreased to 24 kJ mol^{-1}.

4.2.5 IR Spectroscopy to Monitor Reactivity in MOFs

IR spectroscopy can be used to check the stability of metallorganic framework materials in different conditions (by looking for no change in the vibrational framework modes or no appearance of new –OH signals in the >3000 cm^{-1} region), but also to provide evidence of reactions occurring inside the cavities involving the MOF matrices.

4.2.5.1 Adsorption and Reactivity of Molecular Oxygen

Oxygen adsorption and further reaction has been studied in some MOFs characterized by open metal sites with a peculiar affinity towards molecular oxygen, namely Cr$_3$(BTC)$_2$ and CPO-27-Fe.[117–119] Combined structural and spectroscopic studies performed in a controlled atmosphere and at variable temperature followed the formation of species that evolve upon rising the sample temperature from 100 K to 300 K. Due to the strong affinity of Fe^{2+} sites for

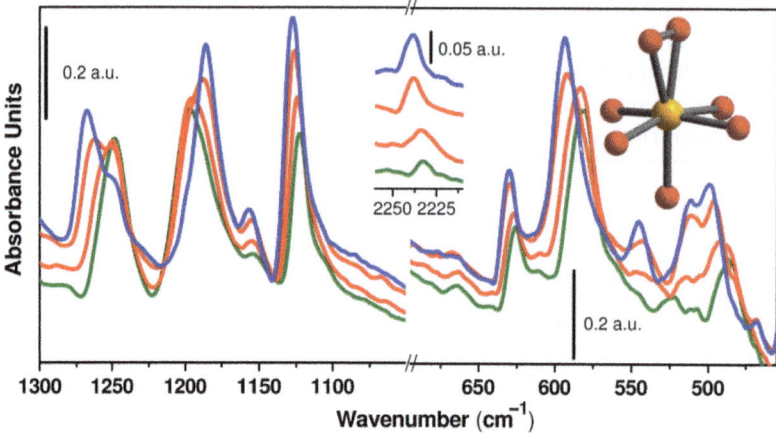

Figure 4.16   IR spectrum of activated CPO-27-Fe (green curve); effect of O_2 dosage at
low temperature (blue curve); progressive outgassing (red curves). The
inset reports the first overtone of the ν(O–O) of superoxo species coordinated to the iron site. On the right side of the picture a schematic draw
of the superoxo species.
Reproduced with permission from Bloch *et al.*,[118] copyright American
Chemical Society, 2011.

oxygen, oxygen is not expected to be adsorbed reversibly, but the formation of
coordinated oxo species is envisaged, at least at room temperature.

Due to the fact that all the bands associated to oxo moieties are observed in
the framework mode region, in order to allow the monitoring of this spectral
range, IR measurements were performed in transmission mode on a thin film.
The series of spectra obtained at ~ 100 K with different dioxygen loadings are
reported in Figure 4.16 and show a total reversibility of the phenomenon. The
blue curve corresponds to the highest equilibrium pressure (30 mbar); the red
curves correspond to intermediate coverages; the green curve corresponds to
the spectrum of the activated sample. New bands at 1129, 541 and 511 cm^{-1} can
be assigned to the formation of superoxo species. In particular, the 1129 cm^{-1}
band is associated with v(O–O) of the superoxo moiety coordinated to the iron
site; 541 cm^{-1} band corresponds to the ν_{asym}(Fe–O$_2$) of the superoxo adduct,
while the band at 511 cm^{-1} is due to the framework mode (Fe–O linker) that
shows a change upon oxygen adsorption. The inset shows the overtone band of
the v(O–O) of superoxo. Moreover, significant shifts are seen in most of the
framework bands, originally at 1250, 1198 cm^{-1} and at 580 cm^{-1} now at 1266,
1186 and 595 cm^{-1}, respectively. Animation of the vibrational modes of CPO-
27-Ni homologue on the CRYSTAL[86,87] optimized structure reveal that all of
them are related to the C–O bonds of the linker, with the bands at 1250 and
580 cm^{-1} strongly correlated.[44,89]

Superoxo species are reversible only at low temperature. Conversely, if
oxygen loaded CPO-27-Fe is warmed up at room temperature, the IR spectrum
evolves significantly showing new fingerprints, as shown in Figure 4.17, that can

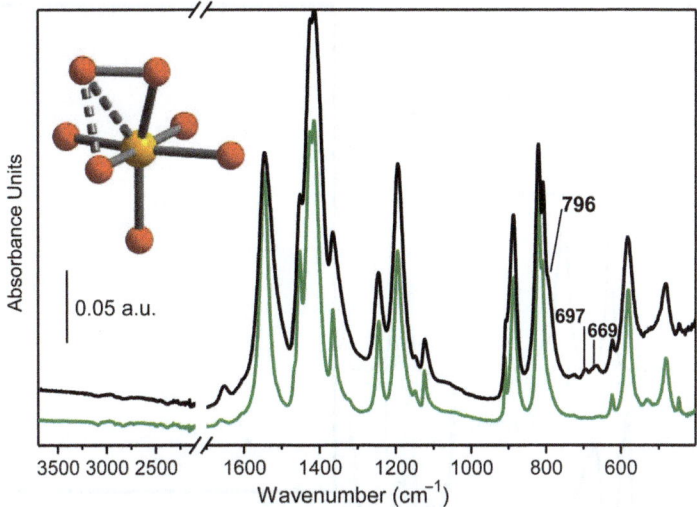

Figure 4.17 ATR spectra of activated CPO-27-Fe (green curve); partially oxidized sample (black curve). Inset: schematic draw of the peroxo species. Reproduced with permission from Bloch *et al.*,[118] copyright American Chemical Society, 2011.

be easily interpreted by the formation of peroxo species: shoulder at $796 \, \text{cm}^{-1}$ assigned to $v(\text{O–O})$ and doublet at 697 and $669 \, \text{cm}^{-1}$ due to the peroxo ring modes of the Fe-$(\eta^2\text{-O}_2)$ unit. Due to the fact that the sample was air sensitive, the spectra have been collected with an ATR instrument inside a glove box.

4.2.5.2 Encapsulation of Organometallic Units and Further Reactivity

Simple organometallic molecules, in particular $M(\text{CO})_6$, have been entrapped and further reacted inside MOFs ($M = \text{Cr, Mo}$). IR spectroscopy performed in a controlled atmosphere, in combination with molecular modeling, demonstrated to be powerful tools to describe the molecular adsorption and successive reactivity of $M(\text{CO})_6$ with the benzene units of the linker to give rise to $(\eta^6\text{-C}_6\text{H}_6)M(\text{CO})_3$.[75] See Section 4.3.3.1 and Figure 4.23 for a discussion on the electronic properties of the functionalized material.

Figure 4.18(c,d) compares the data obtained in the case of Cr(CO)_6 entrapped inside UiO-66 and in CPO-27-Ni and their subsequent evolution to give $(\eta^6\text{-C}_6\text{H}_6)\text{Cr(CO)}_3$; in the same Figure, IR spectra of both molecular units in THF solution (part b) and computed unscaled IR spectra for the Cr(CO)_6 (grey), $(\eta^6\text{-C}_6\text{H}_6)\text{Cr(CO)}_3$ (black) and $[\eta^6\text{-C}_6\text{H}_4(\text{COONa})_2]\text{Cr(CO)}_3$ eclipsed (red) conformers (part a), are reported for the sake of comparison.[120] Cr(CO)_6 in the gas phase is characterized by four IR-active modes of T_{1u} symmetry: CO stretching $v(\text{CO})$, metal–carbon–oxygen bending $\delta(\text{Cr–C–O})$, metal–carbon stretching $v(\text{Cr–C})$ and carbon–metal–carbon deformation $def(\text{C–Cr–C})$, that give rise to four IR absorption bands centered at 2000, 668,

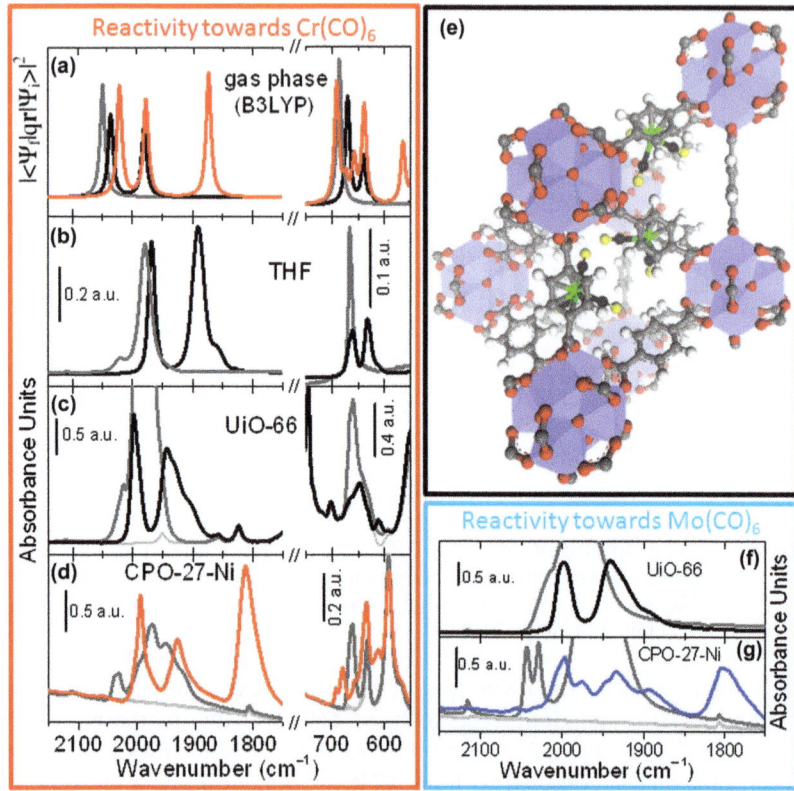

Figure 4.18   Part (a): theoretical unscaled IR spectra of Cr(CO)$_6$ (dark grey), (η^6-C$_6$H$_6$)Cr(CO)$_3$ (black) and [η^6-C$_6$H$_4$(COONa)$_2$]Cr(CO)$_3$ eclipsed (red) conformers at B3LYP level of theory. Part (b): FTIR spectra of Cr(CO)$_6$ (dark grey), (η^6-C$_6$H$_6$)Cr(CO)$_3$ (black) diluted in THF solvent. Part (c): FTIR spectra of UiO-66 before (light gray) and after (gray) Cr(CO)$_6$ dosage and after reaction at 150 °C (black). Part (d): FTIR spectra of CPO-27-Ni before (light gray) and after (gray) Cr(CO)$_6$ dosage and after reaction at 120 °C (red). Left and right panels of parts (a–d) display the ν(CO) and δ(Cr–C–O) regions, respectively. Part (e): 3D representation of UiO-66 framework with some benzene rings of the linkers functionalized with M(CO)$_3$ complexes (M = Cr or Mo). Part (f): FTIR spectra of UiO-66 before (light gray) and after (gray) Mo(CO)$_6$ dosage and after reaction at 150 °C (black). Part (g): FTIR spectra of CPO-27-Ni before (light gray) and after (gray) Mo(CO)$_6$ dosage and after reaction at 120 °C (red).
Previously unpublished figure reporting spectra and schemes from ref. 75,121.

440 and 98 cm^{-1}, respectively. Due to instrumental limitations, only the ν(CO) and δ(Cr–C–O) regions are accessible with standard mid-IR instruments. When Cr(CO)$_6$ is dissolved in THF (dark grey curve in Figure 4.18b), its molecular symmetry is slightly perturbed. As a consequence, the IR absorption bands slightly shift in frequency (ν(CO) at 1979 cm^{-1} and δ(Cr–C–O) at 665 cm^{-1})

and also the Raman active $v(CO)$ vibration of E_g symmetry becomes visible (weak band at 2020 cm^{-1}, not reported). A similar "solvent effect" is observed when $Cr(CO)_6$ is adsorbed inside UiO-66 matrix (dark grey curves in Figure 4.18c). IR spectra are dominated by a very intense, out of scale, IR absorption band centered around 1980 cm^{-1} (IR-active mode) and by a weaker band around 2020 cm^{-1} (Raman active). Similar bands have been reported in the past in case of $Cr(CO)_6$ adsorbed on oxides. In the low frequency range, the band associated to $\delta(Cr–C–O)$ is observed at around 664 cm^{-1} in both cases.

The FT-IR spectrum of $Cr(CO)_6$ adsorbed on CPO-27-Ni at room temperature (grey spectrum in Figure 4.18d) is even more complex.[121] In fact, in this case, in addition to the main IR absorption bands at 1994, 1974 and 1951 cm^{-1} due to the IR-active CO stretching vibrations of T_{1u} symmetry, a weak IR absorption band at 2110 cm^{-1} and a doublet at 2040–2030 cm^{-1} are observed. These bands are assigned to the Raman active CO stretching vibrations of A_{1g} and E_g (non degenerate) symmetry, respectively. In the low frequency region a vibration associated to $\delta(Cr–C–O)$ is observed around 662 cm^{-1} (~ 6 cm^{-1} red shifted from the gas value, 668 cm^{-1}). Both the splitting of the $v(CO)$ band in several components and the presence of intense bands due to Raman active modes, provide evidence that adsorption induced a distortion of $Cr(CO)_6$ from the perfect octahedral symmetry. Although evident also for $Cr(CO)_6$ in MOF-5[122] and UiO-66,[75] the distortion is greater in this last case and is induced by the presence of open nickel sites. The strong interaction observed between $Cr(CO)_6$ and the CPO-27-Ni can be explained by considering the polarizing nature of the surface.

Thermal treatment at 150 °C for 12 h in a dynamic vacuum of UiO-66/$Cr(CO)_6$ gives rise to a coloured (dark yellow) sample, suggesting the formation of $(\eta^6\text{-}C_6H_6)Cr(CO)_3$ species. Indeed, the FT-IR spectra obtained after the thermal decomposition of $Cr(CO)_6$ is dominated by two new IR absorption bands in the $v(CO)$ region (black curve in Figure 4.18c), at a lower frequency with respect to the $Cr(CO)_6$ bands. These two bands, very similar to those observed in the case of $(\eta^6\text{-}C_6H_6)Cr(CO)_3$ in THF (black curve in Figure 4.18b), are easily assigned to the non-degenerate total symmetric stretching $v(CO_{total\text{-}sym})$ and to the doubly degenerate total asymmetric stretching $v(CO_{asym})$ of the highly symmetric (arene)$Cr(CO)_3$ species. A qualitative picture of the bonding in transition metal π-complexes involves a donation of charge from the arene to the metal and then, in turn, from the metal to the carbonyl groups. This results in an increase of the strength of the Cr-CO bond as compared with $Cr(CO)_6$ and in a decrease of the C–O bond strength as compared with $Cr(CO)_6$ and CO. This explains why the vibrational modes of the $(\eta^6\text{-}C_6H_6)Cr(CO)_3$ species are at lower frequency with respect to those of $Cr(CO)_6$. In the low frequency range, the disappearance of the sharp component around 664 cm^{-1} testifies to the absence of unreacted $Cr(CO)_6$, while the attendant appearance of weak bands in the $\delta(Cr–C–O)$ region, at 659 and 610 cm^{-1} for UiO-66/$Cr(CO)_3$, is associated with the formation of $(\eta^6\text{-}C_6H_6)Cr(CO)_3$ species (compare the same region for $(\eta^6\text{-}C_6H_6)Cr(CO)_3$ in THF, Figure 4.18b). These spectroscopic evidences give direct proof that the post-synthesis functionalization of some linkers has occurred, as depicted in Figure 4.18e.

The formation of $(\eta^6$-arene$)Cr(CO)_3$ complexes inside CPO-27-Ni was achieved by thermal decomposition of the $Cr(CO)_6$ precursor at 120 °C (a temperature dictated by the limited thermal stability of the framework) and maximized by removing the CO released during the reaction (*i.e.* performing the thermal decomposition in dynamic vacuum). The successful functional-ization of CPO-27-Ni with $Cr(CO)_3$ is demonstrated by the disappearance in the IR spectrum, of the IR absorption bands due to adsorbed $Cr(CO)_6$ and by the concomitant growth of new bands in the $v(CO)$ region, as shown in Figure 4.18d (black curve). Three IR absorption bands are present in the $v(CO)$ region: a sharp and intense band at 1993 cm^{-1} showing a low frequency shoulder at 1979 cm^{-1}, a quite broad band at 1930 cm^{-1} and an intense and broader band at 1812 cm^{-1}. The IR spectrum of the highly symmetric $(\eta^6$-$C_6H_6)Cr(CO)_3$ (Figure 4.18b, black curve) shows two IR absorption bands in the $v(CO)$ region, at 1968 and 1890 cm^{-1}, which are assigned to the non-degenerate total symmetric stretching $v(CO_{totsym})$ and to the doubly degenerate total asymmetric stretching $v(CO_{asym})$, respectively. The presence of three IR adsorption bands in the $v(CO)$ region for $Cr(CO)_3$ grafted on CPO-27-Ni suggests the rupture of the quasi-degeneracy of the asymmetric stretching modes, due to the interaction of the CO molecules with the exposed Ni^{2+} ions. These results are supported by calculations on $[\eta^6C_6H_4(COONa)_2]Cr(CO)_3$, where the Na^+ cations were chosen to simulate the effect of open metal sites nearby the $Cr(CO)_3$ tripod. Among the two possible conformers (staggered and eclipsed) for $[\eta^6$-$C_6H_4(COONa)_2]Cr(CO)_3$ the minimum conformation was found to be the eclipsed one. The higher stability of the eclipsed conformer would result from the interaction of the Na^+ cation with the negatively charged O atoms of the $Cr(CO)_3$ unit.[120]

On both UiO-66 and CPO-27-Ni a similar chemistry has been observed upon dosing $Mo(CO)_6$: grafted $(\eta^6$-$C_6H_6)Mo(CO)_3$ species are obtained on UiO-66, while a more complex situation is observed for CPO-27-Ni because of the presence of coordinatively unsaturated Ni^{2+}, see parts (f) and (g) of Figure 4.18.

4.3 Electronic Spectroscopies: UV-Vis, Luminescence, XPS, XANES and XES

4.3.1 Basic Consideration on the Techniques

Figure 4.19 reports a sketch on the electronic structure of a general transition metal atom embedded in the cornerstone of a MOF framework. Explicitly reported are the 1s, 2s, $2p_{1/2}$ and $2p_{3/2}$ core levels and the ns, np, nd ($n = 3$ or 4) and molecular orbitals (MO) valence level, that are represented as partially overlapped owing to the fact that hybridizations of metal atomic orbitals (AO) occur because of the ligand effect and that MO are obtained by linear combi-nation of metal and ligand AO.

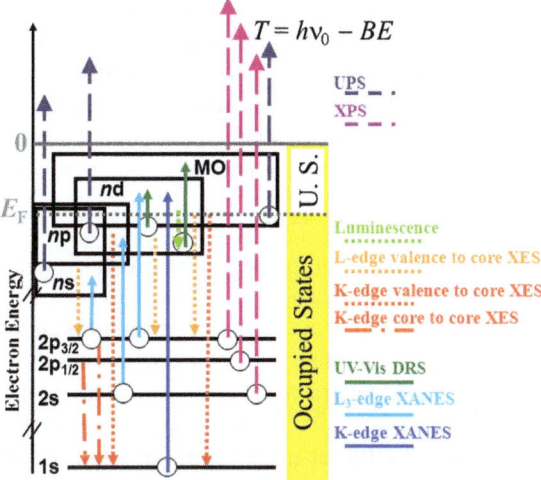

Figure 4.19   Schematic representation of the core and valence electronic levels of a *n*d-transition atom embedded inside a generic MOF framework. The valence states *n*s, *n*p, *n*d and the molecular orbitals (MO) formed with the linker atoms are represented as partially overlapped owing to the fact that the hybridization of metal atomic orbitals (AO) occurs because of ligand effect, in that MO are obtained by linear combination of metal and ligand AO. The Fermi energy (E_F, dotted gray line) divides occupied from unoccupied states (U.S.), while the 0-level of electron energy (solid gray line) divides bound from free electrons. Solid arrows indicate phenomena where a photon (not shown) is adsorbed, promoting the electron into an unoccupied bound state (E < 0): this is the case of XANES and UV-Vis spectroscopies. Dashed arrows indicate phenomena where a photon (not shown) is adsorbed, promoting the electron into a free state (E > 0): this is the case of XPS and UPS spectroscopies. Both full and dashed line phenomena leave a hole in the electronic configuration, that can be filled by a radiative process (emitted photon, not shown). This is the case for disexcitation spectroscopies (dotted and dot dashed arrows), such as XES and luminescence. Dotted and dot dashed lines refer to disexcitation phenomena involving valence and core electrons, respectively. The emission of an Auger electron is a three-level process alternative to XES and has been omitted in this scheme. Previously unpublished figure.

Electronic spectroscopies are aimed to gain information on occupied or on unoccupied electronic states or on both. The Fermi energy (E_F) is defined as the border between occupied and unoccupied states, see horizontal dotted gray line in Figure 4.19. The knowledge of occupied and unoccupied states allows for extraction of important information on the transition metal such as its oxidation state, its local geometry (tetrahedral, octahedral, square planar, *etc.*...) and the chemical nature of its neighbors.[123]

Photoelectron spectroscopies[124–126] (PES) are named UPS (exciting with $h\nu_0$ in the UV region) or XPS (exciting with $h\nu_0$ in the X-ray region) depending on whether they probe valence or core occupied states (dashed up arrows in

Figure 4.19). In some cases, the acronym ESCA (electron spectroscopy for chemical analysis) is used instead of XPS. These experiments are performed using monochromatic photons of sufficiently high energy, $h\nu_0$, to promote one electron to the continuum. Knowing $h\nu_0$ and measuring the electron kinetic energy $T = \frac{1}{2}m_e v_e^2$, according to the Einstein equation of the photoelectric effect $(T = h\nu_0 - BE)$,[127] the electron binding energy BE can be obtained as:

$$BE = h\nu_0 - T \qquad (4.13)$$

The intensity of the photoelectron peak will be proportional to the occupied electron density of states. The measure of the kinetic energy of an electron requires ultrahigh vacuum conditions in the experimental chamber. There are few synchrotron-based XPS facilities worldwide where a complex differential pumping system allows XPS spectra to be collected in a few mbar in the close vicinity of the sample surface.[128,129] However, besides these remarkable facilities, standard XPS and UPS set-ups remain (and will remain in the future) ultrahigh vacuum instrumentations. Moreover, the low free mean path of electrons inside condensed matter implies that PES techniques are surface sensitive techniques: typically 20–100 Å depending on the experimental conditions (average Z of the sample, adopted X-rays energy ($h\nu_0$) and their incidence angle *etc.*....). This characteristic provides to PES strong potentialities as in most of the techniques the response of the surface is overshadowed by the response of the bulk. So PES can be of great power if PES are combined with bulk techniques. If not, the scientist must be aware that the information coming from the surface may not be representative of the whole material, usually dominated by bulk atoms.

UV-Vis and XANES spectroscopies promote into unoccupied states valence and core electrons, respectively (full up arrows in Figure 4.19). According to the Fermi Golden rule, see eqn (4.9) and (4.10), K- and L_1-edge XANES, promoting an s-electron, probes mainly unoccupied p states. Consequently, L_2- and L_3-edge XANES, promoting a p-electron, probe mainly unoccupied s and d states.

The processes UPS, XPS, UV–Vis (CT part) and XANES leave a hole behind (empty circles in Figure 4.19) that is filled by an electron of the higher shells. The radiative decay of such processes can be followed by luminescence or X-ray emission spectroscopy (XES) depending whether we are dealing with valence or core holes. These spectroscopies are called disexcitation spectroscopies and are informative on the occupied states, as is the case for UPS and XPS. Some examples of the use of XPS on MOFs can be found in ref. 130–140.

Because of the low efficiency of the process (particularly for valence to core transitions) and because of the high $\Delta E/E$ required ($<10^{-4}$), the potential of XES spectroscopy can be fully exploited only at high brilliance beamlines hosted in III-generation synchrotrons. With this experimental set-up, XES is combined with XANES (that creates the core hole) and can be seen as a two photon process or a photon in/photon out process. As an example, K-edge XANES creates a 1s core hole (blue up arrow) that can be filled by a $2p_{1/2}$ or $2p_{3/2}$ electron (core to core XES, red dot-dashed down arrows) or by a valence electron (valence to core XES, red dotted down arrows). Having in mind the photon in/photon out scheme described so far, it becomes evident why XES has also been called inelastic X-ray scattering or Raman X-ray scattering. Core to core XES provides information that is equivalent to XPS, while valence to core XES provides information that are equivalent to UPS and UV-Vis spectroscopies. The disadvantage of XES is that a III generation synchrotron radiation facility is required. Conversely core (valence) to core XES is able to obtain the same information as XPS (UPS) but using hard X-rays: this means that ultra high vacuum conditions are not required and that information on the occupied electrons states can be obtained on MOF materials in presence of solvent or in interaction with gas or liquid phases. This opens great possibilities for investigating MOF materials under reaction conditions.[141] Also, the penetration depth of XES and PES are completely different. When comparison is made between valence to core XES and UV-Vis spectroscopies, the main differences are that UV-Vis has a better absolute energy resolution, while XES is able to detect d–d and charge transfer transitions also element selectively,[142] *vide infra* Section 4.3.5.

Examples of possible use of UV-Vis and photoluminescence spectroscopies for MOFs characterization are detailed of the next sections. In fact, in most of the cases, the metal atoms present in MOFs are transition metal atoms. In this section, the possibility of characterizing the optical properties of the MOF materials by means of UV-Vis and photoluminescence is exemplified by reporting the results of CPO-27-Ni.

However, any change not only of the nature and number of the ligands but also of the coordination geometry around the metal atom affects the optical properties of the material. Thus, desolvatation (Section 4.3.2.1), gas adsorption (Section 4.3.3) and also more complex reactions such as ligand and metal node exchange (Section 4.3.2.2) can be easily monitored by means of UV-Vis spectroscopy.

4.3.2 Optical Properties of MOFs upon Desolvation

4.3.2.1 CPO-27-Ni

CPO-27-Ni (see section 4.2.2.1 for the description of the structure) has a strong yellow–green colour due to the combination of electronic transitions associated with the organic linker (2,5-dihydroxyterephthalic acid) with those related to the presence of Ni(II) cations. Figure 4.20a compares the DRS-UV-Visible-NIR spectra of CPO-27-Ni sample as synthesized (hydrated; black curve) with that obtained after dehydration (gray curve).[69] The spectrum of the hydrated sample is characterized by the presence of three main features: a band at 1100 nm, a feature constituted by a double peak at 655 and 750 nm and a component at about 500 nm. The last one is partially hidden by the intense absorption with an edge at 460 nm, associated with the lowest π–π* energy transition due to the organic linker (dashed line); the elements' selectivity of RIXS spectroscopy (Section 4.3.5) allows clear detection of all metal d-d transitions as it does not measure the π–π* transition of the ligand. The other three main bands can be easily assigned by considering the UV-Visible-NIR spectrum of a Ni(II)

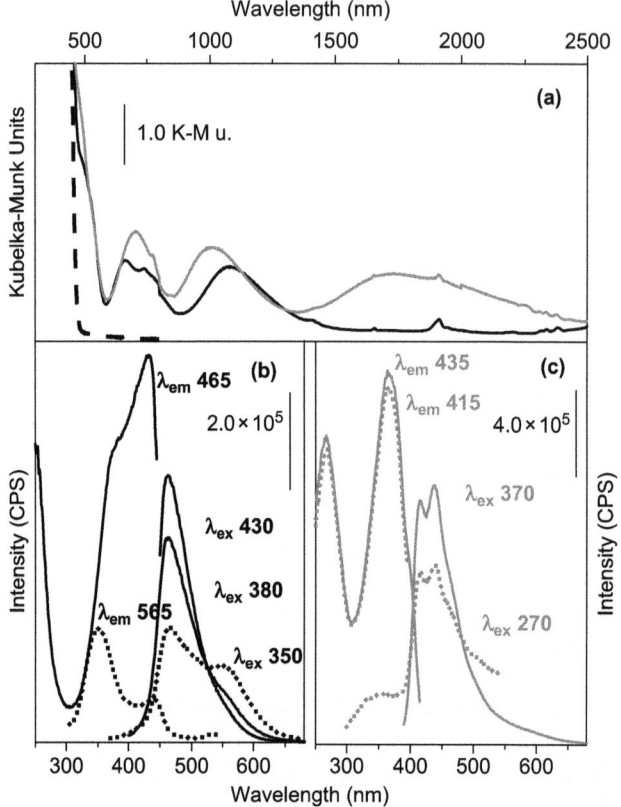

Figure 4.20 DRUV-visible-NIR spectra of CPO-27-Ni and 2,5-dihydroxytereph-
thalic acid linker (solid and dashed lines, respectively). Black and gray
curves are used for as synthesized and dehydrated samples, respectively.
(b, c) Photoluminescence of as synthesized (black curves) and dehydrated
(gray curves) CPO-27-Ni, respectively. Solid and dotted curves are used
to identify corresponding excitation and emission spectra.
Adapted with permission from Bonino *et al.*,[69] copyright American
Chemical Society, 2008.

aqueous solution as model, where Ni(II) has an O_h-like symmetry. In this case,
the three components, observed at 1150, 690 and 425 nm, are assigned to spin-
allowed d–d transitions: $^3A_{2g}(^3F) \rightarrow {}^3T_{2g}(^3F)$, $^3A_{2g}(^3F) \rightarrow {}^3T_{1g}(^3F)$, and
$^3A_{2g}(^3F) \rightarrow {}^3T_{1g}(^3P)$, respectively. Similar features were observed in the case of
$MgGa_2O_4$ or phosphate glasses doped with Ni(II).[143,144] Thermal treatment *in
vacuo* at 393 K produces deep changes in the optical spectrum: a blue shift for
all the components, and above all, the appearance of a new band at 1780 nm.
This profile is extremely similar to what Ciampolini[145] observed in the case of a
five-coordinated Ni(II) species in a squared-pyramidal geometry; the
assignment of the electronic transitions is as follows: $^3B_1(^3F) \rightarrow {}^3E_1(^3F)$,
$^3B_1(^3F) \rightarrow {}^3A_2(^3F)$, $^3B_1(^3F) \rightarrow {}^3B_2(^3F)$, $^3B_1(^3F) \rightarrow {}^3E_2$ (^3P), $^3B_1(^3F) \rightarrow {}^3E(^3P)$,
$^3B_1(^3F) \rightarrow {}^3A_2(^3P)$. This result indicates that the removal of water molecules

directly coordinated to Ni(II) species causes a substantial modification of the electronic structure of the metal. The change in symmetry and ligand field causes a rearrangement of the metal states, as reported above.

The combination of aromatic linkers with transition metal ions suggests the possibility to observe in fact some relevant features associated to electronic emissions by means of photoluminescence, as also shown in the case of many MOFs,[146–150] where the metal coordination to luminescent organic ligands had been able to enhance or quench and shift luminescent emission of the pure organic ligands.

Parts (b) and (c) of Figure 4.20 reports photoluminescence results obtained for CPO-27-Ni on as synthesized (black curves) and dehydrated samples (gray curves), respectively.[69] For both of the samples, excitation and emission spectra, obtained by fixing the emission or the excitation wavelength respectively, are reported in the same graph. Fixed emission and excitation wavelengths are indicated for each spectrum. Different mark styles refer to the corresponding excitation–emission couple of spectra. The hydrated sample (Figure 4.20b) is characterized by strong luminescence in the visible region. The most favorable excitation was found across the UV and visible regions (350–450 nm), where the lowest energy electronic transition associated to the linker and the strongest d–d transition of Ni^{2+} cations are present (see Figure 4.20) and correspond to the dark green arrows in Figure 4.19.

In particular, excitations at 380 and 430 nm give rise to an intense emission peak at 465 nm. An excitation at 350 nm causes a two peak emission spectrum with maxima at 465 and 565 nm. By comparing the photoluminescence of CPO-27-Ni and 2,5-dihydroxyterephthalic acid (spectra not shown for brevity), it is observed that the linker is significantly more fluorescent than the MOF. The absence of a double emission peak for the linker excited at 350 nm implies that the emission peak observed for CPO-27-Ni at 565 nm can be assigned to a charge transfer from the ligand to the metal (LMCT). In the literature, many examples of emission spectra due to LMCTs and not to the aromatic ligands are reported.[146,149,150] CPO-27-Ni luminescence is perturbed upon water removal and change of Ni(II) coordination sphere (see Figure 4.20c). In particular, upon dehydration (thermal treatment at 393 K), emission spectra shift to lower wavelengths and are split (maxima at 415 and 435 nm), because of degeneration removal by symmetry reduction. Excitation both at 270 and 370 nm cause the same type of emission.

4.3.2.2 Metal Node Exchange and Reactivity

The use of the inorganic nodes in MOF-5 as unusual chelating ligands illustrates a potentially rich area of exploration in coordination chemistry. As a proof-of-principle, Dinca *et al.*[151] reconceived the secondary building unit (SBU) of the iconic $Zn_4O(BDC)_3$ (MOF-5, BDC = 1,4-benzenedicarboxylate) as a tripodal ligand for metals that are typically incompatible with tetrahedral oxygen ligand fields, such as Ni^{2+}. Usually, Ni^{2+} (d^8) prefers octahedral coordination in oxygen ligand fields and assumes tetrahedral geometry only when trapped in condensed lattices such as ZnO, or when surrounded by bulky supporting ligands. Dinca's results indicated that spontaneous substitution of Ni^{2+} into MOF-5 (partially replacing Zn^{2+} sites) is thermodynamically favorable and suggested that Ni^{2+}-substituted MOF-5 may also be accessible by direct synthesis. The authors provide a hypothesis for this surprising observation: the yellow colour of as-synthesized Ni^{2+}-substituted MOF-5 is indicative of octahedral Ni^{2+}.

It seems that in the $Zn_4O(carboxylate)_6$ SBUs one Zn^{2+} is hexa-coordinate and binds two DMF molecules, resulting in the formulation of Ni-substituted MOF-5 as $(DMF)_2Ni$-MOF-5 $((DMF)_{2x}Ni_xZn_{4-x}O(BDC)_3$ with $0<x<1)$, see Figure 4.21a. Remarkably, heating $(DMF)_2Ni$-MOF-5 under vacuum afforded deep blue–purple crystals of $Ni_xZn_{4-x}O(BDC)_3$ (Ni-MOF-5), a new analogue of MOF-5 that contains pseudo-tetrahedral Ni^{2+} supported only by oxygen

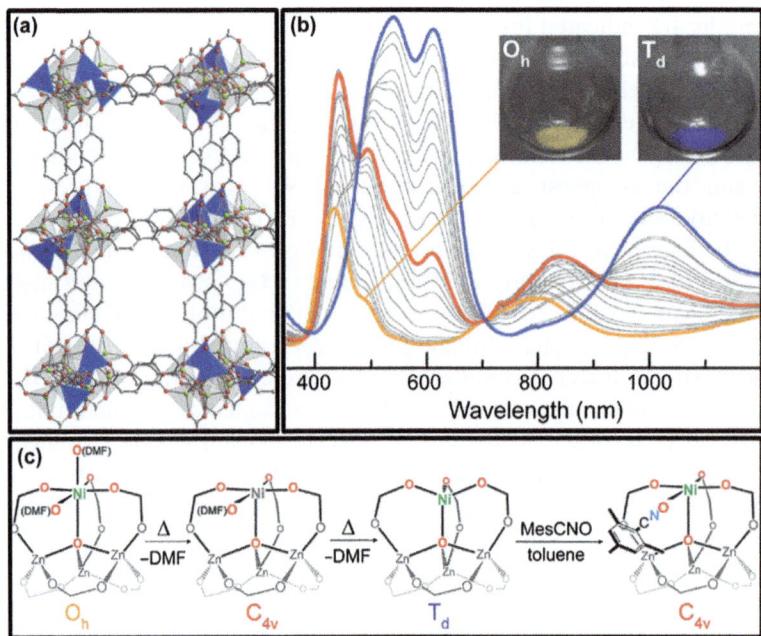

Figure 4.21   Part (a): part of the crystal structure of $Ni_1Zn_3O(BDC)_3$. Due to crystallographically-imposed symmetry, the position of Ni^{2+} centers (blue tetrahedra) within individual NiZn$_3$ clusters could not be identified unambiguously by the authors, who depicted them at random. Green, red and grey spheres represent Zn, O and C atoms, respectively. Hydrogen atoms are omitted for clarity. Part (b): *in situ* UV-Vis DRS spectra depicting the colour progression from yellow $(DMF)_2Ni$-MOF-5 to blue Ni-MOF-5 *via* a putative pentacoordinated Ni^{2+} intermediate (red trace). The inset shows optical images of the yellow and blue crystals. Part (c): the first 3 schemes report the local coordination of Ni^{2+} cations along the desolvation process followed by UV-Vis in part (b), the fourth scheme shows the effect of interaction with MesCNO in toluene solvent (UV-Vis spectrum not reported).
Adapted with permission from Brozek and Dinca,[151] copyright Royal Society of Chemistry, 2012.

ligands. The pseudo-tetrahedral geometry around the Ni^{2+} and the homogeneity of Ni-MOF-5 was quantified by diffuse-reflectance UV-Vis-NIR spectroscopy (blue curve in Figure 4.21b). Despite the slight deviation from tetrahedral geometry around Ni^{2+}, Ni-MOF-5 exhibited a spectrum that resembled solution-phase spectra of strictly tetrahedral Ni^{2+} complexes.[151] Thus, a peak at 1020 nm can be assigned to the $^3T_1(F) \rightarrow {}^3A_2$ transition of a d^8 tetrahedral ion, while the doublet of peaks at 540 and 608 nm can be assigned to the $^3T_1(F) \rightarrow {}^3T_1(P)$ transition.

If *in situ* DRUV-Vis-NIR spectra are recorded upon heating the starting $(DMF)_2$Ni-MOF-5 compound (see the evolution of the curves in Figure 4.21b, from yellow to blue through the red one), an isosbestic point around 700 nm is clearly detected, which suggested that DMF loss occurred in two kinetically independent processes *via* a well-defined five-coordinate Ni^{2+} species, as described in the first three schemes reported in Figure 4.21c.

Brozek and Dinca[151] also tested with UV-Vis spectroscopy the reaction of desolvated Ni-MOF-5 with small ligands such as PMe_3, THF and MeCN, that rapidly produced octahedral Ni^{2+}, indicated by a colour reversal to yellow. In contrast, the sterically-demanding MesCNO afforded an orange adduct, whose spectrum matched that of the putative pentacoordinate (DMF)Ni-MOF-5 adduct reported in the fourth scheme of Figure 4.21c.

4.3.3   Effect of Adsorbates

4.3.3.1   *UV-Vis Spectroscopy*

Electronic spectroscopies have been widely used to study the adsorbate–adsorbent interaction, monitoring the change in the optical properties either of the adsorbate or of the adsorbent. Particularly for MOFs with coordinatively unsaturated metal sites (the most promising for gas storage and separation, sensors and catalysis applications) adsorption/desorption of guest species at the co-ordinatively unsaturated metal site (transition metal) causes a strong change in the MOF optical properties.[70,71,75,117,118,148]

In CPO-27-M desolvated frameworks, metal sites are in square pyramidal geometry (see insets in Figure 4.9) and upon adsorption of gases are converted to distort octahedral leading to change in colour of sample.[69,70] In the case of reversible adsorption this change is shown to be reversed by degassing the sample under vacuum. This fantastic optical response shown by MOFs is promising for developing gas sensors and transparent cartridges for gas purification. An example of that is reported in Figure 4.22a where it is evident that the adsorption of different gases (CO, NO and H_2O) causes strong subsequent changes in the UV-vis absorption spectra of CPO-27-Ni. In particular the presence (H_2O, NO or CO) or the absence (desolvated material) of a sixth ligand in the first shell of Ni^{2+} cations changes its local symmetry, and thus the Ni^{2+} d–d transitions, see dark green arrow in Figure 4.19.

The strong change observed in the UV-Vis spectra (Figure 4.22b) upon the O_2 adsorption on the air unstable CPO-27-Fe suggests that the fact that this MOF reacts with O_2 at RT is the main reason of its air instability.[118]

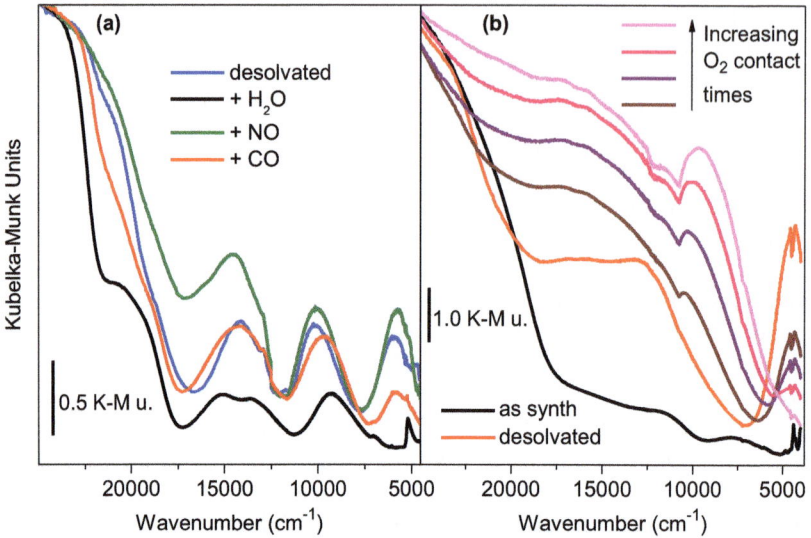

Figure 4.22   Part (a): effect of the adsorption of different probes on the electronic properties of CPO-27-Ni (Ni^{2+} d–d transitions) as monitored from DRS UV-Vis. Part (b): effect of O_2 contact (40 mbar) on the electronic properties of CPO-27-Fe as a function of time (last spectrum corresponds to 5 min contact). The spectra obtained after contact with O_2 are affected by artifacts because of the high Kubelka–Munk values; fine features cannot be trusted but the general trend is reliable.
Previously unpublished figure; in part (a) some spectra already published in ref. 69,70,90.

One of the most promising and challenging applications in the MOF field is represented by the use of MOFs as catalysts.[141,152–162] However, this application is limited by the lack of functionalities and limited coordination unsaturation of the metal sites. Different routes have been envisaged to introduce functionality into MOFs by exploiting both in-synthesis and post-synthesis approaches.[74] In some cases, this type of modification of the framework has altered the electronic properties of MOFs. The best examples of this are IRMOF-3 and UiO-66-NH_2 with the amino terephthalic acid linker; the presence of the amino group shifts the absorption towards the visible light region giving a yellow colour compared to the white parent MOF samples. This electronic modification of the framework is utilized by Silva *et al.*,[160] demonstrating the enhanced photocatalytic activity of UiO-66-NH_2 over UiO-66 in the production of hydrogen from the methanol/water system. This enhanced activity was attributed to the increase in light absorption by the presence of –NH_2 group in the linker.

Chavan *et al.* studied the post-synthetic functionalization of UiO-66 by grafting $Cr(CO)_3$ at the linker.[75] The grafting of $Cr(CO)_3$ at the organic linker changes the colour of the sample from white to yellow, extending absorption into the visible region, see Figure 4.23 (see also Section 4.2.5.2 and Figure 4.18, where we discussed the vibrational properties of the functionalized material).

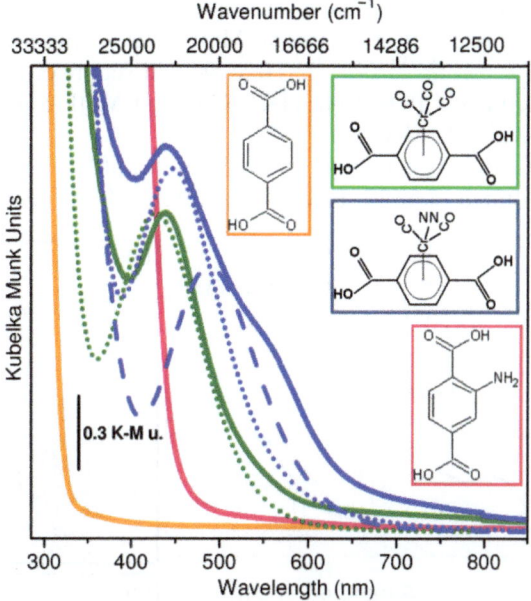

Figure 4.23   Full lines: experimental UV-Vis DRS spectra of UiO-66 (orange curve), UiO-66-NH$_2$ (pink curve), UiO-66 with grafted Cr(CO)$_3$ (green curve) and UiO-66-Cr(CO)$_2$N$_2$ (blue curve) obtained by photo substitution of CO by N$_2$. With the same colour as the spectra, the boxes report the schemes of the corresponding functionalized ligands. Green dotted line: simulated optical spectra for (C$_6$H$_4$)(COOH)$_2$Cr(CO)$_3$; blue dashed and dotted curves: simulated optical spectra for conformers A and B of (C$_6$H$_4$)(COOH)$_2$Cr(CO)$_3$N$_2$, respectively.
Previously unpublished figure reporting spectra from ref. 75.

For UiO-66/Cr(CO)$_3$, the optical spectrum displays an intense, out-of-scale band at $\sim$30 000 cm^{-1} and a well-defined one at 23 000 cm^{-1}. In close analogy with (η^6-C$_6$H$_6$)Cr(CO)$_3$, these absorption bands are associated with (CO)$_3$Cr$\rightarrow$arene charge-transfer transitions (green full curve in Figure 4.23). Furthermore, the photo substitution of CO by N$_2$ was also demonstrated by using UV-vis spectroscopy. After irradiation for a few minutes, the sample slightly changes its colour; as a consequence, the overall intensity of the UV-visible spectrum increases and a new band at 18 450 cm^{-1} appears (blue full curve in Figure 4.23). These changes are nicely predicted by theoretical calculations. See dotted and dashed curves in Figure 4.23. Note that the presence of two components in the experimental spectrum (blue full curve) is due to the simultaneous presence of two conformers of the (C$_6$H$_4$)(COOH)$_2$Cr(CO)$_3$N$_2$ complex.[75]

4.3.3.2   XANES Spectroscopy

Investigation of the XANES region of the X-ray absorption spectrum can provide detailed information on both the oxidation state and the coordination symmetry of the absorber atom, and was successfully applied in the structural and electronic characterization of several MOFs, measuring the corresponding metal K-edge.[43,68–70,77,81,90,142,163–167] Here we will report on the evolution of the XANES spectra of CPO-27-Ni, HKUST-1 and MOF-5 upon desolvation and interaction with adsorbates.

In Figure 4.24a, the experimental XANES spectra of HKUST-1 are reported as-prepared (hydrated sample, black line), after solvent removal at 453 K

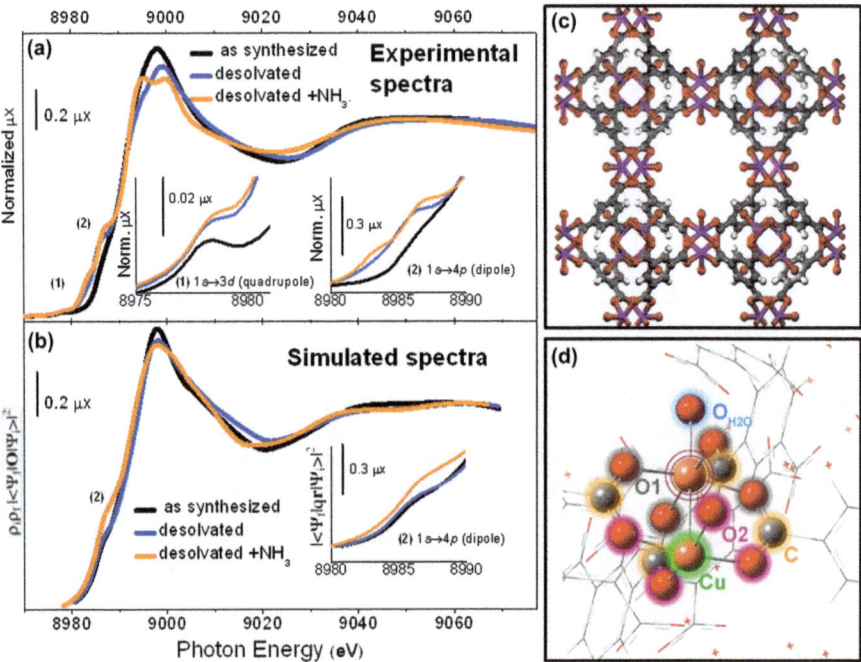

Figure 4.24   Part (a): experimental XANES spectra of HKUST-1 as synthesized (black line), after solvent removal (blue line) and successive NH₃ dosage (orange line). The insets report magnifications of the 1s → 3d quadupolar transition (1) and of the shakedown 1s → 4p dipolar transition (2). Part (b): simulated XANES spectra as obtained with FEFF8.4 code,[169] cutting a cluster centered on one of the Cu atoms with a radius of 6 Å around it, see part (d), from the XRD refined structure,[170] see part (c). For further improvement of the simulations, an optimization of geometrical parameters was performed exploiting Fitit software.[171] Calculations for the absorption spectra and densities of states have been done using the muffin-tin potential and the Hedin–Lundqvist exchange correlation, taking note of the presence of a core hole in a self-consistency scheme (for the self-consistent calculations a radius of 4 Å has been used). To take into account the experimental broadening, an imaginary potential $V_i = 1.0$ eV has been introduced. Part (c) sticks and balls representation of the structure of HKUST-1: Cu violet; O red; C gray; H, white. Part (d): cluster reproducing the local environment of the X-ray-absorbing Cu atom used to compute the simulated XANES spectra.
Previously unpublished figure reporting spectra and schemes published by Borfecchia *et al.*[167]

(dehydrated sample, blue line) and upon contact with 60 mbar of ammonia at room temperature (orange line). The XANES spectra of the as prepared and dehydrated sample are typical of Cu(II) species,[163,167] showing the edge jump at 8990 eV and two characteristic pre-edge peaks at *ca.* 8976 eV and *ca.* 8986 eV, labeled as **(1)** and **(2)**, respectively, in Figure 4.24a, and separately reported in the insets of the same panel. Feature **(1)** is assigned to the very weak 1s→3d quadrupolar transition, while the shoulder **(2)** appearing along the white line profile[168] is related to the dipolar shakedown 1s→4p transition. A quadrupolar transition is a transition where the operator used in the Fermi golden rule is the quadrupolar and not the dipolar moment, see eqn (4.10), a fact that makes the matrix element of much smaller value and so the observed transition of much lower intensity. The simulation of the XANES spectrum of the hydrated material was performed with FEFF8.4 code[169] starting from the XRPD structure,[170] see Figure 4.24c, and cutting the cluster reported in Figure 4.24d. Then the FitIt software[171] was used to optimize five structural parameters: (i) the Cu–Cu and (ii) Cu–H_2O bond length, the distance between the Cu atom and the trimesic acid carboxyl groups with a separate optimization of (iii) Cu–O1 and of (iv) Cu–C and Cu–O2 distances (optimized in a correlated way allowing the use of only one free parameter) and (v) a general overall contraction or elongation of all the other distances ($R_{XANES} = \alpha R_{XRPD}$).[172] See Figure 4.24d for the atom notation.

The XANES spectrum of the activated sample (blue curve in Figure 4.24a) shows: (i) a decrease of the white line[168] intensity, and (ii) an increase in the intensity of feature **(2)**, which appears more as a well separated band than as a white line shoulder. The simulation of the XANES spectrum for the activated sample was performed with the same method previously described for the hydrated sample removing the atoms of the two water molecules, resulting in an optimization with four parameters only. In the simulated spectra, the same trend noticed on the experimental curves is obtained, where the decrease of the white line intensity is correlated to the lower coordination number, and the increase of the intensity of feature **(2)** is ascribable to a lower symmetry of the Cu(II) species. In the simulated curves, this trend is less evident. Unfortunately, a full description of the asymmetric distortion undergone upon activation by the Cu cluster requires the optimization of too many parameters. The optimization of the bond distances resulted in a slight contraction of the distances between the absorbing Cu atom and carbonyl groups and all other distances, in agreement with the results found by EXAFS fit reported in the same work.[3,167]

The XANES spectrum of the activated sample after the interaction with NH_3 (orange curve in Figure 4.24a) shows an additional increase in the intensity and a slight blue-shift of the dipole band **(2)** assigned to the 1s→4p transition. Moreover, a new pre-edge peak is observed at *ca.* 8983 eV. Despite this, the edge position is not modified with respect to the as prepared and activated samples (no change in oxidation state of the metal), the white line is modified towards a more structured appearance, and seems to return towards the shape observed in the case of the hydrated sample. The simulation of the spectrum for the sample after interaction with NH_3 was performed following the same method adopted for the hydrated material and substituting the water molecules with two NH_3 molecules. After the optimization of the geometrical parameters, it has been observed that an increase of the intensity of the pre-edge feature **(2)** was proportional to the Cu-NH_3 distance which was optimized at 2.3 Å with a considerable elongation with respect to the previous position of the water molecules. Moreover, a splitting of the white line feature (although less evident than in the experimental spectrum), has been obtained in the simulation with a distortion of carbonyl groups (slight elongation of Cu-O1 distances and a severe shortening of both Cu-C and Cu-O2 and all other distances). Again, the deformation introduced to simulate the XANES spectra is in agreement with the results of the EXAFS fit.[167]

Figure 4.25 reports the evolution of the XANES spectra of CPO-27-Ni, HKUST-1 and MOF-5 upon desolvation and interaction with probe molecules, parts (a), (b) and (c), respectively. In all cases, the edge of the XANES spectra are observed at significantly higher energies with respect to the corresponding

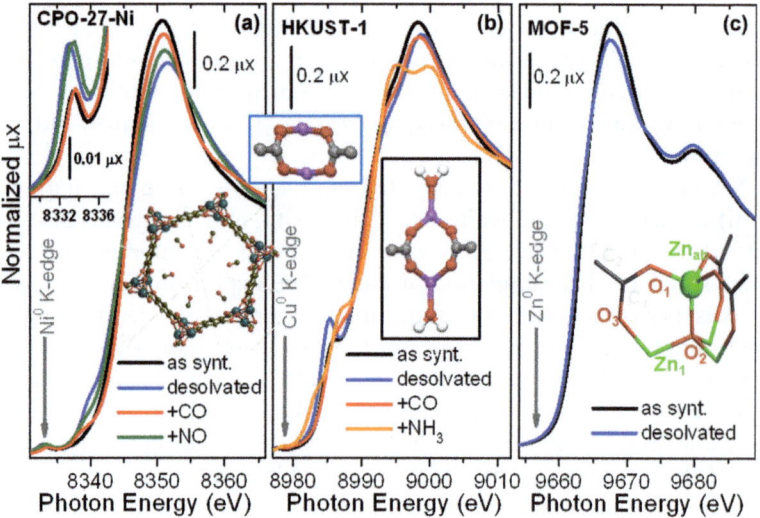

Figure 4.25   Part (a): Ni K-edge XANES spectra of CPO-27-Ni before (black curve) and after (blue curve) solvent removal and after successive dosage of CO (red curve) and of NO (green curve). The top inset reports the magnification of the 1s→3d transitions. The middle inset reports the structure of the desolvated material as optimized by CRYSTAL code[86] with a stoichiometry Ni^{2+} : CO = 1. Part (b): Cu K-edge XANES spectra of HKUST-1 before (black curve) and after (blue curve) solvent removal and after successive dosage of CO (red curve) and of NH_3 (orange curve). Insets report the structure of the $[Cu_2C_4O_8]$ cage in its hydrated (blue) and dehydrated (black) forms, respectively. Part (c): Zn K-edge XANES spectra of MOF-5 before (black curve) and after (blue curve) solvent desolvation. Inset reports the local structure of the inorganic cornerstone showing Zn^{2+} in T_d-like symmetry. In each panel the vertical gray arrow indicates the edge position of the XANES spectrum of the corresponding metal foil. Previously unpublished figure reporting spectra and schemes published in ref. 44,69,70,77,167

metal foils (see vertical gray arrow) reflecting the fact that we are dealing with divalent cations, requiring more energy then the zero-valent atoms of the metals to perform the photoelectric effect from the 1s orbital. In all cases, a white line[168] decrease is observed upon desolvation (from black to blue curves) reflecting a decreased number of the scattering atoms around the absorbing metals. More interestingly is the fact that the overall changes in the XANES spectrum upon desolvation is minimal for MOF-5 (Figure 4.25c), reflecting the fact that the first coordination shell of the metal is not changed in the process. Conversely, for both CPO-27-Ni and HKUST-1, water molecule removal from the first coordination shell of the metal results in a significant change in both pre-edge and edge regions. For these materials, the coordination vacancy left on the metal site upon desolvation can be filled by adsorbed molecules, as shown with CO on both CPO-27-Ni[70] and HKUST-1,[163] with NO on CPO-27-Ni[69] and with NH_3 on HKUST-1.[167]

4.3.4 Investigation of the Electronic Properties of Metal Nanoparticles Encapsulated inside MOFs by XPS

As was the case for other high surface area materials such as *e.g.* oxides,[173–177] carbons,[177] zeolites,[178] zeotypes,[179] and polymers,[180] MOFs have also been used to support metal nanoparticles.[134–136,138–140]

Here we will focus the attention on the paper of Ramos-Fernandez *et al.*[138] reporting on the synthesis, the characterization (FT-IR, XPS, NMR, UV-Vis), and catalytic performances of Pt supported on phosphotungstic acid (PTA) encapsulated in NH_2-MIL-101(Al), synthesized in one step and acting as anchoring sites for the Pt precursor species. The authors have tested the material in the oxidation of CO, the preferential oxidation of CO in the presence of H_2, and the hydrogenation of toluene. They found that the reduction at 473 K results in the formation of small Pt^0 clusters and Pt^{2+} species and that the reduction at 573 K induces the formation of intermetallic $Pt-W^{5+}$ species, which exhibit the best CO oxidation activity and a higher selectivity toward CO_2 than alumina supported Pt, resembling the combination of a noble metal on a reducible support. In toluene hydrogenation, the MOF catalyst performances are worse than Pt on alumina, ascribed to the too small size of the Pt clusters in the MOF catalysts. Hereafter, we will focus on the XPS characterization only, with the aim of elucidating the information that the technique can provide for the characterization of MOF materials. XPS binding energies (*BE*) provides information on the valence state and on the chemical environment of the atoms present on the surface of the sample.[126,181,182] A quantitative speciation of the element present on the sample surface is also possible by measuring the relative intensities of the XPS peaks of all detected species, but requires careful analysis of the collected spectra, which must be analyzed by taking into account several factors such as: the atomic cross sections for X-rays of energy hv_0, see eqn (4.13); the photoelectron cross sections; the escape depths of the electrons; the density of the sample, *etc.*

Ramos-Fernandez *et al.*[138] have determined the surface composition of the NH$_2$-MIL-101(Al), of the PTA-MOF composite, and of the Pt-PTA-MOF catalysts after different pretreatments by means of XPS.[138] By collecting photo-emitted photons in the binding energy *BE*, see eqn (4.13), range of 30–600 eV, the authors where able to extract electrons from O(1s), N(1s), C(1s), Al(2p), W(4f), and Pt(4f) levels. The XPS spectra collected on the different Pt containing samples are reported in Figure 4.26.

From the collected spectra, the following information can be obtained. All analyzed samples exhibit a *BE* value for Al(2p) at 74.5 eV, testifying that we are dealing with Al^{3+} species.[138] This observation indicates that the oxidation state of the metal trimers does not change upon reduction. The *BE* value for O(1s) for the case of NH$_2$-MIL-101(Al) was deconvoluted into three contributions with values of 531.8, 533.0, and 529.9 eV. The main peak, centered at 531.8 eV,

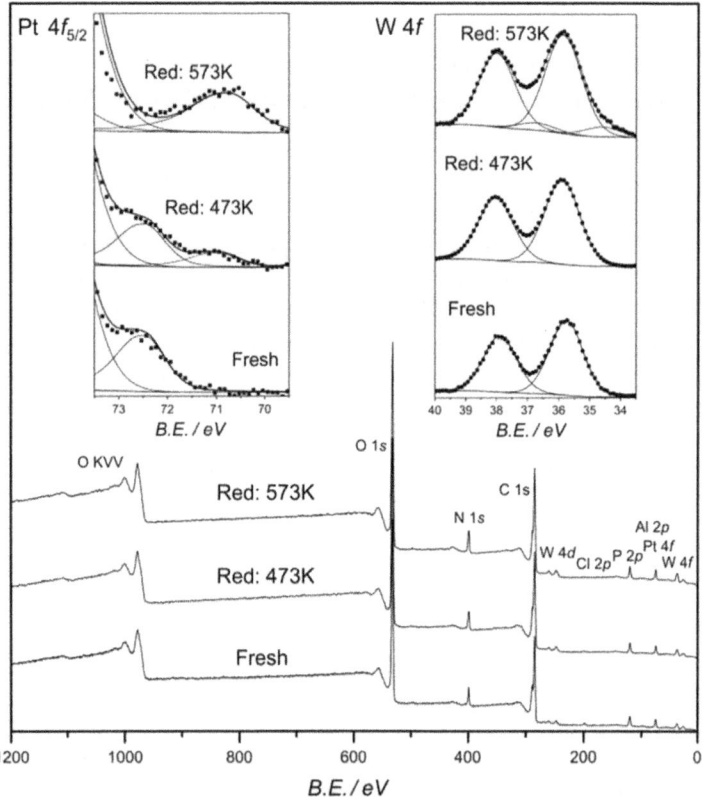

Figure 4.26 XPS survey spectra for the Pt supported on phosphotungstic acid (PTA) encapsulated in NH$_2$-MIL-101(Al) MOF before and after reduction at 473 and 573 K. Left and right insets report a zoom on the Pt(4f$_{5/2}$) and W(4f) levels, respectively.
Reproduced with permission from Ramos-Fernandez *et al.*,[138] copyright Elsevier, 2012.

is due to the terephthalate oxygen.[183] The other two peaks show a small contribution to the O(1s) peak, and they were assigned to a small impurity of uncoordinated terephthalic acid (529 eV) and the hydroxyl groups present in the structure (533 eV).[183]

The Pt(4f) level was analyzed after Pt impregnation and subsequent reductions at 473 and 573 K in H_2 atmosphere (left inset in Figure 4.26). The untreated sample shows a binding energy of 75.7 eV for the Pt(4f$_{7/2}$) and 72.eV for the Pt(4f$_{5/2}$), which are typically assigned to electron deficient Pt^{n+} species. After reduction at 473 K, a new feature appears at 71.0 eV, due to the Pt(4f$_{5/2}$) level for Pt^0, although the peak at 72.3 eV is still present, indicating that Pt is not fully reduced at 473 K. Complete Pt reduction was achieved after reduction at 573 K, as demonstrated by the presence of only one peak for the level Pt(4f$_{5/2}$) centered at 70.5 eV.[138,184]

The binding energy values for the W(4f) level in the PTA-MOF composite and in the non-reduced Pt sample are 35.7 and 37.8 eV for the W(4f$_{5/2}$) and W(4f$_{7/2}$) levels, respectively, see right inset in Figure 4.26. These values are typical of W^{6+} in the Keggin structure. When the Pt sample is reduced at 473 K, the binding energy for the W(4f) levels do not change, but after the treatment at 573 K under hydrogen, W^{6+} is partially reduced and two new peaks appear at 34.5 W(4f$_{5/2}$) and 36.6 eV W(4f$_{7/2}$).[138] These peaks are assigned to W^{n+} with $n<6$, indicating a modification of the W oxidation state. The authors also observed that the sample became blue, and blue PTAs are well known for being reduced.[185]

Ramos-Fernandez *et al.*[138] compared the quantitative atomic ratios calculated from the XPS data with those obtained by inductively coupled plasma-atomic emission spectroscopy (ICP-OES). As XPS is a surface sensitive technique whereas ICP-OES is a bulk one, a comparison of the ratios obtained with the two techniques gives information on the homogeneity of the composition from the surface to the internal regions of the sample. The authors found that in all cases, the XPS atomic ratios are lower than the ones calculated by ICP-OES, suggesting that W and Pt are well inside the MOF cavities. Moreover, the Pt/Al and W/Al XPS ratios can be an indication of the Pt and W dispersion.[138,181,186] The Pt/W XPS atomic ratio of the unreduced sample is identical to the ICP analysis, but the XPS ratio decreases with increasing temperature of reduction. The reduced samples present different atomic ratios, indicating that the Pt surface concentration decreases with increasing reduction temperature, while that of W increases.

4.3.5 X-Ray Emission Spectroscopy: An Element Selective Tool to Investigate d–d Transitions

As discussed in Section 4.3.1, electronic excitation and disexcitation energies and line shapes provide information on both unoccupied and occupied electronic levels, respectively, see Figure 4.19. In particular, the d–d transitions, in which only the occupation of the transition metal 3d orbitals changes, are sensitive to the local coordination at the metal, thus providing information on the electronic levels involved by the coordination. d–d transitions are mostly investigated by means of UV-Vis spectroscopy (see Sections 4.3.2 and 4.3.3.1), XANES (see Section 4.3.3.2), electron energy loss[187] and resonant inelastic X-ray scattering (RIXS).[188–190] In the present section, we will report the RIXS study on the d-d transitions of Ni^{2+} hosted in NiPBP, see section 4.2.3.2 and Figure 4.6 where we discussed the water removal as viewed by IR spectroscopy.

The electronic properties of NiPBP were investigated by Mino *et al.*[81] using UV-Vis DRS, XANES and RIXS spectroscopies and are summarized here in Figure 4.27. The DRS-UV-Vis spectrum of as-synthesized NiPBP (inset of Figure 4.27a) shows a strong band due to the $\pi \rightarrow \pi^*$ electronic transitions of the organic linker around 28 000 cm^{-1} (3.5 eV) and two weaker components in the region of the CF d–d transitions at 9700 cm^{-1} (1.2 eV) and 16 000 cm^{-1} (2.0 eV). Possibly, a third one, expected at higher energy in the presence of distorted octahedral Ni(ii) ions,[123,191] can be seen around 25 000 cm^{-1} (3.1 eV); however, its presence and its exact location cannot be straightforwardly determined from UV-Vis as it is significantly overshadowed by the lower-energy component of the $\pi \rightarrow \pi^*$ transition of the ligand.

The Ni K-edge XANES spectrum, collected in total fluorescence yield (TFY) of NiPBP sample (Figure 4.27a) is very similar to that observed for the CPO-27-Ni MOF (black curve in Figure 4.25a), exhibiting a comparable distorted octahedral local environment. In particular, moving from low to high photon energies, at least four groups of transitions are clearly appreciable: (i) a weak 1s → 3d electronic transition at 8332.8 eV; (ii) a second component at 8337.2 eV; (iii) a strong 1s → 4p dipole-allowed electronic transition near 8340 eV (scarcely visible because it is too close to the edge jump); (vi) a white line[168] at 8348 eV.

The Ni K-edge RIXS spectrum of NiPBP was collected by tuning the excitation energy to the energy maximum of the pre-edge peak at 8332.8 eV (star in Figure 4.27a), which corresponds to the transition between the ground state electronic configuration of Ni(ii), $|g> 1s^23d^8$, to orbitals that have mainly Ni 3d character (we therefore write the intermediate state as $|i> 1s^13d^9$). As evident from Figure 4.27b (see also the scheme in the inset), in moving from lower to higher energies three main features can be identified in the RIXS spectrum: (i) the charge transfer region, extending from 8322 eV up to about 8327 eV. The charge-transfer energy can be estimated as 8325.7 eV (half of the maximum intensity, corresponding to 6.3 eV in energy transfer), in good agreement with published results for Ni(ii) in other systems exhibiting O_h-like symmetry.[191] (ii) The d–d region, where three peaks are clearly visible at 1.25 eV, 1.95 eV and 3.1 eV in energy transfer (see Figure 4.27c). (iii) The elastic peak, due to the elastic de-excitation transition $|i> 1s^13d^9 \rightarrow |g> 1s^23d^8$.

According to previously performed atomic multiplet calculations,[191] Mino *et al.*[81] have assigned the peaks at 1.25 eV, 1.95 eV and 3.1 eV, to the $^3T_{2g}(F) \rightarrow {}^3A_{2g}(F)$, $^3T_{1g}(F) \rightarrow {}^3A_{2g}$ and $^3T_{1g}(F) \rightarrow {}^3A_{2g}(F)$ electronic transitions, respectively. Assuming the maximum of the elastic peak as the relative zero energy, it is possible to compare the d–d region obtained by both RIXS and UV-Vis spectroscopies (see Figure 4.27c). As a result of this comparison, it is

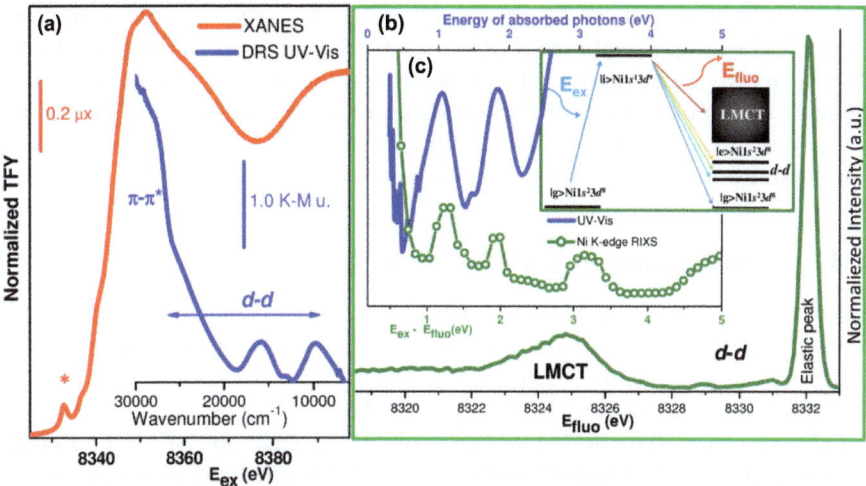

Figure 4.27   Part (a): Ni K-edge XANES spectrum (collected in TFY mode) of as prepared NiPBP (see Figure 4.6b for the MOF structure). The inset reports the corresponding UV-Vis DRS spectrum. Part (b): Ni K-edge RIXS spectrum collected on NiPBP with the excitation energy tuned to the energy maximum of the $1s \rightarrow 3d$ pre-edge peak, red star in part (a). To demonstrate the weakness of such transitions, the intensity of the elastic peak at 8332 eV has also been reported. Note that the latter is technically not a fluorescence emission. Part (c): comparison between the Ni K-edge RIXS (green curve, magnified and reported as a function of the energy transfer $E_{fluo} - E_{ex}$) and UV-Vis (blue curve) spectrum of NiPBP. The cartoon reports a simplified scheme of the two photon process describing the RIXS phenomenon, $|g>$, $|i>$ and $|e>$ representing the ground, the intermediate and the excited state of the valence d–d and/or LMCT transitions. The ground state is formally $|g>Ni1s^23d^8$: the absorption of a photon at 8332 eV promotes a 1s electron into an empty 3d state. From the intermediate $|i>Ni1s^13d^9$ state, the system evolves either into the ground state again (resulting in the elastic peak), or into an excited $|e>Ni1s^23d^8$ state. The excitation can result in (i) the promotion of a Ni 3d electron into an excited 3d state, so that the net transition between initial and final states is the same as that obtained in an UV-Vis experiment for a d–d excitation; or in (ii) the promotion of an electron from the ligands to the Ni, equivalent to a net LMCT transition.
Adapted with permission from Mino *et al.*,[81] copyright Royal Society of Chemistry 2012.

worth noting that RIXS completes the investigation of the d–d region, as it is able to prove the presence and to identify the exact energy location of the third transition at 3.1 eV, which was not clearly observable by means of optical spectroscopy (see inset in Figure 4.27a). Mino *et al.*[81] were able to disclose the whole set of d–d transitions expected for a Ni^{2+} ion in a distorted O_h symmetry because of the atomic selectivity towards Ni, the RIXS technique does not excite the ligand $\pi \rightarrow \pi^*$ transitions that overshadow the d–d transition at higher energy in Ni^{2+} in a standard UV-Vis spectrum. Thus, even if its resolution is much worse than that of UV-Vis spectroscopy (see the scattered points in Figure 4.27a to appreciate the RIXS resolution), due to its element selectivity RIXS can provide us specific information about the desired element. Recently, Gallo *et al.*[192] collected the RIXS spectrum of a hydrated CPO-27-Ni MOF, which was interpreted by combining DFT with multiplets theory approaches. The crystal field splitting and the electron-electron interactions were obtained.

4.4   Conclusions and Perspectives

From the case studies reported in this chapter it becomes evident that the combination of vibrational and electronic spectroscopies provides information on the property of the investigated MOF materials that can not be obtained by the structural refinement of XRPD data, BET and TG analyses.

Laboratories working in the synthesis and characterization of MOFs should be equipped with a FTIR and a DRS-UV-Vis instrument, allowing the provision of the basic, but most informative, information on vibrational and electronic properties. These spectrometers should be equipped with *ad hoc* conceived cells and related set-ups allowing *in situ* solvent removal and dosage of desired probe molecules. Possibly the cell should be also equipped with a temperature control system allowing weakly bonded molecules to be adsorbed on the linkers and metal exposed sites by cooling the sample. Such a set-up opens the way to measuring the site-dependent heat of adsorption of a given probe molecule.[44,103] More expensive (Raman and XPS) and less easily available (INS, XANES and RIXS) spectroscopies should be used in specific cases. Again, the ability to build up specific experimental set-ups allowing *in situ* experiments to be done is unavoidable to investigate this class of materials.

The presence in the group of competencies in computational chemistry is highly recommended as quantum mechanics will allow full understanding of the complex vibrational and electronic features obtained by the experiments.

Acknowledgement

We are indebted to Prof. E. Garrone for helpful discussion on the VTIR method.

References

1. The BET technique stands for a volumetric gas (usually N_2) adsorption at constant temperature (usually 77 K). The acronym BET refer to the names of the three authors that developed a method to obtain the sample surface area from the gas uptakes at different equilibrium pressures: S. Brunauer, P. H. Emmett and E. Teller, *J. Am. Chem. Soc.*, 1938, **60**, 309. Another widely employed method is the Langmuir one: I. Langmuir, *J. Am. Chem. Soc.*, 1916, **38**, 2221.
2. E. A. Pidko and E. J. M. Hensen, *Computational Approach to Chemical Reactivity*, in Metal–Organic Frameworks as Heterogeneous

Catalysts, ed. F. X. Llabrés i Xamena and J. Gascón, RSC, Cambridge, 2013, ch. 7.

3. E. Borfecchia, D. Gianolio, G. Agostini, S. Bordiga and C. Lamberti, *Characterization of MOFs. 2. Long and Local Range Order Structural Determination of MOFs by Combining EXAFS and Diffraction Techniques*, in Metal–Organic Frameworks as Heterogeneous Catalysts, ed. F. X. Llabrés i Xamena and J. Gascón, RSC, Cambridge, 2013, ch. 5.

4. P. R. Griffith, *Fourier Transform Spectroscopy*, Wiley, New York, 1986.

5. S. F. Parker, *Inelastic Neutron Scattering Spectroscopy*, in Handbook of Vibrational Spectroscopy, ed. J. M. Chalmers and P. R. Griffiths, John Wiley & Sons, Chichester, 2002, vol. 1, p. 838.

6. Thermal neutrons are neutrons exhibiting a kinetic energy distribution of Boltzman type in equilibrium with an heavy water bath at $T = 300$ K. They are consequently characterized by an average energy of $<1/2 m_N v_N^2> = 25$ meV and an average speed of $<v_N> = (3kT/m_N)^{1/2} = 2200$ m s^{-1}, where $k = R/N_A = 1.38 \times 10^{-23}$ J K^{-1} is the Boltzman constant and where $m_N = 1.6749 \times 10^{-27}$ kg is the neutron rest mass. Thermal neutrons exhibits an average de Broglie wavelength $<\lambda> = h/(m_N <v_N>) = 2$ Å.[3] As $<\lambda>$ is of the same order of magnitude than the spacings between atoms in crystalline solids, beams of thermal neutrons are ideal for investigating the structure of crystals, particularly for locating positions of hydrogen atoms, which are not well located by X-ray diffraction techniques. For more information see: G. L. Squires, *Introduction to the Theory of Thermal Neutron Scattering*, Dover, New York, 1996.

7. G. Herzberg, *Infrared and Raman Spectra of Polyatomic Molecules*, van Nostrand, Princeton, 1945.

8. K. Nakamoto, *Infrared and Raman Spectra of Inorganic and Coorination Compounds 5th Ed.*, John Wiley & Sons, New York, 1997.

9. H. F. Shurvell, *Spectra-Structure Correlation in the Mid-and Far-infrared*, in Handbook of Vibrational Spectroscopy, ed. J. M. Chalmers and P. R. Griffiths, John Wiley & Sons, Chichester, 2002, vol. 3, p. 1783.

10. L. G. Weyer and S.-C. Lo, *Spectra-Structure Correlation in the Near-infrared*, in Handbook of Vibrational Spectroscopy, ed. J. M. Chalmers and P. R. Griffiths, John Wiley & Sons, Chichester, 2002; vol. 3, p. 1817.

11. H. G. M. Edwards, *Spectra-Structure Correlation in the Raman Spectoscopy*, in Handbook of Vibrational Spectroscopy, ed. J. M. Chalmers and P. R. Griffiths, John Wiley & Sons, Chichester, 2002, vol. 3, p. 1838.

12. K. Nakamoto, *Infrared and Raman Spectra of Inorganic and Coorination Compounds*, in Handbook of Vibrational Spectroscopy, ed. J. M. Chalmers and P. R. Griffiths, John Wiley & Sons, Chichester, 2002, vol. 3, p. 1872.

13. P. Atkins and R. Friedman, *Molecular Quantum Mechanics*, Oxford University Press, Oxford, 1999.

14. M. Alonso and E. J. Finn, *Fundamental University Physics Vol.3 – Quantum and Statistical Physics*, Addison–Wesley, Reading MA, 1968.

15. A. C. Phillips, *Introduction to Quantum Mechanics*, John Wiley & Sons Ltd, New York, 2003.

16. P. R. Griffiths, *Introduction to Vibrational Sprctroscopy*, in Handbook of Vibrational Spectroscopy, ed. J. M. Chalmers and P. R. Griffiths, John Wiley & Sons, Chichester, 2002, vol. 1, p. 33.

17. Please note that in IR spectroscopy a.u. means "absorbance units", and not "arbitrary units". This means that the ordinate values of an IR spectrum must be quantified and care must be done in avoiding a.u. values larger than 2, because in such a case almost all photons are absorbed by the sample and the statistics on I_1 counts will be too low. As an example, when $\mu x = 3$, then $I_1 = I_0 e^{-3} < 0.05 I_0$. In such cases, the experiment has to be repeated using a thinner sample.

18. P. R. Griffiths and J. M. Olinger, *Continuum Theories of Diffuse Reflection*, in Handbook of Vibrational Spectroscopy, ed. J. M. Chalmers and P. R. Griffiths, John Wiley & Sons, Chichester, 2002, vol. 2, p. 1125.

19. J. D. Dahm and K. D. Dahm, *Discontinuum Theories of Diffuse Reflection*, in Handbook of Vibrational Spectroscopy, ed. J. M. Chalmers and P. R. Griffiths, John Wiley & Sons, Chichester, 2002, vol. 2, p. 1140.

20. M. Milosevic, *Internal Reflection and ATR Spectroscopy*, John Wiley & Sons, 2012.

21. D. Wolverson, *Raman spectroscopy*, in Characterization of Semiconductor Heterostructures and Nanostructures, ed. C. Lamberti, Elsevier, Amsterdam, 2008, p. 249.

22. G. Keresztury, *Raman Spectroscopy: Theory*, in Handbook of Vibrational Spectroscopy, ed. J. M. Chalmers and P. R. Griffiths, John Wiley & Sons, Chichester, 2002, vol. 1, p. 71.

23. G. Ricchiardi, A. Damin, S. Bordiga, C. Lamberti, G. Spanò, F. Rivetti and A. Zecchina, *J. Am. Chem. Soc.*, 2001, **123**, 11409.

24. S. Bordiga, A. Damin, F. Bonino, G. Ricchiardi, A. Zecchina, R. Tagliapietra and C. Lamberti, *Phys. Chem. Chem. Phys.*, 2003, **5**, 4390.

25. A. Damin, F. Bonino, S. Bordiga, E. Groppo, C. Lamberti and A. Zecchina, *ChemPhysChem*, 2006, **7**, 342.

26. P. A. M. Dirac, *Proc. Roy. Soc. (London) A*, 1927, **114**, 243.

27. E. Fermi, *Nuclear Physics*, University of Chicago Press, Chicago, 1950.

28. D. Steel, *Infrared Spectroscopy: Theory*, in Handbook of Vibrational Spectroscopy, ed. J. M. Chalmers and P. R. Griffiths, John Wiley & Sons, Chichester, 2002, vol. 1, p. 44.

29. H. Matsuura and H. Yoshida, *Calculation of Vibrational Frequencies by Hartree–Fock-based and Density Functional Theory*, in Handbook of Vibrational Spectroscopy, ed. J. M. Chalmers and P. R. Griffiths, John Wiley & Sons, Chichester, 2002, vol. 3, p. 2012.

30. R. Dovesi, B. Civalleri, R. Orlando, C. Roetti and V. R. Saunders, *Rev. Comput. Chem.*, 2005, **21**, 1.

31. S. Bordiga, E. Garrone, C. Lamberti, A. Zecchina, C. O. Arean, V. B. Kazansky and L. M. Kustov, *J. Chem. Soc., Faraday Trans.*, 1994, **90**, 3367.

32. C. Lamberti, S. Bordiga, F. Geobaldo, A. Zecchina and C. Otero Aréan, *J. Chem. Phys.*, 1995, **103**, 3158.
33. A. Zecchina and C. O. Arean, *Chem. Soc. Rev.*, 1996, **25**, 187.
34. A. Zecchina, C. Lamberti and S. Bordiga, *Catal. Today*, 1998, **41**, 169.
35. A. Zecchina, D. Scarano, S. Bordiga, G. Spoto and C. Lamberti, *Adv. Catal.*, 2001, **46**, 265.
36. E. Payen, J. Grimblot, J. C. Lavallay, M. Daturi and F. Maugé, *Vibrational Spectroscopy in the Study of Oxide (Excluding Zeolites) and Sulfide Catalysts*, in Handbook of Vibrational Spectroscopy, ed. J. M. Chalmers and P. R. Griffiths, John Wiley & Sons, Chichester, 2002, vol. 4, p. 3005.
37. A. Zecchina, G. Spoto and S. Bordiga, *Vibrational Spectroscopy of Zeolites*, in Handbook of Vibrational Spectroscopy, ed. J. M. Chalmers and P. R. Griffiths, John Wiley & Sons, Chichester, 2002, vol. 4, p. 3042.
38. G. Spoto, E. Gribov, G. Ricchiardi, A. Damin, D. Scarano, S. Bordiga, C. Lamberti and A. Zecchina, *Prog. Surf. Sci.*, 2004, **76**, 71.
39. E. N. Gribov, S. Bertarione, D. Scarano, C. Lamberti, G. Spoto and A. Zecchina, *J. Phys. Chem. B*, 2004, **108**, 16174.
40. C. Lamberti, E. Groppo, G. Spoto, S. Bordiga and A. Zecchina, *Adv. Catal.*, 2007, **51**, 1.
41. A. Vimont, F. Thibault-Starzyk and M. Daturi, *Chem. Soc. Rev.*, 2010, **39**, 4928.
42. C. Lamberti, E. Groppo, A. Zecchina and S. Bordiga, *Chem. Soc. Rev.*, 2010, **39**, 4951.
43. L. Valenzano, B. Civalleri, S. Chavan, S. Bordiga, M. H. Nilsen, S. Jakobsen, K. P. Lillerud and C. Lamberti, *Chem. Mater.*, 2011, **23**, 1700.
44. L. Valenzano, J. G. Vitillo, S. Chavan, B. Civalleri, F. Bonino, S. Bordiga and C. Lamberti, *Catal. Today*, 2012, **182**, 67.
45. S. Chavan, J. G. Vitillo, D. Gianolio, O. Zavorotynska, B. Civalleri, S. Jakobsen, M. H. Nilsen, L. Valenzano, C. Lamberti, K. P. Lillerud and S. Bordiga, *Phys. Chem. Chem. Phys.*, 2012, **14**, 1614.
46. F. Geobaldo, C. Lamberti, G. Ricchiardi, S. Bordiga, A. Zecchina, G. T. Palomino and C. O. Arean, *J. Phys. Chem.*, 1995, **99**, 11167.
47. G. L. Marra, A. N. Fitch, A. Zecchina, G. Ricchiardi, M. Salvalaggio, S. Bordiga and C. Lamberti, *J. Phys. Chem. B*, 1997, **101**, 10653.
48. N. L. Rosi, J. Eckert, M. Eddaoudi, D. T. Vodak, J. Kim, M. O'Keeffe and O. M. Yaghi, *Science*, 2003, **300**, 1127.
49. J. L. C. Rowsell, J. Eckert and O. M. Yaghi, *J. Am. Chem. Soc.*, 2005, **127**, 14904.
50. Y. L. Liu, J. F. Eubank, A. J. Cairns, J. Eckert, V. C. Kravtsov, R. Luebke and M. Eddaoudi, *Angew. Chem., Int. Edit.*, 2007, **46**, 3278.
51. F. M. Mulder, T. J. Dingemans, H. G. Schimmel, A. J. Ramirez-Cuesta and G. J. Kearley, *Chem. Phys.*, 2008, **351**, 72.
52. C. M. Brown, Y. Liu, T. Yildirim, V. K. Peterson and C. J. Kepert, *Nanotechnology*, 2009, **20**, 204025.

53. L. Z. Kong, G. Roman-Perez, J. M. Soler and D. C. Langreth, *Phys. Rev. Lett.*, 2009, **103**, 096103.
54. S. A. FitzGerald, J. Hopkins, B. Burkholder, M. Friedman and J. L. C. Rowsell, *Phys. Rev. B*, 2010, **81**, 104305.
55. F. M. Mulder, B. Assfour, J. Huot, T. J. Dingemans, M. Wagemaker and A. J. Ramirez-Cuesta, *J. Phys. Chem. C*, 2010, **114**, 10648.
56. N. R. Stuckert, L. F. Wang and R. T. Yang, *Langmuir*, 2010, **26**, 11963.
57. P. D. C. Dietzel, P. A. Georgiev, J. Eckert, R. Blom, T. Strassle and T. Unruh, *Chem. Commun.*, 2010, **46**, 4962.
58. K. Sumida, C. M. Brown, Z. R. Herm, S. Chavan, S. Bordiga and J. R. Long, *Chem. Commun.*, 2011, **47**, 1157.
59. S. H. Yang, S. K. Callear, A. J. Ramirez-Cuesta, W. I. F. David, J. L. Sun, A. J. Blake, N. R. Champness and M. Schroder, *Faraday Discuss.*, 2011, **151**, 19.
60. L. F. Wang, N. R. Stuckert, H. Chen and R. T. Yang, *J. Phys. Chem. C*, 2011, **115**, 4793.
61. D. J. Tranchemontagne, K. S. Park, H. Furukawa, J. Eckert, C. B. Knobler and O. M. Yaghi, *J. Phys. Chem. C*, 2012, **116**, 13143.
62. I. Matanovic, J. L. Belof, B. Space, K. Sillar, J. Sauer, J. Eckert and Z. Bacic, *J. Chem. Phys.*, 2012, **137**, 014701.
63. M. Z. Xu, Y. S. Elmatad, F. Sebastianelli, J. W. Moskowitz and Z. Bacic, *J. Phys. Chem. B*, 2006, **110**, 24806.
64. P. D. C. Dietzel, B. Panella, M. Hirscher, R. Blom and H. Fjellvag, *Chem. Commun.*, 2006, 959.
65. J. A. Groves, S. R. Miller, S. J. Warrender, C. Mellot-Draznieks, P. Lightfoot and P. A. Wright, *Chem. Commun.*, 2006, 3305.
66. E. Albanese, B. Civalleri, M. Ferrabone, F. Bonino, S. Galli, A. Maspero and C. Pettinari, *J. Mater. Chem.*, 2012, **22**, 22592.
67. C. Serre, F. Millange, C. Thouvenot, M. Nogues, G. Marsolier, D. Louer and G. Ferey, *J. Am. Chem. Soc.*, 2002, **124**, 13519.
68. J. H. Cavka, S. Jakobsen, U. Olsbye, N. Guillou, C. Lamberti, S. Bordiga and K. P. Lillerud, *J. Am. Chem. Soc.*, 2008, **130**, 13850.
69. F. Bonino, S. Chavan, J. G. Vitillo, E. Groppo, G. Agostini, C. Lamberti, P. D. C. Dietzel, C. Prestipino and S. Bordiga, *Chem. Mater.*, 2008, **20**, 4957.
70. S. Chavan, J. G. Vitillo, E. Groppo, F. Bonino, C. Lamberti, P. D. C. Dietzel and S. Bordiga, *J. Phys. Chem. C*, 2009, **113**, 3292.
71. S. R. Miller, G. M. Pearce, P. A. Wright, F. Bonino, S. Chavan, S. Bordiga, I. Margiolaki, N. Guillou, G. Ferey, S. Bourrelly and P. L. Llewellyn, *J. Am. Chem. Soc.*, 2008, **130**, 15967.
72. M. Kandiah, M. H. Nilsen, S. Usseglio, S. Jakobsen, U. Olsbye, M. Tilset, C. Larabi, E. A. Quadrelli, F. Bonino and K. P. Lillerud, *Chem. Mater.*, 2010, **22**, 6632.
73. J. R. During, Utility of Isothopic Data, in *Handbook of Vibrational Spectroscopy*, ed. J. M. Chalmers and P. R. Griffiths, John Wiley & Sons, Chichester, 2002, vol. 3, p. 1935.

74. A. D. Burrows, *Post-synthesis modifications of MOFs*, in Metal-Organic Frameworks as Heterogeneous Catalysts, ed. F. X. Llabrés i Xamena and J. Gascón, RSC, Cambridge, 2013, ch. 3.

75. S. Chavan, J. G. Vitillo, M. J. Uddin, F. Bonino, C. Lamberti, E. Groppo, K. P. Lillerud and S. Bordiga, *Chem. Mater.*, 2010, **22**, 4602.

76. M. Kandiah, S. Usseglio, S. Svelle, U. Olsbye, K. P. Lillerud and M. Tilset, *J. Mater. Chem.*, 2010, **20**, 9848.

77. J. Hafizovic, M. Bjorgen, U. Olsbye, P. D. C. Dietzel, S. Bordiga, C. Prestipino, C. Lamberti and K. P. Lillerud, *J. Am. Chem. Soc.*, 2007, **129**, 3612.

78. W. O. George and R. Lewis, *Hydrogen Bonding*, in Handbook of Vibrational Spectroscopy, ed. J. M. Chalmers and P. R. Griffiths, John Wiley & Sons, Chichester, 2002, vol. 3, p. 1919.

79. C. Pazé, S. Bordiga, C. Lamberti, M. Salvalaggio, A. Zecchina and G. Bellussi, *J. Phys. Chem. B*, 1997, **101**, 4740.

80. N. Masciocchi, S. Galli, V. Colombo, A. Maspero, G. Palmisano, B. Seyyedi, C. Lamberti and S. Bordiga, *J. Am. Chem. Soc.*, 2010, **132**, 7902.

81. L. Mino, V. Colombo, J. G. Vitillo, C. Lamberti, S. Bordiga, E. Gallo, P. Glatzel, A. Maspero and S. Galli, *Dalton Trans.*, 2012, **41**, 4012.

82. U. Ravon, G. Chaplais, C. Chizallet, B. Seyyedi, F. Bonino, S. Bordiga, N. Bats and D. Farrusseng, *ChemCatChem*, 2010, **2**, 1235.

83. A. D. Wiersum, E. Soubeyrand-Lenoir, Q. Yang, B. Moulin, V. Guillerm, M. Ben Yahia, S. Bourrelly, A. Vimont, S. Miller, C. Vagner, M. Daturi, G. Clet, C. Serre, G. Maurin and P. L. Llewellyn, *Chem.–Asian J.*, 2011, **6**, 3270.

84. C. Morterra, L. Orio and C. Emanuel, *J. Chem. Soc., Faraday Trans.*, 1990, **86**, 3003.

85. C. Morterra, G. Cerrato, L. Ferroni and L. Montanaro, *Mater. Chem. Phys.*, 1994, **37**, 243.

86. R. Dovesi, R. Orlando, B. Civalleri, C. Roetti, V. R. Saunders and C. M. Zicovich-Wilson, *Z. Kristallogr.*, 2005, **220**, 571.

87. R. Dovesi, V. R. Saunders, R. Roetti, R. Orlando, C. M. Zicovich-Wilson, F. Pascale, B. Civalleri, K. Doll, N. M. Harrison, I. J. Bush, P. D'Arco and M. Llunell, *CRYSTAL09*, University of Torino, Torino, 2009.

88. Animations of anharmonic vibrational modes of hydroxylated and dehydroxylated UiO-66 and UiO-67 are available at the http://www. crystal. unito. it/ vibs/ uio66_hydro/, http:// www. crystal. unito. it/ vibs/ uio66_dehydro/, http://www.crystal.unito.it/vibs/uio67_hydro/ and http:// www.crystal.unito.it/vibs/uio67_dehydro/ websites, respectively, or upon request to bartolomeo.civalleri@unito.it.

89. L. Valenzano, B. Civalleri, K. Sillar and J. Sauer, *J. Phys. Chem. C*, 2011, **115**, 21777.

90. S. Chavan, F. Bonino, J. G. Vitillo, E. Groppo, C. Lamberti, P. D. C. Dietzel, A. Zecchina and S. Bordiga, *Phys. Chem. Chem. Phys.*, 2009, **11**, 9811.

91. H. Knozinger and S. Huber, *J. Chem. Soc., Faraday Trans.*, 1998, **94**, 2047.

92. N. Drenchev, E. Ivanova, M. Mihaylov and K. Hadjiivanov, *Phys. Chem. Chem. Phys.*, 2010, **12**, 6423.

93. C. Volkringer, H. Leclerc, J.-C. Lavalley, T. Loiseau, G. Ferey, M. Daturi and A. Vimont, *J. Phys. Chem. C*, 2012, **116**, 5710.

94. S. H. Strauss, *J. Chem. Soc., Dalton Trans.*, 2000, 1.

95. V. Bolis, A. Barbaglia, S. Bordiga, C. Lamberti and A. Zecchina, *J. Phys. Chem. B*, 2004, **108**, 9970.

96. L. Valenzano, B. Civalleri, S. Chavan, G. T. Palomino, C. Arean and S. Bordiga, *J. Phys. Chem. C*, 2010, **114**, 11185.

97. S. Chavan, *Characterization of Metal-organic Frameworks for Gas Storage and Catalysis applications*, PhD thesis in Materials Science, University of Turin (I), 2010.

98. S. Bordiga, L. Regli, F. Bonino, E. Groppo, C. Lamberti, B. Xiao, P. S. Wheatley, R. E. Morris and A. Zecchina, *Phys. Chem. Chem. Phys.*, 2007, **9**, 2676.

99. B. Panella and M. Hirscher, *Adv. Mater.*, 2005, **17**, 538.

100. A. G. Wong-Foy, A. J. Matzger and O. M. Yaghi, *J. Am. Chem. Soc.*, 2006, **128**, 3494.

101. M. Dincă and J. R. Long, *Angew. Chem., Int. Ed.*, 2008, **47**, 6766.

102. W. Zhou, H. Wu and T. Yildirim, *J. Am. Chem. Soc.*, 2008, **130**, 15268.

103. J. G. Vitillo, L. Regli, S. Chavan, G. Ricchiardi, G. Spoto, P. D. C. Dietzel, S. Bordiga and A. Zecchina, *J. Am. Chem. Soc.*, 2008, **130**, 8386.

104. S. A. FitzGerald, B. Burkholder, M. Friedman, J. B. Hopkins, C. J. Pierce, J. M. Schloss, B. Thompson and J. L. C. Rowsell, *J. Am. Chem. Soc.*, 2011, **133**, 20310.

105. The H_2 molecule is hardly detectable by standard XRPD experiment because of the low X-ray scattering power of the molecule, bearing only 2 electrons. For this reason neutrons are used in such cases instead of X-rays. However the D_2 molecule is preferred instead of the H_2 one, because the hydrogen nucleus (or proton) has a too high incoherent neutron scattering length that increases the background level of the pattern. Fortunately this inconvenience is not present with the deuterium nucleus. See Section 5.2.2 of Chapter 5 of this book.

106. S. Bordiga, J. G. Vitillo, G. Ricchiardi, L. Regli, D. Cocina, A. Zecchina, B. Arstad, M. Biørgen, J. Hafizovic and K. P. Lillerud, *J. Phys. Chem. B*, 2005, **109**, 18237.

107. C. Otero Areán, S. Chavan, C. P. Cabello, E. Garrone and G. Turnes Palomino, *ChemPhysChem*, 2010, **11**, 3237.

108. E. A. Paukshtis and E. N. Yurchenko, *Russ. Chem. Rev.*, 1983, **52**, 242.

109. G. Spoto, E. Gribov, S. Bordiga, C. Lamberti, G. Ricchiardi, D. Scarano and A. Zecchina, *Chem. Commun.*, 2004, 2768.

110. E. Garrone, M. R. Delgado, B. Bonelli and C. Otero Areán, *Phys. Chem. Chem. Phys.*, 2005, **7**, 3519.

111. E. Garrone and C. Otero Areán, *Chem. Soc. Rev.*, 2005, **34**, 846.

112. C. Otero Areán, G. Turnes Palomino, E. Garrone, D. Nachtigallová and P. Nachtigall, *J. Phys. Chem. B*, 2006, **110**, 395.

113. J. Estephane, E. Groppo, J. G. Vitillo, A. Damin, C. Lamberti, S. Bordiga and A. Zecchina, , 2009, **11**, 2218.

114. G. Spoto, J. G. Vitillo, D. Cocina, A. Damin, F. Bonino and A. Zecchina, *Phys. Chem. Chem. Phys.*, 2007, **9**, 4992.

115. K. Sillar, A. Hofmann and J. Sauer, *J. Am. Chem Soc.*, 2009, **131**, 4143.

116. J. L. C. Rowsell and O. M. Yaghi, *J. Am. Chem. Soc.*, 2006, **128**, 1304.

117. L. J. Murray, M. Dinca, J. Yano, S. Chavan, S. Bordiga, C. M. Brown and J. R. Long, *J. Am. Chem. Soc.*, 2010, **132**, 7856.

118. E. D. Bloch, L. J. Murray, W. L. Queen, S. Chavan, S. N. Maximoff, J. P. Bigi, R. Krishna, V. K. Peterson, F. Grandjean, G. J. Long, B. Smit, S. Bordiga, C. M. Brown and J. R. Long, *J. Am. Chem. Soc.*, 2011, **133**, 14814.

119. M. Marcz, R. E. Johnsen, P. D. C. Dietzel and H. Fjellvag, *Microporous Mesoporous Mater.*, 2012, **157**, 62.

120. J. G. Vitillo, E. Groppo, S. Bordiga, S. Chavan, G. Ricchiardi and A. Zecchina, *Inorg. Chem.*, 2009, **48**, 5439.

121. S. Chavan, J. G. Vitillo, C. Larabi, E. Alessandra Quadrelli, P. D. C. Dietzel and S. Bordiga, *Microporous Mesoporous Mater.*, 2012, **157**, 56.

122. S. S. Kaye and J. R. Long, *J. Am. Chem. Soc.*, 2008, **130**, 806.

123. B. N. Figgis and M. A. Hitchman, *Ligand Field Theory and Its Applications*, Wiley, New York, 2000.

124. C. T. Au, A. F. Carley and M. W. Roberts, *Int. Rev. Phys. Chem.*, 1986, **5**, 57.

125. I. S. Tilinin, A. Jablonski and W. S. M. Werner, *Prog. Surf. Sci.*, 1996, **52**, 193.

126. A. P. Pijpers and R. J. Meier, *Chem. Soc. Rev.*, 1999, **28**, 233.

127. A. Einstein, *Anal. Phys.*, 1905, **322**, 132. An English translation can be found in: D. Ter Haar, *The Old Quantum Theory*, Pergamon, Oxford, 1967, ch. 3: On a Heuristic Point of View about the Creation and Conversion of Light. The text of the original German version can be obtained at http://onlinelibrary.wiley.com/doi/10.1002/andp.19053220607/ abstract while the English translation is available at http://users.physik. fu-berlin.de/~kleinert/files/eins_lq.pdf.

128. M. Salmeron and R. Schlogl, *Surf. Sci. Rep.*, 2008, **63**, 169.

129. A. Knop-Gericke, E. Kleimenov, M. Havecker, R. Blume, D. Teschner, S. Zafeiratos, R. Schlogl, V. I. Bukhtiyarov, V. V. Kaichev, I. P. Prosvirin, A. I. Nizovskii, H. Bluhm, A. Barinov, P. Dudin and M. Kiskinova, *Adv. Catal.*, 2009, **52**, 213.

130. F. Costa, C. J. R. Silva, M. M. M. Raposo, A. M. Fonseca, I. C. Neves, A. P. Carvalho and J. Pires, *Microporous Mesoporous Mater.*, 2004, **72**, 111.

131. X. X. Li, F. Y. Cheng, S. N. Zhang and J. Chen, *J. Power Sources*, 2006, **160**, 542.

132. G. Blanco-Brieva, J. M. Campos-Martin, S. M. Al-Zahrani and J. L. G. Fierro, *Fuel*, 2011, **90**, 190.
133. T. K. A. Hoang, A. Hamaed, G. Moula, R. Aroca, M. Trudeau and D. M. Antonelli, *J. Am. Chem. Soc.*, 2011, **133**, 4955.
134. Y. B. Huang, Z. J. Lin and R. Cao, *Chem.–Eur. J.*, 2011, **17**, 12706.
135. M. Muller, S. Turner, O. I. Lebedev, Y. M. Wang, G. van Tendeloo and R. A. Fischer, *Eur. J. Inorg. Chem.*, 2011, 1876.
136. H. H. Zhao, L. J. Chou and H. L. Song, *React. Kinet. Mech. Catal.*, 2011, **104**, 451.
137. Q. D. Qin, Q. Q. Wang, D. F. Fu and J. Ma, *Chem. Eng. J.*, 2011, **172**, 68.
138. E. V. Ramos-Fernandez, C. Pieters, B. van der Linden, J. Juan-Alcaniz, P. Serra-Crespo, M. Verhoeven, H. Niemantsverdriet, J. Gascon and F. Kapteijn, *J. Catal.*, 2012, **289**, 42.
139. M. M. Zhang, J. C. Guan, B. S. Zhang, D. S. Su, C. T. Williams and C. H. Liang, *Catal. Lett.*, 2012, **142**, 313.
140. Y. B. Huang, S. Y. Gao, T. F. Liu, J. Lu, X. Lin, H. F. Li and R. Cao, *ChemPlusChem*, 2012, **77**, 106.
141. A. Corma, H. Garcıa and F. X. Llabrés i Xamena, *Chem. Rev.*, 2010, **110**, 4606.
142. A. Rossin, B. Di Credico, G. Giambastiani, M. Peruzzini, G. Pescitelli, G. Reginato, E. Borfecchia, D. Gianolio, C. Lamberti and S. Bordiga, *J. Mater. Chem.*, 2012, **22**, 10335.
143. T. Suzuki, G. S. Murugan and Y. Ohishi, *J. Lumines.*, 2005, **113**, 265.
144. R. Ravikumar, J. Yamauchi, A. Chandrasekhar, Y. P. Reddy and P. S. Rao, *J. Mol. Struct.*, 2005, **740**, 169.
145. M. Ciampolini, *Inorg. Chem.*, 1966, **5**, 35.
146. C. A. Kent, D. Liu, T. J. Meyer and W. Lin, *J. Am. Chem. Soc.*, 2012, **134**, 3991.
147. C. C. Wang, C. C. Yang, W. C. Chung, G. H. Lee, M. L. Ho, Y. C. Yu, M. W. Chung, H. S. Sheu, C. H. Shih, K. Y. Cheng, P. J. Chang and P. T. Chou, *Chem.–Eur. J.*, 2011, **17**, 9232.
148. S. Bordiga, C. Lamberti, G. Ricchiardi, L. Regli, F. Bonino, A. Damin, K. P. Lillerud, M. Bjorgen and A. Zecchina, *Chem. Commun.*, 2004, 2300.
149. K. L. Huang and C. W. Hu, *Inorg. Chim. Acta*, 2007, **360**, 3590.
150. L. Wang, M. Yang, G. H. Li, Z. Shi and S. H. Feng, *Inorg. Chem.*, 2006, **45**, 2474.
151. C. K. Brozek and M. Dinca, *Chem. Sci.*, 2012, **3**, 2110.
152. C. D. Wu, A. Hu, L. Zhang and W. B. Lin, *J. Am. Chem. Soc.*, 2005, **127**, 8940.
153. D. N. Dybtsev, A. L. Nuzhdin, H. Chun, K. P. Bryliakov, E. P. Talsi, V. P. Fedin and K. Kim, A, , *Angew. Chem., Int. Ed.*, 2006, **45**, 916.
154. F. X. Llabrés i Xamena, A. Abad, A. Corma and H. Garcia, *J. Catal.*, 2007, **250**, 294.
155. Y. Goto, H. Sato, S. Shinkai and K. Sada, *J. Am. Chem. Soc.*, 2008, **130**, 14354.

156. S. Horike, M. Dinca, K. Tamaki and J. R. Long, *J. Am. Chem. Soc*, 2008, **130**, 5854.

157. M. Savonnet, S. Aguado, U. Ravon, D. Bazer-Bachi, V. Lecocq, N. Bats, C. Pinel and D. Farrusseng, *Green Chem.*, 2009, **11**, 1729.

158. J. Lee, O. K. Farha, J. Roberts, K. A. Scheidt, S. T. Nguyen and J. T. Hupp, *Chem. Soc. Rev.*, 2009, **38**, 1450.

159. L. Ma, C. Abney and W. Lin, *Chem. Soc. Rev.*, 2009, **38**, 1248.

160. C. G. Silva, I. Luz, F. X. Llabrés i Xamena, A. Corma and H. Garcia, *Chem.–Eur. J.*, 2010, **16**, 11133.

161. M. Ranocchiari and J. A. van Bokhoven, *Phys. Chem. Chem. Phys.*, 2011, **13**, 6388.

162. M. Yoon, R. Srirambalaji and K. Kim, *Chem. Rev.*, 2012, **112**, 1196.

163. C. Prestipino, L. Regli, J. G. Vitillo, F. Bonino, A. Damin, C. Lamberti, A. Zecchina, P. L. Solari, K. O. Kongshaug and S. Bordiga, *Chem. Mater.*, 2006, **18**, 1337.

164. K. C. Szeto, K. P. Lillerud, M. Tilset, M. Bjorgen, C. Prestipino, A. Zecchina, C. Lamberti and S. Bordiga, *J. Phys. Chem. B*, 2006, **110**, 21509.

165. K. C. Szeto, C. Prestipino, C. Lamberti, A. Zecchina, S. Bordiga, M. Bjorgen, M. Tilset and K. P. Lillerud, *Chem. Mater.*, 2007, **19**, 211.

166. S. Bordiga, F. Bonino, K. P. Lillerud and C. Lamberti, *Chem. Soc. Rev.*, 2010, **39**, 4885.

167. E. Borfecchia, S. Maurelli, D. Gianolio, E. Groppo, M. Chiesa, F. Bonino and C. Lamberti, *J. Phys. Chem. C*, 2012, **116**, 19839.

168. The white line is a terminology kept from the pioneering X-ray absorption experiments, where photographic plates were used as detectors (see *e.g.* J. D. Hanawalt, *Phys. Rev.* 1931, **37**, 715). It refers to the first resonance after the edge, where the sample absorption is maximum, and thus the photographic plates were less impressed, keeping a white colour.

169. A. L. Ankudinov and J. J. Rehr, *Phys. Rev. B*, 2000, **62**, 2437.

170. S. S. Y. Chui, S. M. F. Lo, J. P. H. Charmant, A. G. Orpen and I. D. Williams, *Science*, 1999, **283**, 1148.

171. G. Smolentsev and A. Soldatov, *J. Synchrot. Radiat.*, 2006, **13**, 19.

172. E. Groppo, C. Prestipino, C. Lamberti, P. Luches, C. Giovanardi and F. Boscherini, *J. Phys. Chem. B*, 2003, **107**, 4597.

173. E. Groppo, S. Bertarione, F. Rotunno, G. Agostini, D. Scarano, R. Pellegrini, G. Leofanti, A. Zecchina and C. Lamberti, *J. Phys. Chem. C*, 2007, **111**, 7021.

174. R. Pellegrini, G. Leofanti, G. Agostini, L. Bertinetti, S. Bertarione, E. Groppo, A. Zecchina and C. Lamberti, *J. Catal.*, 2009, **267**, 40.

175. R. Pellegrini, G. Leofanti, G. Agostini, E. Groppo, M. Rivallan and C. Lamberti, *Langmuir*, 2009, **25**, 6476.

176. G. Agostini, R. Pellegrini, G. Leofanti, L. Bertinetti, S. Bertarione, E. Groppo, A. Zecchina and C. Lamberti, *J. Phys. Chem. C*, 2009, **113**, 10485.

177. G. Agostini, E. Groppo, A. Piovano, R. Pellegrini, G. Leofanti and C. Lamberti, *Langmuir*, 2010, **26**, 11204.

178. D. Scarano, S. Bordiga, C. Lamberti, G. Ricchiardi, S. Bertarione and G. Spoto, *Appl. Catal. A*, 2006, **307**, 3.

179. G. Agostini, S. Usseglio, E. Groppo, M. J. Uddin, C. Prestipino, S. Bordiga, A. Zecchina, P. L. Solari and C. Lamberti, *Chem. Mater.*, 2009, **21**, 1343.

180. E. Groppo, W. Liu, O. Zavorotynska, G. Agostini, G. Spoto, S. Bordiga, C. Lamberti and A. Zecchina, *Chem. Mater.*, 2010, **22**, 2297.

181. F. P. J. M. Kerkhof and J. A. Moulijn, *J. Phys. Chem.*, 1979, **83**, 1612.

182. D. M. Hercules, A. Proctor and M. Houalla, *Acc. Chem. Res.*, 1994, **27**, 387.

183. S. L. Tait, Y. Wang, G. Costantini, N. Lin, A. Baraldi, F. Esch, L. Petaccia, S. Lizzit and K. Kern, *J. Am. Chem. Soc.*, 2008, **130**, 2108.

184. F. A. Debruijn, G. B. Marin, J. W. Niemantsverdriet, W. H. M. Visscher and J. A. R. Vanveen, *Surf. Interface Anal.*, 1992, **19**, 537.

185. A. Troupis, A. Hiskia and E. Papaconstantinou, *Angew. Chem., Int. Ed.*, 2002, **41**, 1911.

186. E. V. Ramos-Fernandez, A. F. P. Ferreira, A. Sepulveda-Escribano, F. Kapteijn and F. Rodriguez-Reinoso, *J. Catal.*, 2008, **258**, 52.

187. B. Fromme, *d-d Excitations in Transition-Metal Oxides A Spin-Polarized Electron Energy-Loss Spectroscopy (SPEELS) Study*, Springer-Verlag, Berlin, 2001.

188. P. Glatzel and U. Bergmann, *Coord. Chem. Rev.*, 2005, **249**, 65.

189. P. Glatzel, M. Sikora, G. Smolentsev and M. Fernandez-Garcia, *Catal. Today*, 2009, **145**, 294.

190. J. Singh, C. Lamberti and J. A. van Bokhoven, *Chem. Soc. Rev.*, 2010, **39**, 4754.

191. S. Huotari, T. Pylkkanen, G. Vanko, R. Verbeni, P. Glatzel and G. Monaco, *Phys. Rev. B*, 2008, **78**, 041102.

192. E. Gallo, P. Glatzel and C. Lamberti, *Inorg. Chem.*, 2013, **52**, 5633.

CHAPTER 5

Characterization of MOFs. 2. Long and Local Range Order Structural Determination of MOFs by Combining EXAFS and Diffraction Techniques

ELISA BORFECCHIA,[a] DIEGO GIANOLIO,[a,b] GIOVANNI AGOSTINI,[a] SILVIA BORDIGA[a] AND CARLO LAMBERTI*[a]

[a] Department of Chemistry, NIS Centre of Excellence and INSTM Reference Center, Via Giuria 7, University of Turin 10125 Torino, Italy; [b] Diamond Light Source Ltd, Harwell Science and Innovation Campus, Didcot, OX11 0DE, United Kingdom
*Email: carlo.lamberti@unito.it

5.1 Introduction

This chapter is divided into two parts. The former, including Sections 5.2, 5.3 and 5.4, provides the basic concepts needed to understand the physics that is behind X-ray and neutron scattering and photoelectron backscattering. This first part has a specific didactic purpose. At the beginning, the scattering process is introduced in a general way and then differentiation between crystalline samples and amorphous samples is made successively, leading to the Bragg equation and to the Debye equation and the Pair Distribution Function

RSC Catalysis Series No. 12
Metal Organic Frameworks as Heterogeneous Catalysts
Edited by Francesc X. Llabrés i Xamena and Jorge Gascon
© The Royal Society of Chemistry 2013
Published by the Royal Society of Chemistry, www.rsc.org

(PDF) approach, respectively. The basics of Extended X-ray Absorption Fine Structure (EXAFS) spectroscopy are also reported. The latter part (Section 5.5) includes recent examples from the literature where the concepts described in the first part have been applied to the understanding of the structure of different MOF materials.

The large unit cells, the enormous flexibility and the variation in structural motifs of MOFs represent a big challenge in the characterization of MOF materials, particularly in cases where single crystal diffraction data are not available. Indeed, solving complex structures with powder data, even if of high quality, is far from trivial.[1–5] The selected cases reported in Section 5.5 will show that in cases where only powder diffraction data are available, additional structural information are often mandatory in order to solve the structure. Three additional and complementary techniques will be discussed in this chapter. (i) Neutron powder diffraction (NPD) data allows better defining of the structure of the organic linkers (formed by low-Z elements only) and to obtain data with higher signal/noise in the high q region: $q = 4\pi \sin(\theta)/\lambda$. (ii) Total scattering (PDF approach) sheds light on the structure of amorphous MOFs and provides additional information on crystalline ones. (iii) The element selectivity of EXAFS spectroscopy, performed at the metal K- or L-edges, provides the local structure of the inorganic cornerstones or backbones. There are cases where the inorganic cornerstones do not follow the symmetry of the overall structure. In such cases diffraction techniques will just "see" an average structure, missing the local structure: a lack that may be critical for understanding the specific properties of the material. In such cases PDF and EXAFS spectroscopy can be fundamental in understanding the actual structure of the material as they do not require long range periodicity. In particular, EXAFS is the tool that provides complementary structural information on the inorganic cluster and the way it binds to the ligand. Selected examples will show how EXAFS and/or PDF will be relevant in: (i) confirming the structure obtained from diffraction refinements; (ii) highlighting that the inorganic cornerstone has a lower symmetry with respect to that of the organic framework; (iii) obtaining structural information on MOF subjected to amorphization processes; (iv) obtaining the local structure of the inorganic cluster in the desolvated material after coordination of a probe (or reactant) molecule, including cluster deformation upon molecule coordination and metal–molecule binding distance; (v) evidencing the presence of impurities in form of amorphous extra-phases.

This chapter is not aimed at reporting an exhaustive review on the most relevant papers that have significantly contributed in understanding MOFs structure, so the reported bibliography is by no way comprehensive and has not been chosen on the basis of the scientific impact of the papers. The chapter is instead aimed at providing general guidelines and to explain the information that can be obtained using the different structural techniques and to stress their complementarity. As detailed in chapter 4 of this book,[6] vibrational and electronic spectroscopies are also extremely informative. In addition, although not directly described in this chapter, theoretical calculations represent an important support to the experimental techniques, allowing checking of, for example, the stability of a structure inferred from Rietveld refinement. For a detailed description of density functional theory (DFT) methods applied to MOFs materials the reader is referred to chapter 6 of the present book.[7]

5.2 X-Ray and Neutron Scattering: Basic Background

5.2.1 X-Ray Scattering: Theoretical Background

In this section, first the physics of the X-ray elastic scattering process will be briefly discussed, showing how to derive very general equations, valid both for crystalline materials and less ordered systems. Subsequently, the peculiarities of the X-ray scattering both from perfect crystals and non-crystalline samples, such as amorphous solids and solutions, will be highlighted. The fundamental mathematical instruments employed for data modeling will be therefore separately discussed depending on the ordering level. In particular, the Laue conditions and the structure factor equation will be introduced for crystalline systems, whereas the Debye equation and the radial Pair Distribution Functions (PDF) formalism will be presented in relation to less-ordered materials. With this background it will be possible to investigate not only the large variety of crystalline MOFs but also the fraction of amorphous MOFs obtained after gas absorption or pressure gradients. In addition, crystallization processes, where the MOFs structure progressively emerges from the precursors in the solvent medium, can be monitored.

X-rays are suitable for structural determination because they are photons having a wavelength λ of the same order of the interatomic distance in condensed matter. The photon energy is linked to the photon wavelength (λ) by the relation $E = h\nu = hc/\lambda$, c being the speed of light ($c = 2.9979 \ 10^8 \ ms^{-1}$), h the Planck constant ($h = 6.626 \times 10^{-34}$ J s), so that: $hc = 12.3984 \ \text{Å} \ keV$. As a consequence, the relationship between photon energy (in keV) and the photon wavelength (in Å) is:

$$E = [12.3984(\text{Å keV})]/\lambda \tag{5.1}$$

5.2.1.1 X-Ray Elastic Scattering: Basic Physical Principles.

When an X-ray beam interacts with the sample, a part of the incoming radiation is elastically scattered. Details on the quantum-mechanical theory developed to describe such processes can be found in the original reports by Waller and Wentzel,[8–11] as well as in later specialized literature.[12–14] Hereinafter, only a synthetic discussion of fundamental concepts and key equations is proposed. Once an X-ray plane wave of wavevector $\mathbf{k}$ is impinging on a sample having an electron density $\rho_e(\mathbf{r})$, the amplitude of the scattered X-ray wave $A(\mathbf{q})$ is expressed by eqn (5.2):

$$A(\mathbf{q}) = \int \rho_e(\mathbf{r}) \, exp \, (-i\mathbf{q} \cdot \mathbf{r}) d\mathbf{r} \tag{5.2}$$

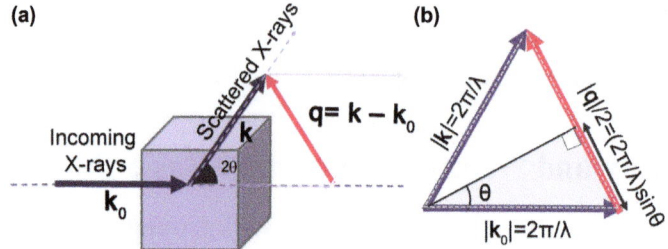

Figure 5.1 Part (a): schematic representation of the elastic scattering process; k_0 and k are the incident and the scattered wave vectors respectively, while $q = k - k_0$ is the vector describing the momentum transfer during the interaction, and 2θ is the scattering angle. In the experiment, a detector is positioned along the k direction to collect scattered intensity $I(q) = |A(q)|^2$, see eqn (5.4). Part (b): geometrical construction for the determination of the modulus of the scattering vector $|q| = 4\pi\sin\theta/\lambda$. Previously unpublished figure.

where the integration on r runs on the space occupied by the sample. In eqn (5.2), $q = k - k_0$ is the wave vector proportional to the momentum transfer (p: $p = h/(2\pi)q$) during the elastic scattering interaction (scattering vector), as schematically represented in Figure 5.1a, where k_0 and k are the incident and the scattered wave vectors respectively, with $|k_0| = |k| = 2\pi/\lambda$. From the simple geometrical construction in Figure 5.1b it is evident that the modulus of the scattering vector is:

$$q = |q| = 4\pi \sin\theta/\lambda \qquad (5.3)$$

The X-ray scattering amplitude $A(q)$ and the sample electron density $\rho_e(r)$, expressed as a function of the 3D coordinate r, are related by a Fourier transform, *i.e.* eqn (5.2). This relation is of key importance and shows how the X-ray scattering signal is intimately dependent on the sample structure. However, we are experimentally limited to measure only the square modulus of the scattering amplitude $A(q)$, that is the scattered intensity $I(q)$ expressed by eqn (5.4):

$$I(q) = |A(q)|^2 = \left| \int \rho_e(r) exp\left(-iq \cdot r\right) dr \right|^2 \qquad (5.4)$$

As will be discussed in details in Section 5.2.1.4.4, any information on the X-ray phase is unavoidably lost: this long-standing complication in crystallography is well-known as the "phase problem".[15] As a consequence, it is impossible to directly apply the Fourier transform relation between $\rho_e(r)$ and $I(q)$ and therefore determine $\rho_e(r)$ from the measured $I(q)$.

However, we are allowed to express the global electron density $\rho_e(r)$ as a superimposition of the individual atomic electron densities centered in the nuclear positions r_n, eqn (5.5), where ρ_n is the electron density of the nth atom and the vector r describes a generic position from the origin of the reference system.

$$\rho_e(\mathbf{r}) = \sum_n \rho_n(\mathbf{r}) \otimes \delta(\mathbf{r} - \mathbf{r}_n) = \sum_n \rho_n(\mathbf{r} - \mathbf{r}_n) \tag{5.5}$$

where $\otimes$ represents the convolution product of two functions and where δ is the Dirac delta-function. By combining eqn (5.2) and eqn (5.5) it is possible to express the scattering amplitude $A(\mathbf{q})$ in the framework of the $\mathbf{r}_n$ positions as follows:

$$A(\mathbf{q}) = \int \left[\sum_n \rho_n(\mathbf{r} - \mathbf{r}_n) \right] e^{(-i\mathbf{q}\cdot\mathbf{r} - i\mathbf{q}\cdot\mathbf{r}_n + i\mathbf{q}\cdot\mathbf{r}_n)} d\mathbf{r}$$

$$= \sum_n \left[\int \rho_n(\mathbf{r} - \mathbf{r}_n) e^{-i\mathbf{q}\cdot(\mathbf{r} - \mathbf{r}_n)} d(\mathbf{r} - \mathbf{r}_n) \right] e^{-i\mathbf{q}\cdot\mathbf{r}_n} = \sum_n f_n(\mathbf{q}) e^{-i\mathbf{q}\cdot\mathbf{r}_n} \tag{5.6}$$

where the index n runs over all atoms included in the sample (*i.e.* on a number of atoms comparable with the Avogadro number!) and where $f_n(\mathbf{q})$ is the so-called atomic form factor for the nth atom,[13,14] which is the Fourier transform of its electron density, as evidenced in eqn (5.7):

$$f_n(\mathbf{q}) = \int \rho_n(\mathbf{r}) exp(-i\mathbf{q} \cdot \mathbf{r}) d\mathbf{r} \tag{5.7}$$

Please note that eqn (5.7) mirrors for the single atom eqn (5.2), that holds for the whole sample. Considering that the atoms can be approximated as spheres, the atomic form factor $f_n(\mathbf{q})$ can be expressed as a function of the modulus of $\mathbf{q}$, *i.e.* $f_n(q)$. The scattered intensity can be therefore expressed as in eqn (5.8):

$$I(\mathbf{q}) = \left| \sum_n f_n(q) exp(-i\mathbf{q} \cdot \mathbf{r}_n) \right|^2 \tag{5.8}$$

Atomic form factor (also referred to as atomic scattering factor) plays a crucial role in the X-ray diffraction theory. A brief description of their principal properties is therefore proposed in the next section.

5.2.1.2 X-Ray Atomic Form Factors

The scattering properties of a specific atom are determined by the shape of its electron density, *i.e.* by the spatial charge distribution in the atomic orbitals. Here we are discussing the atomic form factor of a single, isolated, atom: the index n in eqn (5.7) can be hence removed in both $f(q)$ and $\rho(r)$. Using Euler's formula[16] we can write:

$$f(q) = \int \rho(\mathbf{r}) exp(-i\mathbf{q} \cdot \mathbf{r}) d\mathbf{r} = \int \rho(\mathbf{r}) cos(\mathbf{q} \cdot \mathbf{r}) d\mathbf{r} - i \int \rho(\mathbf{r}) sin(\mathbf{q} \cdot \mathbf{r}) d\mathbf{r} \tag{5.9}$$

The integrals in eqn (5.9) formally run over the whole space, while actually they run in the small region where the atomic electron density is significantly

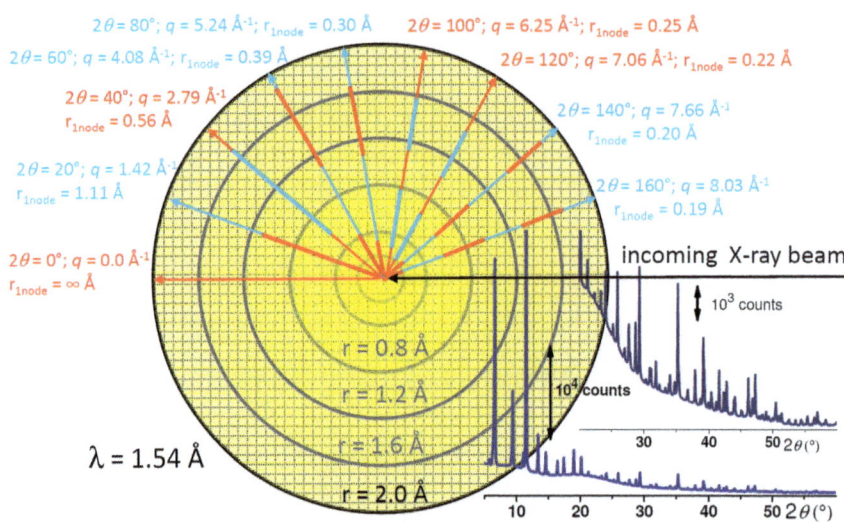

Figure 5.2   Main part: schematic representation of the electron density $\rho(r)$ of a high
Z atom such as Hf or Ta. Concentric circles (drawn at $r = 0.2, 0.4, 0.8, 1.2$,
1.6 and $2.0\,\text{Å}$) represent regions of decreasing $\rho(r)$, supposing $\rho(r) \sim 0$ for
$r > 2\,\text{Å}$. The corresponding atomic form factor f is obtained integrating
$\rho(r)$ modulated by the oscillatory function $\exp(-i\mathbf{q} \cdot \mathbf{r})$ in the spatial region
$r \leq 2\,\text{Å}$, see second term in eqn (5.9). The black arrow represents the
direction of the incoming X-ray beam, while the colored arrows indicate
the sign of $\mathrm{Re}[\exp(-i\mathbf{q} \cdot \mathbf{r})] = \cos(\mathbf{q} \cdot \mathbf{r})$ along the designed scattering
direction: red and blue colors represent positive and negative regions,
respectively. A value $\lambda = 1.54\,\text{Å}$ is assumed to compute the nodes of
$\cos(\mathbf{q} \cdot \mathbf{r})$. For each 2θ direction the first node occurs at $r_{1\text{node}} = (\pi/2)/q$.
The intensity of the yellow color mimics the increased electron density $\rho(r)$
in proximity of the nucleus. Bottom right corner: typical XRPD pattern
(desolvated HKUST-1 MOF collected with $\lambda = 1.54\,\text{Å}$),[19] showing the
typical decrease of the Bragg peak intensity by increasing 2θ (increasing
q). The inset reproduces a magnification of the high 2θ region of the same
diffractogram. See Figure 5.3 for a direct representation of the atomic form
factors f for a selection of atoms and ions. Previously unpublished figure.

different from zero. So, depending on the specific atomic species considered, the
integrals run over a sphere of radius in the 0.3–$3.0\,\text{Å}$ range,[17] see Figure 5.2. As
already mentioned, the electron density $\rho(\mathbf{r})$ for an isolated atom exhibits
spherical symmetry, so that $\rho(-\mathbf{r}) = \rho(\mathbf{r})$.[18] Under such assumption the sinus-FT
in the third hand term of eqn (5.9) is null (as all integrals over the whole space
of an odd function) and $f(q)$ is a real function.

The fact that atoms have spatially extended orbitals instead of point-charges
ones causes a reduction in the coherently scattered intensity. This reduction will
be more efficient the larger either $\mathbf{q}$ or $\mathbf{r}$ become, because the phase shift among
the $d\mathbf{r}$ regions where the integral is performed is determined by the scalar
product $\mathbf{q} \cdot \mathbf{r}$ in the second term of eqn (5.9), as sketched in Figure 5.2. Indeed,
$\cos(\mathbf{q} \cdot \mathbf{r})$ is an oscillatory function that has, in the r-space, a period $2\pi/q$, having

the nodes at $r_{node}(n) = [(2n + 1)\pi/2]/q$. This means that, integrating along r the function $\rho(r) \cos(\mathbf{q} \cdot \mathbf{r})$, the integral receives a positive contribution in the $0 < r < (\pi/2)/q$ region and a negative contribution (that partially cancels the first one) in the $(\pi/2)/q < r < (3\pi/2)/q$ interval. Positive and negative contributions alternate until we reach a region where r is sufficiently large to assure $\rho(r) \approx 0$; the integration process can be stopped there. This situation is represented in Figure 5.2 for an high Z atom, where the cut-off in r-space indicatively occurs at $r = 2$ Å. Reported values refer to an X-ray diffraction experiment performed with Cu Kα radiation ($\lambda = 1.54$ Å). Increasing the scattering angle to 2θ we also progressively increase the modulus of the scattering vector $q = 4\pi\sin\theta/\lambda$, see eqn (5.3). Correspondingly, the period $2\pi/q$ of the $\cos(\mathbf{q} \cdot \mathbf{r})$ function becomes progressively smaller, and the first node occurs at progressively shorter r values: $r_{1node} = (\pi/2)/q$. This means that $\cos(\mathbf{q} \cdot \mathbf{r})$ oscillates in r-space, more and more rapidly upon increasing q (or 2θ), as sketched in Figure 5.2, where red and blue segments represent r-regions where $\cos(\mathbf{q} \cdot \mathbf{r})$ assumes positive and negative values for the different scattering angles. The faster $\cos(\mathbf{q} \cdot \mathbf{r})$ oscillates, the more destructive are the interference in the integrals of eqn (5.9), and the corresponding $f(q)$ value drops down for high q. This is the reason why in XRD the peak intensity progressively drops down upon increasing the scattering vector q, see bottom right inset in Figure 5.2.

Let us consider first the $f(q)$ function behavior at the lower scattering extreme of its domain, *i.e.* $q \to 0$ ($2\theta \to 0$). Here, the electronic cloud scatters the incident X-ray radiation perfectly in phase. $f(q = 0)$ is therefore equal to the total charge of the number of electrons of the atom (atomic number Z) or ion. Indeed, being $\cos(\mathbf{0} \cdot \mathbf{r}) = 1$, the second term of eqn (5.9) simply becomes: $f(0) = \int \rho(\mathbf{r}) d\mathbf{r} = Ze^-$. Let us progressively increase the q value, *e.g.* assuming that the diffraction experiment is performed using Cu Kα radiation ($\lambda = 1.54$ Å). At $2\theta = 20°$, $q = 1.42$ Å^{-1}; at this q value the integral in eqn (5.9) receives a positive contribution up to $r = 1.11$ Å, while in the 1.11–2.00 Å the contribution is negative. Prosecuting the analysis for higher scattering angles, it can be noticed that in the integration range $0 < r < 2$ Å the function $\cos(\mathbf{q} \cdot \mathbf{r})$ shows a number of nodes of: 0, 1, 2, for a scattering angle of $2\theta = 0$, 20 and 40°, respectively; a fourth node occurs at $2\theta = 100°$ and a fifth is observed at $2\theta = 140°$, see Figure 5.2.

Several quantum mechanical methods of increasing sophistication have been employed to evaluate the atomic electron density $\rho(r)$,[20–25] that in turns allowed a better definition of the atomic form factors.[26–31] Tabulated values of $f(q)$ for all chemical elements and some relevant ions can be found in specialized literature.[32] In particular, $f(q)$ is commonly expressed according to eqn (5.10), hence using an approximated 9-parameter model function, introduced by Cromer and Mann.[28] This approach ensures a precision of 10^{-6} in the determination of the atomic form factors, and can be employed up to $q = 4\pi\sin\theta/\lambda \sim 25$ Å^{-1} (*i.e.* up to $\sin\theta/\lambda \sim 2$ Å^{-1}). This q range is able to cover all laboratory experiments. Indeed, at $2\theta = 160°$, the q value reached with Cu ($\lambda = 1.54$ Å), Mo ($\lambda = 0.71$ Å) and Ag ($\lambda = 0.56$ Å) anodes is 8.03, 17.72 and

$22.46 \, \text{Å}^{-1}$, respectively. The coefficients a_i, b_i and c for each chemical element are also tabulated.[32]

$$f(q) = \int \rho(\mathbf{r}) \exp(-i\mathbf{q} \cdot \mathbf{r}) d\mathbf{r} \simeq \int \left[\sum_{i=1}^{4} A_i e^{-B_i r^2} + c\delta(\mathbf{r}) \right] \exp(-i\mathbf{q} \cdot \mathbf{r}) d\mathbf{r}$$

$$= c + \frac{4\pi}{q} \int_0^\infty r \left[\sum_{i=1}^{4} A_i e^{-B_i r^2} \right] \sin(qr) dr = c + \sum_{i=1}^{4} a_i e^{-b_i q^2} \qquad (5.10)$$

where $r = |\mathbf{r}|$ and $c + \Sigma a_i = Ze^-$. The excellent approximation (5.10) is based on the fact that the electron density can be well reproduced by the sum of different Gaussian functions of different standard deviation, defining the different closed shells of the orbitals: $\rho(\mathbf{r}) = \rho(r) = \Sigma_i A_i \exp(-B_i r^2)$. The last equality in (5.10) holds because the Fourier transform of a Gaussian function is still a Gaussian:[33] $FT\{A_i \exp(-B_i r^2)\} = a_i \exp(-b_i q^2)$, where $a_i = (\pi/B_i)^{3/2} A_i$ and where $b_i = (1/4B_i)$. For large Z atoms, the closest shells can be considered sufficiently sharp in r-space that the corresponding Gaussian can be approximated with a Dirac δ-function, which Fourier transform is constant and equal to unit in q-space. A pictorial representation of the Fourier relation between the electron density $\rho(r)$, expressed by the sum of two Gaussian functions of different standard deviation and a Dirac δ-function, and the expression of $f(q)$ according to eqn (5.10) is provided in Figure 5.3a, b.

Figure 5.3c shows the $f(q)$ plots computed using the third term of eqn (5.10), for selected atoms and ions. As discussed above, $f_n(q=0) = 1, 6, 7, 8, 28, 29, 30, 40$ and $72 \, e^-$, for H, C, N, O, Ni, Cu, Zn, Zr and Hf atoms, respectively; *i.e.* $f_n(q=0) = Ze^-$. Although in the very low q-range $f(q)$ maintains values close to the Ze^- (this is true in the small angle regime, where $q < \sim 0.1 \, \text{Å}^{-1}$),[34] a steep decrease with increasing q is always observed for all atoms in the 0.2–2.0 $\text{Å}^{-1} q$-region. The increasing phase differences between the waves scattered by unit volumes of electron density enhance the destructive interference phenomena (see Figure 5.2), rapidly driving $f(q)$ to 0 e^- for the light elements (see the H, C, N and O case in the inset of Figure 5.3c), where the c constant of eqn (5.10) is null. This is not true for the high-Z atoms, which exhibit a constant $f_n(q)$ slope in the whole $4 \, \text{Å}^{-1} < q < 8 \, \text{Å}^{-1}$ region, where: $f_{Zr}(q) \sim 2 \, e^-$ and $f_{Hf}(q) \sim 10 \, e^-$: this corresponds to the c constant in eqn (5.10). This means that the first shell ($1s$ electrons) for Zr and the first 2 shells ($1s$, $2s$ and $2p$ electrons) for Hf are confined in a r-region enough small that, at $q = 8 \, \text{Å}^{-1}$, guarantees $qr \ll \pi/2$, *i.e.* $r \ll 0.2 \, \text{Å}$. Using the values reported by Clementi *et al.*[22] for the radii of maximum charge density we can estimate the spatial extent of the different atomic orbitals. At $q = 8 \, \text{Å}^{-1}$, for zirconium, we obtain $r(1s, Z = 40) = 0.028 \, \text{Å}$ and $\cos(qr) = 0.975$, while for hafnium we have $r(2s, Z = 72) = 0.077 \, \text{Å}$ and $\cos(qr) = 0.816$. These simple considerations qualitatively explain the values of the reported atomic form factors in Figure 5.3: $f_{Zr}(q = 8 \, \text{Å}^{-1}) = 2.07 \, e^-$ and $f_{Hf}(q = 8 \, \text{Å}^{-1}) = 8.58 \, e^-$. For Zr($1s$) $\cos(qr) \sim 1$ and the electrons belonging to the first shell behave as point charges and gives the same contribution to the $f_n(q)$ function in the investigated q domain (the

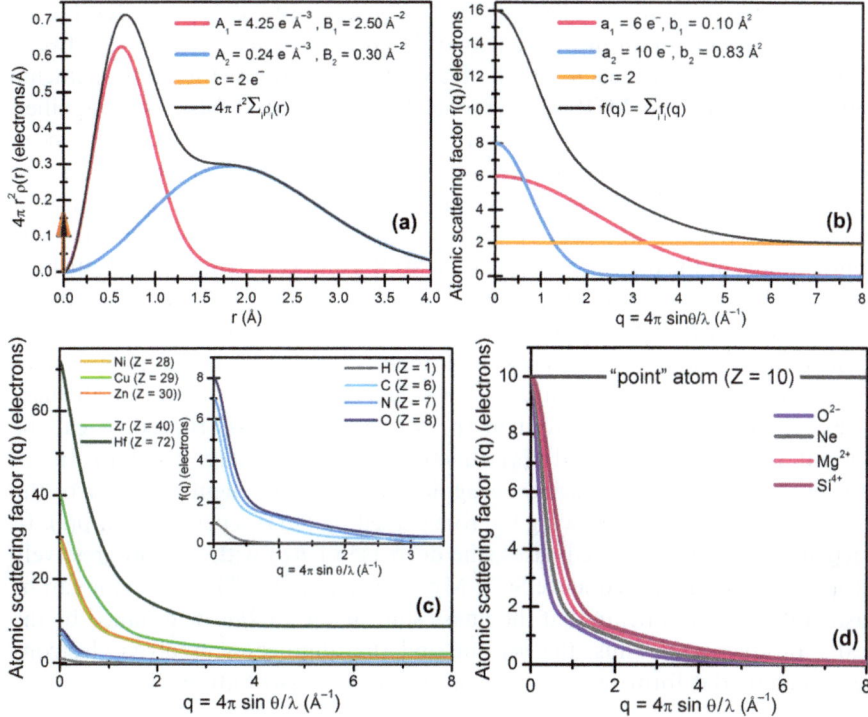

Figure 5.3 Parts (a) and (b): pictorial representation of the Fourier relation between the electron density $\rho(r)$ expressed by the sum of two Gaussian functions of different standard deviation and a Dirac δ-function (part (a), where the quantity $4\pi r^2 \rho(r)$ is plotted), and the expression of $f(q)$ according to eqn (5.10), part (b). Comparing r-space and q-space plots it is evident how the broader Gaussian components in r-space result in sharper Gaussian components in q-space, while the Dirac δ-function in r-space results in a constant function in q-space. Part (c): $f(q)$ functions for four elements commonly present in MOFs linkers, i.e. H, C, N, and O (see also magnified detail in the inset), and for five elements which have been employed as metal centers in MOFs frameworks, i.e. Ni, Cu, Zn, Zr, Hf. Part (d): $f(q)$ functions for three ions, iso-electronic to the Ne atom ($Z = 10$, also reported), i.e. the Mg^{2+} and Si^{4+} cations and the O^{2-} anion. The horizontal grey line $f(q) = 10\,e^-$ highlights how a constant q-independent form factor would be obtained in correspondence of a 10 e^- point charge. Previously unpublished figure.

residual 0.07 e^- represents the contribution of the 38 electrons occupying the higher shells). For Hf($2s$) an higher deviation from the $\cos(qr) \sim 1$ case is observed, and the 8 electrons belonging to the $2s$ and $2p$ orbitals contribute only as $\sim 6.58\ e^-$, i.e. they contribute at $\sim 82\%$ of their total charge (2 e^- coming from the point charge Hf($1s$) orbital).

Note that the rapid decrease of atomic form factors in the high-q range of all atoms is particularly critical when considering the fundamental role of high-q information, *e.g.* to improve the resolution in the reconstruction of the unit cell electron density (*vide infra*, Section 5.2.1.4.3).

The $f(q)$ plots reported in Figure 5.3c demonstrate the reason why X-rays are less sensitive to light atoms, primarily to H, and the reduced contrast between elements adjacent to each other. Here, $f(q)$ functions for four elements commonly present in MOFs linkers, *i.e.* H, C, N and O (see also magnified detail in the inset of the same Figure), and some elements which have been employed as metal centers in MOFs frameworks, *i.e.* Ni, Cu, Zn, Zr, Hf, are reported. The Z-dependent difference in $f(q)$ values between the two groups of elements is striking, and clearly demonstrates the difficulties in locating the lighter atoms of the organic linkers in the MOFs unit cells, and in distinguishing between almost iso-electronic elements, such as C, N, O, or Ni, Cu, Zn, due to the high similarity between their $f(q)$ functions in the whole reported q-range.

In addition, Figure 5.3d reports the $f(q)$ plots for three ions, iso-electronic to the Ne atom, *i.e.* the Mg^{2+} and Si^{4+} cations and the O^{2-} anion: in all these cases, 10 electrons are involved in the scattering process. Indeed, it can be noticed how all the plotted $f(q)$ functions, for $q = 0$, are equal to 10 e^-. However, a trend can be clearly recognized in the subsequent behavior of the different $f(q)$ functions, showing the role played by the r-dependence of $\rho(r)$. In particular, for the proposed cases, the decrease of $f(q)$ with q is progressively steeper considering in sequence Si^{4+}, Mg^{2+}, Ne, and O^{2-}. This behavior can be easily interpreted noticing that the phase shift in eqn (5.9) is determined by the scalar product $\mathbf{q} \cdot \mathbf{r}$. Thus, the q-value and the number of involved electrons being equal, the form factor for a larger charge distribution dies out more rapidly. With respect to the reported examples, cations (Mg^{2+} and Si^{4+}) exhibit a smaller atomic radius with comparison to neutral atoms (Ne), due to the reduced shielding of the nuclear charge, and the consequent enhancement in nuclear attraction towards the remaining electrons. Similarly, a larger radius is found for negatively charged anions (O^{2-}).

Finally, it is worth noting that, for some applications, the inelastic scattering of X-rays has to be accounted. Correspondingly, the definition of the atomic form factor is extended, including a complex anomalous scattering contribution $\Delta f(\lambda) = \Delta f'(\lambda) + i\Delta f''(\lambda)$.[35] The Δf contribution, dependent on the incident X-ray wavelength λ (or energy), is added to the atomic form factors f defined above, with its real and imaginary parts indicating a magnitude variation and a phase shift of the scattered wave, respectively. However, in most cases, anomalous scattering factors $\Delta f'$ and $\Delta f''$ are negligible with comparison to f, as can be noticed from tabulated numerical values.[32]

5.2.1.3   Ordering Levels and Structural Information Level

The physical phenomena hitherto described and the related sets of equations have a general valence, and can be indistinctly applied to any kind of sample: gases, liquids, solids of amorphous or crystalline nature. In the case of crystalline materials (crystallography), the periodic long-range ordering arrangement of atoms inside the crystal lattice allows further elaboration of the general form in eqn (5.8), leading for instance to Laue conditions.[12,14,36] In such a case the elastic scattering process is commonly referred to as diffraction, and the scattered intensity $I(\mathbf{q})$ is characterized by sharp, well-defined Bragg peaks. This case will be discussed in more details in the Section 5.2.1.4.

When the ordering level of the investigated sample is lowered (single molecules in the gas or liquid phase or amorphous materials), the scattering signals are diffused in the entire q-space and, due to random molecular orientation, the 3D information is reduced to 1D, and can be extracted from isotropic scattering patterns. The resulting Debye equation, and the commonly employed radial pair distribution function (PDF) formalism are discussed in Section 5.2.1.5.

A fundamental conceptual difference can be envisaged between the perfectly-ordered crystal case and systems characterized by a lower ordering level. As will be detailed in Section 5.2.1.4, in a diffraction experiment, once the phases and amplitudes of *hkl* reflections (*vide infra* eqn (5.12)) have been obtained, we can directly reconstruct the 3D electron density $\rho(\mathbf{r})$ from the experimental data set.

Conversely, such a direct and univocal relationship (*diffraction pattern ↔ 3D structure*) is no longer valid for non-crystalline systems, due to lowering in information content. In this case, the data analysis strategy requires a conceptual inversion of the previous relation.[37] The relation changes into *hypothetical 3D structures → simulated scattering patterns*. The best-fit of experimental 1D scattering curves then provides the most plausible 3D structure. These considerations emphasize the crucial role of theoretical modeling for the interpretation of scattering data from non-crystalline environments.

5.2.1.4   X-Ray Diffraction from Perfect Crystals

5.2.1.4.1   Laue Equations.   Let us focus the analysis on a perfect crystal, where the unit cell is periodically repeated in the three dimensions of the space. In particular, consider the simple case of a cubic lattice, including N_1, N_2 and N_3 atoms in the three spatial dimensions, respectively. Here, the position $\mathbf{r}_n$ of the nth atom can be expressed in the form $\mathbf{r}_n = n_1\mathbf{a} + n_2\mathbf{b} + n_3\mathbf{c}$, with $(n_1; n_2; n_3)$ integer numbers and $(\mathbf{a}; \mathbf{b}; \mathbf{c})$ primitive vectors of the crystal lattice.

The periodic arrangement of atoms allows further elaboration of the term $|\Sigma_n \exp(-i\mathbf{q}\cdot\mathbf{r}_n)|^2$, appearing in eqn (5.8). Substituting the expression of $\mathbf{r}_n$ previously introduced, we obtain:

$$\left|\sum_n e^{-i\mathbf{q}\cdot\mathbf{r}_n}\right|^2 = \left|\sum_{n_1, n_2, n_3} e^{-i\mathbf{q}\cdot(n_1\mathbf{a}+n_2\mathbf{b}+n_3\mathbf{c})}\right|^2$$

$$= \left|\sum_{n_1=0}^{N_1-1} e^{-i\mathbf{q}\cdot(n_1\mathbf{a})}\right|^2 \cdot \left|\sum_{n_2=0}^{N_2-1} e^{-i\mathbf{q}\cdot(n_2\mathbf{b})}\right|^2 \cdot \left|\sum_{n_3=0}^{N_3-1} e^{-i\mathbf{q}\cdot(n_3\mathbf{c})}\right|^2 \quad (5.11)$$

If the system contains enough atoms and the wavelength of the incident X-ray beam is smaller enough with comparison to interatomic distances, it can be demonstrated that each of the factors of general form $|\Sigma_j e^{-i\mathbf{q}\cdot(jl)}|^2$ (with $j = n_1, n_2, n_3$ and $\mathbf{l} = \mathbf{a}, \mathbf{b}, \mathbf{c}$) in eqn (5.11) is null unless scalar product $\mathbf{q}\cdot\mathbf{l}$ is equal to an integer multiple of 2π. Hence, the coherent scattered intensity is $\neq 0$ only along specific directions, *i.e.* particular values of $\mathbf{q}$, depending on atomic positions within the crystal. This condition on $\mathbf{q}$-values allows constructive interference, and therefore observable scattered intensity is expressed by the following Laue equations:[36]

$$\mathbf{q} \cdot \mathbf{a} = h2\pi, \mathbf{q} \cdot \mathbf{b} = k2\pi, \mathbf{q} \cdot \mathbf{c} = l2\pi \tag{5.12}$$

where (h, k, l) are integer numbers (also known as Miller indices), allowed to vary in the $[-\infty, +\infty]$ range.

Laue equations can be read in several different ways, hereinafter summarized (further details and a more comprehensive mathematical description can be found in the specialized literature).[14,38]

The Miller indexes $(h\ k\ l)$ appearing in eqn (5.12) identify a family of parallel lattice planes, with inter-planar distance d_{hkl}. In particular, the indexes denote a plane that intercepts the three points $\mathbf{a}/h$, $\mathbf{b}/k$, and $\mathbf{c}/l$, or some multiple therein. Furthermore, a general solution that simultaneously verifies the Laue conditions can be expressed in the form:

$$\mathbf{q} = h\mathbf{a}^* + k\mathbf{b}^* + l\mathbf{c}^* = \mathbf{g} \tag{5.13}$$

where $(\mathbf{a}^*, \mathbf{b}^*, \mathbf{c}^*)$ define a new set of lattice vectors, related to the $(\mathbf{a}, \mathbf{b}, \mathbf{c})$ vectors by the following relations:

$$\mathbf{a}^* = 2\pi \frac{\mathbf{b} \wedge \mathbf{c}}{\mathbf{a} \cdot (\mathbf{b} \wedge \mathbf{c})}, \mathbf{b}^* = 2\pi \frac{\mathbf{c} \wedge \mathbf{a}}{\mathbf{b} \cdot (\mathbf{c} \wedge \mathbf{a})}, \mathbf{c}^* = 2\pi \frac{\mathbf{a} \wedge \mathbf{b}}{\mathbf{c} \cdot (\mathbf{a} \wedge \mathbf{b})} \tag{5.14}$$

where $\mathbf{a} \cdot (\mathbf{b} \wedge \mathbf{c}) = \mathbf{b} \cdot (\mathbf{c} \wedge \mathbf{a}) = \mathbf{c} \cdot (\mathbf{a} \wedge \mathbf{b})$ represent the cell volume V_{cell}.

This new lattice is referred to as reciprocal lattice. Each point in the reciprocal lattice corresponds to a set of lattice planes of index $(h\ k\ l)$ in the direct lattice. Furthermore, the direction of each reciprocal lattice vector $\mathbf{g}$ is normal to a family of real-space lattice planes, while its magnitude is proportional to the reciprocal of the interplanar spacing, *i.e.* $|\mathbf{g}| = 2\pi / d_{hkl}$. Laue conditions can be therefore reformulated in the implication: $I(\mathbf{q}) \neq 0 \leftrightarrow \mathbf{q} \equiv \mathbf{g}$, *i.e.* a non-zero diffracted intensity is measured only when the scattering vector $\mathbf{q}$ coincides with a vector of the reciprocal lattice.

Finally, notice that the Laue conditions and the well-known Bragg law[39–43] are just two different perspectives on the same physical phenomenon. Their equivalency can be easily demonstrated considering that:

(i) the general expression for the modulus of the scattering vector previously introduced, see Figure 5.1b and eqn (5.3), is given by $|\mathbf{q}| = 4\pi \sin\theta/\lambda$;
(ii) the modulus of a generic reciprocal lattice vector is given by $|\mathbf{g}| = 2\pi/d_{hkl}$.

Then, starting from the above mentioned formulation of the Laue conditions $\mathbf{q} = \mathbf{g}$, the simple passages eqn (5.15) reported below yield the Bragg law:[39]

$$\mathbf{q} = \mathbf{g} \to |\mathbf{q}| = |\mathbf{g}| \to \frac{4\pi \sin\theta}{\lambda} = \frac{2\pi}{d_{hkl}} \to 2d_{hkl} \sin\theta = \lambda \tag{5.15}$$

5.2.1.4.2 The Structure Factor. The crystallographic unit cell is the smallest unit by which the periodic order in the crystal is repeated: once the structural information is obtained within the cell, it can be simply extended to the whole crystal by periodical replication. Consequently, hereinafter the analysis will focus on the atoms included in the unit cell.

In particular, let us introduce the so-called structure factor $F(\mathbf{q})$, defined as:

$$F(\mathbf{q}) = \sum_n f_n(q)e^{-i\mathbf{q}\cdot\mathbf{r}_n} \tag{5.16}$$

where $f_n(q)$ is the atomic form factor of the n-atom in the unit cell, see eqn (5.7), and where the index n now runs over the atoms within the unit cell. Please note the huge difference between eqn (5.16) and eqn (5.8), where n was running over all the atoms of the sample! $F(\mathbf{q})$ includes all the effect of internal interference due to the geometric phase relationships between the atoms in the crystal unit cell, with each nth atomic contribution weighted by its form factor $f_n(q)$.

Once the atomic form factors $f_n(q)$ for all the involved elements are known, the positions $\mathbf{r}_n$ of the N atoms within the unit cell can be specified by their dimensionless fractional coordinates (x_n, y_n, z_n), *i.e.* $\mathbf{r}_n = x_n\mathbf{a} + y_n\mathbf{b} + z_n\mathbf{c}$. When considering the case of detectable diffracted intensity, the Laue condition, eqn (5.13), can be employed to simplify the $\mathbf{q}\cdot\mathbf{r}_n$ phase factor, as follows:

$$\mathbf{q}\cdot\mathbf{r}_n = (h\mathbf{a}^* + k\mathbf{b}^* + l\mathbf{c}^*)(x_n\mathbf{a} + y_n\mathbf{b} + z_n\mathbf{c}) = 2\pi(hx_n + ky_n + lz_n) \tag{5.17}$$

as, according to eqn (5.14), $\mathbf{a}^*\cdot\mathbf{a} = \mathbf{b}^*\cdot\mathbf{b} = \mathbf{c}^*\cdot\mathbf{c} = 2\pi$ and $\mathbf{a}^*\cdot\mathbf{b} = \mathbf{a}^*\cdot\mathbf{c} = \mathbf{b}^*\cdot\mathbf{c} = 0$. Substituting (5.17) in (5.16), the structure factor, now indexed using the Miller index $(h\ k\ l)$, can be consequently expressed as in eqn (5.18), commonly referred to as the structure factor equation:

$$F_{h,k,l} = \sum_{n=1}^{N} f_n e^{-2\pi i(hx_n + ky_n + lz_n)} \tag{5.18}$$

where the atomic form factors f_n have to be computed for all atoms at the q value corresponding to the scattering conditions of the $(h\ k\ l)$ plane: $q = 2\pi/d_{hkl}$, see eqn (5.15). The square modulus of the structure factor is proportional to the intensity $I_{h,k,l}$ of the reflection measured in the diffraction pattern, corresponding to the Bragg condition being satisfied for the particular $(h\ k\ l)$ plane:

$$I_{h,k,l} \propto |F_{h,k,l}|^2 \tag{5.19}$$

The use of the proportionality symbol instead of the simple equality in eqn (5.19) is due the presence of other contributions to the reflection intensity, such as the Lorentz-polarization factor, the Debye-Waller factor, and the absorption factor, not discussed here for the sake of brevity. The same holds for instrumental parameters such as incoming X-ray beam intensity and detector efficiency and angular acceptance. It is very important to notice that the structure factor $F_{h,k,l}$ is a complex quantity, which can be expressed in terms of amplitude and phase, as evidenced in eqn (5.20), and graphically represented as a vector in the Argand plane:

$$F_{h,k,l} = |F_{h,k,l}|e^{i\phi(h,k,l)} = \sqrt{I_{h,k,l}}e^{i\phi(h,k,l)} \tag{5.20}$$

where $\phi(h, k, l)$ is the phase associated with the point in reciprocal space of coordinates (h, k, l).

5.2.1.4.3 Inversion of the Structure Factor Equation and Electron Density Reconstruction.

The end-task in diffraction crystallography is to get reliable information on the electron density function within the unit cell. For this purpose, using the Fourier relation between atomic form factor and atomic electron density, see eqn (5.7), we can rewrite eqn (5.18) as:

$$F_{h,k,l} = \int_{cell} \rho(x, y, z)e^{-2\pi i(hx+ky+lz)} dxdydz \tag{5.21}$$

where $\rho(x,y,z) = \Sigma_n \rho_n(x,y,z)$ represents the electron density value in the point of coordinates (x, y, z) within the unit cell volume (obtained summing the atomic electron densities $\rho_n(x, y, z)$ within the unit cell), where the integration is performed. Assume, for the moment, that the structure factor $F_{h,k,l}$ is fully determined, both in its amplitude and phase (this assumption is not fully correct, as will be discussed in Section 5.2.1.4.4, introducing the "phase problem"). It is hence possible to invert eqn (5.21) as follows:

$$\rho(x, y, z) = \frac{1}{V_{cell}} \sum_{h,k,l} F_{h,k,l} e^{2\pi i(hx+ky+lz)} \tag{5.22}$$

where V_{cell} is the unit cell volume, assuring that both hands of eqn (5.22) are measured in $e^-\text{Å}^{-3}$.

Due to the discrete nature of the structure factors set collected in a diffraction experiment (one structure factor $F_{h,k,l}$ for each experimentally determined and indexed *hkl* reflection, with a discrete ensemble of reflections), in eqn (5.22) the integral is replaced by the sum over all the collected (h, k, l) values, *i.e.* a discrete Fourier transform[44] is performed.

What we obtain is therefore a discrete approximation of the continuous electron density function. Consequently, the summation has to be performed on the finest possible grid of (x, y, z) points to obtain a smooth electron density distribution in the unit cell. Subsequently, connecting the points with equal electron density values, electron density isosurfaces can be obtained (see Figure 5.4). The reconstructed electron density is then fitted to the effective atomic model, typically by least-squares minimization procedures.

Eqn (5.21)–(5.22) highlight some fundamental issues, hereinafter briefly discussed. First, the "holistic" character of the diffraction technique, directly deriving from the Fourier relation between the scattering amplitude and the spatial distribution of the scatterers, is evidenced. Indeed, to determine the electron density in an individual (x, y, z) point within the unit cell, the whole diffraction dataset, *i.e.* $F_{h,k,l}$ structure factors for all the measured reflections, is

decreasing d_{min} or increasing q_{max} $\Longrightarrow$ increasing $\rho(r)$ resolution level

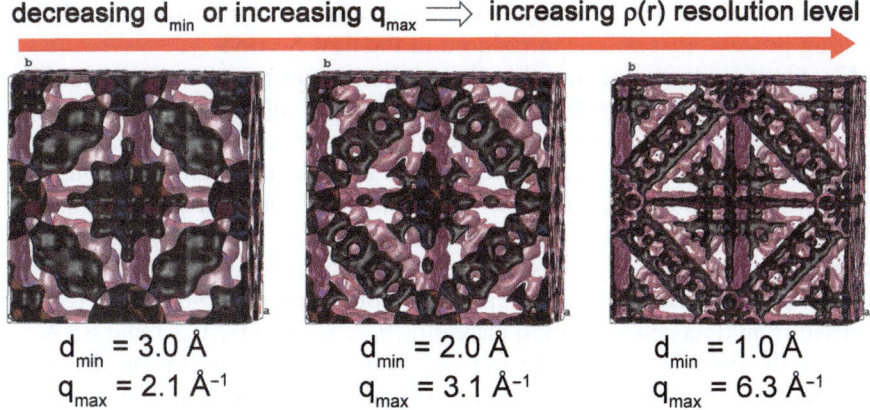

$$d_{min} = 3.0 \text{ Å} \qquad d_{min} = 2.0 \text{ Å} \qquad d_{min} = 1.0 \text{ Å}$$
$$q_{max} = 2.1 \text{ Å}^{-1} \qquad q_{max} = 3.1 \text{ Å}^{-1} \qquad q_{max} = 6.3 \text{ Å}^{-1}$$

Figure 5.4 Effect of the resolution level on electron density map $\rho(x,y,z)$ caused by using eqn. (5.22) on a progressively larger set of measured $I_{h,k,l}$. From left to right: $\rho(x,y,z)$ obtained using $I_{h,k,l}$ collected up to $q_{max} = 2.1$, 3.1 and 6.3 Å^{-1}, corresponding to $d_{min} = 3.0$, 2.0 and 1.0 Å, respectively. The image refers to single crystal XRD data collected on UiO-67 MOF, *vide infra* Section 5.5.3. Increasing q_{max}, the improvement in reconstruction of both the 4,4′ biphenyl-dicarboxylate (BPDC) linker and the $Zr_6O_4(OH)_4$ inorganic cornerstones is remarkable. Image kindly provided by the courtesy of Prof. K.P. Lillerud (Oslo University, NO).

simultaneously employed. As a consequence, the achievable spatial resolution in the determination of $\rho(x,y,z)$ is directly connected with the quality of the diffraction dataset employed, in terms of the number of structure factors used in the sum, and their degree of observability. In particular, the resolution level of an electron density map is commonly quantified using the minimum distance "visible" in the map, d_{min}, which can be simply deduced from Bragg's equation:

$$d_{min} = \frac{\lambda}{2 \sin \theta_{max}} = \frac{2\pi}{|\mathbf{q}_{max}|} \tag{5.23}$$

where θ_{max} is the maximum angle at which a Bragg peak is detected above the noise level. Therefore, to improve the resolution level, we can either reduce the incident wavelength, or extend the collection of diffraction data up to the highest possible angles, *i.e.* collect data at higher q-values. Unfortunately, on the experimental grounds, the steep decrease of atomic form factors (see Figure 5.3 in Section 5.2.1.2.) significantly lowers the signal-to-noise ratio of diffraction data in the high-q region, with respect to the low-q range. This means that high spatial resolution electron density maps can be obtained with a data collection strategy devoting much more acquisition time on the high q-part of the diffractograms in order to increase the statistics.

In addition, the reliability of the reconstructed electron density is also influenced by the precision in experimental determination of diffraction amplitudes (intensities) and, even more critically, by the correctness of the phasing process, that will be introduced in the next Section.

Finally, it is worth noticing that the minimum number of structure factors necessary to obtain a satisfactory reconstruction of the electron density depends on the structural complexity of the specific sample investigated. For macromolecules, and in particular for protein crystals, the available amount of experimental data is a crucial point. The same holds for MOFs. Furthermore, Figure 5.4 clearly evidences the crucial role played by the high q-region data collection in the ability of reconstructing the charge density *via* eqn (5.22). From left to right we observe the $\rho(x,y,z)$ distributions for the UiO-67 MOF reconstructed from data collected up to increasing q_{max} values of 2.1, 3.1 and 6.3 Å^{-1}, respectively: the differences in the accuracy of the reconstructed charge densities are evident. Generally speaking, MOFs, with unit cell volumes as large as some tens of thousands of Å^3 indicatively containing up to 10^3 atoms, are among the more complex inorganic structures investigated via X-ray diffraction,[45,46] and require consequently high quality data collection in the high q-region. For this reason, neutron or synchrotron radiation data are welcome.[47–49] Indeed synchrotron data are characterized by a much higher incident photon flux, while neutron scattering lengths are q-independent, *vide infra* Section 5.2.1.2.

5.2.1.4.4 The Phase Problem. Eqn (5.22) can be rewritten explicitly expressing the complex nature of the structure factors, highlighted in eqn (5.20). We therefore obtain:

$$\rho(x,y,z) = \frac{1}{V_{cell}} \sum_{h,k,l} |F_{h,k,l}| e^{i\phi(h,k,l)} e^{2\pi i(hx+ky+lz)} \tag{5.24}$$

where only $|F_{h,k,l}| = (I_{h,k,l})^{1/2}$ can be obtained from the experiment. Assume now to have collected a complete set of diffraction data on a completely unknown crystal, to the best of experimental possibilities. An almost obvious question can be asked: does the diffraction data contain all the information to fully determine the structure of the crystal? Looking at eqn (5.24), we can easily realize that a crucial piece of the puzzle is missing. The amplitude of the structure factor $|F_{h,k,l}|$ can be obtained by simply calculating the square root of the intensity $I_{h,k,l}$, as expressed in eqn (5.19). However, any information about the phase $\phi(h, k, l)$ is unavoidably lost. As depicted in Figure 5.5, this lack of information has dramatic implications, because the phases carry the most of information.

Notwithstanding this inherent limitation, in the most cases the information contained in the diffraction dataset is enough to obtain a satisfactory level in electron density reconstruction. Indeed, several methods have been developed to determine the $\phi(h, k, l)$ values, thus bypassing the phase problem. Among the commonest phasing approaches we can mention *ab initio* or direct methods,[15] Patterson methods,[51–53] molecular replacement,[53,54] anomalous dispersion-based methods,[35,55–57] and charge flipping methods.[58] A detailed discussion of

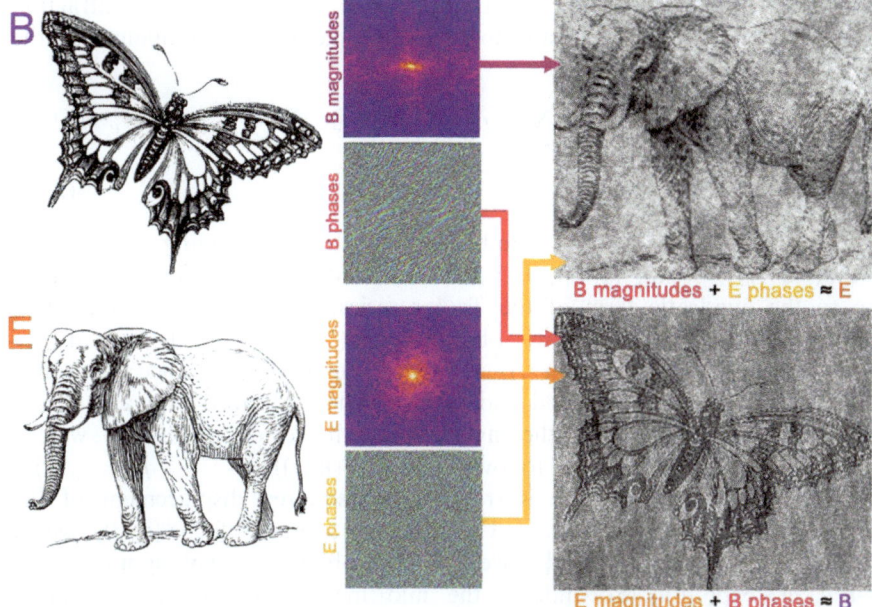

Figure 5.5 Visual representation of the phase problem. Left panel: pictures of a butterfly (B) and an elephant (E). These images can be treated as density maps and compute their Fourier transforms, obtaining for each picture the reported amplitudes (or magnitudes) and phases (middle panels). In right panels are reported the images obtained combining the B amplitudes and E phases (top panel) or, oppositely, E amplitudes and B phases (bottom panel). Comparing the two resulting pictures, it is striking that it is the phase that carries the major part of the information. The pictures have been elaborated using the FTL-SE program.[50] Previously unpublished figure.

these methodologies is beyond the scope of the present work, and can be found elsewhere in the specialized literature.[59–61]

5.2.1.4.5 The Effect of the Crystal Vibrations on the Diffracted Intensity. All the equations written so far in Sections 5.2.1.4.1 to 5.2.1.4.4 hold for ideally perfect crystals where the atoms of the unit cell occupy defined r_n positions that do not change with time, *i.e.* neglecting the phenomenon of thermal vibrations in solids. If one would be able to totally eliminate the thermal effects, the Bragg reflections would be uniformly sharp with no dependence on scattering angle to the distribution of intensities.[62] Actually, even in the 0 Kelvin limit, atoms vibrate inside a lattice, so that the atomic positions are not time independent but can be represented as time-dependent displacements $u_n(t)$ around the atomic equilibrium positions r_n used so far: $r_n(t) = r_n + u_n(t)$. This fact leads to a modification of the definition of the structure factor $F(q)$, see eqn (5.16), that takes into account the atomic displacements $u_n(t)$. As the time dependence of $u_n(t)$ is, by several orders of magnitude, faster than the acquisition time of a standard diffraction experiment, then eqn (5.16) still holds if we make a time averaging over the atomic positions:

$$F(\mathbf{q}) = <\sum_n f_n(q)e^{-i\mathbf{q}\cdot\mathbf{r}_n(t)}> = \sum_n f_n(q)<e^{-i\mathbf{q}\cdot\mathbf{r}_n(t)}> = \sum_n f_n(q)e^{-i\mathbf{q}\cdot\mathbf{r}_n}<e^{-i\mathbf{q}\cdot\mathbf{u}_n(t)}>$$

(5.25)

where the symbol $<\cdots>$ indicates the time averaging and where $<\exp(-i\mathbf{q}\cdot\mathbf{r}_n)> = \exp(-i\mathbf{q}\cdot\mathbf{r}_n)$, as $\mathbf{r}_n$ are time-independent. Now, developing up to the quadratic term of the Taylor series, the exponential containing the atomic displacements, we obtain: $<\exp(-i\mathbf{q}\cdot\mathbf{u}_n(t))> \sim 1 - i<\mathbf{q}\cdot\mathbf{u}_n(t)> -\frac{1}{2}<[\mathbf{q}\cdot\mathbf{u}_n(t)]^2> = 1 - \frac{1}{2}<[\mathbf{q}\cdot\mathbf{u}_n(t)]^2>$ because $<\mathbf{q}\cdot\mathbf{u}_n(t)> = 0$ since the atomic displacements are uncorrelated and random when averaged over a sufficient long time.[38,62] Defining $\alpha_n(t)$ the angle between the vectors $\mathbf{q}$ and $\mathbf{u}_n(t)$, we can rewrite the scalar product as follows: $<\exp(-i\mathbf{q}\cdot\mathbf{u}_n(t))> \sim 1 - \frac{1}{2} q^2 \cdot <u_n^2(t)> <\cos^2[\alpha_n(t)]>$. Defining the isotropic mean squared displacement of the nth atom as $u_n^2 = <u_n(t)>^2$, and geometrical average of $\cos^2[\alpha_n(t)]$ over a sphere being equal to $\frac{1}{3}$, we finally obtain: $<\exp(-i\mathbf{q}\cdot\mathbf{u}_n(t))> \sim 1 - q^2u_n^2/6 \sim \exp(-q^2u_n^2/6)$. Indeed, the mid member of the last equation corresponds with the first term of the Taylor series of the exponential in the third member. Consequently, eqn (5.25) can be rewritten as:

$$F(\mathbf{q}) \simeq \sum_n f_n(q)e^{-i\mathbf{q}\cdot\mathbf{r}_n}e^{-\frac{1}{6}q^2u_n^2} = \sum_n f_n(q)e^{-i\mathbf{q}\cdot\mathbf{r}_n}e^{-W_n}$$

(5.26)

where the last exponential is a damping factor that further reduces the scattering amplitudes in the high q (or high $\sin(\theta)/\lambda$) part of the diffraction patterns. Figure 5.6a reports pictorially a 1D-representation of the effect that the thermal motion has on the spread of the atomic electron density. The atomic motion is usually defined using the Debye–Waller factor[62–64] $W_n = 1/6q^2u_n^2$ or using the isotropic temperature factor $B_n = 8\pi^2u_n^2$. According to the different conventions, the exponential that accounts for the dumping of the diffracted intensities due to the thermal motion of the atoms can be written as $\exp(-q^2u_n^2/6)$ or as $\exp(-2W_n)$ or as $\exp[-\frac{1}{3}B_n \sin^2(\theta)/\lambda^2]$. According to eqn (5.26), previously written for the form factor $F(\mathbf{q})$, the measured intensity $I(\mathbf{q})$ can be expressed by:

$$I(\mathbf{q}) = |F(\mathbf{q})|^2 \simeq \sum_n f_n^2(q)\left|e^{-i\mathbf{q}\cdot\mathbf{r}_n}\right|^2 e^{-2W_n}$$

(5.27)

where each nth term of the sum is damped by a factor $\exp(-2W_n)$. On the experimental ground, to limit the dumping effect described so far, low temperature data collections are welcome, in order to keep the mean squared displacement u_n^2 of the atoms as low as possible.

In the simplified discussion provided in this section we supposed that all atoms have undergone an isotropic displacement along the three directions $\mathbf{a}$, $\mathbf{b}$ and $\mathbf{c}$; this allowed us to write $<\mathbf{q}\cdot\mathbf{u}_n(t)> = \frac{1}{3}q^2u_n^2$. Actually, the different strength of the chemical bonds around the nth atom can result in different atomic displacements along the three independent directions. An ellipsoid is a convenient way of visualizing the anisotropic vibration of atoms inside a lattice and therefore their 3D time-averaged position. We thus speak about *thermal ellipsoids*, more formally termed *atomic displacement parameters* (adp), that are ellipsoids used to indicate the magnitudes and directions of the thermal vibration of the different atoms in the unit cell. Thermal ellipsoids are actually tensors (*i.e.* mathematical objects which allow the definition of magnitude and orientation of vibration with respect to three mutually independent axes). The three principal axes of the thermal vibration for the nth atom are denoted $u^2_{n,a}$, $u^2_{n,b}$ and $u^2_{n,c}$, and the corresponding thermal ellipsoid is based on these axes. The size of the ellipsoid is scaled so that it occupies the space in which there is a particular probability of finding the electron density of the atom.[65] An example of structure reporting the refined thermal ellipsoid is shown in Figure 5.6b. The

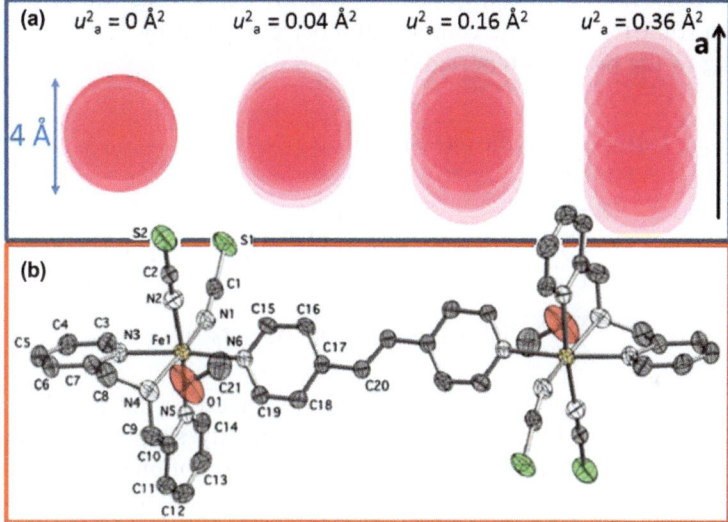

Figure 5.6 Part (a): pictorial representation on how the atomic displacement is able to spread the electron density in space. The same atom used in Figure 5.2, having the appreciable atomic electron density confined within a sphere of radius 2 Å, is here represented assuming a mean squared displacement u^2_a of, from left to right, 0, 0.04. 0.16 and 0.36 Å^2, respectively. For simplicity, only the atomic displacement along the a axis is here represented, assuming $u^2_b = u^2_c = 0$ Å^2. From what has been discussed in Sections 5.2.1.2 and 5.2.1.4.5, it is evident that the larger is the electron density spread, the larger will be the dumping of the scattered intensity at high q values. Part (b): molecular structure of [{Fe(dpia)(NCS)$_2$}$_2$(bpe)] · 2CH$_3$ OH MOF at 290 K where the thermal ellipsoids obtained after a structural refinement are reported (H atoms are omitted for clarity). Ellipsoids enclose 30% probability. From the refined structure it emerges how the atoms S1, S2 and O1 exhibit quite anisotropic displacement parameters. Reproduced with permission from Matouzenko *et al.*,[66] copyright (2011) Royal Society of Chemistry.

reconstruction, for all atoms in the unit cell of the thermal ellipsoids requires a very high quality data set. In most of the cases only isotropic thermal factors can be independently refined.

Finally, please note that, as the measured scattered intensity, $I(\mathbf{q})$, is the squared modulus of the structure factor $F(\mathbf{q})$, see eqn (5.8), then the thermal motion affects the measured intensity according to the following law: $I(\mathbf{q}) = I_0(\mathbf{q}) \exp(-\frac{1}{3}q^2u^2)$, where $I_0(\mathbf{q})$ is the expected intensity in the case of an ideal crystal characterized by non-vibrating atoms.

5.2.1.5 X-Ray Scattering from Non-crystalline Samples

In non-crystalline samples (gases, liquids, and amorphous materials), where only short range ordering is present, the probed volume is constituted by a statistical ensemble of sub-systems (molecules) randomly oriented. We can start from the general form in eqn (5.8) and explicitly express the square modulus operation (that for a generic complex number z is given by $|z|^2 = zz^*$, $z \in C$), according to the passages reported below:

$$
\begin{aligned}
I(\mathbf{q}) &= \left| \sum_n f_n(q) exp(-i\mathbf{q} \cdot \mathbf{r}_n) \right|^2 \\
&= \left[\sum_n f_n(q) exp(-i\mathbf{q} \cdot \mathbf{r}_n) \right] \left[\sum_m f_m(q) exp(i\mathbf{q} \cdot \mathbf{r}_m) \right] \qquad (5.28) \\
&= \sum_n \sum_m f_n(q) f_m(q) exp(-i\mathbf{q} \cdot (\mathbf{r}_n - \mathbf{r}_m))
\end{aligned}
$$

We are therefore allowed to isotropically average over all possible orientations the resulting form,[13,14,67] obtaining a greatly simplified formula in which the scattered intensity is expressed as a function of one instead of three spatial dimensions. The resulting form, eqn (5.29), is named the Debye equation:[68]

$$
I(q) = \sum_n \sum_m f_n(q) f_m(q) \frac{\sin(q r_{nm})}{q r_{nm}} \qquad (5.29)
$$

where r_{nm} is the distance between n^{th} and m^{th} atom.

The Debye equation represents the key theoretical instrument in interpreting results from X-ray scattering experiments involving non-crystalline samples. However, the direct use of eqn (5.29) is associated with some critical issues, in particular passing from gas-phase scattering to experiments involving solution-phase or solid non-crystalline systems. Indeed, when dealing with many-atom systems (more than 500 atoms), calculation using the Debye equation is extremely resource-consuming, due to the extremely high number of possible combinations of n and m indices.

For these reasons, it is useful to express the Debye formula in terms of radial pair distribution functions (PDF) $g_{nm}(r)$.[69–71] Radial PDFs are defined in such a way that the probability of finding an m-type atom at distance r from an n-type atom is equal to $4\pi r^2 g_{nm}(r)$ (Figure 5.7 reports a schematic representation of how radial PDFs are obtained).

Using the formalism of radial PDF, eqn (5.29) can be reformulated as in eqn (5.30):[14]

$$I(q) = \sum_n N_n^2 f_n^2(q) + \sum_m \sum_{n \neq m} \frac{N_n N_m}{V} f_n(q) f_m(q) \cdot \int_0^\infty g_{nm}(r) \frac{\sin(qr)}{qr} 4\pi r^2 dr$$

$$(5.30)$$

where the indexes n and m run over all the atoms types included in the sample, N_n and N_m are the numbers of n-type and m-type atoms respectively and V the volume of the sample probed by X-rays.

In the case of gas-phase scattering, the radial PDF $g_{nm}(r)$ can be approximated to a Dirac delta function and consequently eqn (5.30) reduces again to eqn (5.29). Eqn (5.30) highlights the possibility of obtaining a simulated $I(q)$ curve directly from the $g_{nm}(r)$ functions for the system of interest, to be used for comparison with the measured scattering data. This, hence, represents the theoretical basis for modeling and interpretation of scattering experiments, providing a direct connection between acquired scattered intensity and structural features of the sample. More sophisticated analysis approaches will be then applied depending on sample peculiarities and the specific

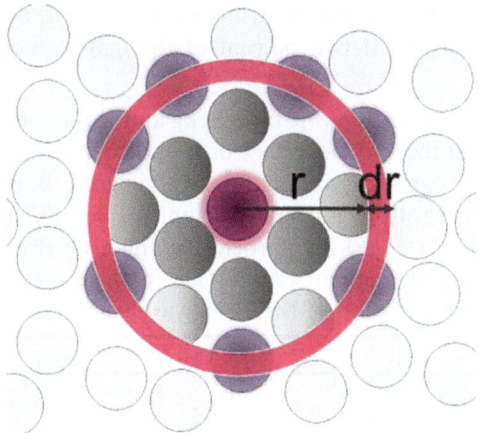

Figure 5.7 Scheme of the radial pair distribution function (PDF) $g(r)$. The radial PDF is a measure to determine the correlation between particles within a system. Specifically, it is an average measure of the probability of finding a particle at a distance of r away from a given reference particle. The general algorithm involves the determination of the number of particles within r and $r + dr$ (circular pink shell in the figure) from the reference particle (represented in violet).

experimental conditions. In particular, depending on the investigated angular range, the SAXS (Small Angle X-ray Scattering)[34,72,73] and WAXS (Wide Angle X-ray Scattering)[14] techniques can be distinguished, providing structural information of different nature, and on different length-scales. At the highest q-exchanged values we have the total scattering or Pair Distribution Function (PDF) approach,[74] which will be discussed in more details in Section 5.4, once the basic concepts of EXAFS have been introduced in Section 5.3, while relevant examples in MOF characterization will be reported in Section 5.5. For space limitation, the SAXS theory will not be discussed, consequently no example on the use of SAXS in MOFs characterization will be reported in Section 5.5. Here we refer the reader to some relevant papers on this subject.[75–80]

5.2.2 Neutron Diffraction

5.2.2.1 Historical Background

In his PhD thesis dating back to 1924, de Broglie assigned to any massive particle P a wavelength λ given by the ratio between the Plank constant ($h = 6.626068 \times 10^{-34}$ Js) and its momentum p:[81]

$$\lambda = \frac{h}{p} = \frac{h}{m_P v_P \sqrt{(1 - v_P^2/c^2)}} \simeq \frac{h}{m_P v_P} \text{ for } v_p/c \ll 1 \qquad (5.31)$$

later named the de Broglie's wavelength. De Broglie's intuition opened the possibility to perform diffraction experiments using as incident beam, instead of X-rays, any kind of particle beam of defined momentum **p** having a modulus p in the order of 7×10^{-24} kg m s^{-1}, so that the corresponding λ will be in the range of 1 Å. This value is obtained for neutrons having a speed of about 4000 ms^{-1}, and for electrons moving at two percent of the light speed c, see Table 5.1,

Table 5.1   Top part: rest mass (m_P, both in kg and in mc^2 units), charge (q) and spin (s) of the particles ($P = e$ or N), that represent the two most used particle beams for diffraction experiments. Bottom part: particle speed (v_P, both in m/s and in c units: $c = 2.9979 \times 10^8$ m/s) and the particle energy (E) for particles having a de Broglie wavelength of $\lambda = 1$ Å. For comparison, the last row reports the corresponding values for a photon of the same wavelength, see eqn (5.1).

Particle	m_P (kg)	$m_P c^2$ (MeV)	q (C)	s
electron	9.1094×10^{-31}	0.511	1.6022×10^{-19}	1/2
neutron	1.6749×10^{-27}	939.565	0	1/2
photon (X-ray)	0	0	0	1

Particle	*Values for $\lambda = 1$ Å*			
	v_P (m/s)	v_P/c	E (eV)	p (kg m s^{-1})
electron	7.2739×10^6	2.4263×10^{-2}	150.412	6.626×10^{-24}
neutron	3.9560×10^3	1.3196×10^{-5}	8.180×10^{-2}	6.626×10^{-24}
photon (X-ray)	2.9979×10^8	1	$1.23984 \times 10^{+4}$	6.626×10^{-24}

so that relativistic corrections are useless for neutrons and in a first approximation are negligible for electrons.[82]

As the technology needed to produce X-ray tubes had already been available for some years,[83] electron sources of sufficient intensity became available in the twenties. The natural particle candidate to experimentally test de Broglie's theory was therefore the electron. The experimental proof of de Broglie's idea, providing a remarkable validation of the particle-wave dualism, arrived in 1927 by Davisson and Germer. The scientists, at Bell Telephone laboratories, obtained an electron diffraction pattern from a thin Ni single crystal.[84] Independently at the University of Aberdeen, Thomson and Reid passed a beam of electrons first through a thin celluloid[85] film and subsequently through a thin Pt[86] film. In both laboratories diffraction patterns explainable combining the Bragg equation with the de Broglie assumption were detected. Several other observations followed these pioneering works,[87–89] and some historical perspectives are available in the literature.[90,91]

The use of neutrons, discovered by Chadwick in 1932,[92–94] in diffraction experiments was postponed until the first nuclear reactors were available.[95] The first neutron diffraction experiments were carried out in the second half of the forties by Wollan and Shull, using the Graphite Reactor at Oak Ridge,[96–103] and by Fermi and Marshall, using the heavy water pile at the Argonne National Laboratory.[104–107]

In the following, electron diffraction will not be discussed further. This choice is related to the difficulties in collecting electron diffraction patterns on non-damaged MOF crystals.[108] The main problem is the pronounced instability of MOFs under the electron beam, leading to loss of long range order information after mere seconds of illumination by the electron beam. In particular, experimental setups such as cryogenic sample holder stages and very low electron illumination conditions are mandatory to hope to obtain significant results.[109,110] It is so not surprising to realize that the literature contains very few contributions where electron scattering in Trasmission Electron Microscope (TEM) instruments has been used to characterize MOFs structures.[108–115]

5.2.1.2 Analogies and Differences between Neutrons and X-Ray Scattering

Although electron, neutron, and X-ray scattering interactions with matter are based on different physical processes, the resulting diffraction patterns are analyzed using the same coherent scattering approach described above for X-rays.[116,117] Basically, once the concept of de Broglie wavelength (5.31) is established, the Bragg equation can be rewritten as:

$$2d_{hkl} \sin(\theta) = h/(m_P v_P) \tag{5.32}$$

where P stands for e or N, and applied to describe experiments performed with electrons or neutrons, as eqn (5.15) is used for X-rays.

Neutrons interact with the nuclei, *i.e.* with objects having a size of few fm ($1\,\text{fm} = 10^{-5}\,\text{Å}$). This implies that the nuclei with the same number of protons but with a different number of neutrons (isotopes) will behave as two completely different scattering objects in neutron diffraction: this was not true for X-rays where the scattering comes from the electrons. The interaction with nuclei implies that, in eqn (5.9), both $\exp(-\mathbf{q} \cdot \mathbf{r})$ and $\cos(\mathbf{q} \cdot \mathbf{r})$ can be considered as ~ 1, for any q value practically attainable in diffraction experiments (even those where the data collection is extended at very high angles, employing a very short λ at synchrotron sources). Consider now a nucleus with $r = 2$ fm, at $q = 50\,\text{Å}^{-1}$. In this case $qr = 10^{-3}$ and $\cos(qr)$ differs from units at the ninth decimal digit only. This implies that the neutron scattering intensity is constant at any θ (or q) and does not drop off at high θ (or q) values as occurs for the X-ray scattering, see Figure 5.2 and related discussion. Neutron diffraction is consequently intrinsically superior to XRD in the collection of high quality data at high q, which is a key advantage in the structure refinement of complex structures like MOFs, see Section 5.2.1.4.3. This superiority is clearly visible in Figure 5.8a, where the simulated neutron diffraction pattern of desolvated UiO-66 is compared with the corresponding simulated XRPD pattern. *Vide infra* in the examples section Figure 5.18ab for a comparison between experimental neutron and X-ray diffraction patterns of $Cr_3(BTC)_2$ MOF.

The nuclear scattering intensity for neutrons is quantified by the total absorption cross section σ_{tot}, defined as the total number of neutrons scattered per second, normalized by the incident neutron flux Φ. As Φ is given by the number of incident neutrons per second per surface area, σ_{tot} is an area, usually measured in barn ($1\,\text{barn} = 1 \times 10^{-28}\,\text{m}^2 = 100\,\text{fm}^2$, that approximately represents the geometrical cross sectional area of one uranium nucleus). The nuclear scattering intensity for neutrons can be alternatively quantified with the corresponding nuclear scattering length b, defined as the radius of an ideal hard sphere able to provide the same scattering. b is consequently related to σ_{tot} by the simple geometric relationship:

$$\sigma_{\text{tot}} = 4\pi < |b|^2 > \tag{5.33}$$

where 4π comes from the integration over the whole solid angle Ω and where the "$< >$" symbol represents the statistical average over the neutron and nucleus spins. Indeed it has been demonstrated since the earliest studies of Fermi that the neutron–nucleus interaction is spin dependent.[107] Defining $\mathbf{s}$ and $\mathbf{I}$ the neutron and the nucleus spin, respectively, the b parameter can be defined as:[118]

$$b = b_c + 2b_i[\mathbf{I}(\mathbf{I} + 1)]^{-1/2}\mathbf{s} \cdot \mathbf{I} \tag{5.34}$$

where b_c and b_i are the coherent and incoherent neutron scattering lengths of the nucleus, respectively. On this basis, the total cross section σ_{tot} is given by the contribution of a coherent (σ_c) and an incoherent (σ_i) cross section:[118]

$$\sigma_{\text{tot}} = \sigma_c + \sigma_i \quad \text{where } \sigma_c = 4\pi|b_c|^2 \text{ and } \sigma_i = 4\pi|b_i|^2 \tag{5.35}$$

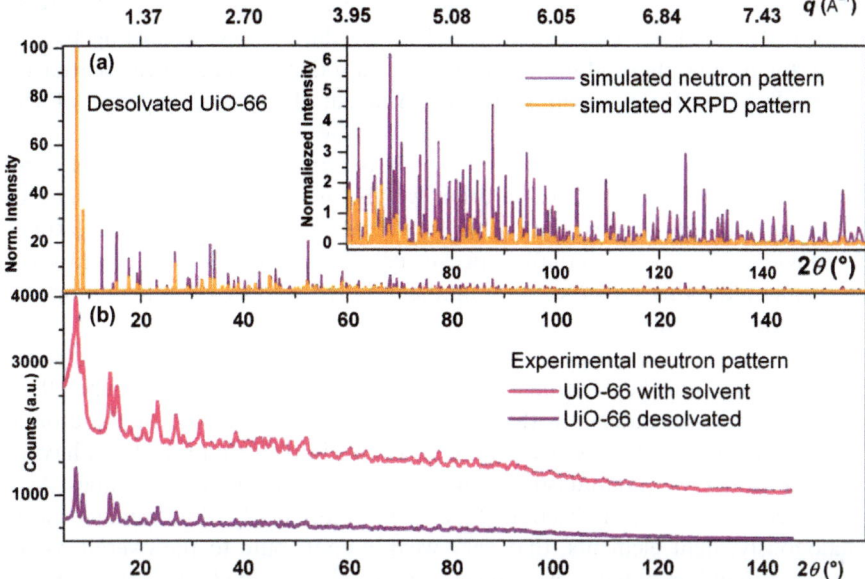

Figure 5.8   Part (a): comparison between the simulation of the X-ray (orange) and neutron (violet) diffraction patterns for UiO-66 MOF in its desolvated form ($\lambda = 1.954$ Å). Previously unpublished figure, by courtesy of A. Piovano (ILL, Grenoble F). Part (b): experimental neutron diffraction patterns of UiO-66 MOF before (pink curve, collected at 300 K) and after (violet curve, collected at 523 K) solvent removal. Previously unpublished data collected at Super-D2B instrument (ILL, Grenoble F) with $\lambda = 1.59$ Å: by courtesy of M. Ceretti (University of Montpellier-2, F).

The coherent scattering depends on the scattering vector **q**, and contains the structural information. The incoherent scattering is isotropic and, in a diffraction experiment, contributes to an overall increase of the background. Samples prepared with isotopic substitutions can be used to minimize σ_i and consequently increase the quality of the neutron diffraction data. In this regard, the most relevant case concerns the hydrogen atom where for ^1H we have $b_c = -3.74$ fm and $b_i = 25.27$ fm, while for ^2H we have $b_c = 6.67$ fm and $b_i = 4.04$ fm. The synthesis of MOFs with deuterated ligands will consequently significantly decrease the neutron incoherent scattering.[119] This effect is clearly visible in the experimental neutron diffraction data reported in Figure 5.8b, where the removal of the solvent inside the UiO-66 porescauses a decrease of the scattering background by a factor larger than three. Indeed the solvent molecule, dimethylformamide (DMF, $(CH_3)_2NC(O)H$), has seven hydrogen atoms so that they are the major responsible for the incoherent neuron scattering of the solvated material. The patterns reported in Figure 5.8b are also informative because they show a clear decrease of the Bragg peak intensities at high 2θ moving from the solvated to the desolvated form of UiO-66. This is due to the thermal vibrations of the lattice. Indeed, the solvated UiO-66 has been measured at 300 K, while the data collection of the desolvated form has been performed at 523 K. This experimental result shows how, according to eqn (5.26), the atomic thermal motion introduces a q-dependence on the scattering amplitudes also for neutron scattering. Indeed, the q-independence of the neutron scattering length is based on the point-dimension of the nuclei (few fm), but the atomic thermal motion moves the atoms (and so the nuclei) by fractions of Å, see Figure 5.6a. In other words, the nuclear density is a Dirac-delta function only in ideal lattices where vibrations are absent. In real lattices the nuclear density is actually spread because of the atomic vibrations. Experimentally, this spread can be significantly reduced by adopting cryogenic sample environments.

Both b_c and b_i are defined by nuclear interaction, that can not be quantified by theory, so they must be determined experimentally.[118] As shown in Figure 5.9, b_c has no specific dependence with Z, as conversely was the case of the atomic scattering factor for X-rays (see Figure 5.3). This implies that for atoms having similar Z values, we can find isotopes having consistently different b_c values. As a consequence, neutron diffraction experiments can easily discriminate them. Analogously, light elements, that very weakly contribute to the overall X-ray scattering signal, may have $|b_c|$ values comparable to some high-Z elements, or even higher. Therefore a neutrons-based analysis multiplies the chances to locate low-Z elements with respect to the X-ray case.

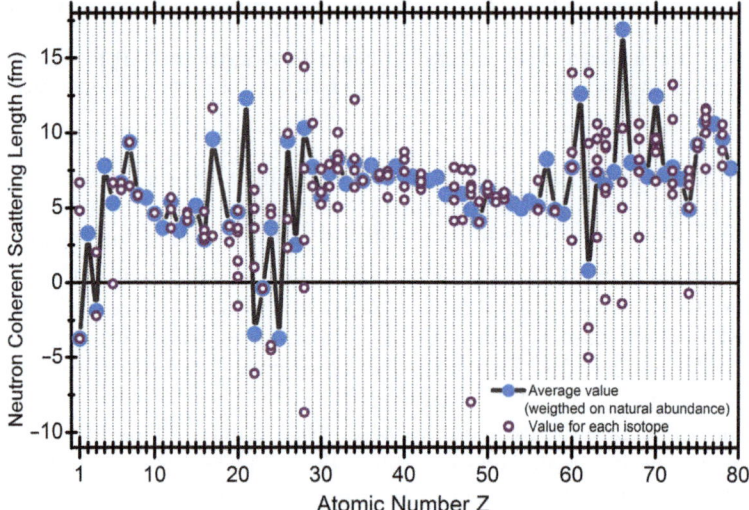

Figure 5.9 Coherent neutron scattering lengths b_c, averaged over the natural isotopes abundance (full cyan spheres), for all elements in the $1 \leq Z \leq 80$ interval. For each element, a selection of specific isotopes is also reported (violet circles). Previously unpublished figure, reporting values from the tables of Sears;[118] the same data are also available online on the NIST web site (http://www.ncnr.nist.gov/resources/n-lengths/).

Care must be taken when comparing the results obtained with X-rays or neutron data collections because the former provides the electron density, while the latter provides the nuclear density. It is so common that the O–H (or C–H) distances refined from X-ray are markedly shorter than those refined from neutron data on the same compound. This is due to the fact that neutron scattering provides the correct nuclear–nuclear distance, while XRD observes that the electron cloud along the O–H (or C–H) bond is localized closer to the O (or C) atom, owing to its higher electronegative character. As an example, for the sucrose molecule ($C_{12}H_{22}O_{11}$), crystallizing with two molecules per cell in the monoclinic space group $P2_1$, the refined O–H (C–H) distance determined by X-rays is 0.79 Å (0.96 Å), while that determined by neutrons is 0.97 Å (1.095 Å).[120]

As reported in Table 5.1, there are six orders of magnitude in energy between a neutron and an X-ray photon with $\lambda = 1$ Å. This means that the energy release of a neutron beam on the sample is negligible with respect to that occurring using an X-ray beam, particularly when a synchrotron source is used. This aspect guarantees that there is no risk of radiation damage on the sample for relatively short acquisitions. Conversely, the sample can become radioactive after exposure to a neutron beam, a risk not present with X-rays. In such cases the samples may not be available for additional experiments for a while. Additional critical aspects of neutron diffraction concern the low neutron cross section and the low flux of neutron sources, forcing the scientists to synthesize huge amounts of sample and requiring long data collections. Moreover, on one hand it is particularly difficult to obtain highly monochromatic neutron beams and on the other hand the $\Delta\theta$ accuracy of the detector is larger with respect to the X-ray diffraction case. Note that, for the same instrument, the $\Delta\theta$ accuracy is λ-dependent. As consequence, as it can be noted that by differentiating the Bragg law, see eqn (5.15), we obtain:

$$\Delta d_{hkl} = \left|\frac{\partial d_{hkl}}{\partial\lambda}\right|\Delta\lambda + \left|\frac{\partial d_{hkl}}{\partial\theta}\right|\Delta\theta = \frac{1}{2\sin(\theta)}\Delta\lambda + \frac{\lambda\cos(\theta)}{2\sin^2(\theta)}\Delta\theta \qquad (5.36)$$

The larger dispersion in the wavelength selected by the monochromator ($\Delta\lambda$) and a poorer $\Delta\theta$ resolution induce a larger incertitude on the lattice inter-planar distance determination. Typical values of $\Delta d/d$ obtained at neutron diffractometer are in the $10^{-2} - 10^{-3}$ range. This problem can be significantly overcome with a sophisticated combination of movable multi-detectors, see *e.g.* the Super-D2B two axis instrument at the ILL reactor (able to reach $\Delta d/d \sim 5 \times 10^{-4}$).[121] Alternatively, the polychromatic time-of-flight (TOF) detection mode[122] can be employed, see *e.g.* the HRPD instrument at the ISIS spallation neutron source able to reach $\Delta d/d \sim 4 \times 10^{-4}$.[123] These values are to be compared with what's available with X-rays at synchrotron sources: $\Delta\theta = 10^{-3}$ and $\Delta E/E = 5 \times 10^{-5}$ at 10 keV corresponding to $\lambda = 1.24$ Å, see eqn (5.1).[124,125]

The TOF approach is based to the fact that neutrons with different λ are traveling at different velocities, so that they will reach the sample (first) and the detector (then) at different times. Working at fixed 2θ angle, with a small $\Delta\theta$ angular acceptance, the d_{hkl} can be obtained measuring the time (t_{TOF}) needed to travel the source-to-detector distance (S), according to the Bragg equation:

$$d_{hkl} = \frac{1}{2\sin(\theta)}\frac{h}{m_N v_N} = \frac{1}{2\sin(\theta)}\frac{h t_{TOF}}{m_N S} \tag{5.37}$$

where the neutron wavelength has been expressed using the de Broglie equation, see eqn (5.32). Differentiating eqn (5.37) with respect to the angular acceptance and the time resolution of the detector, we obtain:

$$\Delta d_{hkl} = \left|\frac{\partial d_{hkl}}{\partial\theta}\right|\Delta\theta + \left|\frac{\partial d_{hkl}}{\partial t_{TOF}}\right|\Delta t_{TOF} = \frac{\cos(\theta)}{2\sin^2(\theta)}\frac{h t_{TOF}}{m_N S}\Delta\theta + \frac{1}{2\sin(\theta)}\frac{h}{m_N S}\Delta t_{TOF} \tag{5.38}$$

It is evident from eqn (5.38) that Δd_{hkl} can be minimized using large S distances. Moreover, a TOF-instrument does not need a monochromator, therefore it exploits the whole λ-spectrum emitted by a pulsed source, with an evident benefit on the total flux. In order to cover a larger d-spacing interval, defined by the acquisition time window between two successive neutron bunches, more than one detector bank can be used at different 2θ. The HRPD instrument at ISIS, for instance, is equipped with two detector banks at $2\theta = 90°$ and $168°$.[123]

Finally, the limited access of neutron sources represents an additional practical problem. Table 5.2 summarizes, in a schematic way, the main properties, advantages and disadvantages of diffraction experiments performed with X-ray or neutron beams.

5.3   XAS Spectroscopy: Basic Background

The aim of this section is to provide the reader with a concise review of the basic physical principles on which the interpretation XAS data is based. For a more detailed description of the theoretical background and experimental aspects of XAS we refer the reader to the extensive specialized literature (e.g. ref. 126–133).

5.3.1   XAS Theoretical Background

XAS measures the variations of the X-ray absorption coefficient μ as a function of the incident X-ray energy E. According to the Fermi Golden Rule,[6,134,135] the XAS signal is proportional to the electron transition probability from the core-state $|i>$ of energy E_i to the unoccupied state $|f>$ of energy E_f, as expressed by eqn (5.39), where the product $\mathbf{e}\cdot\mathbf{r}$ indicates the electronic transition dipole operator,[136] where $\rho_i(occ)$ and $\rho_f(unocc)$ are the

Table 5.2 Summary of the main properties, advantages and disadvantages of diffraction experiments performed with X-ray or neutron beams. RA = rotating anode; BM = bending magnet; U = undulator.

Property	*X-Rays*	*Neutrons*
Interact mainly with	all electrons	nuclei and unpaired electrons
Scattering intensity	decreases at high q values	constant at all q values
Beam penetration depth (m)	$10^{-1} - 10^{-3}$ for $1 < Z < 10$; $10^{-3} - 10^{-5}$ for $10 < Z < 60$; $10^{-5} - 10^{-6}$ for $60 < Z < 92$;	$\sim 10^{-3}$ for ^{10}B, ^{113}Cd, ^{149}Sm, ^{157}Gd; $10^{-1} - 10^{-3}$ for remaining nuclei
Needed sample volume	small, typically a capillary (0.1 < diameter < 1.0 mm)	huge, typically some cm^3
Instrument availability	very high at lab; low at synchrotrons	very low at nuclear reactors or spallation sources
Beam flux (particle s^{-1} m^{-2})	10^{10} (RA); 10^{17} (BM); 10^{24} (U)	10^{11}
Beam divergence (mrad^2)	0.5×10 (RA); 0.1×5 (BM); 0.01×0.1 (U)	10×10
$\Delta E/E$	10^{-4}	10^{-2}
Main advantages	• high $\Delta\theta$ (Δq) resolution • fast data collection, allowing time dependent experiments to be done (sub second resolution for BM and U) • no sample radioactive activation • possibility to use μm beams for space resolved experiments	• ability to detect low Z elements • ability to discriminate elements with similar Z • advantages related to the use of isotopic-substituted samples • no radiation damage • the scattering power is constant at any θ (q) • ability to obtain spin density maps
Main disadvantages	▪ low ability to detect light elements ▪ low ability to discriminate elements with similar Z ▪ inability to discriminate isotopes ▪ risk of radiation damage ▪ the scattering power falls progressively off at increasing θ (or q) ▪ intrinsic severe difficulties to investigate magnetic structures	▪ moderate $\Delta\theta$ (or Δq) resolution ▪ long acquisition times (hampering the possibility of time-resolved studies) ▪ risk of radioactive activation of the sample ▪ space resolved experiments are critical due to the low available flux and the low efficiency in the beam focusing ▪ elements with high incoherent scattering lengths (such as H) should be avoided

densities of initial occupied and final unoccupied states, respectively, and where $\delta(E_f - E_i - E)$ is a Dirac delta function.[132]

$$\mu(E) \propto |<i|\mathbf{e} \cdot \mathbf{r}|f>|^2 \delta(E_f - E_i - E)\rho_i(occ)\rho_f(unocc) \qquad (5.39)$$

The behavior of the $\mu(E)$ function is represented in Figure 5.10b. A general decrease of the absorption with increasing incident energy can be noticed, following approximately the law:

$$\mu(E)/\rho \approx Z^4/AE^3, \qquad (5.40)$$

where ρ is the sample density, Z the atomic number and A the atomic mass. This equation holds for a sample containing a unique chemical species like a metal foil, but can be easily generalized for any sample of known composition. Also evident in Figure 5.10b is the presence of the characteristic saw-tooth like edges, whose energy position is a distinctive feature of each kind of absorbing atom. See also the blue and cyan transitions in Figure 4.19 of chapter 4 in this book.[6] These absorption edges correspond to transitions where a core-orbital electron is excited to (i) the free continuum (*i.e.* when the incident energy is above the ionization energy of the absorber atom) or (ii) unoccupied bond states lying just below the ionization energy. The nomenclature adopted for the edges recalls the atomic orbitals from which the electron is extracted, as shown in Figure 5.10a: K-edges are related to transitions from orbitals with the principal quantum number $n=1$ ($1s_{1/2}$), L-edges refers to electron from the $n=2$ orbitals (L_I to $2s_{1/2}$, L_{II} to $2p_{1/2}$, and L_{III} to $2p_{3/2}$ orbital), and so on for M, N, ... edges.

When the energy of the X-ray photon exceeds the ionization limit (case (i) mentioned above), the excited electron (generally named "photoelectron") has a kinetic energy E_K given by $E_K = h\nu - E_B$, where E_B indicates the electron binding energy, that is typical of the absorption edge (K, L_I, L_{II} or L_{III}) of the selected atomic species.[133,137] Once ejected, the photoelectron propagates through the sample as a spherical wave diffusing from the absorber atom, with a wavevector of modulus k defined by eqn (5.41):

$$k = \frac{1}{\hbar}\sqrt{2m_e E_K} \qquad (5.41)$$

A close zoom on the energy region in the proximity of an absorption edge shows a well defined fine-structure. In particular, only when the absorber is surrounded by neighboring atoms (molecules or crystals) a structure of oscillatory nature modulates the smooth $\mu(E)$ profile at energies above the edge. Figure 5.10c provides an example for the activated $Cr_3(BTC)_2$ MOF, where the energy ranges around the Cr K-edge. Such modulation in the absorption coefficient derives from the interference between the outgoing photoelectron wave diffusing from the absorber and the wavefronts back-scattered by the neighboring atoms.[138,139] In a typical XAS experiment, the energy range probed around the edge is conventionally divided into two different regions (Figure 5.10c):

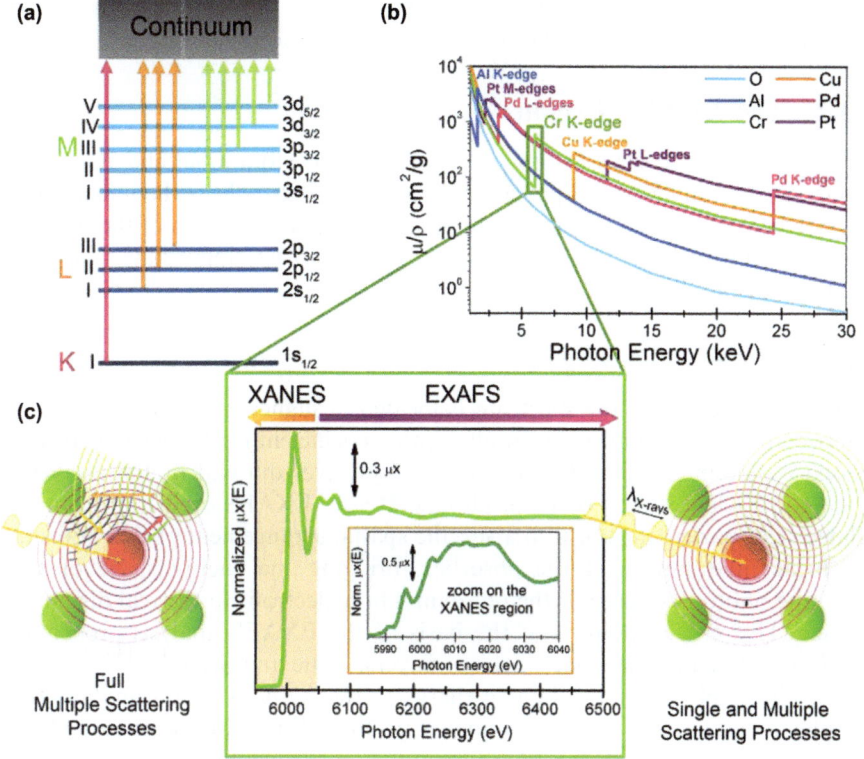

Figure 5.10 Part (a): X-ray absorption edges nomenclature and representation of their relation with the atomic orbitals from which the electron is extracted. Part (b): general behavior of the X-ray absorption coefficient μ/ρ, see eqn (5.40), as a function of the incident X-ray energy E for O ($Z=8$), typically contained in MOFs organic ligands, and for some selected metals present in MOF cornerstones, *i.e.* Al ($Z=13$), Cr ($Z=24$), Cu ($Z=29$), Pd ($Z=46$), and Pt ($Z=78$). Data obtained from NIST web site http://physics.nist.gov/PhysRefData/XrayMassCoef/tab3.html. Note the logarithmic scale of the ordinate axis. Part (c): Cr K-edge XAS of activated $Cr_3(BTC)_2$ MOF[140] (data collected at BM01B beamline of the ESRF). The conventional division between the XANES and EXAFS region and the schematic models of full multiple and single scattering processes, dominating respectively the XANES and EXAFS region, are indicated (color code: absorber atom in magenta; neighbor atoms that backscatter the photoelectron wave outgoing from the absorber in green). Previously unpublished figure.

(i) X-Ray Absorption Near Edge Structure (XANES) region: portions of the XAS spectrum just below and above the edge energy;

(ii) Extended X-ray Absorption Fine Structure (EXAFS) region: portion at higher energies with respect to the edge (from tens to hundreds of eV), characterized by the oscillatory modulation in the absorption coefficient.

Hereby, we will briefly discuss the main information that can be extracted from the analysis of each of the two regions listed above.

5.3.2 The XANES Region

As discussed in chapter 4 of this book,[6] devoted to vibrational and electronic spectroscopies, the XANES part of the XAS spectrum reflects the unoccupied electronic levels of the selected atomic species. The investigation of these levels provides information on oxidation and coordination state of the absorber atom.

5.3.3 The EXAFS Region

The EXAFS region of the spectrum is located at higher energies and is characterized by the modulation of the absorption coefficient $\mu(E)$. Such a feature is caused by the interference between the X-ray waves diffused by the absorber atom and back-scattered by its neighbors. Hence, EXAFS oscillations can be related *via* Fourier transform to a specific spatial arrangement of the atoms in the local environment of the absorber, bridging the energy space of the incoming photon (k-space of the outgoing photoelectron) to the real distances r-space. This crucial point is at the basis of the EXAFS analysis procedure developed after the milestone works of Sayers, Lytle and Stern.[138,139,141]

The higher photoelectron kinetic energy in the EXAFS region implies that the phenomenon is no longer dominated by the full multiple scattering (MS) regime, that prevails in the XANES region;[141] consequently data analysis can be performed using the simpler Fourier transform operation.[138]

The EXAFS signal $\chi(E)$ is generally expressed as the oscillatory part of the $\mu(E)$ function, normalized to the edge-jump, *i.e.* $\chi(E) = [\mu(E) - \mu_0(E)]/\Delta\mu_0(E)$, where $\mu_0(E)$ is the atomic-like background absorption and $\Delta\mu_0(E)$ the normalization factor. Above the absorption edge, the energy E can be substituted with the photoelectron wave-vector k using eqn (5.41), therefore obtaining the EXAFS function $\chi(k)$. The relation between the modulation of the $\chi(k)$ signal and the structural parameters is provided by the EXAFS formula that, in the single scattering (SS) approximation, is reported in eqn (5.42).

$$\chi(k) = S_0^2 \sum_i N_i F_i(k) e^{-2\sigma_i^2 k^2} e^{-\frac{2r_i}{\lambda(k)}} \frac{\sin[2kr_i + 2\delta_l(k) + \theta_i(k)]}{kr_i^2} \qquad (5.42)$$

S_0^2 is the overall amplitude reduction factor; the index i runs over all the different shells of neighboring atoms around the absorber, $F_i(k)$ is the back-scattering amplitude as a function of k for each shell, N_i is the coordination number (number of equivalent scatterers), σ_i is the Debye-Waller factor accounting for thermal and static disorder. The parameter r_i indicates the interatomic distance of the ith shell from the central absorber. The phase shift of the photoelectron is distinguished in two contributions, related to the absorber ($2\delta_l$) and to the scatterer (θ_i).

In eqn (5.42) the electron back-scattering amplitude $F_i(k)$ is measured in Å,[142–144] because the cross section, which is an area, see eqn (5.33), is the modulus squared of the backscattering amplitude $F_i(k)$ and plays a similar role to the atomic form factors $f_n(q)$ in eqn (5.8) or (5.16) for X-ray scattering. Indeed, $F_i(k)$ defines the weight that the ith neighbor has in the overall EXAFS signal. As the electron scattering is mainly performed by the electron clouds of the neighbor atoms, it is evident that $F_i(k)$ will be larger for larger-Z neighbors. Consequently, as was the case for X-ray scattering, EXAFS will be less efficient in the detection of low-Z neighbors and the discrimination among neighbors having similar Z will be critical. When the difference in Z is sufficiently large, then both backscattering amplitude $F_i(k)$ and phase shift functions are markedly different to allow an easy discrimination between the different neighbors, see Figure 5.11.

The term $\lambda(k)$ is the energy-dependent photoelectron mean free path, typically a few Å,[145] determining the local nature of the technique that can investigate only up to 5–8 Å around the photo-excited atom. This apparent

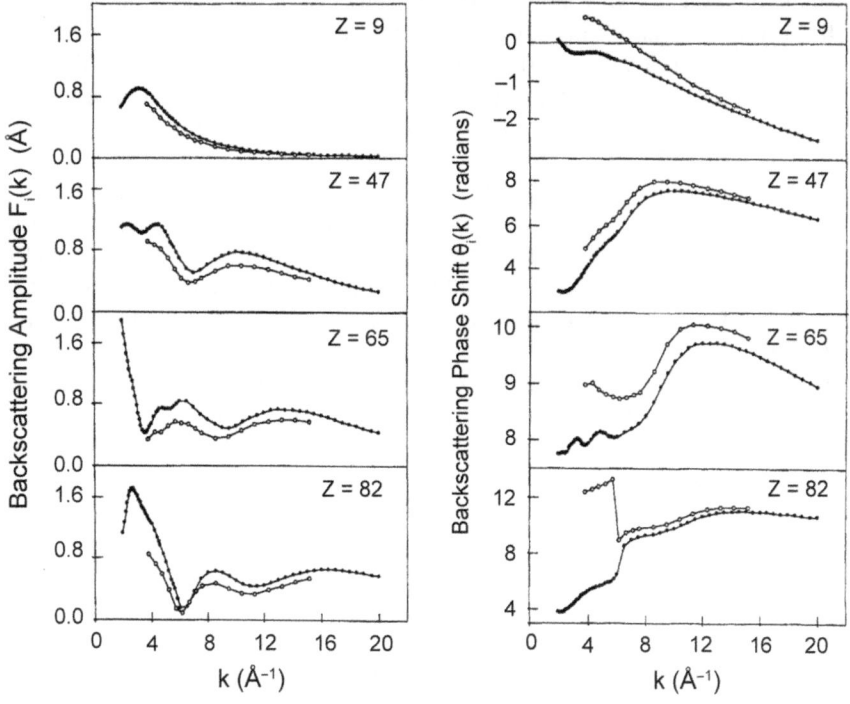

Figure 5.11 Left panels: backscattering amplitude functions $F_i(k)$ (in Å) *versus* k as obtained from plane wave calculations for F ($Z=9$), Ag ($Z=47$), Tb ($Z=65$), and Pb ($Z=82$) scattering atoms. Right panels: as left panels for the backscattering phase shift functions $\theta_i(k)$ (in radians). Open and full circles refer to the works of Teo and Lee[142] and of McKale *et al.*,[143] respectively. Reproduced with permission from ref. 143 (copyright American Chemical Society 1988).

limitation of EXAFS is conversely a big advantage in the investigation of disordered materials like glasses or liquids as the long-range order is not required.[146]

Coming to the Debye-Waller term $\exp(-2\sigma_i^2 k^2)$, it has the same physical origin of the term $\exp(-q^2 u_n^2/6)$ introduced in Section 5.2.1.4.5, for diffraction experiments,[63,64] see eqn (5.26). Actually the thermal parameters σ_i^2, measured in an EXAFS experiment, and $u_{n,a}^2$, $u_{n,b}^2$, $u_{n,c}^2$, measured in a diffraction experiment, represent two different aspects of the atomic vibrations in crystals. The anisotropic mean squared displacements $u_{n,a}^2$, $u_{n,b}^2$, $u_{n,c}^2$, or the isotropic mean squared displacement u_n^2 measure how the nth atom vibrates inside the unit cell along **a**, **b** and **c** directions (u_n^2 being an average of $u_{n,a}^2$, $u_{n,b}^2$, $u_{n,c}^2$). The σ_i^2 parameter measures how the couple of atoms of the absorber – i^{th} shell neighbor vibrates along the direction that connects these two atoms. This means that in case of strongly correlated vibrations, the scatterer and the absorber can vibrate in phase, resulting in a low σ_i^2 parameter and in a larger u_n^2 parameter.

The standard EXAFS formula, eqn (5.42), provides a convenient parameterization for fitting the local atomic structure around the absorbing atom to the experimental EXAFS data.[147] The dependence of the oscillatory structure of the EXAFS signal on interatomic distance and energy is clearly reflected in the $\sin(2kr_i)$ term. The strength of the interfering waves depends on the type and number of neighboring atoms through the backscattering amplitude $F_i(k)$ and the coordination number N_i, and hence is primarily responsible of the magnitude of the EXAFS signal. Once the phase and amplitude functions have been independently measured on model compounds or *ab initio* computed, the structural parameters N_i, r_i, and σ_i^2, can be determined in a least squares approach where the difference between the experimental and the modeled $k^n \chi(k_j)$ function is minimized along all the sampled experimental points k_j. The minimization routine can be done either in k-space, directly on the measured $k^n \chi(k_j)$ function, or in r-space, working on the Fourier-transformed functions. So, for each coordination shell, the coordination number, the atomic distance and the thermal factor can be extracted from an accurate EXAFS study. Extending the Nyquist-Shannon theorem[148,149] (also known as sampling theorem) to the EXAFS case, the maximum number of optimized parameters cannot exceed the number of truly independent points (n_{ind}), where n_{ind} is defined by the product of the sampled interval in k-space (Δk) and the interval in R-space (ΔR) containing the optimized shells:

$$n_{ind} = 2\Delta k \Delta R/\pi. \qquad (5.43)$$

A careful monitoring of the fitting results is fundamental to avoid local or non-physical minima of the minimization process. Analogously, correlation factors between each couple of optimized parameters should ideally be lower than 0.8 in absolute value and should never exceed 0.9. Eqn (5.43) underlines the importance of acquiring the EXAFS spectrum over the largest possible k-interval. Experimental data collected up to a high maximum k-value k_{max}: (i) increases n_{ind}, as Δk increases; (ii) reduces the correlation between N_i and σ_i^2 parameters; and (iii) increases the ability to discriminate between two close distances. The distance resolution (r_{min}) of an EXAFS spectrum is indeed defined from k_{max} according to the relation:

$$r_{min} = \pi/(2\,k_{max}). \tag{5.44}$$

Two equally intense signals generated by the same scatterer located at r_1 and r_2 can indeed be singled out only in the case of two oscillating functions $\sin(2kr_1)$ and $\sin(2kr_2)$ that are able to generate at least a beat in the sampled k-range, and this occurs for $2k(r_1-r_2)=\pi$. Consequently, in order to reach a distance resolution of $r_{min} = 0.1$ Å, the EXAFS spectrum has to be collected up to about $16\,\text{Å}^{-1}$. Eqn (5.44) is the EXAFS-equivalent of Eqn (5.23) discussed for diffraction experiments.

5.4   X-Ray and Neutron Total Scattering: Basic Considerations

The total scattering technique[150–158] is able to provide the overall pair distribution function (PDF) $G(r)$ of the material. The experimental setup needed is that of X-ray or neutron powder diffraction,[159,160] but the scattering pattern has to be collected to much higher exchanged q-values, up to at least $20–30\,\text{Å}^{-1}$. Low-λ sources and high-2θ collections are consequently required for PDF analysis. For standard Cu Kα ($\lambda = 1.54$ Å) and Mo Kα ($\lambda = 0.71$ Å) tubes, a collection up to $2\theta = 140°$ results in $q = 7.7$ and $16.6\,\text{Å}^{-1}$, respectively. Working with synchrotron sources at $\lambda = 0.5, 0.4, 0.3$ and 0.2 Å, q-values as high as $23.8, 29.8$ and 39.7 and $59.0\,\text{Å}^{-1}$, respectively, can be reached for a data collection up to $2\theta = 140°$.

To analyze PDF data, the coherent scattering function $I_C(q)$ has to be extracted from the experimentally collected intensity $I_{exp}(q)$. Before performing this operation, the background intensity due to extrinsic contributions, *e.g.* Compton scattering, fluorescence, scattering from the sample holder, and other experimental artifacts[161–163] has to be removed. $I_C(q)$ has sharp intensities where there are Bragg peaks, and broad features in between from the diffuse scattering. The total-scattering structure function, $S(q)$, is then obtained from $I_C(q)$ as follows:[74,152]

$$S(q) = [I_C(q) - <f(q)^2> + <f(q)>^2]/<f(q)>^2 \tag{5.45}$$

where the symbol $<>$ denotes an average over all the chemical species in the sample and $f(q)$ is the X-ray atomic form factor. As $f(q)$ decreases upon increasing q, see Figure 5.3 and related discussion, very long integration times are needed at high q to obtain a good statistic. For this reason, area detectors are more suitable than point detectors because they allow the integration over a wide region of the diffraction cone. In addition, the poorer angular resolution of area detector is not a significant disadvantage in a q-region where the diffractogram undergoes only smooth variations. Alternatively, PDF studies can be performed using neutrons because the coherent neutron scattering length is constant in the whole q-region of interest, see Section 5.2.2. Both $I_{exp}(q)$ and $I_C(q)$ data appear smooth and featureless in the high-q region (this holds even for crystalline materials where usually no Bragg peaks are observed above $q \approx 10 \,\text{Å}^{-1}$). However, after normalizing and dividing by the square of the atomic form-factor, important oscillations appear in this region of the $S(q)$ function. A similar behavior is observed in EXAFS experiments, comparing $\mu(E)$ and $\chi(k)$ functions at high E (high k or high q) after the edge. Finally, the reduced pair distribution function, $G(r)$, is obtained from $S(q)$ through a sine FT:

$$G(r) = \frac{2}{\pi} \int_{q_{min}}^{q_{max}} q[S(q) - 1] \sin(qr) dq \qquad (5.46)$$

where q_{min} and q_{max} are the limits of the data collection in q-space, being $q_{min} \sim 0 \,\text{Å}^{-1}$ and q_{max} as large as possible. The PDF function (5.46) gives the inter-atomic distance distribution, having peaks at positions r corresponding to the most probable distances between each pair of atoms in the solid. The PDF function therefore contains EXAFS-like information that is, however, not atomically selective: $G(r)$ includes contributions arising from the local environments of all the atomic species present in the sample. In this regard, the intrinsic differences in the nature of the $\chi(k)$ and $S(q)$ signals obtained from EXAFS and PDF experiments on single-component disordered systems were deeply discussed by Filipponi.[164] In that work, particular effort was devoted to connect the $\chi(k)$ signal with quantities commonly employed within the distribution function theory in disordered matter. As the physical phenomenon behind PDF is X-ray scattering and not photoelectron scattering, the PDF signal is not damped by the short photoelectron mean-free path and by the core hole life-time as EXAFS is, see eqn (5.42), so valuable structural information is contained in the pair-correlations extending to much higher values of r, than typically reachable by EXAFS (≈ 5–$8 \,\text{Å}$). In fact, with high q-space resolution data, PDFs can be measured out to tens of nanometres (hundreds of Ångströms) and the structural information remains quantitatively reliable. With respect to EXAFS, the PDF data analysis does not have to deal with MS paths, as only SS signals are present, remarkably simplifying the data modeling. However the PDF signal contains, entangled, the structural information about the local environment around all the atomic species present in the sample, and further complications are related to the isolation of the different contributions.

5.5 Applications

5.5.1 Determination of Possible Interpenetrating Frameworks and of Possible Extra Phases in Some MOF-5 Syntheses by Combining Single Crystal XRD, XRPD and Zn K-edge EXAFS

5.5.1.1 Problems Related to Synthesis Reproducibility of MOF-5

Traditionally, MOF syntheses have been designed to yield high quality single crystals suitable for structural analysis. A diversity of methods have been adopted, and they often involve a slow introduction of the reactants to reduce the rate of crystallite nucleation, such as slow diffusion of one component solution into another through a membrane or an immobilizing gel, slow evaporation of a solution of the precursors, or layering of solutions. Fortunately, solvothermal techniques have been found to be a convenient replacement for these often time-consuming methods. However, if an increased yield is more desirable than a high crystal quality, the reaction times can be significantly reduced by increasing reactant concentration and by employing agitation.[165,166] The product formed under these conditions may or may not be exactly identical to those obtained from methods used to produce highly crystalline MOF materials.

In this regard, MOF-5 represents a case study because of both its high popularity (it is probably the most highly cited MOF) and the large number of different synthesis recipes reported in the literature. Indeed, the initial findings of reversible hydrogen adsorption,[167–169] thermal robustness[170,171]and interesting luminescence properties[172] of MOF-5 have made it one of the most studied MOFs. Moreover, a broad set of synthesis conditions and procedures have been tried to obtain MOF-5 either in large scale quantities or with a particular crystallite size or to reduce synthesis time.[173,174] Yaghi and co-workers have reported several different synthesis methods for MOF-5, and all of them yield a product with a fairly large crystallite size, suitable for structure determination.[170,175] Huang *et al.*[173] reported a synthesis strategy for fast formation of nanocrystalline MOF-5. Naturally, the inherent challenge with a nanocrystalline material is the limited structural information provided by XRPD. Later, Ni *et al.*[174] presented a new microwave-assisted solvothermal synthesis approach, which allows MOF-5 crystals of uniform size (4 ± 1) μm to be synthesized in less than a minute. On the basis of the similar, or at least related, powder XRD patterns the products from all the synthesis procedures have been claimed to be the same MOF-5 phase. However, after a careful scrutiny of the published XRPD patterns, Lillerud and co-workers[166] revealed clear intensity differences, especially in the two first and most intense peaks, see left part of Figure 5.12. Substantial variations in the surface area, ranging from 700 to 3400 m^2g^{-1}) of MOF-5 prepared according to the different procedures have also been reported, but were not explained.[168,173,176,177]

5.5.1.2 XRPD and Single Crystal XRD Studies

To clarify this inconsistency in the literature data, Lillerud *et al.*[166] prepared two set of MOF-5 material, the first repeating Huang's synthesis, resulting in nano-sized crystals (hereafter MOF-5_n), the second modifying slightly Huang's method to obtain micro-sized crystals sufficiently large for single-crystal XRD characterization (hereafter MOF-5_m).

The left part of Figure 5.12 reports the XRPD patterns of (a) MOF-5_m, (b) MOF-5_n, (c) digitalized powder XRD pattern of MOF-5 (from the work of Huang *et al.*[173]), (d) simulated powder XRD pattern of a MOF-5 crystal, and (e) digitalized powder XRD pattern of MOF-5 (from the work of Yaghi *et al.*[175]

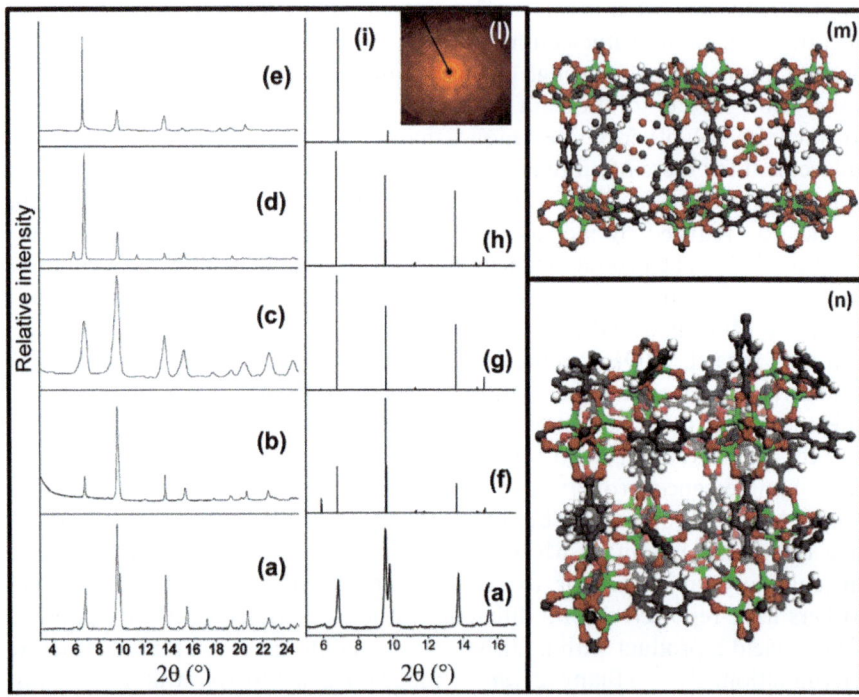

Figure 5.12   Comparison among experimental XRPD patterns of MOF-5 obtained in different labs, from different synthesis and comparison with some simulated XRPD patterns. Patterns have been collected with (or computed for) Cu Kα radiation ($\lambda = 1.5406$ Å). Left part: (a) Experimental XRPD of MOF-5_m by Lillerud *et al.*[166] The pattern has been reported in both left and right parts of the Figure for comparison. (b) Experimental XRPD pattern of MOF-5_n by Lillerud *et al.*[166] (c) Experimental XRPD pattern by Huang *et al.*[173] (d) Simulated XRPD pattern for MOF-5 according to the ideal structure. (e) Experimental XRPD pattern by Yaghi *et al.*[175] Right part: simulated patterns (reported without line broadening) of: MOF-5_m_Zn (f); MOF-5_m_int with solvent in the pores (g); MOF-5_m_int without solvent in the pores (h); and original MOF-5 when transformed from cubic to trigonal symmetry (i). Part (l) reports a 2D diffraction image collected on MOF-5_m at ESRF ID11 beamline ($\lambda = 0.38745$ Å). Parts (m and n): sticks and balls representation of structure refined from single crystal XRD performed on samples selected from MOF-5_m synthesis. Zn atoms are shown in green, O in red, C in gray, and H in white. Part (m): structure of MOF-5_m_Zn with Zn cluster in the small cage and unorganized solvent in the large cage; the electron densities are illustrated with partly occupied oxygen and carbon atoms. Part (n): structure of MOF-5_m_int showing two interpenetrated MOF-5 ideal lattices. Adapted with permission from ref. 166, copyright 2007 ACS.

From this comparison Lillerud *et al.*[166] deduced that the experimental pattern obtained on a high-surface-area MOF-5 reported by Yaghi *et al.*[175] (Figure 5.12e) is highly consistent with the calculated MOF-5 pattern

(Figure 5.12d) and that the patterns of MOF-5_m, MOF-5_n and nanocrystalline MOF-5 reported by Huang *et al.*[173] (Figure 12a,b,c) are similar to each
other and deviate significantly from the pattern of the ideal MOF-5 phase
(Figure 5.12d,e). The main differences in the XRPD patterns concern the
intensities of the first (6.9°, corresponding to a *d* of 12.8 Å) and second (9.7°,
corresponding to a *d* of 9.1 Å) peak. As expected, the larger crystals of
MOF-5_m give less peak broadening, and the peak at 9.7° consists of two
distinct contributions. This peak splitting phenomenon does not occur for
MOF-5_n. On the basis of the reported XRPD patterns and on the Langmuir
surface areas of MOF-5_n (747 m^{2}g^{-1}) and MOF-5_m (1104 m^{2}g^{-1}), Lillerud
et al.[166] concluded that MOF-5_m, MOF-5_n and the nanocrystalline MOF-5
reported by Huang *et al.* (722 m^{2}g^{-1}) represent the same phase.

The microcrystalline nature of the samples allowed Lillerud *et al.*[166] to
perform synchrotron radiation single crystal data collection (ESRF ID11, see
part (l) of Figure 5.12), and refinement to attain a deeper understanding of this
material. The single-crystal data were collected for more than ten different
MOF-5_m crystals. During this procedure two different kinds of MOF-5_m
crystals were discovered, hereafter denoted MOF-5_m_Zn and MOF-5_m_int.
The MOF-5_m_Zn structure was solved in the trigonal *R* $\overline{3}m$ space group (No.
166), while the original MOF-5 structure has a cubic symmetry (*Fm* $\overline{3}m$, No.
225). Its structure, reported in Figure 5.12m, has a framework consisting of the
same building units as the ideal MOF-5, but exhibiting additional electron
density in the center of the cavities (*i.e.*, the half of the cages where the benzene
rings are twisted into the cage center). The best refinement was obtained by
assigning the electron density to a Zn atom with occupancy of 0.5. The coordination sphere of the Zn atom consists of partially occupied oxygen positions.
Figure 5.12n illustrates the structure of MOF-5_m_int after removal of the
solvent. MOF-5_m_int consists of two MOF-5 frameworks interpenetrated
with each other. The frameworks are not physically connected, but there is
sufficient interaction to cause a significant distortion of the cell from cubic to
trigonal. The structure was therefore also refined in the *R* $\overline{3}m$ space group with
an equal occupation of the two frameworks.[166] Summarizing, in MOF-5_m_Zn
(Figure 5.12m) the Zn–O species in the pores cause a change in the axes'
lengths, while in the MOF-5_m_int the interaction between the two interpenetrated frameworks is responsible for the same effect. These results show that
the large MOF-5 cell is highly flexible and allows significant changes in the unit
cell axes without destroying the structure. The models of the two phases
reported in Figure 5.12m,n are also able to take into account for the large
deviation in the surface area of MOF-5 materials (700–3400 m^{2}g^{-1}) reported in
the literature:[168,173,176,177] as for MOF-5_m_Zn, the presence of the nonvolatile
compounds (Figure 5.12m) makes the host cavity inaccessible and may
also block the entrance to the adjacent cavities, for MOF-5_m_int, the
presence of doubly interpenetrated MOF-5 networks will also reduce the
adsorption capacity. These models were also in agreement with the reported
XRPD patterns (Figure 5.12ab), because it is well known from the large

experience on zeolitic materials that when crystalline materials have the pores filled with electron density (from whatever origin: template, adsorbed molecules *etc.*) then the intensity of the low-2θ peaks decreases.[178,179]

Combining single crystal XRD and TG data, Lillerud *et al.*[166] were able to conclude that the MOF-5_m batch is composed by 94% MOF-5_m_Zn and 6% MOF-5_m_int.

5.5.1.3   *Role of EXAFS in Determining Extra Phase Zn Species Foreseen by XRD*

The missing part of the puzzle was a direct, atomically selective proof that the electron density found in the MOF-5_m_Zn phase (representing 94% of the MOF-5_m batch) by single crystal XRD is actually attributable to zinc oxide nanoclusters. Zn K-edge EXAFS is of course the technique of choice to answer to this question. Lillerud *et al.*[166] performed XANES and EXAFS measurements on both MOF-5_m and MOF-5_n batches finding very similar results, indicating that the average Zn local environment is basically the same in both syntheses. The authors concluded that solvent removal does not affect neither the XANES nor the EXAFS spectra, testifying that Zn atoms in MOF-5 do not exhibit any coordination vacancy.

The fit of the EXAFS signal using the path degeneration, expected from the ideal MOF-5 structure obtained by XRD and fixing reasonable σ^2 values, failed because it was not able to correctly reproduce the relative intensities of the first and second shell contributions; with respect to the experimental datum, the fit underestimated the first, while overestimating the second. Allowing σ^2 values to vary independently for the two shells, resulted in physically too low or too high values being obtained for the first and the second shell, respectively. The failure of this fit was the consequence of the presence of a highly disordered nano-structured zinc oxide phase trapped inside the MOF-5 cavities. Disordered nanoclusters are expected to mainly give just a first shell Zn–O contribution, so explaining the inability to reproduce the intensity ratio between the first and second shell contribution using as a model the ideal MOF-5 structure. The quality of the fit was significantly improved by adding an extra-phase Zn–O contribution.[166] The Zn–O distance of this phase was optimized at 2.11 ± 0.01 Å. This value is somewhat stretched with respect to the Zn–O distance in crystalline zinc oxide as expected for highly disordered nanoclusters. So Zn K-edge EXAFS was able to attribute the electron density found in the single crystal XRD study of the MOF-5_m_Zn phase to highly disordered ZnO nano-clusters.[166]

5.5.2   Combined XRPD, EXAFS and *Ab Initio* Study of NO, CO and N_2 Adsorption on Ni^{2+} Sites in CPO-27-Ni

Dietzel *et al.*[180] synthesized a three-dimensional honeycomb-like metallorganic framework (Figure 5.13a) with Ni^{2+} as the metal component: $Ni_2(dhtp)(H_2O)_2 \cdot 8H_2O$ (dhtp = 2,5-dihydroxyterephthalic acid). This new material, named CPO-27-Ni, belongs to the family of CPO-27-M (Mg, Co, Ni) also known as MOF-74 (synthesized by Yaghi and co-workers) and is isostructural to framework materials with Zn^{2+}, Co^{2+} and Mg^{2+} metal component.[181–184]

The structure of CPO-27-Ni was solved by Dietzel *et al.*[180] using synchrotron XRPD data collected at BM01B beamline of the ESRF at $\lambda = 0.50134$ Å. The data were refined up to $2\theta_{max} = 34.5°$ and $28.0°$ for the hydrated an dehydrated forms, resulting in $d_{min} = 0.85$ and 1.04 Å, respectively, see eqn (5.23). The CPO-27 framework contains one-dimensional channels (Figure 5.13a) filled with water that can be removed by a mild thermal treatment. Upon dehydration the crystalline structure is preserved and a material with a high surface area is obtained (about 1100 m^2g^{-1}), which contains unsaturated metal sites organized in helicoidal chains.[180] At the intersections of the honeycomb are helical chains of *cis*-edge connected nickel oxygen octahedra running along the *c* axis. Nearest neighbors helices are of opposite handedness. Each chain is connected by the organic ligand with three adjacent chains, resulting in the honeycomb motif. The channels in the honeycomb have a diameter of ~ 11 Å (see Figure 5.13a). All of the O atoms of the ligand are involved in the coordination of Ni^{2+}; these oxygen atoms account for five out of six ligands for each nickel atom, while the sixth coordinative bond is to a water molecule which points towards the cavity.

CPO-27-Ni, in both its hydrated an dehydrated forms, was studied in detail by Bonino *et al.*[185] The refined structure from XRPD Rietveld refinement[180] was used as input for the EXAFS model, resulting in an excellent agreement between the set of distances optimized with the two different techniques. The EXAFS signal was quite complex because it was constituted by several SS and MS paths. Therefore Bonino *et al.*[185] cross-checked the validity of their EXAFS model analyzing the data collected on dehydrated CPO-27-Ni at 300 and 77 K (see Table 5.3). The model was validated as all optimized distances were comparable in the two datasets, while the thermal parameters σ^2, see eqn (5.42), increased moving from 77 to 300 K. Water removal from CPO-27-Ni significantly changes both its XANES and EXAFS spectra. In particular, the average Ni–O first shell distance decreases from 2.03 ± 0.01 Å down to 1.99 ± 0.01 Å, while an even more impressive contraction was observed for the second shell Ni–Ni distance, that moves from 2.980 ± 0.005 Å down to 2.892 ± 0.005 Å, see Table 5.3. The desolvation process caused the removal of the water molecule coordinated to the metal center, resulting in Ni^{2+} cations with a coordinative vacancy potentially able to coordinate ligand molecules. The interaction of NO, CO and N_2 ligands with desolvated CPO-27-Ni has been deeply investigated by means of Ni K-edge XANES and EXAFS spectroscopies, supported by parallel IR and UV-Vis techniques.[185–187]

High quality data were obtained in transmission mode up to almost $k = 20$ Å^{-1}, see Figure 5.13b: this allowed a high resolution in *r*-space, better than 0.08 Å, see eqn (5.44). The EXAFS data (and corresponding best fits) obtained on dehydrated CPO-27-Ni and after interaction with H_2O, NO, CO

Table 5.3   Summary of the parameters optimized by fitting of the EXAFS data collected at 77 K and 300 K on CPO-27-Ni MOF in interaction with different molecules. The fits were performed in r-space in the 1.0–5.0 Å range over k^3-weighted FT of the $\chi(k)$ functions performed in the 2.0–18.0 Å^{-1} interval. A single ΔE_0 and a single S_0^2 have been optimized for all SS and MS paths. The Ni–O, and Ni–Ni (first and second neighbor) SS paths have been modeled with their own path length and Debye-Waller factors, while a unique σ^2 and a unique path length parameter α, common to all other SS and MS paths, have been optimized. NO, CO and N$_2$ adsorption has been simulated by treating the molecule as a rigid body linearly adsorbed on Ni^{2+}. Consequently only two additional parameters are needed: the Ni-molecule distance (R_{ads}) and the corresponding Debye-Waller factor (σ^2_{ads}). N_{ind} = number of independent points ($\pi \Delta R \Delta k/2$); N_{var} = number of optimized parameters. Previously unpublished table summarizing data from ref. 185–187.

Sample condition	Dehydrated, from ref. 185	In vacuo, from ref. 185	Hydrated, from ref. 185	+ NO, from ref. 185	+ CO, from ref. 186	+ N$_2$, from ref. 187
T/K	300	77	300	300	77	77
R_{factor}	0.043	0.033	0.027	0.045	0.018	0.011
N_{ind}	40	40	40	40	40	40
N_{var}	10	10	7	7	12	12
ΔE_0/eV	-2.5 ± 1.0	-1.7 ± 1.0	0.5 ± 0.5	-2.5 ± 1.0	0.8 ± 0.7	2.1 ± 1.0
S_0^2	1.17 ± 0.09	1.20 ± 0.08	1.17	1.17	1.18 ± 0.06	1.18 ± 0.05
$\langle R_O \rangle$/Å	1.99 ± 0.01	2.00 ± 0.01	2.03 ± 0.01	1.99 ± 0.01	2.024 ± 0.005	2.012 ± 0.005
$\sigma^2(O)$/Å^2	0.0049 ± 0.0005	0.0042 ± 0.0004	0.0049	0.0049	0.0044 ± 0.0004	0.0038 ± 0.0003
R_{Ni1}/Å	2.892 ± 0.005	2.889 ± 0.005	2.980 ± 0.005	2.95 ± 0.01	2.973 ± 0.005	2.937 ± 0.005
$\sigma^2(Ni_1)$/Å^2	0.0055 ± 0.0006	0.0045 ± 0.0004	0.0055	0.0055	0.0038 ± 0.0004	0.0035 ± 0.0003
R_{Ni2}/Å	4.82 ± 0.02	4.87 ± 0.01	4.78 ± 0.03	4.79 ± 0.02	4.89 ± 0.02	4.86 ± 0.02
$\sigma^2(Ni_2)$/Å^2	0.0059 ± 0.0018	0.0052 ± 0.0014	0.0059	0.0059	0.008 ± 0.002	0.006 ± 0.002
α	-0.008 ± 0.009	-0.003 ± 0.007	-0.021 ± 0.005	-0.024 ± 0.007	-0.013 ± 0.009	-0.003 ± 0.007
σ^2/Å^2	0.009 ± 0.004	0.007 ± 0.003	0.009	0.009	0.009 ± 0.005	0.009 ± 0.004
R_{ads}/Å	—	—	2.10 ± 0.04	1.85 ± 0.02	2.11 ± 0.02	2.27 ± 0.03
σ^2_{ads}/Å^2	—	—	0.01 ± 0.01	0.0065 ± 0.002	0.006 ± 0.002	0.010 ± 0.005
$-\Delta H_{ads}^{expt}$/kJ mol^{-1}	—	—	100	92	58	17

and N_2 are reported in *r*-space in Figure 5.13c–g and Table 5.3. The higher intensity of the EXAFS oscillation in the case of the CPO-27-Ni/CO and CPO-27-Ni/N_2 is evident and is due to the fact that corresponding spectra were collected at 77 K,[186,187] while the spectra of CPO-27-Ni contacted by H_2O and NO were collected at 300 K.[185] This implied that authors were forced to fix many more parameters in the analysis of the this last case, see Table 5.3. Independent IR experiments allowed the coordination number of the adsorbed molecules (NO, CO and N_2) to be fixed to 1,[185–187] see also in this book Section 4.2.4.4. of Chapter 4.[6]

Adsorption of molecules on Ni^{2+} sites strongly modifies the whole framework structure inducing elongation in Ni–O and Ni–Ni distances. Figure 5.14a–d summarizes experimental structural data (XRPD and EXAFS) on the adsorption of H_2O, NO, CO, CO_2, and N_2 molecules on CPO-27-Ni material. Data are reported as a function of the enthalpy of adsorption measured *via* standard microcalorimetric techniques[185,186] or *via* temperature-dependent IR desorption[187] or *via* isosteric heat of adsorption.[190] The figure summarizes data collected at both 77 and 300 K (open and full symbols, respectively). Comparison with the analogous values obtained from a theoretical study performed at the B3LYP-D*/TZVP level of theory (using a periodic boundary conditions) is reported in parts (e)–(h) of Figure 5.14.

From the reported set of data, it clearly emerges that the computed framework distances and computed adsorption distances are systematically overestimated by the theory. Notwithstanding this fact, the trends observed in the experimental data are clearly mirrored by the theoretical data. In particular, it emerges from both experimental and theoretical data that the larger the adsorption energy, the larger is the perturbation induced by the adsorbed molecule to the MOF framework in terms of elongation of the $<R_O>$, R_{Ni1} and R_{Ni2} distances, see parts (a)–(c) and (e)–(g) of Figure 5.14, respectively. As far as the adsorption distance is concerned, it follows an opposite trend: the larger the $-\Delta H_{ads}(-\Delta E^c_{ads})$, the shorter is R_{ads}, see Figure 5.14d and Figure 5.14h, respectively.

Regarding the EXAFS results, it is worth noticing that $<R_O>$, and R_{Ni2} increase by decreasing the temperature from room temperature (300 K, filled triangles) to 77 K (empty triangles) while R_{Ni1} does not change. An increase in the cell volume by decreasing the temperature indicates a negative thermal expansion coefficient; this rare property is shared by some other MOF structures, as determined by temperature dependent diffraction experiments on MOF-5[191–193] or foreseen by force-field calculations on the IRMOF-1/-10/-16 family,[72,194,195] and on HKUST-1.[196,197]

The calculations predict that upon adsorption there will be an increase of all the framework distances considered and an almost linear relationship between the adsorption energy and the distance elongation, Figure 5.14(e)–(g). These findings have been confirmed by the experiments: as a general statement, both XRD and EXAFS indicate an increase of all the framework distances upon molecular adsorption. However, in the experiments, a larger spread of the data is observed due to the different coverages adopted in the different experiments. In fact, whereas in the calculations the coverage ratio was fixed to

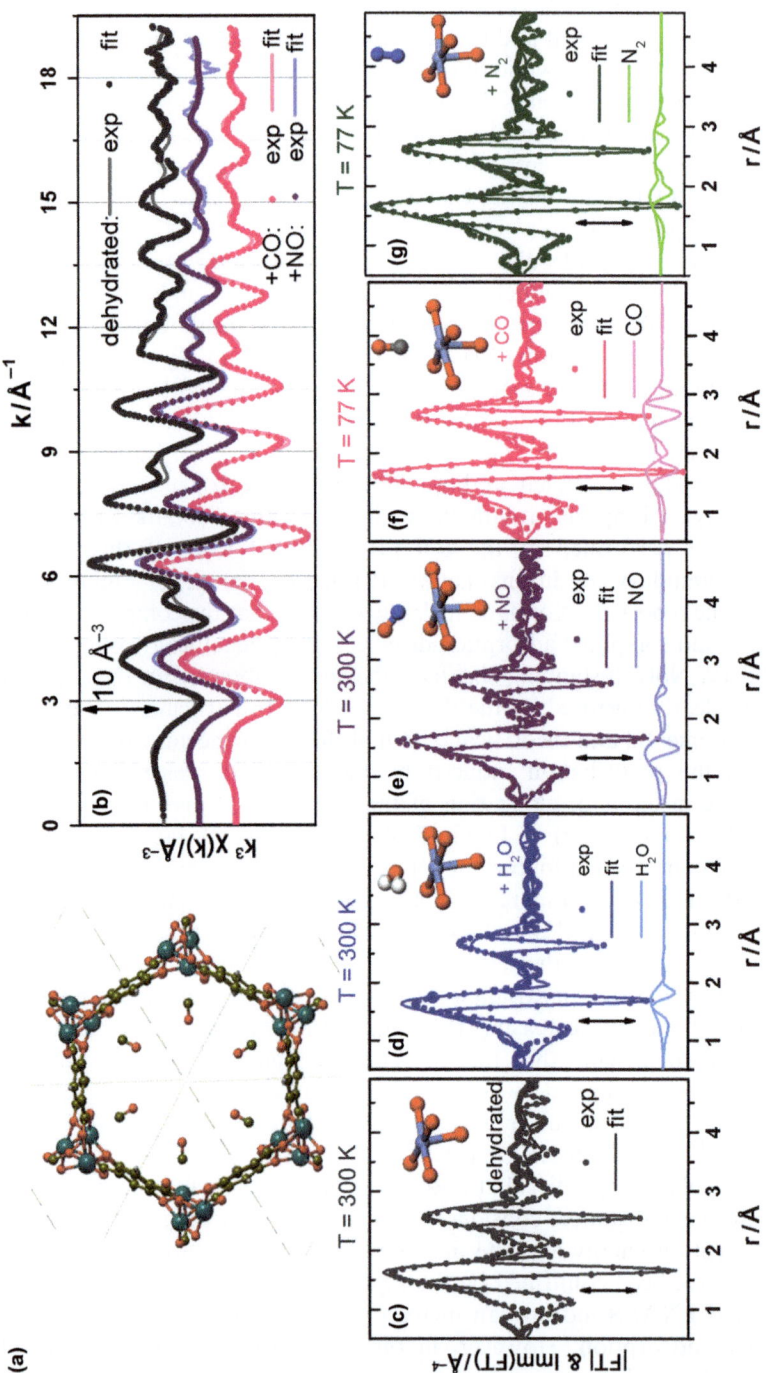

Ni : molecule $= 1 : 1$, in the XRD for CO_2 and H_2O a $Ni : CO_2 = 1 : 0.5–0.6$ and $Ni : H_2O = 1 : 5$ were adopted. The different coverages are likely ascribable to the different behaviors observed for H_2O adsorption of R_{Ni2} obtained with EXAFS and XRPD, which indicate, respectively, a shortening and a lengthening of this distance. In fact, whereas the XRPD data have been recorded for the highest coverage, the EXAFS measurements have been recorded at a lower $Ni : H_2O$ ratio and then the Ni_{ads}–Ni_2 shortening is a reflection of the high interaction energy. In fact, a shortening of R_{Ni2} has been also observed for NO, the second in interaction energy among molecules considered in Figure 5.14. Coming to the distance between the Ni atom and the adsorbed molecules (R_{ads}), in this case an opposite trend is observed in both experiments and calculations as expected: in fact this distance shortens by increasing the energy of the interaction, the shortest distances being observed for the larger interacting molecules that is for H_2O and NO.

The here reviewed multitechnique approach[180,184–187,189,190] requiring XRPD and EXAFS for structural determination and micro-calorimetry or temperature-dependent IR desorption or isosteric heat of adsorption for adsorption enthalpies determination, and supported by periodic DFT calculations, is relevant in understanding and foreseeing applications to a potential practical uses of MOF materials. Indeed, the understanding of the molecular adsorption on a given surface site is the first step in understanding whether the site may have a potential catalytic reactivity or not.[198] On the other hand, measuring (and/or computing) adsorption enthalpies of different molecules establishes an adsorption strength scale that is relevant in determining a selective adsorption ranking useful for gas separation and selective adsorption purposes. In more detail: (i) the significant difference in the $-\Delta H_{ads}$ (and $-\Delta E^c_{ads}$) for the adsorption of H_2 and CO implies that CPO-27-Ni is an interesting material for the purification of a H_2/CO mixture used to feed fuel cells. (ii) The material can clearly also play a role in the CO_2 capture, even at

---

Figure 5.13   Part (a): the structure of CPO-27-Ni/CO ($Ni^{2+} : CO = 1$) optimized by *ab initio* periodic approach with CRYSTAL code[184,189] and viewed along the *c* axis. Part (b): $k^3\chi(k)$ of CPO-27-Ni after: desolvation (gray curves); interaction with NO (violet curves); and interaction with CO (pink curves). Parts (c)–(g): modulus and imaginary part of the k^3-weighted, phase uncorrected, FT of the EXAFS spectra collected on dehydrated CPO-27-Ni (c); and after interaction with H_2O (d); NO (e); CO (f) and N_2 (g). Where adsorbates are present, the vertically translated contribution (in both modulus and imaginary parts) of the adsorbed molecule optimized in the fits is also reported. The models used in the fits adopted a $Ni^{2+}/adsorbate = 1 : 1$ stoichiometry and assumed a linear adsorption geometry for CO and N_2 and a Ni–N–O angle of 130° for the NO (only the O atom of the H_2O molecule has been included in the fit). Insets report the local environment of Ni^{2+} in its dehydrated form, part (a), and upon molecular adsorption, parts (b–e), as optimized by *ab initio* calculations. In parts (b)–(g) scattered and continuous curves refer to the experimental data and the best fit, respectively. Previously unpublished figure: the EXAFS spectra have been adapted from ref. 185–187.

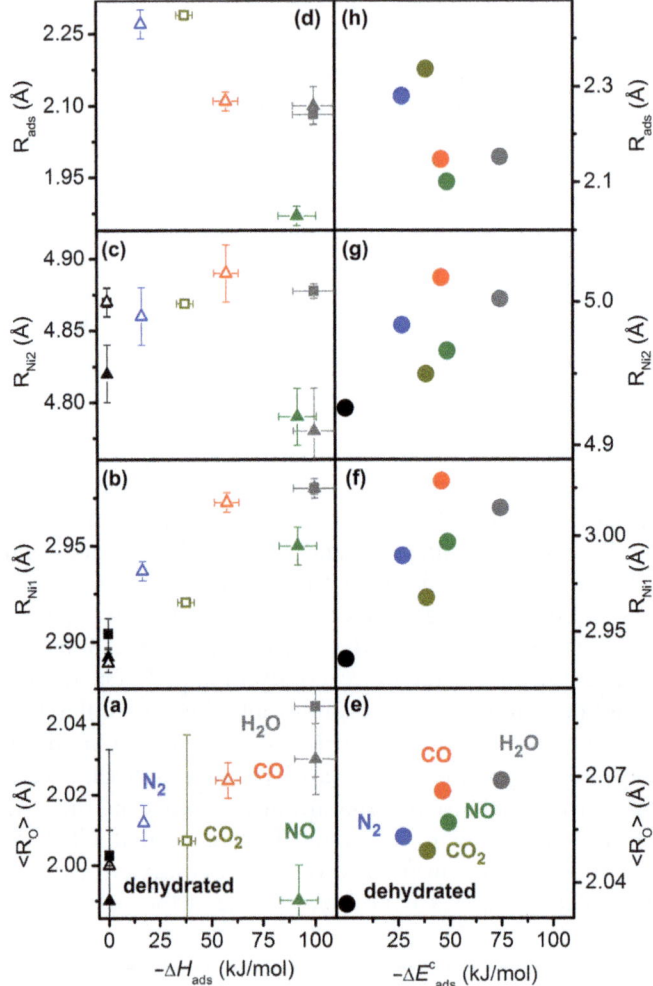

Figure 5.14 Correlation of the different structural parameters upon molecular adsorption on Ni^{2+} site with the corresponding adsorption energy. Parts (a)–(d): experimental values (high-resolution XRPD squares, EXAFS triangles) and corresponding uncertainty. Full and open symbols refer data collected at 300 and 77 K, respectively. Parts (e)–(h): theoretical values. Note that left and right parts do not have the exactly the same ordinate intervals. This reflects the systematic overestimation of the theoretical distances. Beside this fact, all trends are well reproduced. Adapted with permission from ref. 189, copyright 2012 Elsevier.

relatively high temperatures, *i.e.* for *post-combustion* capture as demonstrated by the work of Dietzel *et al.*[190] (iii) Finally, the ability of H_2O to progressively displace NO from the Ni^{2+} sites,[185] makes CPO-27-Ni a good candidate for a controlled NO drug delivery inside the human body, similarly to the HKUST-1 MOF investigated by the group of Morris.[199,200]

5.5.3 Combined XRPD, EXAFS and *Ab Initio* Studies of Structural Properties on MOFs of the UiO-66/UiO-67 Family: Same Topology but Different Linkers or Metal

The recently discovered UiO-66/67/68 class of isostructural MOFs[201] has attracted great interest because of its remarkable stability at high temperatures, high pressures and in the presence of different solvents acids and bases.[188] UiO-66 is obtained by connecting $Zr_6O_4(OH)_4$ inorganic cornerstones with 1,4-benzene-dicarboxylate (BDC) as linker, while the isostructural UiO-67 material, obtained using the longer 4,4′ biphenyl-dicarboxylate (BPDC) linker[202] (Figure 5.15a) and Hf-UiO-66 is obtained by keeping the UiO-66 linker (BDC) and substituting the $Zr_6O_4(OH)_4$ blocks with $Hf_6O_4(OH)_4$ corners (inset in Figure 5.15e). XRPD, see Figure 5.15b,e testifies to the quality of the synthesis. Due to the rigidity of the framework, several isostructural UiOs has been prepared and tested for the stability and gas adsorption. Kandiah *et al.*[203] studied the thermal and chemical stabilities of isostructural UiO-66-X ($X = NH_2$, Br and NO_2) and observed the lower stability of this analogue with respect to parent UiO-66. Conversely, as documented by the thermogravimetry studies reported in Figure 5.15c,f UiO-67[202] and Hf-UiO-66[204] show thermal and chemical stability similar to that of UiO-66 and exhibit the expected surface area, as determined by low temperature volumetric N_2 adsorption isotherms (Figure 5.15d,g). Such a high stability is related to the fact that each Zr- (Hf-) octahedron is 12-fold connected to adjacent octahedra. This connectivity is very common for metals, resulting in the highly packed *fcc* structure, but it is still almost unique in MOF topologies.

The desolvation process left the XRPD pattern of such materials (Figure 5.15b) almost unchanged: besides a gain of intensity of the basal reflections (due to the removal of the electron density inside the pores),[166,178,179] all peaks remain in almost the same 2θ position with small intensity changes. Conversely, a huge modification of the EXAFS spectra is obtained in all cases, see Figure 5.16.

In the three hydroxylated materials, the structure determined from the Rietveld refinement of the XRPD corresponding patterns resulted in a straightforward interpretation of the complex EXAFS signals, see first three columns in Table 5.4. The dramatic modification undergone by the EXAFS spectrum upon dehydroxylation (see Figure 5.16) makes the data analysis not so straightforward. In the case of UiO-66 (see Figure 5.16b, but similar effects are observed in the two other cases) the changes are basically explained in terms of three main effects: (i) small contraction of the first metal-oxygen (M–O) shell accompanied by a small decrease in coordination (erosion of the shoulder around 1.9 Å); (ii) relevant distortion of the second shell contribution showing a maximum that moves from 3.17 Å to 2.91 Å, with a shoulder at 3.41 Å, thus reflecting an important splitting of the R_{M1} distances of the octahedron sides; (iii) the almost complete disappearance of the weak contribution around 4.7 Å, due to the M–M SS signal of the octahedron diagonal (R_{M2}). For the three

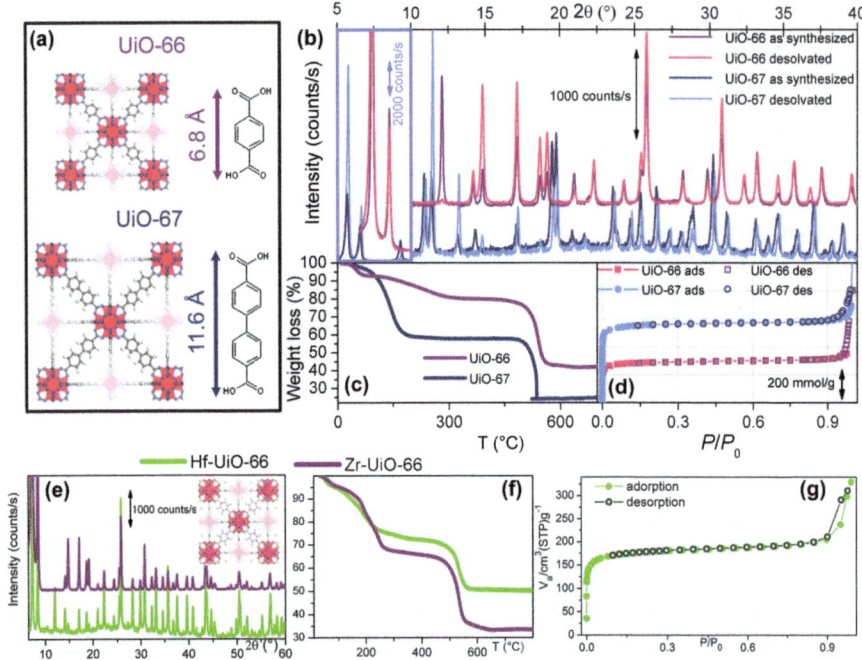

Figure 5.15 Part (a): from top to bottom: comparison of the dimension of linker and structure for the isostructural UiO-66 and UiO-67 MOFs. Part (b): comparison of the XRPD patterns ($\lambda = 1.540$ Å) for as prepared UiO-66 (UiO-67), violet (dark blue) curve and activated at 300 °C (pink and light blue curves). The patterns in the 10–40 2θ region have been amplified by a factor 4. Patterns related to UiO-66 have been vertically translated for clarity. Part (c): TGA curve of UiO-66 and UiO-67 samples, violet and dark blue curves, respectively. In both cases, the heating ramp was 5 °C min^{-1} in a N$_2$ flow (100 mL min^{-1}). Part (d): volumetric N$_2$ adsorption isotherms recorded at 77 K on UiO-66 (violet squares) and UiO-67 (blue circles). Filled and empty symbols refer to the adsorption and desorption branches, respectively. Part (e): XRPD pattern ($\lambda = 1.540$ Å) of Hf-UiO-66 (green) and Zr-UiO-66 (violet) in their solvated forms. The inset reports the MOF structure. Part (f): weight loss of Hf-UiO-66 and Zr-UiO-66 relative to the start mass (green and violet curve, respectively). Since hafnium is 41% heavier than zirconium, the Zr-UiO-66 shows both higher initial and breakdown losses. Part (g): N$_2$ adsorption/desorption isotherm for Hf-UiO-66 at 77 K. Previously unpublished figure reporting data from ref. 188, 201, 202, 204, 205.

cases, differently to the hydroxylated cases, the 3D model obtained from the Rietveld refinement of XRPD data in the highly symmetric *Fm-3m* space group was inadequate to simulate the experimental EXAFS data. The origin of this failure was obviously due to the inability of the model to account for two different R_{M1} and R_{M2} distances. For both the UiO-66[188] and UiO-67[202] cases, the failure of the XRPD model was overcome by using the optimized geometry obtained by *ab initio* periodic calculations.

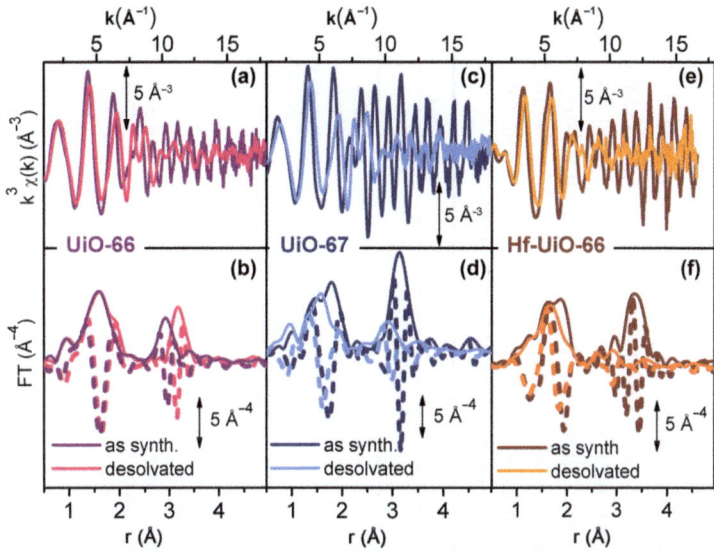

Figure 5.16 *k*- (top panels) and *r*-space (bottom panels) EXAFS data collected on UiO-66, UiO-67 and Hf-UiO-66, parts (a,b), (c,d) and (e,f), respectively. Both as synthesized (or hydroxylated) and desolvated (or dehydroxylated) forms of the three different isostructural MOFs have been measured. With the exception of desolvated Hf-UiO-66 sample (collected at 573 K) remaining spectra were collected at 300 K. Previously unpublished figure reporting data from ref. 188, 201, 202, 204, 205.

The inorganic cornerstones of the as synthesized materials are perfect $M_6(OH)_4O_4$ octahedra (see model in Figure 5.17b), with 6 equivalent M at the vertex, 12 equivalent M–M1 sides and 3 equivalent M–M2 diagonals. Upon desolvation, 2 structural water molecules are lost per cornerstone unit (Figure 5.17a), causing evolution from $M_6(OH)_4O_4$ to M_6O_6.[188,202,204,205] The new M_6O_6 octahedron showed a compression (2 opposite vertexes approaching, see model in Figure 5.17c) resulting in the shortening of 8 of the 12 edges, and the elongation of the other 4 edges. To take into account this variation the EXAFS contribution was simulated with two independently parameterized paths, fixing the degeneration with a ratio of 1/3 and 2/3 with respect to the case of the single contribution. For the three systems, this combined XRPD, EXAFS and DFT approach allowed a full interpretation of the EXAFS data in both hydroxylated and dehydroxylated forms.[188,202,204,205] Please note that IR spectroscopy was used to confirm this model and allowed to observe the disappearance of the O–H stretching band in these materials, see also Section 4.2.3.3 of chapter 4 in the present book.[6]

In particular, EXAFS spectroscopy allowed for the detection of the evolution from $M_6(OH)_4O_4$ to M_6O_6 (M = Zr or Hf) of the inorganic cornerstones of UiO-66, UiO-67 and Hf-UiO-66 MOFs occurring in the desolvation process, that escaped XRPD detection. On Zr-UiO-66 and Zr-UiO-67, period

Table 5.4 Summary of the EXAFS refinement obtained on the hydroxylated and dehydroxylated forms of UiO-66, UiO-67 and Hf-UiO-66. Parameters without indication of the errors were not optimized. The EXAFS refinement of the hydroxylated materials was obtained using as input model the optimized structure from Rietveld refinement of the corresponding XRPD patterns. The EXAFS refinement of the dehydroxylated materials was obtained using as input model optimized *ab initio* calculations for the hydroxylated UiO-66. With this approach, the coordination number (N) of each contribution is fixed by the model stoichiometry. Refinement of the experimental amplitude is done by optimizing the overall amplitude factor S_0^2 only. The fitting of the higher shells was possible only adopting the axial compressed model of the M_6O_6 octahedron represented where eight octahedron sides R_{M1} are split into eight short prismatic distances (R_{M1}, $N = 8/3$) and four long planar ones (R_{M1b}, $N = 4/3$) and where the three diagonals R_{M2} are split into a short axial diagonal (R_{M2}, involving two M atoms out of six; $N = 1/3$) and two long planar diagonals (R_{M2b}, involving four M atoms out of six; $N = 2/3$): Figure 5.17c. For the dehydroxylated Hf-UiO-66 sample (last column), the second distances R_{M2} and R_{M2b} were not refined due to the low S/N ratio at high *r*. Previously unpublished table, reporting data from ref. 188, 201, 202, 204, 205.

	hydroxylated			dehydroxylated		
	UiO-66	UiO-67	Hf-UiO-66	UiO-66	UiO-67	Hf-UiO-66
T/K	300	300	300	300	300	573
R-factor	0.01	0.04	0.02	0.02	0.04	0.05
Δk/Å^{-1}	2.0–18.0	2.0–16.2	2.0–16.0	2.0–15.0	2.0–15.0	2.0–15.0
ΔR/Å	1.0–5.3	1.0–5.3	1.0–5.5	1.0–5.3	1.0–5.3	1.0–3.9
Indep. points	43	38	40	35	35	25
N. variables	14	13	10	15	15	9
ΔE_0/eV	5±1	1±1	2.2±0.6	5	1	2.4±0.7
S_0^2	1.17±0.08	1.17	0.91±0.06	1.17	1.17	0.91
$R_{\mu3-O}$/Å	2.087±0.008	2.12±0.02	2.12±0.01	2.06±0.01	2.096±0.007	2.06±0.01
$\sigma^2(\mu3-O)$/Å^2	0.0036±0.0009	0.005±0.002	0.005±0.002	0.008±0.003	0.006±0.001	0.009±0.002
R_{O1}/Å	2.235±0.008	2.26±0.01	2.25±0.01	2.221±0.007	2.249±0.007	2.19±0.01
$\sigma^2(O1)$/Å^2	0.0074±0.0008	0.006±0.001	0.005	0.007±0.002	0.004±0.001	0.009
R_C/Å	3.19±0.02	3.40±0.06	3.23±0.06	3.17±0.04	3.15±0.04	3.22±0.05
$\sigma^2(C)$/Å^2	0.004±0.002	0.012±0.002	0.014±0.011	0.009±0.009	0.004±0.003	0.016±0.013
R_{M1}/Å	3.511±0.007	3.512±0.006	3.510±0.005	3.35±0.01	3.365±0.015	3.31±0.03
$\sigma^2(M1)$/Å^2	0.007±0.001	0.004±0.001	0.0042±0.0004	0.009±0.001	0.009±0.002	0.009±0.004
R_{M1b}/Å	—	—	—	3.74±0.02	3.80±0.03	3.45±0.06
$\sigma^2(M1b)$/Å^2	—	—	—	0.009±0.002	0.008±0.003	0.009
R_{M2}/Å	4.99±0.04	4.95±0.03	4.964	4.14±0.07	4.15±0.07	—
$\sigma^2(M2)$/Å^2	0.010±0.006	0.004±0.002	0.008±0.002	0.008±0.006	0.006±0.004	—
R_{M2b}/Å	—	—	—	5.30±0.04	5.46±0.05	—
$\sigma^2(M2b)$/Å^2	—	—	—	0.008	0.006	—

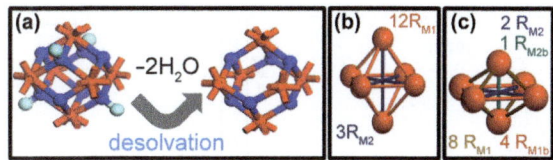

Figure 5.17 Part (a): stick and ball representation of the dehydroxylation undergone by the inorganic $M_6O_4(OH)_4$ cornerstone upon thermal treatment at 300 °C *in vacuo* resulting in a distorted M_6O_6 cluster (M = Zr or Hf). Red, blue and cyan colors refer to M, O and H atoms, respectively. Part (b): stick and ball representation of the perfect M_6 octahedron, showing 12 equivalent R_{M1} sides and 3 equivalent $R_{M2} = \sqrt{2}R_{M1}$ diagonals. Part (c): stick and ball representation of a squeezed M_6 octahedron. The 12 sides are now split into 4 in-plane long R_{M1b} sides and 8 prismatic short R_{M1} sides, while the 3 diagonals evolve into 2 in-plane long R_{M2b} and 1 orthogonal short R_{M2} diagonals. For clarity, O atoms are omitted in parts (b) and (c). Previously unpublished figure reporting schemes published in ref. 188, 202.

calculations performed with CRYSTAL code[206] at DFT level of theory supported EXAFS data.

5.5.4 Molecular Adsorption Inside MOFs: Determination of Adsorption Geometries by Neutron Diffraction

For the reasons outlined above (see Section 5.2.2 and in particular the discussions on Figure 5.9 and Table 5.2), neutron diffraction is an excellent structural technique to determine the location of adsorbed molecules (particularly deuterated ones) inside MOFs structures.[207] In this regard, neutron diffraction is complementary to EXAFS (see Sections 5.5.2 and 5.5.3) in the determination of the molecular adsorption on the metal sites, with the additional advantage of being able to also locate the molecules adsorbed on the organic part of the framework.

Interesting results have been reported by several groups, among them we mention the adsorption experiments of: D_2 on MOF-5,[119] HKUST-1,[208] ZIF-8,[209] CPO-27-Zn,[210] $Y(BTC)(H_2O)_4$,[211] and Cr MIL-53[212] frameworks; of CD_4 in sodalite-type Mn-MOF,[213,214] HKUST-1,[215] PCN-11,[215] PCN-14,[215] and CPO-27-M (M = Mg, Mn, Co, Ni, Zn)[216] frameworks; of O_2 on $Cr_3(BTC)_2$[140] and CPO-27-Fe;[217] of N_2 on CPO-27-Fe;[217] and of CO_2 on CPO-27-Mg[218] and HKUST-1.[218]

The selected example to describe the potential of neutron diffraction *versus* X-ray diffraction is the study of O_2 loading on $Cr_3(BTC)_2$, the Cr^{2+} analogue of HKUST-1 MOF, reported by the group of Long in Berkley.[140] First these authors used neutron powder diffraction to prove that $Cr_3(BTC)_2$ crystallizes in the same *Fm-3m* space group as the Cu^{2+}-homologue,[219] see Figure 5.18a (compare top and bottom patterns collected for the desolvated materials). Insertion of Cr^{2+}, substituting Cu^{2+}, implies a small cell expansion from

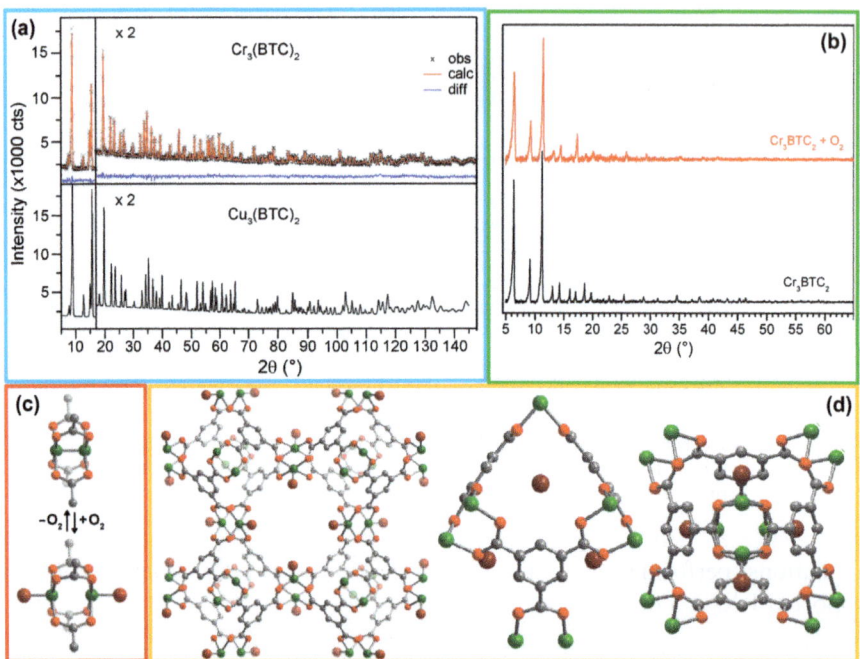

Figure 5.18 Part (a) top: neutron powder diffraction pattern of desolvated $Cr_3(BTC)_2$ MOF (× symbols) Rietveld refinement (red curve) and residual (blue curve). Part (a) bottom: neutron powder diffraction pattern of desolvated $Cu_3(BTC)_2$ MOF (black curve). Data have been collected on the high resolution BT-1 diffractometer at the NIST Centre for Neutron Research (US) with $\lambda = 2.0785$ Å. Part (b): XRPD pattern of desolvated $Cr_3(BTC)_2$ MOF before (black curve) and after (red curve) O_2 dosage. $\lambda = 1.54$ Å. Part (c): modification of the $[Cu_2C_4O_8]$ cage upon O_2 coordination. Note that O_2 absorption causes an increase of the Cr–Cr distance 2.06(2) to 2.8(1) Å. Part (d): different views of the structure of the O_2 loaded $Cr_3(BTC)_2$ as refined from the neutron powder diffraction. In parts (c) and (d) green, red, and grey spheres represent Cr, O, and C framework atoms, respectively, while large dark red spheres represent the centroid of the bound O_2 molecules. Reproduced with permission form ref. 140 (copyright American Chemical Society 2010).

$a = 26.2243(5)$ Å, $V = 18035(1)$ Å^3 of HKUST-1 to $a = 26.6652(3)$ Å, $V = 18959.8(6)$ Å^3 of $Cr_3(BTC)_2$, which is accompanied by a significant shrinking of the metal–metal distance in the dimer from $R_{Cu-Cu} = 2.50 \pm 0.02$ Å, see ref. 19, $R_{Cr-Cr} = 2.06 \pm 0.02$ Å.[140] The impressive change in structure of the $[Cr_2C_4O_8]$ cages is as expected since Cr^{2+} centers can form a strong (quadruple) metal–metal bond upon loss of the axial solvent[220] whereas Cu^{2+} does not. The neutron diffraction data were definitive on this point.[140] The data collection was done with $\lambda = 2.0785$ Å up to a $2\theta_{max} = 140°$, where Bragg peaks were clearly present, resulting in a $d_{min} = 1.11$ Å or a $q_{max} = 5.68$ Å^{-1}, see eqn (5.23).

In this regard, comparison between parts (a) and (b) of Figure 5.18 is striking in terms of the higher potential of neutrons compared to X-rays in obtaining high signal to noise data for the high 2θ Bragg peaks.

The interesting aspect of this new $Cr_3(BTC)_2$ MOF material is that it is able to fix O_2 molecules from air with high selectivity and in a reversible way.[140] This peculiarity is relevant because the separation of O_2 from air is carried out in industry using cryogenic distillation on a scale of 100 Mtons per year, as well as using zeolites in portable devices for medical applications. Moreover, in the near future O_2 may be needed in large scale for CO_2-free energy production inside fuel cells. Thus, there is a clear benefit to developing materials that might enable this process to be carried out with a lower energy cost.

Long *et al.*[140] followed the interaction of $Cr_3(BTC)_2$ with O_2 by neutron powder diffraction, collected at 4 K, IR, UV-Vis-NIR and XANES spectroscopies. Neutron powder diffraction revealed a decrease of the unit cell from $a = 26.6652(3)$ Å, $V = 18959.8(6)$ Å^3 to $a = 25.956(2)$ Å, $V = 17487(4)$ Å^3, accompanied by an impressive elongation of the distance in the dimer, that moves from $R_{Cr-Cr} = 2.06 \pm 0.02$ Å to $R_{Cr-Cr} = 2.8 \pm 0.1$ Å (Figure 18c). The Rietveld refinement afforded a model in which 0.87(3) O_2 molecules are coordinated to the axial sites of each paddle-wheel unit (Figure 5.18d), while 0.197(7) occupy the smallest pore openings within the framework (middle structure in Figure 5.18d). The authors concluded that, although the resolution of the data was insufficient to determine the orientation of O_2, the observed metal–centroid distance, $R_{Cr-O2} = 1.97(5)$ Å, is consistent with a side-on coordination mode.

In Table 5.5 structural data on desolvated and on O_2-interacting $Cr_3(BTC)_2$ are summarized and compared with similar data obtained on the isostructural $Cu_3(BTC)_2$, *i.e.* HKUST-1, in its dehydrated form and after interaction with H_2O or NH_3 ligands. The removal of the water molecule coordinated to Cu^{2+} during the desolvation process in $Cu_3(BTC)_2$ results in a small contraction of the unit cell $\Delta a/a = -0.5\%$ that is accompanied by an important shrinking of the $[Cu_2C_4O_8]$ cage of $\Delta R_{Cu-Cu}/R_{Cu-Cu} = -5.0\%$. It is interesting to underline how $Cr_3(BTC)_2$ behaves differently when the coordinated O_2 molecule is removed from the Cr^{2+} site: the MOF lattice undergoes a much larger contraction $\Delta a/a = -2.7\%$, which is however accompanied by a huge deformation of the $[Cr_2C_4O_8]$ cage, opposite to what observed for the Cu^{2+}-homologue MOF, of $\Delta R_{Cr-Cr}/R_{Cr-Cr} = +35\%$. Such impressive behavior can be explained only on the basis of an extremely high flexibility of the $[Cr_2C_4O_8]$ cage. Structural values on the effect of molecular adsorption (H_2O, NH_3 or O_2) on the two isostructural systems are summarized in Table 5.5.

It is finally worth recalling that to evaluate possible oxidation state changes of Cr upon oxygenation of the activated framework, Long *et al.*[140] have collected the Cr K-edge XANES spectra (see chapter 4 of this book[6]) of the as prepared, desolvated and O_2-contacted material. They concluded that the observed shift was consistent with partial charge transfer from the Cr metal center to the bound O_2 molecule.

Table 5.5 Summary on the structural data of isostructural $M_3(BTC)_2$, obtained with different techniques. $M = Cu$ or Cr. Previously unpublished table reporting data published in quoted refs.

Material	Coordination on M site	a/Å	V/Å³	R_{M-M} (Å)	$R_{M-adsorbate}$
$Cu_3(BTC)_2$	H_2O	26.343(5)[a]	18280(7)[a]	2.628(2)[a]; 2.64(2)[b]; 2.65(2)[d]	2.19(2)[b]; 2.24(3)[d]
$Cu_3(BTC)_2$	—	26.2243(5)[c]	18035(1)[c]	2.50(2)[b]; 2.58(2)[d]	—
$Cu_3(BTC)_2$	NH_3	—	—	2.80(3)[d]	2.31(1)[d]
$Cr_3(BTC)_2$	—	26.6652(3)[e]	18959.8(6)[e]	2.06(2)[e]	—
$Cr_3(BTC)_2$	O_2	25.956(2)[e]	17487(4)[e]	2.8(1)[e]	1.97(5)[e]

[a]Single crystal X-ray diffraction;[219] [b]Cu K-edge EXAFS;[19] [c]XRPD;[19] [d]Cu K-edge EXAFS;[221] [e]neutron powder diffraction.[140]

5.5.5 Trapping Guest Molecules Within Nanoporous MOFs Through Pressure-induced Amorphization: a PDF Approach

The final section of this chapter is devoted to discussing the potential of the total scattering, or PDF, approach. The peculiarity of the PDF approach is to provide a local range structural information based on the X-ray (neutron) scattering process (see Section 5.4), that makes PDF an intermediate between EXAFS and X-ray (neutron) diffraction choices in terms of local *vs.* long range order of materials. Concerning the application of the technique to the structural determination of MOFs materials, PDF certainly represents an ideal complementary technique to support diffraction data (as was the case of EXAFS studies discussed in sections 5.5.1 to 5.5.3); it is however evident that its peculiarity becomes essential in the study of processes that imply a partial or total amorphization of the framework.

The example selected to show the potentialities of the PDF approach is the work of Chapman *et al.*[222,223] who investigated the impact of modest, industrially accessible pressures (~ 1 GPa) on the structure and porosity of $Zn(2\text{-methylimidazole})_2$ (ZIF-8),[224,225] a high-surface area MOF with expanded zeolite topologies where the bidentate imidazolate-based ligand replicates the characteristic T–O–T angle of zeolites. The topology of ZIF-8, with imidazolate-bridged zinc tetrahedra, corresponds to the one of the high symmetry sodalite zeolite. The cubic framework $I\overline{4}3m$ space group with $a \sim 17.0$ Å) can be described by a space-filling packing of regular truncated octahedra, defining 12.0 Å diameter pores connected *via* 3.5 Å diameter apertures (6-rings), with the 4-rings being too small to transmit guests (see Figure 5.19a). In a first work, Chapman *et al.*[222] have demonstrated that ZIF-8 exhibits an irreversible pressure-induced amorphization, which starts at moderate pressures (see Figure 5.19b). The authors succeeded in the generation of a new type of non-crystalline MOF still exhibiting nanoporosity, that has been modified with respect to that of the pristine crystalline phase, as proven by

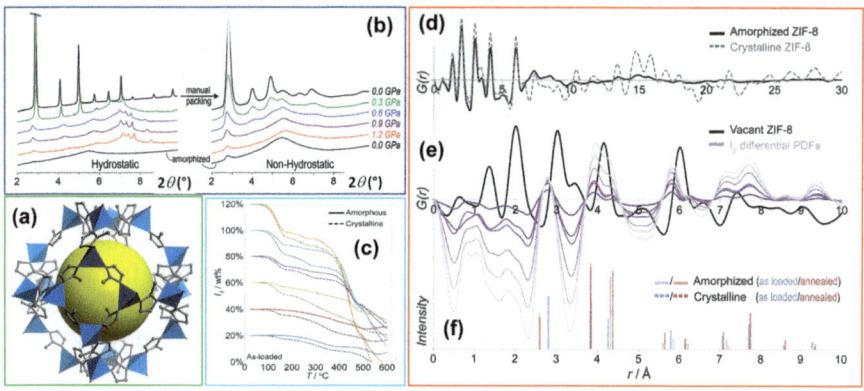

Figure 5.19 Part (a): representation of the ZIF-8 structure, showing the 12 Å pores. Part (b): XRPD data, collected with $\lambda = 0.60511$ Å at the 1-BM beamline at the APS synchrotron (Argonne, IL, USA), showing irreversible amorphization of the ZIF-8 framework under both hydrostatic (left) and non-hydrostatic (right) compressions (1 atm $= 1.01 \times 10^5$ Pa). Part (c): TGA curves reporting the differences between crystalline (dashed curves) and amorphized (continuous curves) ZIF-8 materials during the I_2 release process for different starting I_2 loadings. The mass losses for the vacant ZIF-8 materials have been subtracted to the reported curves. Part (d): Representative PDFs, $G(r)$ see eqn (5.46), for the crystalline and amorphous materials. Part (e): differential PDFs corresponding to I − I and I − framework interactions in the pressure-amorphized as-loaded series (120 wt% loading). Part (f): intensity and position of features in differential PDFs for crystalline and amorphous, and as-loaded and annealed samples. Parts (a),(b) reproduced with permission from ref. 222 (copyright 2009, American Chemical Society). Parts (c)–(f) reproduced with permission from ref. 223 (copyright 2011, American Chemical Society).

the differences in the absorption/desorption of nitrogen[222] and iodine[223] molecules. In particular, the TGA curves reported in Figure 5.19c, show how the retention of I_2 is enhanced in the amorphized ZIF-8 compared to the crystalline one. The mass losses were shifted to higher temperatures for the amorphized materials, by up to 150 °C. These gains were most pronounced for the intermediate I_2 loadings. The retained nanoporosity property of the amorphous phase is a consequence of the fact that the structure retained some structural order after the amorphization process. In a successive work, these authors decided to use the PDF approach to investigate such structural order.[223]

The well-defined long-range correlations, evident in PDFs for the crystalline materials, are absent for the amorphized ZIF-8 systems (Figure 5.19d). However, the shorter range features, including those up to 6 Å, which correspond to the Zn − imidazolate − Zn links, are entirely preserved in the amorphous materials. The combined retention of guests, porosity, and the Zn · · · Zn′ connectivity in the pressure-amorphized materials suggests that the sodalite topology of ZIF-8 is preserved, despite the local structural changes responsible for destroying the long-range order, that is, the crystallinity. The authors concluded that structural changes are likely to involve symmetry-reducing distortions of the 6-ring apertures, eliminating the well-defined features in the PDF beyond ~ 6 Å and hampering diffusion of guest molecules through the framework.[223] While the long-range framework order is eliminated upon amorphization, the short-range $I-I$ and $I-$ framework interactions remain unchanged. Indeed, a larger change in local structure is associated with the annealing and surface-desorption compared to the amorphization itself (Figure 5.19e). Specifically, the nearest neighbor $I-I$ peak shifts from 2.8 to 2.6 Å while simultaneously narrowing, indicating less disorder (dynamic or static). This is accompanied by an increase in the relative intensity of the second and third peaks at 3.85 and $4.3-4.4$ Å, associated with intermolecular inter-actions within pores (Figure 5.19e). These changes reflect a refinement of the I_2 arrangement within the pores upon annealing, which enhances the retention of I_2 by the framework, as documented by the TGA curves reported in Figure 5.19c.

We briefly close this section mentioning two additional examples of the use of a PDF analysis in the understanding of the structure of complex MOFs. The first example comes from the group of Hupp,[226] who recently reported on the synthesis of a new, twofold interpenetrated framework that demonstrates impressive perforrmances in the absorption of CO_2 and hysteresis in its desorption. Combining XRPD with PDF analysis, authors were able to demonstrate that the remarkable structural changes undergone by the framework upon CO_2 sorption involve the interpenetrated frameworks moving with respect to each other.[226]

The second example comes from the Cheetham group.[114] The authors have followed the reversible amorphization processes undergone by ZIF-4 MOF upon heating to 300 °C by neutron and X-ray total scattering. The collected high q-data were used as a basis for reverse Monte Carlo refinement of an atomistic model of the structure of a-ZIF. The authors were able to describe the amorphous structure in terms of a continuous random network analogous to that of a-SiO$_2$.[114] On top of this, optical microscopy, electron diffraction and nanoindentation measurements revealed the amorphous ZIF phase to be an isotropic glasslike phase capable of plastic flow on its formation. The authors concluded by suggesting the possibility of designing broad new families of amorphous and glasslike materials that exploit the chemical and structural diversity of MOFs.[114]

5.6 Conclusions and Perspectives

From the case studies reported in this chapter, it becomes evident that besides the unavoidable standard laboratory XRPD investigation, the possibility of extending the structural characterization to *less common* techniques, such as neutron powder diffraction, metal K-or L-edge EXAFS and neutron or X-ray PDF, will allow better characterization of complex materials such as MOFs. Such information should be coupled with a spectroscopic investigation of the vibrational and electronic properties of the material.[6] Finally, the presence in a research group of competencies in computational chemistry[227] is highly recommended, as quantum mechanics will allow verification of the stability of the structures inferred from Rietveld refinement.

Acknowledgements

We are deeply indebted to F. Camara (Department of Earth Science University of Turin, Italy) for the long and constructive discussions that contributed to a significant improvement of the contents of this chapter. F. Camara and M. Ceretti (Institut de Chimie Moléculaire et des Matériaux Université de Montpellier 2, France) are acknowledged for a critical reading of the manuscript. M. Ceretti is also acknowledged for providing the neutron powder diffractograms reported in Figure 5.8b. We thank A. Piovano (ILL, Grenoble, France) for computing the neutron and X-ray simulated patterns reported in Figure 5.8a and with Prof. K. P. Lillerud (Univerity of Oslo, NO) for providing Fig. 5.4. Finally, CL and EB are grateful to the students of the *Advanced Crystallography* class of the MaMaSELF European Master in Materials Science[228] that have contributed to finding several typos in a first draft of Sections 5.1–5.4 used as lecture notes.

References

1. P. E. Werner, L. Eriksson and M. Westdahl, *J. Appl. Crystallogr.*, 1985, **18**, 367.
2. A. Altomare, C. Giacovazzo, A. Guagliardi, A. G. G. Moliterni and R. Rizzi, *J. Appl. Crystallogr.*, 2000, **33**, 1305.
3. A. Altomare, C. Giacovazzo, A. Guagliardi, A. G. G. Moliterni, R. Rizzi and P. E. Werner, *J. Appl. Crystallogr.*, 2000, **33**, 1180.
4. C. Giacovazzo, A. Altomare, C. Cuocci, A. G. G. Moliterni and R. Rizzi, *J. Appl. Crystallogr.*, 2002, **35**, 422.
5. A. Altomare, R. Caliandro, M. Camalli, C. Cuocci, I. da Silva, C. Giacovazzo, A. G. G. Moliterni and R. Spagna, *J. Appl. Crystallogr.*, 2004, **37**, 957.
6. F. Bonino, C. Lamberti, S. Chavan, J. G. Vitillo and S. Bordiga, *Characterization of MOFs. 1. Combined vibrational and electronic spectroscopies*, in Metal-Organic Frameworks as heterogeneous catalysts, ed. F. X. Llabrés i Xamena and J. Gascón, RSC, Cambridge, 2013, ch. 4.
7. E. A. Pidko and E. J. M. Hensen, *Computational approach to chemical reactivity of MOFs*, in Metal-Organic Frameworks as heterogeneous catalysts, ed. F. X. Llabrés i Xamena and J. Gascón, RSC, Cambridge, 2013, ch. 6.
8. I. Waller, *Phil. Mag.*, 1927, **4**, 1228.
9. I. Waller, *Nature*, 1927, **120**, 155.
10. G. Wentzel, *Zeit. f. Physik*, 1927, **43**, 1.
11. I. Waller, *Zeit. f. Physik.*, 1928, **51**, 213.

12. R. W. James, *The Optical Principles of the Diffraction of X-rays*, G. Bell and Sons Ltd, London, 1948.
13. A. Guinier, *X-Ray Diffraction*, Dover, New York, US, 1963.
14. B. E. Warren, *X-Ray Diffraction*, Dover, New York, 1969.
15. (a) C. Giacovazzo, *Direct Phasing in Crystallography: Fundamentals and Applications*, Oxford University Press: Oxford, 1998, vol. 8. (b) C. Giacovazzo, *Fundamentals of Crystallography*, Oxford University Press, New York, US, 2002.
16. The Euler's formula teaches us that the exponential of a pure imaginary argument is a periodic function in both its real and imaginary parts: $\exp(\mathbf{i}x) = \cos(x) + \mathbf{i}\sin(x)$.
17. R. Shannon, *Acta Cryst A*, 1976, **32**, 751.
18. Note that $\rho(-\mathbf{r}) = \rho(\mathbf{r})$ is no more fully true when the atom will be part of a molecule or embedded inside a crystal as the formation of chemical bonds breaks the spherical symmetry.
19. C. Prestipino, L. Regli, J. G. Vitillo, F. Bonino, A. Damin, C. Lamberti, A. Zecchina, P. L. Solari, K. O. Kongshaug and S. Bordiga, *Chem. Mater.*, 2006, **18**, 1337.
20. J. C. Slater, *Phys. Rev.*, 1930, **36**, 57.
21. E. Clementi and D. L. Raimondi, *J. Chem. Phys.*, 1963, **38**, 2686.
22. E. Clementi, D. L. Raimondi and W. P. Reinhardt, *J. Chem. Phys.*, 1967, **47**, 1300.
23. E. Clementi and C. Roetti, *Atom. Data Nucl. Data Tables*, 1974, **14**, 177.
24. T. Koga, H. Tatewaki and A. J. Thakkar, *Phys. Rev. A*, 1993, **47**, 4510.
25. T. K. Ghanty and S. K. Ghosh, *J. Phys. Chem.*, 1996, **100**, 17429.
26. D. T. Cromer and J. T. Waber, *Acta Cryst.*, 1965, **18**, 104.
27. P. A. Doyle and P. S. Turner, *Acta Crystallogr. A*, 1968, **24**, 390.
28. D. T. Cromer and J. B. Mann, *Acta Crystallogr. A*, 1968, **24**, 321.
29. D. T. Cromer and J. T. Waber, *Mean atomic scattering factors in electrons for free atoms and chemically significant ions. in International Tables for X-ray Crystallography, Volume IV*, Kynock Press, Birmingham, 1974.
30. A. G. Fox, M. A. O'Keefe and M. A. Tabbernor, *Acta Crystallogr. A*, 1989, **45**, 786.
31. P. J. Brown, A. G. Fox, E. N. Maslen, M. A. O'Keefe and B. T. M. Willis, *Intensity of diffracted intensities*, in International Tables for Crystallography, ed. E. Prince, International Union of Crystallography, on-line edition, 2006, vol. C, p. 554.
32. E. N. Maslen, A. G. Fox and M. A. O'Keefe, *X-ray Scattering, in International Tables for Crystallography*, ed. E. Prince, Kluwer Academic, Dordrecht, 2004, vol. C, p. 554.
33. R. N. Bracewell, *The Fourier Transform and Its Applications*, McGraw-Hill, New York, 3rd edn, 1999.
34. G. Portale and A. Longo, *Small Angle X-ray scattering for the study of nanostructures and nanostructured materials, in Characterization of Semiconductor Heterostructures and Nanostructures*, ed. C. Lamberti and G. Agostini, Elsevier, Amsterdam, 2nd edn, 2013, p. 289.

35. J. L. Hodeau, V. Favre-Nicolin, S. Bos, H. Renevier, E. Lorenzo and J. F. Berar, *Chem. Rev.*, 2001, **101**, 1843.
36. W. Friedrich, P. Knipping and M. von Laue, *Bayerische Akademie der Wissenschaften*, 1912, 303.
37. H. Ihee, *Acc. Chem. Res.*, 2009, **42**, 356.
38. C. Kittel, *Introduction to Solid State Physics*, John Wiley & Sons, 2004, ch. 2.
39. W. L. Bragg, *Proc. Cambridge Phil. Soc.*, 1913, **17**, 43–57.
40. W. H. Bragg and W. L. Bragg, *Proc. R. Soc. Lond. A*, 1913, **88**, 428.
41. W. H. Bragg, *Proc. R. Soc. Lond. A*, 1913, **89**, 246.
42. J. R. Helliwell, A. J. Blake, J. Blunden-Ellis, M. Moore and C. H. Schwalbe, *Crystallogr. Rev*, 2012, **18**, 3.
43. J. R. Helliwell, *Crystallogr. Rev*, 2012, **18**, 280.
44. T. Cormen, H., C. E. Leiserson, R. L. Rivest and C. Stein, *Introduction to Algorithms*, MIT Press and McGraw-Hill, Boston, 3rd edn, 2009, section 30.2: The DFT and FFT, pp. 906–914.
45. D. J. Tranchemontagne, J. L. Mendoza-Cortes, M. O'Keeffe and O. M. Yaghi, *Chem. Soc. Rev.*, 2009, **38**, 1257.
46. A. Bacchi, M. Carcelli and P. Pelagatti, *Crystallogr. Rev*, 2012, **18**, 253.
47. A. K. Cheetham and A. P. Wilkinson, *Angew. Chem., Int. Ed. Engl.*, 1992, **31**, 1557.
48. R. J. Harrison, *Rev. Mineral. Geochem.*, 2006, **63**, 113.
49. J. R. Helliwell, *Crystallogr. Rev.*, 2012, **18**, 33.
50. E. Aubert and C. Lecomte, *J. Appl. Cryst.*, 2007, **40**, 1153.
51. A. L. Patterson, *Phys. Rev*, 1934, **46**, 372.
52. A. L. Patterson, *Z. Kristallogr.*, 1934, **90**, 517.
53. M. G. Rossmann and E. Arnold, *Patterson and molecular-replacement techniques*, in International Tables for Crystallography, 2006, vol. B, p. 235.
54. M. G. Rossmann and D. M. Blow, *Acta Cryst.*, 1962, **15**, 24.
55. W. A. Hendrickson, *Science*, 1991, **254**, 51.
56. S. E. Ealick, *Curr. Opin. Chem. Biol.*, 2000, **4**, 495.
57. J. L. Smith, W. A. Hendrickson, T. T. C. and J. Berendzen, *MAD and MIR*, in International Tables for Crystallography, 2006, vol. F, p. 299.
58. G. Oszlanyi and A. Suto, *Acta Crystallogr. Sect. A*, 2008, **64**, 123.
59. D. E. Sands, *Introduction to Crystallography*, Dover Publications, New York, 1994.
60. M. F. C. Ladd and R. A. Palmer, *Structure Determination by X-Ray Crystallography*, Kluwer Academic/Plenum Publishers, New York, 2003.
61. G. Taylor, *Acta Crystallogr. D*, 2003, **59**, 1881.
62. C. C. Wilson, *Crystallogr. Rev*, 2009, **15**, 3.
63. P. Debye, *Ann. Phys.*, 1913, **348**, 49.
64. I. Waller, *Z. Phys.*, 1923, **17**, 398.
65. W. Massa, *Crystal Structure Determination*, Springer-Verlag, Berlin, 2004.
66. G. S. Matouzenko, E. Jeanneau, A. Y. Verat and A. Bousseksou, *Dalton Trans.*, 2011, **40**, 9608.

67. G. A. Martynov, *J. Mol. Liq.*, 2003, **106**, 123.
68. P. Debye, *Ann. der Physik*, 1915, **46**, 809.
69. J. G. Kirkwood and E. M. Boggs, *J. Chem. Phys.*, 1942, **10**, 394.
70. B. R. A. Nijboer and L. Van Hove, *Phys. Rev.*, 1952, **85**, 777.
71. B. Widom, *J. Chem. Phys.*, 1964, **41**, 74.
72. O. Glatter and O. Kratky, *Small angle X-ray scattering*, Academic Press, London, 1982.
73. L. A. Feigin and D. I. Svergun, *Structure analysis by small angle X-ray and neutron scattering*, Plenum Press, New York, 1987.
74. E. S. Bozin, P. Juhás and S. J. L. Billinge, *Local structure of bulk and nanocrystalline semiconductors using total scattering method*, in Characterization of Semiconductor Heterostructures and Nanostructures, ed. C. Lamberti and G. Agostini, Elsevier: Amsterdam, 2nd edn, 2013, p. 289.
75. C. S. Tsao, M. S. Yu, T. Y. Chung, H. C. Wu, C. Y. Wang, K. S. Chang and H. L. Chent, *J. Am. Chem. Soc.*, 2007, **129**, 15997.
76. S. H. Yeon, S. Osswald, Y. Gogotsi, J. P. Singer, J. M. Simmons, J. E. Fischer, M. A. Lillo-Rodenas and A. Linares-Solanod, *J. Power Sources*, 2009, **191**, 560.
77. M. Klimakow, P. Klobes, A. F. Thunemann, K. Rademann and F. Emmerling, *Chem. Mater.*, 2010, **22**, 5216.
78. C. S. Tsao, C. Y. Chen, T. Y. Chung, C. J. Su, C. H. Su, H. L. Chen, U. S. Jeng, M. S. Yu, P. Y. Liao, K. F. Lin and Y. R. Tzeng, *J. Phys. Chem. C*, 2010, **114**, 7014.
79. J. Juan-Alcaniz, M. Goesten, A. Martinez-Joaristi, E. Stavitski, A. V. Petukhov, J. Gascon and F. Kapteijn, *Chem. Commun.*, 2011, **47**, 8578.
80. A. Mallick, E. M. Schon, T. Panda, K. Sreenivas, D. D. Diaz and R. Banerjee, *J. Mater. Chem.*, 2012, **22**, 14951.
81. L. de Broglie, *Ann. Phys. (Paris)*, 1925, **2**, 22.
82. Relativistic corrections can not be neglected when working with SEM or TEM machines. In a TEM working at 300 keV, the relativistically corrected de Broglie wavelength is 0.0197 Å, while a value of 0.0223 Å results from adopting the $\lambda = h/(m_e v_e)$ equation. See *e.g.* D. B. Williams and C. B. Carter, *Transmission Electron Microscopy*, Plenum, New York, 1996; M. De Graef, *Introduction to Conventional Transmission Electron Microscopy*, Cambridge University Press, Cambridge, 2003.
83. M. Siegbahn, *Phil. Mag.*, 1919, **37**, 601.
84. The Davisson and Germer experiment consisted of firing an electron beam from an electron gun perpendicular to the Ni(111) phase. As the potential V was fixed, then the energy of the electrons ($|e|V$) was fixed, and so was the de Broglie wavelength. The experiment included an electron gun consisting of a heated filament that released thermally excited electrons, which were then accelerated through a potential difference V giving them a defined kinetic energy ($E = eV$) towards the nickel crystal. To avoid collisions of the electrons with other molecules on their way towards the surface, the experiment was obviously conducted in

a vacuum chamber. To measure the number of electrons that were scattered at different angles, an electron detector that could be moved on an arc path about the crystal was used. The detector was designed to accept only elastically scattered electrons. When the electrons hit the surface, they were scattered by atoms that originated from crystal planes inside the nickel crystal at precise angular positions that were in agreement with the Bragg law previously demonstrated for X-rays. When the voltage was changed from 54 to 174 V, the angular position where scattered electrons were observed changed according to the electron wavelength change foreseen by de Broglie equation.

85. G. P. Thomson and A. Reid, *Nature*, 1927, **119**, 890.
86. G. P. Thomson, *Nature*, 1927, **120**, 802.
87. G. P. Thomson, *Proc. R. Soc. Lond. A*, 1928, **117**, 600.
88. G. P. Thomson, *Proc. R. Soc. Lond. A*, 1928, **119**, 651.
89. G. P. Thomson, *Proc. R. Soc. Lond. A*, 1929, **125**, 651.
90. G. P. Thomson, *Am, J. Phys.*, 1961, **29**, 821.
91. G. P. Thomson, *Contemporary Phys.*, 1968, **9**, 1.
92. J. Chadwick, *Nature*, 1932, **129**, 312.
93. J. Chadwick, *Proc. R. Soc. Lond. A*, 1932, **136**, 692.
94. J. Chadwick, *Proc. R. Soc. Lond. A*, 1933, **142**, 1.
95. C. G. Shull, *Rev. Mod. Phys.*, 1995, **67**, 753.
96. E. O. Wollan, C. G. Shull and M. C. Marney, *Phys. Rev.*, 1948, **73**, 527.
97. C. G. Shull, E. O. Wollan, G. A. Morton and W. L. Davidson, *Phys. Rev.*, 1948, **73**, 842.
98. E. O. Wollan and C. G. Shull, *Phys. Rev.*, 1948, **73**, 830.
99. E. O. Wollan, W. L. Davidson and C. G. Shull, *Phys. Rev.*, 1949, **75**, 1348.
100. W. C. Koehler, E. O. Wollan and C. G. Shull, *Phys. Rev.*, 1950, **79**, 395.
101. C. G. Shull, E. O. Wollan and W. A. Strauser, *Phys. Rev.*, 1951, **81**, 483.
102. C. G. Shull, E. O. Wollan and W. C. Koehler, *Phys. Rev.*, 1951, **84**, 912.
103. C. G. Shull and E. O. Wollan, *Phys. Rev.*, 1951, **81**, 527.
104. E. Fermi and L. Marshall, *Phys. Rev.*, 1947, **71**, 666.
105. E. Fermi and L. Marshall, *Phys. Rev.*, 1947, **71**, 915.
106. E. Fermi, W. J. Sturm and R. G. Sachs, *Phys. Rev.*, 1947, **71**, 589.
107. E. Fermi and L. Marshall, *Phys. Rev.*, 1947, **72**, 408.
108. R. J. T. Houk, B. W. Jacobs, F. El Gabaly, N. N. Chang, A. A. Talin, D. D. Graham, S. D. House, I. M. Robertson and M. D. Allendorf, *Nano Lett.*, 2009, **9**, 3413.
109. H. F. Greer and W. Z. Zhou, *Crystallogr. Rev*, 2011, **17**, 163.
110. C. Wiktor, S. Turner, D. Zacher, R. A. Fischer and G. Van Tendeloo, *Microporous Mesoporous Mat.*, 2012, **162**, 131.
111. O. I. Lebedev, F. Millange, C. Serre, G. Van Tendeloo and G. Ferey, *Chem. Mat.*, 2005, **17**, 6525.
112. J. Cravillon, S. Munzer, S. J. Lohmeier, A. Feldhoff, K. Huber and M. Wiebcke, *Chem. Mat.*, 2009, **21**, 1410.
113. M. Yamada and S. Yonekura, *J. Phys. Chem. C*, 2009, **113**, 21531.

114. T. D. Bennett, A. L. Goodwin, M. T. Dove, D. A. Keen, M. G. Tucker, E. R. Barney, A. K. Soper, E. G. Bithell, J. C. Tan and A. K. Cheetham, *Phys. Rev. Lett.*, 2010, **104**, Art. n. 115503.

115. J. Hermannsdofer and R. Kempe, *Chem.-Eur. J.*, 2011, **17**, 8071.

116. L. H. Schwartz and J. B. Cohen, *Diffraction from Materials*, Academic Press, New York, 1977.

117. S. W. Lovesey, *Theory of Neutron Scattering from Condensed Matter, Volume 1: Neutron Scattering*, Clarendon Press, Oxford, 1984.

118. V. F. Sears, *Neutron News*, 1992, **3/3**, 26.

119. T. Yildirim and M. R. Hartman, *Phys. Rev. Lett.*, 2005, **95**, 215504.

120. P. Atkins, *Physical Chemistry*, Oxford University Press, Oxford, 3rd edn, 1986.

121. E. Suard and A. Hewat, *Neutron News*, 2001, **12**(4), 30.

122. E. Steichele and P. Arnold, *Phys. Lett. A*, 1973, **44**, 165.

123. R. M. Ibberson, *Nucl. Instrum. Methods Phys. Res. Sect. A*, 2009, **600**, 47.

124. A. N. Fitch, *Nucl. Instrum. Methods Phys. Res. Sect. B*, 1995, **97**, 63.

125. A. N. Fitch, *J. Res. Natl. Inst. Stand. Technol.*, 2004, **109**, 133.

126. D. R. Sandstrom and F. W. Lytle, *Ann. Rev. Phys. Chem.*, 1979, **30**, 215.

127. P. A. Lee, P. H. Citrin, P. Eisenberger and M. Kincaid, *Rev. Mod. Phys.*, 1981, **53**, 769.

128. E. A. Stern, Theory of EXAFS, in X-Ray Absorption: Principles, Applications, Techniques of EXAFS, SEXAFS and XANES, ed. D. C. Koningsberger and R. Prins, John Wiley & Sons, New York, 1988, vol. 92, p. 3.

129. A. Filipponi, A. Di Cicco and C. R. Natoli, *Phys. Rev. B*, 1995, **52**, 15122.

130. A. Filipponi and A. Di Cicco, *Phys. Rev. B*, 1995, **52**, 15135.

131. J. J. Rehr and R. C. Albers, *Rev. Mod. Phys.*, 2000, **72**, 621.

132. F. Boscherini, *X-ray absorption fine structure in the study of semiconductor heterostructures and nanostructures*, in Characterization of Semiconductor Heterostructures and Nanostructures, ed. C. Lamberti, Elsevier, Amsterdam, 2008, p. 289.

133. S. Bordiga, E. Groppo, G. Agostini, J. A. van Bokhoven and C. Lamberti, *Chem. Rev.*, 2013, **113**, 1736.

134. P. A. M. Dirac, *Proc. Roy. Soc. (London) A*, 1927, **114**, 243.

135. E. Fermi, *Nuclear Physics*; University of Chicago Press: Chicago, 1950.

136. B. K. Teo and D. C. Joy, *EXAFS Spectroscopy: Techniques and Applications*, Plenum, New York, 1981.

137. E. Borfecchia, C. Garino, L. Salassa and C. Lamberti, *Phil. Trans. R. Soc. A*, 2013, doi: 10.1098/rsta.2012.0132.

138. D. E. Sayers, E. A. Stern and F. W. Lytle, *Phys. Rev. Lett.*, 1971, **27**, 1204.

139. F. W. Lytle, D. E. Sayers and E. A. Stern, *Phys. Rev. B*, 1975, **11**, 4825.

140. L. J. Murray, M. Dinca, J. Yano, S. Chavan, S. Bordiga, C. M. Brown and J. R. Long, *J. Am. Chem. Soc.*, 2010, **132**, 7856.

141. E. A. Stern, *Phys. Rev. B*, 1974, **10**, 3027.

142. B.-K. Teo and P. A. Lee, *J. Am. Chem. Soc.*, 1979, **101**, 2815.

143. A. G. McKale, B. W. Veal, A. P. Paulikas, S. K. Chan and G. S. Knapp, *J. Am. Chem. Soc.*, 1988, **110**, 3763.
144. M. Vaarkamp, I. Dring, R. J. Oldman, E. A. Stern and D. C. Koningsberger, *Phys. Rev. B*, 1994, **50**, 7872.
145. For a compilation of electron inelastic mean free path lengths (λ) in solids for energies in the range 0–10 000 eV above the Fermi level, see *e.g.* M. P. Seah and W. A. Dench, *Surf. Interface Anal.*,1979, **1**, 2.
146. A. Filipponi, *J. Phys.-Condes. Matter*, 2001, **13**, R23.
147. K. Asakura, *Analysis of EXAFS*, in X-ray absorption fine structure for catalysts and surfaces, ed. Y. Iwasawa, World Scientific, Singapore, 1996, vol. 2, p. 33.
148. H. Nyquist, *Trans. AIEE,* 1928, **47**, 617–644. Reprinted in *Proc. IEEE*, 2002, **90**, 280.
149. C. E. Shannon, *Proc. Institute Radio Engin.*, 1949, **37**, 10–21. Reprinted in *Proc. IEEE*, 1988, **86**, 447.
150. B. H. Toby and T. Egami, *Acta Crystallogr. A*, 1992, **48**, 336.
151. D. A. Keen, *J. Appl. Crystallogr.*, 2001, **34**, 172.
152. T. Egami and S. J. L. Billinge, *Underneath the Bragg peaks: structural analysis of complex materials*, Pergamon Press, Oxford, 2003.
153. S. J. L. Billinge and M. G. Kanatzidis, *Chem. Commun.*, 2004, 749.
154. M. Fernandez-Garcia, A. Martinez-Arias, J. C. Hanson and J. A. Rodriguez, *Chem. Rev.*, 2004, **104**, 4063.
155. S. J. L. Billinge and I. Levin, *Science*, 2007, **316**, 561.
156. F. M. Michel, L. Ehm, S. M. Antao, P. L. Lee, P. J. Chupas, G. Liu, D. R. Strongin, M. A. A. Schoonen, B. L. Phillips and J. B. Parise, *Science*, 2007, **316**, 1726.
157. J. L. Hodeau and R. Guinebretiere, *Appl. Phys. A-Mater. Sci. Process.*, 2007, **89**, 813.
158. E. S. Bozin, P. Juhás and S. J. L. Billinge, *Local structure of bulk and nanocrystalline semiconductors using total scattering methods*, in Characterization of Semiconductor Heterostructures and Nanostructures, ed. G. Agostini and C. Lamberti, Elsevier, Amsterdam, 2nd edn, 2013, pp. 229–257.
159. P. J. Chupas, X. Y. Qiu, J. C. Hanson, P. L. Lee, C. P. Grey and S. J. L. Billinge, *J. Appl. Crystallogr.*, 2003, **36**, 1342.
160. P. J. Chupas, K. W. Chapman, H. L. Chen and C. P. Grey, *Catal. Today*, 2009, **145**, 213.
161. T. Proffen and S. J. L. Billinge, *J. Appl. Cryst.*, 1999, **32**, 572.
162. X. Qiu, J. W. Thompson and S. J. L. Billinge, *J. Appl. Cryst.*, 2004, **37**, 678.
163. C. L. Farrow, P. Juhas, J. W. Liu, D. Bryndin, E. S. Bozin, J. Bloch, T. Proffen and S. J. L. Billinge, *J. Phys.-Condes. Matter*, 2007, **19**, 335219.
164. A. Filipponi, *J. Phys.-Condes. Matter*, 1994, **6**, 8415.
165. J. L. C. Rowsell and O. M. Yaghi, *Microporous Mesoporous Mater.*, 2004, **73**, 3.
166. J. Hafizovic, M. Bjorgen, U. Olsbye, P. D. C. Dietzel, S. Bordiga, C. Prestipino, C. Lamberti and K. P. Lillerud, *J. Am. Chem. Soc.*, 2007, **129**, 3612.

167. N. L. Rosi, J. Eckert, M. Eddaoudi, D. T. Vodak, J. Kim, M. O'Keeffe and O. M. Yaghi, *Science*, 2003, **300**, 1127.
168. J. L. C. Rowsell, A. R. Millward, K. S. Park and O. M. Yaghi, *J. Am. Chem. Soc.*, 2004, **126**, 5666.
169. S. Bordiga, J. G. Vitillo, G. Ricchiardi, L. Regli, D. Cocina, A. Zecchina, B. Arstad, M. Bjorgen, J. Hafizovic and K. P. Lillerud, *J. Phys. Chem. B*, 2005, **109**, 18237.
170. H. Li, M. Eddaoudi, M. O'Keeffe and O. M. Yaghi, *Nature*, 1999, **402**, 276.
171. M. Eddaoudi, H. L. Li and O. M. Yaghi, *J. Am. Chem. Soc.*, 2000, **122**, 1391.
172. S. Bordiga, C. Lamberti, G. Ricchiardi, L. Regli, F. Bonino, A. Damin, K. P. Lillerud, M. Bjorgen and A. Zecchina, *Chem. Commun.*, 2004, 2300.
173. L. M. Huang, H. T. Wang, J. X. Chen, Z. B. Wang, J. Y. Sun, D. Y. Zhao and Y. S. Yan, *Microporous Mesoporous Mater.*, 2003, **58**, 105.
174. Z. Ni and R. I. Masel, *J. Am. Chem. Soc.*, 2006, **128**, 12394.
175. O. M. Yaghi, M. Eddaoudi, H. Li, J. Kim and N. Rosi, *Patent WO 02/088148 A1*, 2002.
176. M. Eddaoudi, D. B. Moler, H. L. Li, B. L. Chen, T. M. Reineke, M. O'Keeffe and O. M. Yaghi, *Accounts Chem. Res.*, 2001, **34**, 319.
177. B. Panella and M. Hirscher, *Adv. Mater.*, 2005, **17**, 538.
178. M. Milanesio, G. Artioli, A. F. Gualtieri, L. Palin and C. Lamberti, *J. Am. Chem. Soc.*, 2003, **125**, 14549.
179. G. Agostini, C. Lamberti, L. Palin, M. Milanesio, N. Danilina, B. Xu, M. Janousch and J. A. van Bokhoven, *J. Am. Chem. Soc.*, 2010, **132**, 667.
180. P. D. C. Dietzel, B. Panella, M. Hirscher, R. Blom and H. Fjellvag, *Chem. Commun.*, 2006, 959.
181. P. D. C. Dietzel, Y. Morita, R. Blom and H. Fjellvag, *Angew. Chem., Int. Ed.*, 2005, **44**, 6354.
182. N. L. Rosi, J. Kim, M. Eddaoudi, B. L. Chen, M. O'Keeffe and O. M. Yaghi, *J. Am. Chem. Soc.*, 2005, **127**, 1504.
183. P. D. C. Dietzel, R. E. Johnsen, R. Blom and H. Fjellvåg, *Chem.-Eur. J.*, 2008, **14**, 2389.
184. L. Valenzano, B. Civalleri, S. Chavan, G. T. Palomino, C. O. Arean and S. Bordiga, *J. Phys. Chem. C*, 2010, **114**, 11185.
185. F. Bonino, S. Chavan, J. G. Vitillo, E. Groppo, G. Agostini, C. Lamberti, P. D. C. Dietzel, C. Prestipino and S. Bordiga, *Chem. Mater.*, 2008, **20**, 4957.
186. S. Chavan, J. G. Vitillo, E. Groppo, F. Bonino, C. Lamberti, P. D. C. Dietzel and S. Bordiga, *J. Phys. Chem. C*, 2009, **113**, 3292.
187. S. Chavan, F. Bonino, J. G. Vitillo, E. Groppo, C. Lamberti, P. D. C. Dietzel, A. Zecchina and S. Bordiga, *Phys. Chem. Chem. Phys.*, 2009, **11**, 9811.
188. L. Valenzano, B. Civalleri, S. Bordiga, M. H. Nilsen, S. Jakobsen, K.-P. Lillerud and C. Lamberti, *Chem. Mater.*, 2011, **23**, 1700.

189. L. Valenzano, J. G. Vitillo, S. Chavan, B. Civalleri, F. Bonino, S. Bordiga and C. Lamberti, *Catal. Today*, 2012, **182**, 67.

190. P. D. C. Dietzel, V. Besikiotis and R. Blom, *J. Mater. Chem.*, 2009, **19**, 7362.

191. J. L. C. Rowsell, E. C. Spencer, J. Eckert, J. A. K. Howard and O. M. Yaghi, *Science*, 2005, **309**, 1350.

192. W. Zhou, H. Wu, T. Yildirim, J. R. Simpson and A. R. H. Walker, *Phys. Rev. B*, 2008, **78**, Art. n. 054114.

193. N. Lock, Y. Wu, M. Christensen, L. J. Cameron, V. K. Peterson, A. J. Bridgeman, C. J. Kepert and B. B. Iversen, *J. Phys. Chem. C*, 2010, **114**, 16181.

194. D. Dubbeldam, K. S. Walton, D. E. Ellis and R. Q. Snurr, *Angew. Chem., Int. Ed.*, 2007, **46**, 4496.

195. S. S. Han and W. A. Goddard, *J. Phys. Chem. C*, 2007, **111**, 15185.

196. M. Tafipolsky, S. Amirjalayer and R. Schmid, *J. Phys. Chem. C*, 2010, **114**, 14402.

197. S. Amirjalayer, M. Tafipolsky and R. Schmid, *J. Phys. Chem. C*, 2011, **115**, 15133.

198. A. Corma, H. Garcıa and F. X. Llabres i Xamena, *Chem. Rev.*, 2010, **110**, 4606.

199. B. Xiao, P. S. Wheatley, X. B. Zhao, A. J. Fletcher, S. Fox, A. G. Rossi, I. L. Megson, S. Bordiga, L. Regli, K. M. Thomas and R. E. Morris, *J. Am. Chem. Soc.*, 2007, **129**, 1203.

200. S. Bordiga, L. Regli, F. Bonino, E. Groppo, C. Lamberti, B. Xiao, P. S. Wheatley, R. E. Morris and A. Zecchina, *Phys. Chem. Chem. Phys.*, 2007, **9**, 2676.

201. J. H. Cavka, S. Jakobsen, U. Olsbye, N. Guillou, C. Lamberti, S. Bordiga and K. P. Lillerud, *J. Am. Chem. Soc.*, 2008, **130**, 13850.

202. S. Chavan, J. G. Vitillo, D. Gianolio, O. Zavorotynska, B. Civalleri, S. Jakobsen, M. H. Nilsen, L. Valenzano, C. Lamberti, K. P. Lillerud and S. Bordiga, *Phys. Chem. Chem. Phys.*, 2012, **14**, 1614.

203. M. Kandiah, M. H. Nilsen, S. Usseglio, S. Jakobsen, U. Olsbye, M. Tilset, C. Larabi, E. A. Quadrelli, F. Bonino and K. P. Lillerud, *Chem. Mater.*, 2010, **22**, 6632.

204. J. Jakobsen, D. Gianolio, D. Wragg, M. H. Nilsen, H. Emerich, S. Bordiga, C. Lamberti, U. Olsbye, M. Tilset and K. P. Lillerud, *Phys. Rev. B*, 2012, **86**, art. n. 125429.

205. D. Gianolio, J. G. Vitillo, B. Civalleri, S. Bordiga, U. Olsbye, K. P. Lillerud, L. Valenzano and C. Lamberti, *J. Phys.: Conf. Ser.*, 2013, **430**, art. n. 012134.

206. R. Dovesi, R. Orlando, B. Civalleri, C. Roetti, V. R. Saunders and C. M. Zicovich-Wilson, *Z. Kristallogr.*, 2005, **220**, 571.

207. S. V. Kolotilov and V. V. Pavlishchuk, *Theor. Exp. Chem.*, 2009, **45**, 277.

208. V. K. Peterson, Y. Liu, C. M. Brown and C. J. Kepert, *J. Am. Chem. Soc.*, 2006, **128**, 15578.

209. H. Wu, W. Zhou and T. Yildirim, *J. Am. Chem. Soc.*, 2007, **129**, 5314.

210. Y. Liu, H. Kabbour, C. M. Brown, D. A. Neumann and C. C. Ahn, *Langmuir*, 2008, **24**, 4772.
211. J. H. Luo, H. W. Xu, Y. Liu, Y. S. Zhao, L. L. Daemen, C. Brown, T. V. Timofeeva, S. Q. Ma and H. C. Zhou, *J. Am. Chem. Soc.*, 2008, **130**, 9626.
212. F. M. Mulder, B. Assfour, J. Huot, T. J. Dingemans, M. Wagemaker and A. J. Ramirez-Cuesta, *J. Phys. Chem. C*, 2010, **114**, 10648.
213. M. Dincă, A. Dailly, Y. Liu, C. M. Brown, D. A. Neumann and J. R. Long, *J. Am. Chem. Soc.*, 2006, **128**, 16876.
214. M. Dincă, W. S. Han, Y. Liu, A. Dailly, C. M. Brown and J. R. Long, *Angew. Chem., Int. Ed.*, 2007, **46**, 1419.
215. H. Wu, J. M. Simmons, Y. Liu, C. M. Brown, X. S. Wang, S. Ma, V. K. Peterson, P. D. Southon, C. J. Kepert, H. C. Zhou, T. Yildirim and W. Zhou, *Chem.-Eur. J.*, 2010, **16**, 5205.
216. H. Wu, W. Zhou and T. Yildirim, *J. Am. Chem. Soc.*, 2009, **131**, 4995.
217. E. D. Bloch, L. J. Murray, W. L. Queen, S. Chavan, S. N. Maximoff, J. P. Bigi, R. Krishna, V. K. Peterson, F. Grandjean, G. J. Long, B. Smit, S. Bordiga, C. M. Brown and J. R. Long, *J. Am. Chem. Soc.*, 2011, **133**, 14814.
218. H. Wu, J. M. Simmons, G. Srinivas, W. Zhou and T. Yildirim, *J. Phys. Chem. Lett.*, 2010, **1**, 1946.
219. S. S. Y. Chui, S. M. F. Lo, J. P. H. Charmant, A. G. Orpen and I. D. Williams, *Science*, 1999, **283**, 1148.
220. F. A. Cotton, E. A. Hillard, C. A. Murillo and H. C. Zhou, *J. Am. Chem. Soc.*, 2000, **122**, 416.
221. E. Borfecchia, S. Maurelli, D. Gianolio, E. Groppo, M. Chiesa, F. Bonino and C. Lamberti, *J. Phys. Chem. C*, 2012, **116**, 19839.
222. K. W. Chapman, G. J. Halder and P. J. Chupas, *J. Am. Chem. Soc.*, 2009, **131**, 17546.
223. K. W. Chapman, D. F. Sava, G. J. Halder, P. J. Chupas and T. M. Nenoff, *J. Am. Chem. Soc.*, 2011, **133**, 18583.
224. K. S. Park, Z. Ni, A. P. Cote, J. Y. Choi, R. D. Huang, F. J. Uribe-Romo, H. K. Chae, M. O'Keeffe and O. M. Yaghi, *Proc. Natl. Acad. Sci. U. S. A.*, 2006, **103**, 10186.
225. X. C. Huang, Y. Y. Lin, J. P. Zhang and X. M. Chen, *Angew. Chem., Int. Ed.*, 2006, **45**, 1557.
226. K. L. Mulfort, O. K. Farha, C. D. Malliakas, M. G. Kanatzidis and J. T. Hupp, *Chem.-Eur. J.*, 2010, **16**, 276.
227. G. M. Day, *Crystallogr. Rev*, 2011, **17**, 3.
228. MaMaSELF is the European Master in Materials Science focused on the use of Large Scale facilities (synchrotrons and neutron sources) developed within the Erasmus-Mundus frame. It involves French (Montpellier-2 and Rennes-1), German (TUM and LMU, München) and Italian (Turin) Universities as full partners and has several associated partners in Swiss, France, Germany, Italy, Russia, Japan, US and India. http://www.mamaself.eu/.

CHAPTER 6

Computational Approach to Chemical Reactivity of MOFs

EVGENY A. PIDKO*[a,b] AND EMIEL J. M. HENSEN[a]

[a] Inorganic Materials Chemistry group, Eindhoven University of Technology, P.O. Box 513, 5600 MB Eindhoven, The Netherlands; [b] Institute for Complex Molecular Systems, Eindhoven University of Technology, P.O. Box 513, 5600 MB Eindhoven, The Netherlands
*Email: e.a.pidko@tue.nl

6.1 Introduction

Nowadays, progress in fundamental understanding of chemical reactivity heavily relies on computations. Molecular modeling is considered a valuable approach to understand experimental observations and rationalize them into novel concepts about chemistry and reactivity. Ultimately, this can lead to a detailed molecular-level description of chemical reactivity. As a result, computational methods are currently recognized as an indispensable tool in chemical, physical, biomedical and engineering sciences. Besides being of value in the interpretation of experimental results, these methods are also increasingly used for the prediction of physicochemical properties of chemical systems at the molecular level. In some cases, computational chemistry may even replace experiments, so that ultimately the experiment is only used to validate predictions.

The computational chemistry toolbox includes a wide variety of methods which can be classified according to their accuracy and the dimensions (both length- and time-scale) of chemical systems they can simulate. There is usually a trade-off between the dimensions of the simulated system and the achievable accuracy. To illustrate this, the hierarchy of computational methods applicable

RSC Catalysis Series No. 12
Metal Organic Frameworks as Heterogeneous Catalysts
Edited by Francesc X. Llabrés i Xamena and Jorge Gascon
© The Royal Society of Chemistry 2013
Published by the Royal Society of Chemistry, www.rsc.org

to different system sizes and allowing for different accuracy is schematically shown in Figure 6.1. The behavior of chemical systems containing thousands of atoms can be routinely simulated by using interatomic potentials (force fields) within the framework of classical mechanics. The corresponding methods are usually empirical and do not consider explicitly the electrons in the system. The possibility to treat extended systems with such methods is achieved at the expense of a substantial loss of generality and accuracy. The resulting energies are then used to optimize structures by means of minimization methods to calculate ensemble averages using Monte Carlo simulations or to model dynamic processes *via* molecular dynamics simulations using classical Newton's law of motion.

When appropriate force fields are available, such methods can be successfully used to simulate and predict various properties of metal organic frameworks, the subject of this book, as long as no bond breaking or making processes need to be described. These include, for example, framework stability, textural and adsorption properties.[1,2]

The second class of computational methods is based on the quantum mechanical description of a chemical system. Such methods are particularly important for processes that involve bond breaking or making, which include, of course, catalytic reactions. Quantum mechanics represents the highest level in the hierarchy of computational methods. It in principle allows the Schrödinger equation to be solved for any chemical system to provide us an energy and wavefunction. The latter can be used to derive all properties of all atoms in the system. In practice, however, the Schrödinger equation cannot be solved exactly for any molecular system containing more than one electron. Thus, a number of simplifying approximations must be made that lead to different levels of quantum chemical methods. The most drastic approximations are assumed in the so-called semiempirical methods. These methods

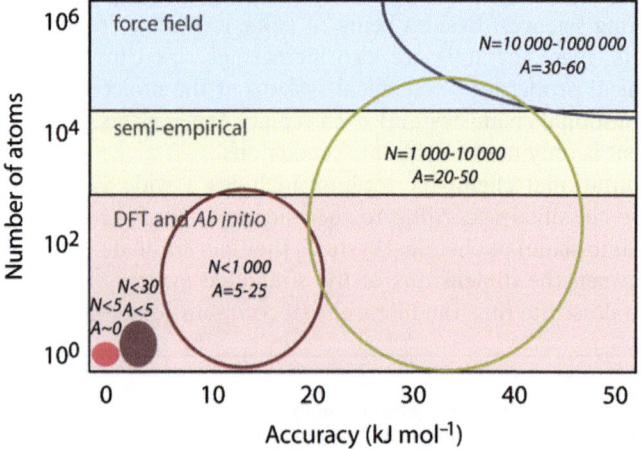

Figure 6.1   Hierarchy of computational methods.[3]

are relatively fast. Still they are roughly 100–1000 times slower than the classical force-field methods. Although semiempirical methods are based on the Schrödinger equation, they are extensively parametrized with experimental values. The accuracy of such methods is low. The major limitation, however, is that they give quite unreliable results for systems that are different from those used in the parametrization.

The most accurate computational methods provide an approximate solution of the Schrödinger equation without direct fitting to experiment. There are two general approaches to do this. One way is to solve the quantum mechanical equations directly to derive a many-electron wavefunction. This approach gives rise to a family of so-called *ab initio* methods ("from the begining", in other words, "from first principles") in which only mathematically sound approximations are made to derive the solution of the Schrödinger equation. A different approach is represented by density functional theory (DFT), in which the exact one-electron density is derived instead of the many-electron wavefunction. These methods calculate the energy of a system from the electron density using approximate exchange-correlation functionals. DFT methods are usually more efficient and scale better compared to *ab initio* methods, yet retain a high accuracy. Both approaches have their advantages and limitations and can be efficiently used for the description of different chemical processes in MOFs. Nevertheless, because of their robustness and general availability, the DFT methods have predominated in recent applications.

This chapter presents an introductory overview of important theoretical concepts and practical tools essential for computational modeling of chemical reactivity of metal organic frameworks using quantum chemical calculations. The text is organized as follows. After describing the concept of the potential energy surface that is central to the subject of computational chemistry, we will discuss the many-body problem in quantum mechanics. This will be followed by presenting an overview of the basic elements of *ab initio* and DFT methods that are used to tackle this problem. Then we briefly review some practical aspects of quantum chemical modeling of extended molecular systems. Finally, the power and capabilities of modern quantum chemical techniques will be illustrated by discussing relevant examples from the recent literature dealing with computational modeling of chemistry of metal organic frameworks.

6.2 The Concept of the Potential Energy Surface

The potential energy surface (PES) represents one of the core concepts in computational chemistry. It represents a relationship between the energy of a chemical system and its geometry. One of the primary tasks of computational chemistry is to determine the structure and energy of chemical structures including transition states of chemical transformations that "link" different stable states, *i.e.* reaction intermediates. The knowledge about the geometrical properties of different reaction intermediates and their energy creates a framework for predicting chemical reactivity of catalytic systems. These essential structures corresponding to the stable and transition states represent stationary points on PES. The first derivative of the potential energy at these points with respect to every degree of freedom in the system is zero:

$$\frac{\partial E}{\partial q_1} = \frac{\partial E}{\partial q_2} = \cdots = 0 \tag{6.1}$$

Stable reaction intermediates correspond to energy minima on PES. The second derivative $\left(\frac{\partial^2 E}{\partial q^2}\right)$ of the potential energy at such points is then positive for all degrees of freedom q.

The minima are linked by the lowest-energy path, the reaction coordinate or intrinsic reaction coordinate (IRC). Upon a reaction, the chemical system generally follows this path towards another energy minimum if it possesses sufficient energy to overcome the activation barrier. The second derivative of the potential energy along the reaction coordinate is negative for all degrees of freedom. For realistic chemical systems the degree of freedom corresponding to the reaction coordinates is usually composed of many individual geometrical parameters such as bond distances, angles and dihedrals.

An energy maximum along the reaction coordinate linking two minima corresponds to the transition state. In all other directions this stationary point represents an energy minimum. The transition state represents a saddle point on the PES for which the second derivative of the potential energy with respect to a single degree of freedom, which however may be composed of several geometrical parameters, is negative, while it is positive for the degrees of freedom along all other directions.

Determining structures of stable reaction intermediates, finding the minimum energy paths and locating the transition states on the potential energy surface are non-trivial processes. There are different methods available for accomplishing these tasks that differ in their accuracy and efficiency. Some of the associated methods will be briefly reviewed in Section 6.8. Generally, to find the stationary points and the reaction paths, the knowledge of the potential energy of a chemical system and its first derivative with respect to the atomic coordinates is required.

6.3 The Many-body Problem

A wide variety of chemical and physical properties, in the first place the energy, of a chemical system can be obtained by solving the many-body Schrödinger equation:

$$\widehat{H}\Psi = E\Psi \tag{6.2}$$

Where E is the total energy of the N-particle system, $\widehat{H}$ is the Hamiltonian operator and Ψ is the many-body wavefunction. The non-relativistic Hamiltonian for a system consisting of a set of nuclei and electrons is written as a sum of different kinetic and potential energy contributions arising from the interacting nuclei and electrons. In atomic units its form is:

$$\widehat{H} = \widehat{T}_N + \widehat{t}_e - \sum_{n=1}^{N}\sum_{i=1}^{M} \frac{Z_n}{|\mathbf{r}_i - \mathbf{R}_n|} + \frac{1}{2}\sum_{i \neq j}^{M} \frac{1}{|\mathbf{r}_i - \mathbf{r}_j|} + \frac{1}{2}\sum_{n \neq v}^{N} \frac{Z_n Z_v}{|\mathbf{R}_n - \mathbf{R}_v|} \quad (6.3)$$

Where Z_n is the charge of a nucleus, and $\mathbf{r}_i$ and $\mathbf{R}_n$ are, respectively, the spatial coordinates of electron i and nucleus n. Electrons and nuclei interact in a chemical system with each other and themselves. The first term $\widehat{T}_N$ and the second term $\widehat{t}_e$ are the kinetic energy operator for nuclei and electrons, respectively. The third, fourth and fifth terms are the potential energy operators of the electron–nuclei attraction, electron–electron and nuclei–nuclei repulsion, respectively. For any chemical system with the number of electrons exceeding 1, the exact solution of the equation 6.2 with the Hamiltonian 6.3 is not possible. Therefore, a number of simplifying approximations resulting in different quantum chemical methods are required to solve it for realistic systems that are invariably composed of many particles.

6.4   Born–Oppenheimer Approximation

The Born–Oppenheimer approximation (introduced in 1927 by Max Born and J. Robert Oppenheimer[4]) decouples the electron and nuclear motion. As the nuclei are much heavier than the electrons, the former move relatively slow and therefore are assumed to be stationary with the electrons moving in their field. Mathematically, this implies that the total wavefunction can be divided into an electronic ψ_e and a nuclear ψ_n part. The electronic wavefunction is solved then for a fixed set of nuclear positions. From the practical perspective, this implies that the total internal energy of a chemical system can be computed by solving the electronic Schrödinger equation

$$\widehat{H}_e\psi_e = \left(\sum_{i=1}^{M} -\frac{1}{2}\nabla_i^2 - \sum_{n=1}^{N}\sum_{i=1}^{M} \frac{Z_n}{|\mathbf{r}_i - \mathbf{R}_n|} + \frac{1}{2}\sum_{i \neq j}^{M} \frac{1}{|\mathbf{r}_i - \mathbf{r}_j|}\right)\psi_e = E_e(\mathbf{R}_N)\psi_e$$

$$(6.4)$$

and adding the resulting electronic energy E_e calculated for a set of nuclear positions R_N to the value of the internuclear repulsion, which is computed trivially. For most of chemical systems in the ground state, for which nonadiabatic effects are negligible, this approximation introduces very small error and it decreases for heavier elements. Note that the Born–Oppenheimer approximation lies in the foundation of the PES concept. Indeed, PES correlates the energy of a collection of nuclei and electrons only against the nuclear positions. The nature of stationary points is determined by the energy response on the perturbations of the positions of nuclei and not electrons. Because of the much faster motion of electrons, they form a "smeared-out cloud" of negative charge that "fixes" the relative position of the nuclei. The geometrical parameters of a molecular system are therefore determined only by the nuclear positions.

Thus, by solving the eqn (6.4), the total energy of a chemical system and its wavefunction can in principle be determined. The exact solution of this equation is however only possible for one-electron systems. While the first two terms corresponding to the kinetic energy of electrons and the electron-nuclei attraction are intrinsically monoelectronic, the third term describes the electron–electron repulsion that cannot be analytically resolved. Further approximations are necessary to describe many-electron systems.

6.5 *Ab Initio* Methods

The many-body problem is reduced within the Born–Oppenheimer approximation to an electronic problem. Although the positions of nuclei are now only a parameter, the electronic wavefunction depends on the coordinates of all electrons in the system and their spin. A traditional approach to derive a reasonable electronic wavefunction and determine the energy of the system is to follow a consecutive procedure, in which crucial physical phenomena are introduced step-by-step. The basis of all *ab initio* methods is the Hartree–Fock (HF) approximation that allows derivation of an approximate zeroth-order electronic wavefunction.

6.5.1 Hartree–Fock Approximation

In the HF method, the electron–electron interaction within an N-electron system is taken into account by a meanfield approximation where each electron is treated as moving in the field generated by the nuclei and the average electron distribution due to the $N - 1$ other electrons. As a result, each electron in the system moves independently, in other words the electron motion is uncorrelated. In the original method by Hartree,[5] an approximate wavefunction for the N-electron system was written as the product of one-electron wavefunctions:

$$\psi_0 = \varphi_0(\mathbf{r}_1)\varphi_0(\mathbf{r}_2) \ldots \varphi_0(\mathbf{r}_N) \tag{6.5}$$

Where ψ_0 is a function of coordinates of all electrons and $\varphi_0(\mathbf{r}_1),\ldots, \varphi_0(\mathbf{r}_N)$ are functions of the coordinates of electrons $1,\ldots, N$, respectively. These one-electron wavefunctions are orbitals. The Hartree process involves the refining of the initial guess wavefunctions *via* an iterative process, in which a one-electron Schrödinger equation is solved for each electron interacting with an average field due to all other electrons. At each step of the iterative procedure an updated N-electron wavefunction is generated. The process is continued until the energy calculated from the wavefunction ψ_k does not change (within a particular threshold) with respect to that obtained from the wavefunction ψ_{k-1} generated at the previous stage. In essence, the iterative procedure converges when the one-electron wavefunctions and the electrostatic field used to describe the electron–electron interaction change very little from one cycle to the next one. The field becomes consistent with the one used at the previous stage. This procedure is called the self-consistent field (SCF) procedure.

The fermionic character of electrons and their spin are not taken into account in the resulting wavefunction. These crucial deficiencies of the Hartree SCF method were corrected by Fock and by Slater in 1930.[6] The resulting average-field approach in which electron spin is considered and the fermionic nature of electrons is introduced by using the antisymmtric sum of one-electron wavefunctions gives rise to the Hartree–Fock (HF) method. Unlike Hartree wavefunctions that are a product of spatial orbitals, Slater wavefunctions used in the HF method are actually determinants of one-electron wavefunctions including spin variables as coordinates (spin orbitals). A Slater determinant for an N-electron system is given as:

$$\psi_e = \frac{1}{\sqrt{N!}} \begin{vmatrix} \varphi_1(\mathbf{r}_1\sigma_1) & \varphi_2(\mathbf{r}_1\sigma_1) & \cdots & \varphi_N(\mathbf{r}_N\sigma_N) \\ \varphi_1(\mathbf{r}_2\sigma_2) & \varphi_2(\mathbf{r}_2\sigma_2) & \cdots & \varphi_N(\mathbf{r}_N\sigma_N) \\ \vdots & \vdots & \cdots & \vdots \\ \varphi_N(\mathbf{r}_N\sigma_N) & \varphi_2(\mathbf{r}_N\sigma_N) & \cdots & \varphi_N(\mathbf{r}_N\sigma_N) \end{vmatrix} \tag{6.6}$$

where $\varphi_i(\mathbf{r}_j\sigma_j)$ are one-electron wavefunctions describing electron i at the position of electron j. The $1/\sqrt{N!}$ coefficient is the normalization factor. The ground state of a system is the one that has the lowest energy. The Hartree–Fock equations are derived by minimizing the energy with respect to the orbitals $\varphi_i(\mathbf{r}_j\sigma_j)$ under the constraint of the orbitals remain orthonormal:

$$\delta\left(E_e - \sum_{i,j=1}^{N} \varepsilon_{ij}\langle\psi_i|\psi_j\rangle \right) = 0 \tag{6.7}$$

where ε_{ij} are the Langrangian multipliers.

The well-known Hartree–Fock equations are then:

$$\widehat{F}_i|\varphi_i\rangle = \sum_{j=1}^{N} \varepsilon_{ij}|\varphi_j\rangle \tag{6.8}$$

where $\widehat{F}$ is the Fock operator given as

$$\widehat{F}_i = \widehat{h}_i + \sum_{j=1}^{N} (2\widehat{J}_{ij} - \widehat{K}_{ij}) \tag{6.9}$$

where operator $\widehat{h}$ comprises the kinetic energy operator and the electron-nuclear attraction, $\hat{J}$ and $\widehat{K}$ are the Coulomb and exchange operators, respectively. The latter arises from the terms in the Slater determinant that differ in electron exchange. Both operators describe the average repulsion of each electron with a charge cloud due to other electrons. It is usually postulated that the exchange operator $\widehat{K}$ does not have a simple physical interpretation that could be expressed by comparing with any classical analogue. Nevertheless, $\widehat{K}$ can be viewed as term introducing a correction to $\hat{J}$ and therefore reducing the electrostatic repulsion between two interacting electron clouds.

This effect originates directly from the Pauli principle, *i.e.* two electrons with antisymmetric wavefunctions cannot occupy the same spin-orbital, but they can populate the same spatial orbital if they have opposite spins. Thus, the repulsion between two electrons having the same spin is stronger than follows solely from the electrostatic considerations included in $\hat{J}$. Therefore, the summed $2\hat{J} - \hat{K}$ term describes the electrostatic repulsion within the meanfield approximation corrected for the Pauli exclusion principle. After the iterative procedure, the HF SCF method results in a set of molecular orbitals φ_j and their respective energy levels ε_j, that can be used to calculate the total energy, and the Slater determinant of φ_j that represents a total wavefunction Ψ. The latter can in principle be used for computing any property of a chemical system, as an expectation value of some operator.

6.5.2 Post Hartree–Fock Methods

Within the HF approximation, each electron is considered to be moving in an average electrostatic field due to all remaining electrons in the system. This description implies that the motion of electrons is not correlated, *i.e.* the probability of an electron having a particular set of spatial coordinates at a particular time does not depend on the coordinates of other electrons at that moment. In reality, the motion of electrons is correlated and they can avoid each other better. As a result, the electron–electron repulsion in the HF treatment is substantially overestimated and the total energy of the chemical system under consideration is too high. The correlation energy is then defined as the energy that the HF procedure fails to account for. Two components of electron correlation are often considered: static correlation and dynamic correlation. Dynamic correlation energy is associated with the fact that within HF electrons are not kept sufficiently apart, in other words it relates to the interaction between electrons in the system. Static correlation is associated with the use of a single determinant. The underestimation of the static correlations arises from the fact that there may be several almost degenerate frontier orbitals for which the assignment of an electron pair cannot be unambiguously done within HF. The HF approach cannot adequately describe either of these. This results in substantial errors in predicting energetics of chemical reactions and in the description of chemical transformations. Thus, a range of post Hartree-Fock methods (post-HF) has been developed that allow for the treatment of the correlated motion of electrons better by introducing additional corrections to the results of HF calculations. These include such methods as CI (configurational interaction), Møller–Plesset perturbation theory (MPn), complete active space (CAS) and coupled cluster (CC) (an extensive and in depth description of these methodologies as well as other computational techniques can be found in ref. 7, 8). The dynamic correlation energy can be efficiently recovered by using the MPn, CI and CC methods, whereas the static correlations can be taken into account by basing the wavefunction on more than one determinant within multireference configurational interaction methods, such as complete active space SCF (CASSCF).

One of the most widely used approaches to recover correlation energy in wavefunction-based methods is based on perturbation theory. The basic idea is that if an idealized simplified system can be described with a mathematical model, then a more complex and more realistic version of the system can be described as a perturbed version of the simple one. In the Møller–Plesset methods, the configurational interactions are treated as small perturbations to the Hamiltonian. Using this expansion, the HF method can be denoted as MP1 and the HF energy is equal to the sum of the zeroth and the first terms. The second-order Møller–Plesset perturbation theory (MP2) is the first MP level that goes beyond HF and is able to describe electron correlation. The MP2 energy (E_{MP2}) can be expressed as a sum of HF energy (E_{HF}) and the perturbation correction E^2. The latter term is purely electrostatics and represents a sum of terms describing double excitations from occupied to formally unoccupied molecular orbitals. The second order Møller–Plesset perturbation theory (MP2) typically recovers 80–90% of the correlation energy, while MP4 provides a reliably accurate solution to most systems.

Another approach to treat correlated motion of electrons is based on the proposal that the HF wavefunction and, therefore, the HF energy can be improved by introducing additional terms representing electron excitations from occupied to virtual molecular orbitals. In the configurational interactions (CI) methods a trial wavefunction is constructed as a linear combination of the ground-state wavefunction and excited-state wavefunctions. Similar to the MP case, the partial population of the excited states makes it easier for electrons to avoid each other and thus decreases the electron–electron repulsion. Although conceptually the treatment of electron correlation in MP and CI methods is very similar, the underlying mathematics is substantially different. The first term in the trial wavefunction represents the ground state configuration and is just the single-determinant HF wavefunction. Each subsequent term represents additional determinants due to excited state configurations originated from the initial HF configuration. If all possible electronic configurations available for a chemical system are considered, the trial wavefunction is the full CI wavefunction. The corresponding full CI calculations give an exact solution for the electronic problem, but can be applied only for very small molecules. In reality, the CI expansion has to be truncated and one has to consider only a limited number of determinants for the description of the electronic wavefunction. The trial wavefunction can include the exchange of 1,2 or 3 electrons from the valence band into unoccupied orbitals; these are known as CI singles (CIS), CI doubles (CID) and CI triples and allow for single, single/double and single/double/triple excitations. A number of mathematical treatments have been developed to facilitate the recovery of correlation energy within CI calculations despite the necessity of truncation of the CI expansion. Currently, the most widely-used modifications of CI are multiconfigurational SCF (MCSCF), complete active space SCF (CASSCF) and coupled cluster (CC) methods. The widely used CASSCF method involves a careful choice of the orbitals creating the active space to be used for constructing the various CI determinants. The active space includes only the orbitals that are assumed to be crucial for the investigated chemical process.

The coupled cluster (CC) method involves both the approaches from the perturbation methods and the CI treatment. The basic idea in the CC approach is to express the trial wavefunction as a sum of deteminants by allowing a series of excitation operators to act on the ground-state HF wavefunction promoting thus electrons into virtual spin orbitals. Depending on how many excitations are considered, one distinguishes the CC doubles (CCD), CC singles and doubles (CCSD) or CC singles, doubles and triples (CCSDT). Because the latter method is extremely demanding, the contribution of triple excitations is usually considered in an approximate way. The resulting CCSD(T) method represents the current golden standard of computational chemistry. With this method, an absolute accuracy within 1 kcal mol^{-1} can be achieved for practical chemical systems up to a moderate size.

Although when using these methods a very high accuracy can be achieved, almost all of the post-HF methods, with the exception of MP2, are prohibitively expensive for calculation of realistic models of heterogeneous catalysts including of course metal organic frameworks.

6.6   Density Functional Theory

6.6.1   Theoretical Background

A more attractive method to address the problem of electronic structure of complex chemical systems such as metal organic frameworks is density functional theory (DFT). The wavefunction contains all the information about the electronic structure. However, it cannot be measured for any molecule or atom. There is no agreement among physicists on the question of whether a wavefunction is just a convenient way to mathematically represent physical properties or a true physical entity itself. DFT is not based on wavefunctions, but on electron density function (electron probability density function, electron density, charge density), $\rho(\mathbf{r})$. The electron density can directly be measured by, for example, X-ray diffraction. Besides being experimentally measurable, the electron density has an additional significant advantage over wavefunction to be used as the core property in calculations. It is a function of position only, *i.e.* it depends on only three spatial coordinates, whereas a wavefunction depends on the three spatial coordinate plus spin coordinate for *each electron*. Thus, for an N-electron system a wavefunction will have $4N$ variables, while the electron density remains a function of only 3 spatial coordinates independently of the size of the system. Thus, compared to the wavefunction, the electron density has three important advantages: it is easily comprehensible, directly measurable and more mathematically tractable.

DFT rests on the work of Hohenberg and Kohn,[9] who formally proved that any ground-state property of a system is a functional of its electron density. DFT can in principle be regarded "*ab initio*" in the sense that it is derived from the first principles and does not necessarily require the use of adjustable parameters. One of the main reasons against this view is that the exact mathematical form of the DFT functional is not known, unlike the mathematically correct fundamental Schrödinger equation that is in the core of the wavefunction-based *ab initio* methods. This implies that whereas in the latter methods, the quality of wavefunction can be improved by employing higher-level correlation energy corrections, the systematic improvement of DFT functional is not straightforward.

Currently DFT calculations on chemical systems are based on the Kohn–Sham (KS) approach, the basis for which was set by the works of Hohenberg and Kohn in 1964,[9] where it was demonstrated that in the ground electronic state all properties of a system including energy (E_0) are a functional of its ground state electron density function, ρ_0. Kohn and Sham[10] extended the theory to practice by demonstrating that a variational approach can be used to calculate the energy and electron density. Thus, if an accurate electron density function ρ was available and if the exact form of energy functional was known, it could have been possible to directly compute the energy of a chemical system. Unfortunately, an accurate ρ is not readily available for any chemical system and, more importantly, the correct energy functional is not known. The latter is the key problem of DFT. The KS approach has been developed to address both these problems.

The ground state of a chemical system can be represented as the sum of the kinetic energy of the electron motion (T_E), nuclear–electron external potential (E_{NE}) and electron–electron interactions (E_{EE}):

$$E_0[\rho_0] = T_E[\rho_0] + E_{NE}[\rho_0] + E_{EE}[\rho_0] \tag{6.10}$$

The nuclear–electron term can be expressed as a classical electrostatic attraction potential energy expression that can be easily calculated:

$$E_{NE}[\rho_0] = \int \rho_0(\mathbf{r}) V_{ext}(r) d\mathbf{r} \tag{6.11}$$

The situation with the remaining two terms is much more difficult, because the kinetic and potential energy functionals in the respective energy terms are not known. A practical solution is to separate the terms that can be readily calculated and place all other terms in a separate approximated functional. Thus, the electron–electron interaction term E_{EE} can be defined as a sum of the classical coulomb repulsion between charge clouds and an additional term (E_{EE-cl}) that includes all the contributions to the "real" electron–electron interaction energy, but missing in the classical description.

$$E_{EE}[\rho_0] = \frac{1}{2} \int \frac{1}{r_{12}} \rho_0(\mathbf{r}_1) \rho_0(\mathbf{r}_2) d\mathbf{r}_1 d\mathbf{r}_2 + E_{EE-cl}[\rho_0] \tag{6.12}$$

A similar trick can be used to deal with the kinetic energy term. It can be represented as a sum of the electronic kinetic energy of a reference system (T_{ref})

and its deviation from the "real" electronic kinetic energy ($T_{E\text{-}ref}$). The equation 6.10 then becomes

$$E_0 = \int \rho_0(\mathbf{r}) V_{ext}(\mathbf{r}) d\mathbf{r} + \frac{1}{2} \int\int \frac{1}{r_{12}} \rho_0(\mathbf{r}_1)\rho_0(\mathbf{r}_2) d\mathbf{r}_1 d\mathbf{r}_2 + T_{ref}[\rho_0]$$
$$+ E_{EE\text{-}cl}[\rho_0] + T_{E\text{-}ref}[\rho_0]$$
(6.13)

The last two terms represent the main problem of the DFT. Their sum is usually defined as the exchange-correlation energy that is also a functional of the electron density function:

$$E_{XC}[\rho_0] = T_{E\text{-}ref}[\rho_0] + E_{EE\text{-}cl}[\rho_0]$$
(6.14)

The $T_{E\text{-}ref}$ term represents the kinetic correlation energy of the electrons and the $V_{EE\text{-}cl}$ describes the potential correlation and exchange energy.

The optimal ρ_0 is then computed following the variational principle. Kohn and Sham proposed that a real electron density can be represented by a "fictious" non-interacting reference system. Electrons in the latter system do not interact, but its ground state electron density distribution ρ_r is exactly the same as ρ_0 corresponding to the real system under consideration. The deviation in the behavior of non-interacting electrons from that of the real ones is then taken into account by the unknown XC functional. The electron density is then represented by a set of non-interacting one-electron wavefunctions:

$$\rho_0 = \rho_r = \sum_{i=1}^{2N} \varphi_i^{KS}(r)^2$$
(6.15)

where φ_i^{KS} are the Kohn–Sham spatial orbitals. The KS equations are then obtained by differentiating the energy with respect to these non-interacting KS orbitals similarly to the procedure used for the derivation of the HF equations:

$$\left[-\frac{1}{2}\nabla^2 + V_{NE} + \frac{1}{2}\int \frac{1}{r_{12}} \rho(r_2) dr_2 + V_{XC} \right] \varphi_i^{KS}(r) = \varepsilon_i \varphi_i^{KS}(r)$$
$$i = 1, \ldots, N$$
(6.16)

The exchange-correlation potential V_{XC} is defined as

$$V_{XC} = \frac{\delta E_{XC}[\rho(\mathbf{r})]}{\delta \rho(\mathbf{r})}$$
(6.17)

The optimal set of KS orbitals and therefore the ground state density can be obtained by solving these equations using the SCF procedure. This method could be exact if the functional for the exchange-correlation energy was known. As we have already indicated above, this is not the case and it has to be represented in an approximate form. The development of such an approximate functional is a very active research topic in theoretical chemistry.

6.6.2 Exchange-correlation Functionals

Here we will present only a brief overview of the most popular approximate exchange-correlation functionals. A good functional should not only account for errors in electron exchange and correlation, but also for self-repulsion errors in the average electron density cloud and kinetic energy error. The simplest approximation to E_{XC} is the local density approximation (LDA) that is based on an assumption that at every point in a chemical system the energy density corresponds to that of a homogeneous electron gas with the same electron density at that point. The exchange-correlation energy for such a model can be analytically computed for any chosen value of uniform density. The LDA exchange-correlation energy is given by

$$E_{XC}^{LDA}(\rho) = \int p(r)\varepsilon_{XC}[\rho(r)]\,dr \qquad (6.18)$$

The LDA approximation is in a sense exact, but only for those systems, in which electron density changes very smoothly in space. Therefore, LDA is able to produce reasonable results only for limited classes of systems. This approximation leads to large errors in bond energies and cannot adequately describe weak bonds such as hydrogen bonds.

The description of LDA can be substantially improved by taking into account not only the density at a given point in space but also its gradient, the first derivative of ρ with respect to its position, $\nabla\rho(r)$. This however cannot be performed by a simple gradient expansion and a number of constraints have to be imposed to derive the exchange-correlation functional, which are included in the F_{XC} functional:

$$E_{XC}^{GGA}(\rho) = \int \rho(r)F_{XC}[\rho(r), \nabla\rho(r)]\,dr \qquad (6.19)$$

A wide variety of F functions and therefore different GGA functional has been proposed so far. Among the most popular GGA functional are PW91,[11] PBE,[12] B88.[13] The use of these functionals allow drastic improvement of the accuracy of DFT calculations compared to the LDA approach. The accuracy of calculations of the atomization energy improves to about $20\,\mathrm{kJmol^{-1}}$ compared to an accuracy of about $180\,\mathrm{kJ\,mol^{-1}}$ obtained with LDA.[14]

Because the inclusion of the information about the spatial variation of density results in such a dramatic increase in accuracy, one can expect that the approximation of the exchange-correlation energy term can be further improved by introducing the dependency on the second derivative of ρ, $\nabla^2\rho(r)$. The corresponding functionals are usually denoted as meta-GGA (MGGA). These functionals depend on kinetic energy density instead of ρ itself. This approach allows some improvements of the results of calculations, although the difference in this case is much less pronounced compared to the difference between LDA and GGA approximations.

The exchange-correlation energy can be expressed as a sum of the exchange and correlation energies. The exchange energy is the dominant term. A very popular way to improve GGA functional is to introduce the exact HF exchange into the exchange-correlation functional. The applicability of this approach follows from the adiabatic connection method (ACM). The ACM implies that the exchange-correlation energy can be expressed as a linear combination of the DFT exchange-correlation energy and HF exchange energy. As a result, hybrid DFT functionals are contributed by a HF-type electron exchange energy that is calculated from the KS wavefunctions for non-interacting electrons. This allows an effective correction to the classical coulomb repulsion for Pauli repulsion. The crucial parameter for hybrid functional is the amount of exact HF exchange used that can be expressed in mixing coefficient. Most of the popular hybrid functionals were constructed by fitting these mixing coefficients to various parameters of large sets of molecules. One of the most popular hybrid functional is B3LYP, for which the exchange-correlation energy is given as:

$$E_{XC}^{B3LYP} = (1-a)E_X^{LDA} + aE_X^{HF} + bE_X^{GGA} + cE_c^{LYP} + (1-c)E_c^{LDA} \qquad (6.20)$$

where $a = 0.2$, $b = 0.72$, and $c = 0.81$.

However, despite its popularity, better functionals are currently available for almost any particular application. Nevertheless, this method remains a very efficient and quite reliable tool to tackle an "average" quantum chemical problem.[15] From the practical point of view, the attractiveness of DFT methods is associated with their high accuracy and at the same time very high efficiency. DFT calculations produce results of a comparable quality to those obtained using the MP2 method, while requiring much less computational time. DFT methods formally scale with the increase in the number of basis functions (electrons) as N^3 and therefore allow investigations of more realistic structural models compared to the higher-level post-HF methods, which usually scale as N^5 for MP2 and up to N^7 for CCSD(T). Nevertheless, because of the approximate nature of the exchange-correlation functional employed in DFT methodologies, their direct application to chemical problems dominanted by effects due to electron correlation can result in substantial errors. Although so far rather reliable parametrization schemes for DFT have been developed, the systematic improvement of the functionals especially for exchange and correlation is a challenge. Note that the associated phenomena can be crucial for the adequate description of chemical processes occurring within microporous voids of MOFs.

6.6.3 Beyond Exchange-correlation Functionals

Standard density functionals generally fail to describe chemical phenomena dominated by effects due to the electron correlation. These include, for example, the correct prediction of the electronic structure of chemical systems with strong electron correlation such as some transition/rare-earth metal compounds and adequate description of molecular systems dominated by weak London dispersion interactions. The latter is particular crucial for the correct represen- tation of the behavior of MOFs and will be considered below in more detail.

One of the main challenges in modeling chemical processes in MOFs is to predict accurately energies of adsorption and reaction profiles for transformations of different molecules in their porous space. As has been outlined above, the current method of choice for modeling reactivity of such complex catalytic materials as MOFs is density functional theory (DFT). However, commonly used density functionals fail to describe correctly the long-range dispersion interactions.[16,17] Because the pore walls of MOFs are predominantly composed of organic fragments, the dominant interactions between confined molecules and MOF pores correspond to weak van der Waals interactions of dispersive nature, which cannot be correctly computed within conventional DFT. All semilocal and conventional hybrid functionals cannot adequately describe the asymptotic $-C_6/R^6$ behavior of the dispersive interactions with increasing interatomic distance. This failure is due to the nature of these dispersion interactions, which originate from the interaction between induced dipoles in the interacting fragments. The induced dipoles are formed due to "charge fluctuations" that are, in essence, the instantaneous electron correlations. The inability of conventional DFT approaches to adequately account for such effects may not only result in inaccurate energetics for chemical reactions but even in wrong predictions about the stability or reactivity trends for systems where the influence of dispersive interactions on the total stabilization energy of the reaction intermediates and transition states is not uniform along the reaction coordinates. This is in particular important for the description of chemical reactions that are accompanied with substantial structural transformations of the reaction environment, such as appears to occur in the pores of MOFs.

Dispersion is an intermolecular correlation effect. The simplest electronic structure method that explicitly describes electron correlation is Møller–Plesset perturbation theory (MP2). However, MP2 calculations for large chemical systems are very demanding. To describe periodic systems they are presently feasible only for very small unit cells containing only few atoms. Recently, an embedding scheme to introduce local corrections at post-HF level to DFT calculations on a periodic structural model of porous materials has been proposed.[18] This approach allows an accurate modeling of structural and electrostatic properties of the microporous reaction environment by using the periodic DFT calculations. The refinement for the self-interaction effects and van der Waals interactions between the adsorbed reactants and the reactive site is achieved by applying resolution of identity implementation of the MP2 method (RI-MP2) to a cluster model representing the essential part of the framework that is embedded into the periodic model of the MOF. The resulting MP2:DFT approach is suited to studying reactions between small or medium sized substrate molecules and very large chemical systems, such as MOF crystals, and allows quantitative computing reaction energy profiles for transformations of hydrocarbons in microporous matrices with near chemical accuracy.

Nevertheless, although the proposed DFT:MP2 scheme allows for the very accurate calculations of adsorption and reaction energies in microporous space, the associated computations are still too demanding to be used for comprehensive studies and for an in depth theoretical analysis of chemical transformations in MOFs. Thus, there is still a strong desire for a robust computational tool capable of a reliable description of chemical reactions in MOFs that must combine efficiency and chemical accuracy of DFT methods along with aproper account for van der Waals dispersive interactions. This is reflected by the fact that the improvement of DFT towards a better description of nonbonding interactions is currently an active research area in theoretical chemistry. An in depth description and limitations of different available dispersion-corrected DFT methods has been recently a subject of an excellent review by Grimme.[17]

The most pragmatic solution of this problem is to include a semiclassical dispersion correction into the quantum chemical approach. The corresponding computational scheme is usually denoted as DFT-D. It consists of adding a semiempirical term $E(D)$ to the DFT energy $E(DFT)$ resulting in the dispersion-corrected energy $E(DFT–D)$. $E(D)$ in this case is expressed as a sum over pairwise interatomic interactions computed using a force-field-like potential truncated after the first term:

$$E(D) = - s_6 \sum \frac{c_{ij}}{r_{ij}^6} f_D(r_{ij}) \tag{6.21}$$

where c_{ij} are the dispersion coefficients, the damping function $f_D(r_{ij})$ removes contributions for short-range interactions, while the global scaling parameter s_6 depends on the particular choice of the exchange-correlation functional. The DFT-D approach has been parameterized for many atoms and a wide variety of functionals and can be used in a combination with popular quantum chemical programs. When applied to chemical processes in microporous materials, this approach has been shown to provide realistic energies of adsorption of different gases in MOFs (see for examples ref. 19, 20, 21). The DFT-D approach represents a powerful, efficient and robust computational tool for investigating different chemical processes and accounts for London dispersion interactions in molecular systems and solids.

The performance of DFT itself may also be substantially improved by parameterization of the exchange-correlation functionals. Zhao and Truhlar have recently reported a family of meta-GGA functionals (M05,[22] M06,[23] and related functionals) which allow for a substantially more accurate description of non-bonding interactions and prediction of reaction energies and activation barriers compared to the conventional GGA and hybrid functionals. The hybrid methods involving a combination of such density functionals and well-parameterized force fields are anticipated to be very efficient and accurate for the investigations of adsorption phenomenon and chemical reactions in MOFs.

Nevertheless, the simplifications involved in the above methods, such as the assumption of pairwise additivity of van der Waals interactions, the presence of empirically fitted parameters both in the force fields and in the parameterized density functionals can lead to unreliable results for systems different from the training set. The recently proposed nonlocal van der Waals density functional (vdW-DF)[24,25] is derived completely from first principles. It describes dispersion in a general and seamless fashion, and predicts correctly its asymptotic behavior.

6.7 Basis Sets

The energy in the DFT methods is formally a function of the electron density. However, in practice the density of the system $\rho(r)$ is represented *via* the Kohn–Sham orbitals (eqn (6.15)). This leads to another approximation usually done both in DFT and wavefunction-based methods. It involves the representation of molecular orbitals by a specific orthonormal basis set. The real electronic structure of a system can be in principle correctly described by an infinite number of basis functions. However, because of computational limitations, in practice these functions are usually truncated and the orbitals are described by a finite number of basis sets. Obviously such an approximation also results in some potential loss in accuracy. A wide variety of different basis sets is currently available. The actual choice of a certain basis set for a quantum chemical calculation strongly depends on the solution method used, the type of the problem considered and the accuracy required in each particular case. These functions can take on one of several forms. The most commonly used approaches use either a linear combination of local atomic orbitals, usually represented by Gaussian-type functions (GTO), or a linear combination of plane-waves (PW) as basis sets. GTO basis sets are mostly used for modeling of molecular systems, which also include cluster models of MOFs discussed in more detail below. GTO basis sets are implemented in various available quantum chemical software (Gaussian,[26] Tubomole,[27] *etc*). Plane-waves are more suitable for simulating solid materials (*e.g.* MOF crystals). This is mainly due to the fact that their application to periodic systems is straightforward. Such calculations are usually faster both for computing energies and gradients as compared to the approaches employing GTO basis sets. The PW approach is implemented in such computer programs as CASTEP,[28] CPMD[29] and VASP,[30] *etc.*, which are often used for studying various periodic systems usually by means of 'pure' DFT (without exact HF exchange). In addition to the PW codes developed to model solids, the CRYSTAL[31] program that utilizes GTO basis sets can be used for studying both periodic and molecular systems within the same formalism irrespectively of the dimensionality of the system. It is important to note that in general, when a sufficiently large number of basis functions is employed, the computational results obtained using either GTO or PW basis sets are essentially the same.[32]

When the PW basis set expansion of the electronic wavefunction is used, the number of plane-wave components needed to correctly describe the behavior of the wavefunction near the nucleus is prohibitively large. To solve this problem, the core electrons are described within the PW approach using the pseudopotential approximation, which assumes that the core electrons do not significantly influence the electronic structure and properties of atoms. Thus, the ionic potential that arises from the nuclear charge and core electron density can be safely replaced by an effective pseudopotential. Although within the GTO approach core electrons can be treated explicitly, the pseudopotential approximation can also be employed to reduce number of basis functions in calculations without dramatic loss of accuracy. This is in particular useful for the description of heavy atoms, such as transition metal ions within the MOF structures.

6.8   Practical Aspects of Modeling Chemistry of MOFs

6.8.1   Geometry Optimization

Using the quantum chemical methodologies described above, the electronic structure of a given molecular configuration can be resolved and its energy can be calculated. All of them have their strengths and weaknesses and the actual choice of the computational method is determined by the type of the scientific question to be answered, the amount of already available information and the amount of available computational resources. The simulation usually begins by constructing molecular models representing a specific site in a MOF structure, an adsorption complex or reaction intermediate, the properties of which have to be investigated. Such an initial guess structure usually does not correspond to an actual local minimum on the potential energy surface (PES). Therefore, its geometry has to be optimized, in other words the energy of the guess structure has to be minimized with respect to the positions of the nuclei. After an electronic energy for the initial structure is computed, the first derivative of the potential energy, that is the gradient, can be computed analytically or numerically. There are numerous geometry optimization procedures available that allow minimizing the gradient until it reaches a small value (below the pre-defined convergence threshold) by changing the positions of atoms in a chemical system. The most common procedures for geometry optimization include the steepest descent, conjugate gradient and Newton–Raphson algorithms. Some of the methods require only the energy and its gradient as the input, whereas the others also use the Hessian matrix, *i.e.* second-order partial derivatives of the potential energy. The use of force constants that can directly be obtained from the Hessian matrix substantially accelerates the geometry optimization procedure. However, the calculation of Hessians is very time-consuming and is very seldom performed at every iteration of the geometry optimization. In practice, often approximate Hessians are used that are constructed based on the results of the gradient calculations.

6.8.2   Transition State Search

Predicting catalytic reactivity requires the knowledge of the energy and geometry of the first-order saddle points (transition state) located on the reaction path linking two minima (reaction intermediates). Whereas energy minima can routinely be located, finding transition states on PES determined from quantum-chemical calculations is usually much more challenging because it requires the knowledge of the local topology of a PES. In principle, one can explicitly compute the PES by manually varying relevant constrained coordinates with the remaining coordinates of a chemical systems being relaxed using standard geometry optimization procedure. Despite its high effectiveness, an automated procedure for a TS search is desired. A number of different automated procedures for locating saddle points on PES have been developed so far. Nevertheless, the task of locating TSs especially for chemical reactions within complex systems is far from being routine. In general, the algorithms for saddle-point search can be divided into two groups, namely, those based on interpolation between two minima and those using only local information such as the local gradient and second derivatives of the potential energy.

The most popular algorithm of the latter family is the Newton–Raphson method. In principle, this method can be initiated anywhere on the PES. When the starting point is close to a saddle point the convergence will be rapidly achieved. The optimization procedure updates the atomic positions along the normal mode of the Hessian with the lowest magnitude aiming at locating the point of maximum negative curvature. Unfortunately, this method performs very poorly when the search is started far from a transition state and when the utilized Hessian contains several negative modes. Sometimes the initial transition state guess structure can be successfully constructed from an intuitive guess. However, often such an approach results in a configuration with a Hessian containing a number of unstable models, none of which corresponds to the motion along the desired reaction coordinate. Furthermore, even if the TS search is successful, there is a high probability that the transition state structure does not connect the reactant and product states.

The dimer method, developed by Henkelman and Jónsson,[33] does not involve the computation of the full Hessian matrix. Instead, only the lowest eigenvalue and the respective eigenvector are calculated in this algorithm. The dimer method aims at determining which activated transitions are possible from a given state and attempts to find a set of low-energy saddle points at the boundary of the potential energy basin associated with the initial energy minimum state.

The second class of methods necessitates additonal information about the structures of the reactant and product. The algorithm then generates a sequence of atomic configurations (images)connecting the given energy minimum states. Such interpolation methods basically convert a saddle-point search to a minimization problem. One of the most popular and highly efficient interpolation algorithms is nudged-elastic band (NEB) method. This method does not require the calculation of a Hessian matrix. The NEB procedure involves the linear interpolation of a set of images between the known initial and final states followed by their geometry optimization. Each intermediate configuration corresponds to a specific structure along the reaction coordinate. Thus, when the geometry optimization of all images is successfully completed, the information about the minimum energy reaction path is obtained.

6.8.3 Frequency Analysis and Thermodynamics

Once geometry optimization is completed and a stationary point is found, its nature has to be verified. This is done by analyzing vibrational frequencies calculated for the optimized structure (normal mode analysis). The vibrational frequencies (ω_v) are usually calculated using the harmonic oscillator approximation. In the harmonic approximation, the potential energy is a quadratic function of the normal coordinate. Thus, the force contstant of a vibration mode equals the second derivative of the potential energy surface. The Hessian is a matrix of the force contacts. The vibrational modes and their frequencies can be computed by matrix diagonalization of the Hessian. The sign of the force constant shows the curvature of the PES along the particular vibrational mode. The correspondence of a particular optimized structure to a stable reaction intermediate or a TS is ensured by analyzing the calculated vibrational frequencies that have to be computed at the end of every geometry optimization. One should, however, note that the calculated frequencies usually substantially deviate from the experimental values determined from infrared spectroscopy because this method does not take into account the anharmonicity of molecular vibrations. Furthermore, the errors due to the computational method itself can also substantially contribute to the mismatch between the computed and experimental vibrational frequencies.

In addition to indicating the IR spectrum and allowing the nature of the stationary point to be checked, the computed vibrational frequencies provide zero-point energies (ZPE). Because even at 0 K molecules vibrate, there is a nonzero vibrational contribution to its energy. This first thermodynamic correction to the energy of a system is defined as a half of the sum of all vibrational frequencies, ω_v:

$$E_{ZPE} = -\frac{1}{2}\sum \hbar\omega_v \qquad (6.22)$$

Note that the approximate Hessian that is built in the course of the geometry optimization is not accurate enough for the calculation of frequencies and, accordingly, ZPE and all other energy contributions. An accurate Hessian matrix should be calculated in a separate step. This can be done either numerically or analytically.

The results of the normal mode analysis can also be used to compute macroscopic thermodynamic properties. This can be done by using statistical thermodynamics within the ideal gas approximation. The macroscopic properties of a system are described by the partition function. By using this approach, finite temperature and entropic energy corrections can be computed at a given temperature and pressure. By summing the electronic energy obtained from quantum chemical calculations with the zero point, finite temperature and entropic energy contributions computed using the results of the normal mode analysis, free energy values at the given temperature and pressure can be calculated.

6.8.4 Structural Models of Metal Organic Frameworks

The approximations done in order to compute energies by different quantum chemical methods as well as the use of a finite basis set for the description of molecular orbitals are not the only factors leading to limited accuracy. When modeling complex chemical systems such as MOFs, one can seldom treat all its atoms at the highest computational level. Often, only a limited subset of the atoms of the MOF structure is used to construct an atomistic model representative of the reaction center. This approach is usually referred to as the cluster modeling approach. The size of a cluster model is critical for obtaining of reliable results. Typically, a representative model of an active site in MOF should include the reactive site or the adsorption site and a part of the structure near the active site to mimic the local properties of the MOF confinement space and to ensure a sufficient structural flexibility of the coordination environment at the active center. Although this approach results in some loss of "model" accuracy, it can be very useful for the analysis of different local properties of porous materials such as elementary reaction steps, adsorption, *etc.* In addition, in the case of cluster modeling the higher level *ab initio* methods as well as hybrid density functionals can be successfully used.

The current progress in computational chemistry has also made it possible to use rather efficiently periodic boundary conditions in DFT calculations of solids. This allows theoretical DFT studies of structure and properties of some MOF structures, usually with relatively small unit cells using a real crystal structure as a model (Figure 6.2). Such periodic DFT calculations of MOFs are however mostly limited to the use of LDA and GGA density functionals.

Hybrid quantum chemical embedding schemes form a wide range of popular methods for computational modeling of different porous materials including

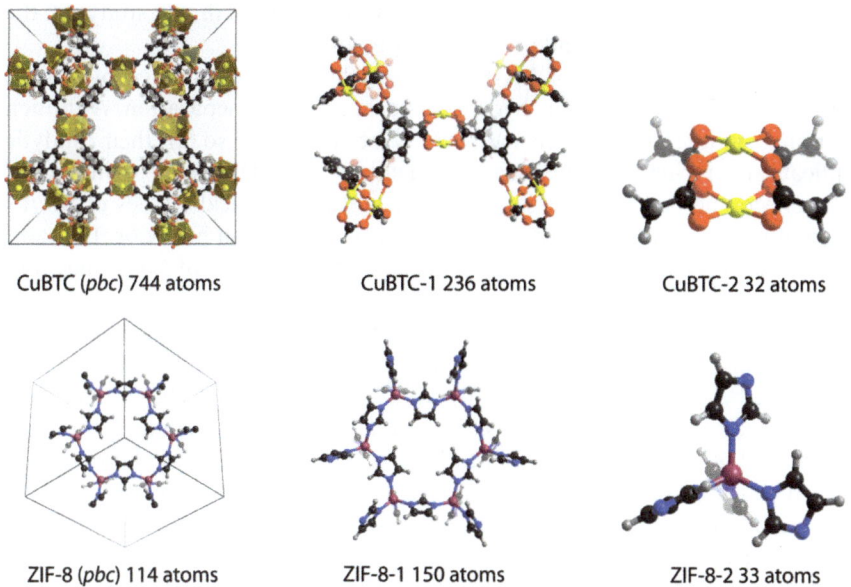

CuBTC (*pbc*) 744 atoms   CuBTC-1 236 atoms   CuBTC-2 32 atoms

ZIF-8 (*pbc*) 114 atoms   ZIF-8-1 150 atoms   ZIF-8-2 33 atoms

Figure 6.2   Examples of periodic and cluster structural models for CuBTC and ZIF-8 metal organic frameworks at decreasing level of description (from left to right; pbc refers to a simulation cell with periodic boundary conditions).

MOFs. They allow the combination of two or more computational techniques in one calculation (see *e.g.* the MP2:DFT method discussed in Section 6.6.3). The region of the system where the chemical process takes place (similar to that used for cluster modeling) is described with an appropriately accurate method, while the remainder of the system is treated at a lower level of theory.

The energy for this system is then calculated as

$$E_{\text{hybrid}}(\text{System}) = E_{\text{high}}(\text{Model}) + E_{\text{low}}(\text{System}) - E_{\text{low}}(\text{Model}) \qquad (6.23)$$

where E_{high} (Model) is the energy calculated for the inner core region at a higher level of theory, E_{low} (System) – E_{low} (Model) is the energy difference between the full system and the core region calculated at the low level of theory. Usually quantum chemical methods (QM) are used for the description of the core region, while the rest of the atoms of the system are treated by molecular mechanics.

6.9 Computational Modeling of MOFs

Quantum chemical methods are based on the calculation of the electronic structure of a chemical system. This defines the main area of their application for the investigation of metal organic frameworks that is modeling and prediction of their chemical properties and reactivity. The formation of new chemical bonds between adsorbates and adsorption sites as well as chemical transformations taking place in the micropores of MOFs can be viewed as a rearrangement of electrons within the interacting species. Thus, electronic structure calculations are the only method that allows direct study of the fundamental phenomena underlying such processes.

So far, research in field of MOF chemistry has been dominated by studies related to such applications as hydrogen storage and gas separation. Although thousands of different MOF structures have been reported so far, their catalytic applications are limited. The catalytic chemistry of MOFs is in its infancy.[34] This reflects the fact that the majority of quantum chemical studies of MOFs predominantly focus on unraveling their unique adsorption behavior. In recent years, however, computational studies on catalytic reactivity of MOFs begin to emerge.

Usually, upon adsorption gas molecules interact with the framework atoms of MOF surfaces through dispersive, repulsive, and Coulombic interactions, which can be accurately described by appropriate force field-based methods. The application of the respective molecular simulation techniques to gas storage and adsorption in MOFs has been recently reviewed by Snurr *et al.*[35] Nevertheless, quantum chemical modeling represents a powerful tool to analyze the properties of individual adsorption complexes, accurately predict interactions between adsorbates and different structural units of MOFs, as well as to investigate such phenomena as adsorbate-induced structural trans-formations of MOFs.

Nachtigall *et al.*[36] employed a combined DFT/CC method to rationalize the experimental isotherms for carbon dioxide adsorption on CuBTC metal organic frameworks at the molecular level. This methodology was also successfully used for the investigation of the adsorption of water[37] and carbon monoxide[38] on coordinatively unsaturated sites in CuBTC. It was concluded that the DFT/CC scheme allows an accurate description of MOFs and can be used for the verification and improvement of the force field-based simulation methods. A conceptually similar MP2:DFT-D method was used by Sauer *et al.*[19] to study CO and CO_2 adsorption on CPO-27-M (M = Mg, Ni, Zn) MOFs. The hybrid method yielded heats of adsorption within $2 \, kJmol^{-1}$ compared to the experimental values.

An interesting property of MOFs that distinguishes them from other classical molecular sieve-type materials such as zeolites is their ability to undergo structural transitions upon adsorption of gases.[39] Quantum chemical studies shed light on the mechanism of the adsorbate-induced structural trans- formations and the effect of such phenomena on the adsorption properties of MOFs. The interaction of CO_2 and N_2 in a flexible Zn_2-(bpdc)$_2$(bpee) MOF (where bpdc = 4,4'-biphenyl dicarboxylate and bpee = 1,2-bis(4-pyridyl) ethylene) was studied by a combination of Raman and IR spectroscopies complemented by quantum chemical calculations using the vdW-DF2 method.[40] It was demonstrated that the ligand connectivity in the framework is the key factor determining the flexibility of the MOF and the pore-opening phenomenon. Framework flexibility was also proposed as the key factor determining the exceptional performance of amino-modified MIL-53(Al) materials (NH_2-MIL-53(Al)) in the separation of CO_2 from different gas mixtures.[41] Originally, it was proposed that the preferred adsorption of CO_2 is associated with its enhanced binding to basic NH_2 sites in NH_2-MIL-53(Al) walls.[42] However, periodic DFT-D calculations clearly showed that the amine moieties play only an indirect role in this process. Upon adsorption, the CO_2 molecules do not directly interact with the NH_2 fragments. The functional- ization of the MIL-53(Al) material modulates its "breathing" behavior thus affecting the flexibility of the MOF framework and therefore its ability to undergo structural transformations induced by the interaction with specific adsorbates. Another example is the recent study on selective alkane adsorption on ZIF-7 from alkane/alkene mixtures.[43] The interaction of ethane and ethylene with a cluster model representing a pore-opening of ZIF-7 was studied using meta-GGA M06-L functional. It was demonstrated that the main difference in the behavior of saturated and unsaturated hydrocarbons towards adsorption on ZIF-7 is related to their ability to form stable adsorption complexes at the external surface.

The molecular details underlying the high catalytic activity of nonfunc- tionalized ZIF-8 in the transesterification of vegetable oils with alcohols were unraveled in a FTIR spectroscopic study complemented by periodic and cluster DFT calculations by Chizallet *et al.*[44] A range of cluster models representing different Zn moieties that can be formed at the external surface of ZIF-8 under the catalytic conditions was considered. It was demonstrated that the catalytic reactivity in this case is predominantly associated with the presence of strongly Lewis acidic undercoordinated Zn sites conjugated with basic N^- and OH groups at the external surface or possibly in bulk defects of ZIF-8. A similar conclusion has been drawn in a recent work by Broadbelt *et al.*[45] reporting a DFT study of ethylbenzene hydroperoxide decomposition over under-coordinated Cu and Co sites in metal organic frameworks. The catalytic reaction was modeled using cluster models representative of structural motifs common for different MOF structures (ZIF-9, HKUST-1, NOTT) but with substantially simplified ligand structures. On the basis of the computational results obtained, it was argued that the catalytic potential of metal sites in Co- and Cu-MOFs is limited. Only the undercoordinated sites on the external catalyst surface can promote the decomposition of bulk hydroperoxide molecules.

6.10   Conclusions

Computational modeling is becoming one of the key contributors to catalysis and material sciences. Quantum chemical calculations are currently widely utilized to assist the interpretation of experimental data, and in developing the molecular-level understanding of adsorption processes and catalytic reactions in metal organic frameworks. In this chapter we presented a short overview of computational methods that can be used for simulating chemical processes in metal organic frameworks. We did not aim to review all of the available methodologies in view of the limited space available here, but instead limited ourselves to highlighting the basic principles underlying different quantum chemical methods, their capabilities and limitations as applied to MOF chemistry. The power of quantum chemical techniques in developing molecular-level understanding of complex chemical phenomena in metal organic frameworks is highlighted with relevant examples from recent literature. We hope that this chapter will provide an initial orientation for the reader into the field of computational chemistry and its application to MOF science.

Acknowledgements

E.A.P. thanks the Netherlands National Science Foundation (NWO) and Technology Foundation STW for his personal VENI grant.

References

1. O. K. Farha, A. O. Yazaydın, I. Eryazici, C. D. Malliakas, B. G. Hauser, M. G. Kanatzidis, S. T. Nguyen, R. Q. Snurr and J. T. Hupp, *Nature Chem.*, 2010, **2**, 944.
2. C. E. Wilmer, M. Leaf, C. Y. Lee, O. K. Farha, B. G. Hauser, J. T. Hupp and R. Q. Snurr, *Nature Chem.*, 2012, **4**, 83.
3. S. Fantacci, A. Amat and A. Sgamellotti, *Acc. Chem. Res.*, 2010, **43**, 802.
4. M. Born and J. R. Oppenheimer, *Ann. Phys.*, 1927, **84**, 457.
5. D. R. Hartree, *Proc. Cambridge Philos. Soc.*, 1928, **24**, 89.
6. (a) J. C. Slater, *Phys. Rev.*, 1930, **35**, 210; (b) V. Fock, *Z. Phys.*, 1930, **61**, 126.

7. E. G. Lewars, *Computational Chemistry*, Springer, Springer Science + Business Media B.V., 2nd edn, 2011.
8. D. C. Young, *Computational Chemistry: A Practical Guide for Applying Techniques to Real-World Problems*, Wiley-Interscience, New York, 2001.
9. P. Hohenberg and W. Kohn, *Phys. Rev.*, 1964, **136**, 864.
10. W. Kohn and L. Sham, *Phys. Rev.*, 1965, **140**, 1133.
11. J. P. Perdew, J. A. Chevary, S. H. Vosko, K. A. Jackson, M. R. Pederson, D. J. Singh and C. Fiolhais, *Phys. Rev. B*, 1992, **46**, 6671.
12. J. P. Perdew, K. Burke and M. Ernzerhof, *Phys. Rev. Lett.*, 1996, **77**, 3865.
13. A. D. Becke, *Phys. Rev. A*, 1988, **38**, 3098.
14. P. Sautet, *Quantum Chemistry Methods*, in Characterization of Solid Materials and Heterogeneous Catalysts: From Structure to Surface Reactivity, ed. M. Che and J. C. Védrine, Wiley-Interscience, New York, 2012.
15. S. F. Sousa, O. A. Fernandes and M. J. Ramo, *J. Phys. Chem. A*, 2007, **111**, 10439.
16. K. E. Riley, M. Pitoňák, P. Jurečka and P. Hobza, *Chem. Rev.*, 2010, **110**, 5023.
17. S. Grimme, *WIREs: Comput. Mol. Sci.*, 2011, **1**, 211.
18. (a) C. Tuma and J. Sauer, *Chem. Phys. Lett.*, 2004, **387**, 388; (b) C. Tuma and J. Sauer, *Phys. Chem. Chem. Phys.*, 2006, **8**, 3955.
19. L. Valenzano, B. Civalleri, K. Sillar and J. Sauer, *J. Phys. Chem. C*, 2011, **115**, 21777.
20. K. Sillar, A. Hofman and J. Sauer, *J. Am. Chem. Soc.*, 2009, **131**, 4143.
21. D. Peralta, G. Chaplais, A. Simon-Masseron, K. Barthelet, C. Chizallet, A.-A. Quoineaud and G. D. Pirngruber, *J. Am. Chem. Soc.*, 2012, **134**, 8115.
22. Y. Zhao, N. E. Schultz and D. G. Truhlar, *J. Chem. Phys.*, 2005, **123**, 161103.
23. Y. Zhao and D. G. Truhlar, *Theor. Chem. Acc.*, 2008, **120**, 215.
24. (a) M. Dion, H. Rydberg, E. Schröder, D. C. Langreth and B. I. Lundqvist, *Phys. Rev. Lett.*, 2004, **92**, 246401; (b) M. Dion, H. Rydberg, E. Schröder, D.C. Langreth and B. I. Lundqvist, *Phys. Rev. Lett.*, 2005, **95**, 109902(E).
25. V. R. Cooper, L. Kong and D. C. Langreth, *Phys. Proc.*, 2010, **3**, 1417.
26. M. J. Frisch, G. W. Trucks, H. B. Schlegel, G. E. Scuseria, M. A. Robb, J. R. Cheeseman, G. Scalmani, V. Barone, B. Mennucci, G. A. Petersson, H. Nakatsuji, M. Caricato, X. Li, H. P. Hratchian, A. F. Izmaylov, J. Bloino, G. Zheng, J. L. Sonnenberg, M. Hada, M. Ehara, K. Toyota, R. Fukuda, J. Hasegawa, M. Ishida, T. Nakajima, Y. Honda, O. Kitao, H. Nakai, T. Vreven, J. A. Montgomery, Jr., J. E. Peralta, F. Ogliaro, M. Bearpark, J. J. Heyd, E. Brothers, K. N. Kudin, V. N. Staroverov, R. Kobayashi, J. Normand, K. Raghavachari, A. Rendell, J. C. Burant, S. S. Iyengar, J. Tomasi, M. Cossi, N. Rega, J. M. Millam, M. Klene, J. E. Knox, J. B. Cross, V. Bakken, C. Adamo, J. Jaramillo, R. Gomperts, R. E. Stratmann, O. Yazyev, A. J. Austin, R. Cammi, C. Pomelli, J. W. Ochterski, R. L. Martin, K. Morokuma, V. G. Zakrzewski, G. A. Voth, P. Salvador, J. J. Dannenberg, S. Dapprich, A. D. Daniels, Ö. Farkas, J. B.

Foresman, J. V. Ortiz, J. Cioslowski, and D. J. Fox, *Gaussian 09*, Gaussian, Inc., Wallingford CT, 2009.

27. R. Ahlrichs, M. Bär, M. Häser, H. Horn and C. Kölmel, *Chem. Phys. Lett.*, 1989, **162**, 165.

28. M. D. Segall, P. J. D. Lindan, M. J. Probert, C. J. Pickard, P. J. Hasnip, S. J. Clark and M. C. Payne, *J. Phys.: Condens. Matter*, 2002, **14**, 2717.

29. (a) D. Marx and J. Hutter, in *Modern Methods and Algorithms of Quantum Chemistry*, NIC, FZ Jülich, 2000, p. 301; (b) W. Andreoni and A. Curioni, *Parallel. Comput.*, 2000, **26**, 819.

30. (a) G. Kresse and J. Hafner, *Phys. Rev. B*, 1994, **49**, 14251; (b) G. Kresse and J. Furthmüller, *Comput. Mater. Sci.*, 1996, **6**, 15; (c) G. Kresse and J. Furthmüller, *Phys. Rev. B*, 1996, **54**, 11169.

31. R. Dovesi, V. R. Saunders, C. Roetti, R. Orlando, C. M. Zicovich-Wilson, F. Pascale, B. Civalleri, K. Doll, N. M. Harrison, I. J. Bush, Ph. D'Arco and M. Llunell, CRYSTAL2006 User's Manual, Universita di Torino, Torino, 2006, http://www.crystal.unito.it/.

32. S. Tosoni, C. Tuma, J. Sauer, B. Civalleri and P. Ugliengo, *J. Chem. Phys.*, 2007, **127**, 154102.

33. G. Henkelman and H. Jónsson, *J. Chem. Phys.*, 1999, **111**, 7010.

34. M. Ranocchiari and J. A. van Bokhoven, *Phys. Chem. Chem. Phys.*, 2011, **13**, 6388.

35. R. B. Getman, Y. S. Bae, C. E. Wilmer and R. Q. Snurr, *Chem. Rev.*, 2012, **112**, 703.

36. L. Grajciar, A. D. Wiersum, P. L. Llewellyn, J.-S. Chang and P. Nachtigall, *J. Phys. Chem. C*, 2011, **115**, 17925.

37. L. Grajciar, O. Bludský and P. Nachtigall, *J. Phys. Chem. Lett.*, 2010, **1**, 3354.

38. M. Rubeš, L. Grajciar, O. Bludský, A. D. Wiersum, P. L. Llewellyn and P. Nachtigall, *Chem Phys Chem*, 2012, **13**, 488.

39. (a) A. J. Fletcher, K. M. Thomas and M. J. Rosseinsky, *J. Solid State Chem.*, 2005, **178**, 2491; (b) G. Férey and C. Serre, *Chem. Soc. Rev.*, 2009, **38**, 1380.

40. N. Nijem, P. Thissen, Y. Yao, R. C. Longo, K. Roodenko, H. Wu, Y. Zhao, K. Cho, J. Li, D. C. Langreth and Y. J. Chabal, *J. Am. Chem. Soc.*, 2011, **133**, 12849.

41. E. Stavitski, E. A. Pidko, T. Remy, E. J. M. Hensen, B. M. Weckhuysen, J. Denayer, J. Gascon and F. Kapteijn, *Langmuir*, 2011, **27**, 3970.

42. S. Couck, J. F. M. Denayer, G. V. Baron, T. Rémy, J. Gascon and F. Kapteijn, *J. Am. Chem. Soc.*, 2009, **131**, 6326.

43. J. van den Bergh, C. Gücüyener, E. A. Pidko, E. J. M. Hensen, J. Gascon and F. Kapteijn, *Chem.–Eur. J*, 2011, **17**, 8832.

44. C. Chizallet, S. Lazare, D. Bazer-Bachi, F. Bonnier, V. Lecocq, E. Soyer, A. A. Quoineaud and N. Bats, *J. Am. Chem. Soc.*, 2010, **132**, 12365.

45. P. Ryan, I. Konstantinov, R. Q. Snurr and L. J. Broadbelt, *J. Catal.*, 2012, **286**, 95.

Part B
Catalysis by MOFs

CHAPTER 7

Strategies for Creating Active Sites in MOFs

FRANCESC X. LLABRÉS I XAMENA,* IGNACIO LUZ
AND FRANCISCO G. CIRUJANO

Instituto de Tecnología Química UPV-CSIC, Universidad Politécnica de
Valencia, Consejo Superior de Investigaciones Científicas, Avda. de los
Naranjos, s/n, 46022, Valencia, Spain
*Email: fllabres@itq.upv.es

7.1 Introduction

In the first part of the book, we have described the different existing methods to
prepare metal organic frameworks, and to modify their properties and
introduce new functionalities through post-synthesis modification of a starting
MOF. We have also shown that quite often the determination of their
structural, textural, electronic and vibrational properties can only be achieved
satisfactorily when a number of characterization techniques are combined
together, sometimes with the essential assistance of computational methods.
Thus, we have provided the necessary tools for preparing new MOFs, for
tailoring their composition, porosity and structure, for assessing and evaluating
their properties and, hopefully, for predicting and understanding their behavior
and potential. We should now be ready to address another important issue,
which is the design and preparation of MOFs with a view to their application.
We will deal with this in the rest of the book. To delimit the scope, we have
confined ourselves to describe only the use of MOFs as heterogeneous catalysts.
For details on other applications of MOFs, such as in gas separation[1,2] and

RSC Catalysis Series No. 12
Metal Organic Frameworks as Heterogeneous Catalysts
Edited by Francesc X. Llabrés i Xamena and Jorge Gascon
© The Royal Society of Chemistry 2013
Published by the Royal Society of Chemistry, www.rsc.org

storage,[3–6] luminescence,[7–9] medicine,[10] sensing[11] or magnetism,[12] the readers should turn to other specialized readings.

The second part of the book, devoted to catalytic applications of MOFs, starts with this introductory chapter, in which we will first try to summarize the main properties of MOFs that make them very appealing for applications in heterogeneous catalysis. In fact, great efforts have been directed in the last decade to study the potential of MOFs in catalysis, with an ever-increasing number of papers appearing each year. We will now see what is behind this "MOF rush". Next, we will present some general considerations that should be taken into account when planning the use of MOFs as heterogeneous catalysts, such as stability, recovery and reusability. And finally, we will revise the different strategies that can be used to introduce the desired catalytic centers in the MOFs. We will show how it is possible by using these strategies to engineer the material for catalysis, and to fine tune the properties of the MOF to influence the catalytic performance.

7.1.1   Strengths of MOFs as Heterogeneous Catalysts

As we have seen in previous chapters, one of the most interesting features of MOFs is that, in certain cases, they can be prepared in the form of crystalline porous materials with extremely *high surface areas and pore volumes* (up to 6240 $m^2 \, g^{-1}$ and 3.6 $cm^3 \, g^{-1}$ have been reported in the case of MOF-210).[13] When this is so, the material combines simultaneously two properties that are highly desirable for the design of heterogeneous catalysts: high porosity and crystallinity. On one hand, the elevated porosity facilitates the diffusion of reactants and products and the interfacial contact with the active sites of the catalyst. On the other hand, given the crystalline nature of these materials, the position of each atom in the space will be precisely known, as well as the coordination and oxidation states of the metal ions forming the MOF. This will largely facilitate establishing structure–activity relationships of the catalytic centers of the MOF. At the same time, the crystallinity of the MOF also determines that the pore system will be strictly periodic and regular in size and shape, as for other crystalline porous solids such as zeolites. An almost endless variety of pore dimensions and topologies are already known for MOFs, ranging from ultra-microporous to mesoporous, and from monodimensional channels to tridimensionally connected cages or cavities, and even chiral porous structures can be readily prepared.

When considering the porosity of a MOF, one can run into different situations. Following the nomenclature introduced by Kitagawa,[14] we can distinguish between *first, second,* and *third generation* porous systems. *First generation* MOFs are materials having a pore system sustained by guest molecules that irreversibly collapses when they are removed. *Second generation* materials have a robust pore system, featuring *permanent porosity* upon evacuation of the guest molecules. Finally, *third generation* MOFs have a flexible pore system, which may change reversibly depending on the presence of guest molecules or by the action of certain external stimuli, such as light, temperature or the application of an electric field. While first generation MOFs would find very limited (or no) use in heterogeneous catalysis, second and third generation MOFs show a very high potential for this and other applications. Thus, this chapter is mainly focused on second and third generation MOFs and their use in catalysis.

Besides the possibility of having permanent porosity, another important feature of MOFs is that their pore size, shape, dimensionality and chemical environment (*e.g.*, *hydrophilic/hydrophobic* pores) can be finely tuned by either selecting the appropriate components (metal and organic ligand) and network connectivity, or by post-synthesis modification of a pre-formed MOF. Note that this is in contrast with other crystalline porous solids, for which the diversity of framework types is significantly limited due to the use of a rather reduced number of building units (*e.g.*, only [SiO_4] and [AlO_4] tetrahedral units are used in zeolites, while AlO_4 and PO_4 units are used in aluminophosphates). The high tunability of MOFs allows for controlling which molecules can diffuse within the pores, since only those molecules smaller than the pore openings will penetrate the inner space of the solid. Thus, MOFs can show *molecular sieving* and *shape-selective properties*, which are highly desired for heterogeneous catalysis. Meanwhile, both the metallic and organic components can modulate the *host–guest interactions* experienced by adsorbed molecules or by the transition states formed during a reaction occurring within the pores, especially when the dimensions of the substrates are similar to the pores in which they are confined. These interactions can thus *activate or orient the substrates* around a catalytic center and, in general, can modify the reactivity of adsorbed substrates and the selectivity of the reaction towards specific products.

Since Robson introduced the concept that MOF networks can be described by using appropriate molecular building blocks and metal ions (the modular approach),[15–17] people realized that one possible strategy for tuning the pore dimensions is the so-called *isoreticular expansion*. This concept consists of expanding the pore size of a known structure by using "expanded" but geometrically analogous organic bridging ligands, so that the new material will have the same framework topology as the original compound, but with larger pores. Using this strategy, it has been possible to expand the domain of porous MOFs from the microporous to the *mesoporous range*. The best known example of the use of this strategy is the so-called IRMOF isoreticular series prepared by the group of Yaghi.[18] All the materials belonging to this series are constructed with the same inorganic building block, Zn_4O (analogous to that found in basic zinc acetate), while the organic linker is selected among various linear dicarboxylic molecules of different dimensions, such as in IRMOF-0, IRMOF-1, IRMOF-10 and IRMOF-16, which contain alkyne, phenyl, biphenyl and triphenyl moieties, respectively. Similar series of isoreticular compounds having increasing pore sizes were prepared by Lillerud *et al.* containing hexameric $Zr_6O_4(OH)_4$ clusters as the inorganic building blocks and linear dicarboxylic acids, *viz.* UiO-66, UiO-67 and UiO-68.[19] Besides linear dicarboxylic ligands, other "expanded ligands" have also been used in this context, such as tricarboxylic molecules geometrically analogous to trimesic acid, leading to the isoreticular series of materials known as MOF-177, MOF-180 and MOF-200.[13] Some representative examples of "expanded ligands" used to control systematically the pore size of the isoreticular MOFs are shown in Figure 7.1.

This high versatility of MOF design provides further advantages for catalysis, since it is possible to introduce the catalytic active site either at the metal or at the organic component. Alternatively, the pore system of the MOF can also serve as the physical space in which a catalytic species (*e.g.*, metal or metal oxide nanoparticle, metallic coordination complex or other discrete molecular species) is encapsulated,[20,21] or as the confined space in where a chemical reaction is taking place. In these cases, the MOF may act as a simple spectator or as a passive dispersing medium for the catalytic species, or it can also intervene in the catalytic reaction, either by stabilizing transition states or by introducing additional active sites.

Following with this list of advantages, the possibility of designing MOFs containing chiral catalytic centers should also be considered, which can lead to *asymmetric (enantioselective) catalysis*. Asymmetric catalysis is routinely practiced in the synthesis of enantiomerically pure drugs, usually relying on the use of expensive homogeneous chiral catalysts. The use of chiral organic ligands in the synthesis of MOFs can lead to a vast number of new chiral

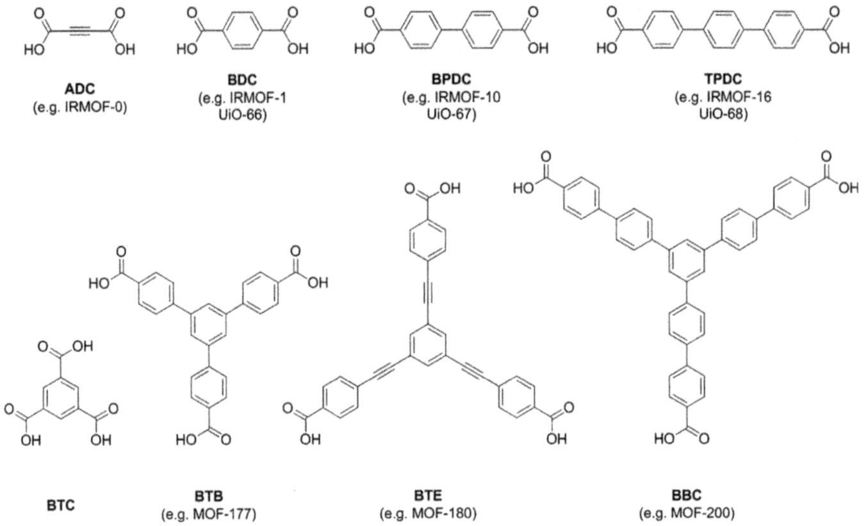

Figure 7.1 Representative examples of "expanded ligands" used for preparing isoreticular MOFs of increasing pore size. Some examples of the isoreticular materials prepared with the ligands are given in parentheses. ADC = acetylenedicarboxylate; BDC = benzenedicarboxylate; BPDC = biphenyl-dicarboxylate; TPDC = triphenyldicarboxylate; BTC = benzenetricar-boxylate; BTB = 4,4'4''-benzene-1,3,5-triyl-tribenzoate; BTE = 4,4',4''-[benzene-1,3,5-triyl-tris(ethyne-2,1-diyl)]tribenzoate; BBC = 4,4',4''-[benzene-1,3,5-triyl-tris(benzene-4,1-diyl)]tribenzoate.

heterogeneous catalysts to be tested in this field. However, the preparation of chiral ligands often requires arduous and expensive laboratory synthesis or separation from racemic mixtures, which can complicate the preparation of a chiral MOF. Fortunately, other synthesis strategies exist for preparing homochiral MOFs from achiral components, either *via* self-resolution during crystal growth or by using chiral co-ligands, as described in detail in Chapter 11. Alternatively, chiral centers can be introduced *a posteriori* in the MOF by post-synthesis modification.[22] The possibility to design chiral MOFs with relative ease make them privileged candidates with respect to other crystalline porous solids, such as zeolites and other zeotypes. In these latter compounds, introduction of chiral moieties is often prevented by the need of harsh conditions during the synthesis of the inorganic solid, usually carried out in strong basic or acidic media or requiring calcination at high temperatures to remove organic templates. In fact, currently only a few chiral zeolites are known, *viz.* SU-32,[23] ITQ-37,[24] or the zeolite mineral goosecreekite,[25] which contrasts with the large number of homochiral MOFs described so far.

When dealing with third generation porous MOFs, rational control over the *framework flexibility* of the material can also have a large impact on its final catalytic properties. Although framework flexibility has been largely overlooked when designing catalytic applications of MOFs, it can be expected that these properties will be considered in future developments. Taking the enzymatic systems as source of inspiration, the objective has to be the preparation of materials capable of adapting the pore space by conformational changes of their building units.

Finally, the rational design of MOFs affords a means for readily preparing *multi-functional catalysts*, which allows one-pot procedures to be performed involving multiple catalytic events with only one catalyst and avoiding unnecessary (and costly) steps of separation and purification of intermediate products. These transformations known as *tandem*, *domino* or *cascade reactions* can decrease the energy consumption of the catalytic process and may increase the selectivity of the whole process, avoiding the generation of waste byproducts. Although the use of MOFs as enantioselective and multi-functional catalysts is still in its infancy, a fast evolution of the field can be anticipated in the next few years, which will surely demonstrate the brilliant potential of these materials as advanced heterogeneous catalysts. Figure 7.2 summarizes the most relevant strengths that MOFs can offer when considering their application as heterogeneous catalysts.

7.1.2 Weaknesses of MOFs as Heterogeneous Catalysts: Framework Stability and Leaching Tests

At the other end of the scale, we need to be aware that MOFs can also have severe limitations, which could limit or even prevent their use for certain reactions or under certain conditions. Particularly relevant in this sense are the relatively low thermal and chemical stability of MOFs, when compared with

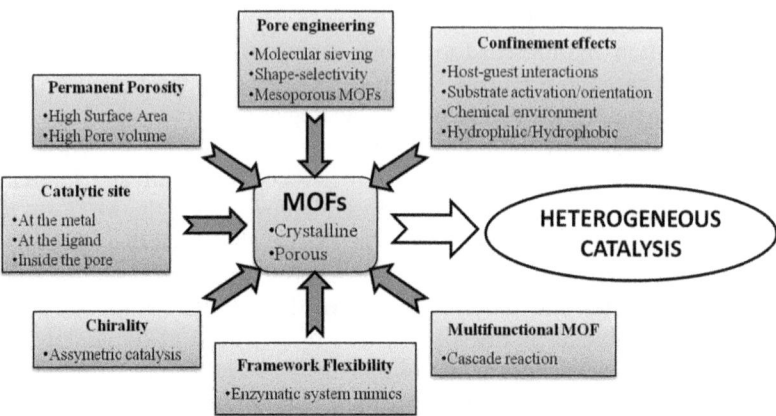

Figure 7.2   Summary of the most relevant properties of MOFs with a view to their potential application in catalysis.

inorganic porous solids. Although a handful of remarkably robust compounds have been prepared,[19,26,27] other MOFs are highly sensitive to moisture and they cannot withstand being in the open air for long, even at room temperature, undergoing amorphization or phase transformation into other crystalline structures.[28,29]

The thermal stability of a MOF is frequently established by thermogravimetric (TGA) or by thermo-diffraction analysis, usually carried out under a flowing inert gas (N_2 or He) or under a vacuum while progressively raising the temperature up to the complete destruction of the framework (typically up to 850–1000 K). Under these conditions, several MOFs are found to be thermally stable at temperatures higher than 573 K. This temperature in principle would be sufficiently high to envisage the applicability of the material in many liquid-phase reactions, which are typically performed at temperatures lower than 473 K. However, the data coming from thermogravimetric or thermo-diffraction analysis have to be taken with caution, since structures that are stable over 573 K for the limited time of the TGA measurement may undergo extensive damage when heated at lower temperature for much longer times. Also, degradation and framework dissolution in certain solvents, in acidic or basic media, or in the presence of certain functional groups, are often drawbacks that can severely delimit the scope of MOFs as heterogeneous catalysts. In general, it is very difficult to anticipate the stability of a given MOF under certain reaction conditions, even when its thermal stability under vacuum or N_2 is very high. For this reason, it is always necessary to check that the crystalline structure of the MOF is preserved after the catalytic reaction. Common tests for assessing MOF stability include the comparison of the X-ray powder patterns, the specific surface area and pore volume, and the elemental analysis of the fresh MOF and the solid recovered after the catalytic reaction by filtration or centrifugation. In some cases, minor changes in the XRD pattern upon catalytic use, particularly in the relative intensity of some peaks, have been attributed to the presence of organic species retained inside the void space of the MOF and not to a destruction of the crystallinity.[30,31] This can be assessed by determining the porosity of the used material after the reaction and by comparing it with that of the fresh material, or by comparing the fraction of organic components burnt during a TGA experiment.

When performing a reaction catalyzed by a MOF it is always advisable to determine the possible occurrence of leaching of the metal (or organic components) to the reaction medium. This is of paramount importance, since it is necessary to know the true origin of the catalytic process. The metal species leached from the solid to the liquid phase, even in minute amounts, could be responsible totally or in part for the observed catalytic activity. In these cases, the solid MOF could be the precursor of the real catalytic species and will act by dispensing a certain amount of metal into the solution. In principle, leaching is an undesirable process in heterogeneous catalysis because it produces a long term the decay of the catalytic activity of the material and reveals the instability of the solid catalyst. Furthermore, in those cases in which a shape-selective process could be expected on the basis of substrate dimensions relative to pore size of the MOF, the existence of leaching could ruin this possibility.

Extensive metal leaching of a MOF during a catalytic reaction is often easily detected by a simple chemical analysis of the filtrate of the reaction medium (once the MOF catalyst has been removed by filtration or centrifugation), or directly from the comparison of the XRD pattern of the recovered solid. Note that, since the metal is an integral part of the crystalline framework, a massive release of the metal would necessary cause a collapse of the MOF structure. However, when the leaching occurs to a much lower extent (at the level of traces) it can be very difficult to detect by chemical analysis or X-ray diffraction. In these cases, a common method for determining the occurrence of leaching is by performing a hot filtration test. This consists of performing the reaction in normal conditions, but quickly removing the catalyst at partial conversion by filtration at the reaction temperature and following the reaction in the filtrate. Then, if no traces of the leached metal have passed to the liquid medium, no further conversion should be observed in the filtrate. Figure 7.3 shows a pictorial representation of the typical curves obtained for a filtration test in the cases of positive or negative metal leaching.

Another important issue that can delimit the stability (and thus, the potential application) of MOFs is their mechanical resistance. Some MOFs are known to collapse when submitted to mechanical compression beyond a certain pressure, thus losing their crystallinity and specific surface area. This is relevant for those (catalytic) applications requiring compressing or molding of the MOF into pellets of various shapes or dimensions to be fed into the reactor (see Chapter 14). In these applications, it would be necessary to asses also the mechanical stability of the MOF upon shaping, which can simply consist of comparing the XRD pattern and the textural properties of the fresh MOF powder and the pelletized solid.

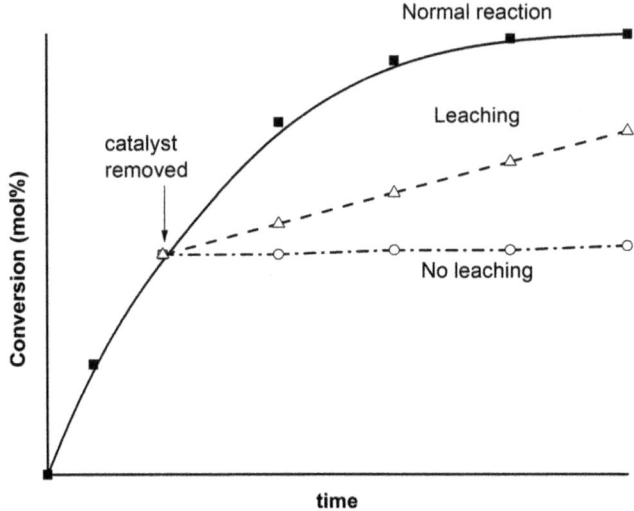

Figure 7.3 Typical curves obtained for hot filtration test experiments in the cases of positive or negative leaching. When there is no leaching, no further conversion is observed in the filtrate after removing the catalyst. If leaching is occurring, catalyst filtration does not stop the reaction, though a decrease in the conversion rate is sometimes observed.

7.1.3 Bridging Homogeneous and Heterogeneous Catalysis with MOFs

Homogeneous and heterogeneous catalysis have traditionally been considered as different disciplines, and indeed they have been usually studied separately. However, the tendency nowadays is to try to design and develop new materials to bring together the best advantages of both disciplines into a catalytic system, while avoiding the limitations of "conventional" homogeneous and heterogeneous catalysts. As we will now see, MOFs offer an excellent opportunity to bring the two worlds together.

Metals in solution, either as metal salts or transition metal complexes, are widely used to effect chemical transformations in two completely distinct scenarios. One consists in the use of stoichiometric amounts of the metal salts as oxidizing reagents (such as $KMnO_4$, $K_2Cr_2O_7$ or $Pb(OAc)_4$), as well as Lewis or Brønsted acids or bases (*e.g.*, $AlCl_3$ or NaOH); while a second approach consists of using more sophisticated, specifically designed metal coordination or organometallic complexes as efficient and selective catalysts. The use of stoichiometric metal salts has many associated problems, such as the generation of large amounts of wastes, either directly generated in the chemical process or formed during the necessary neutralization of acids and bases during waste treatment. The current tendency is whenever possible to replace these compounds by more benign heterogeneous catalysts, such as zeolites or clays. For instance, 4-methoxyacetophenone, an important chemical intermediate, has been traditionally produced in the liquid phase, using dichloromethane as solvent and $AlCl_3$ as Lewis acid reagent. But this process produces up to 4.5 kg of inorganic salts per kg of final product. Nowadays, Rhodia produces this molecule in a continuous process in fixed bed reactors and without any solvent, using zeolite beta as the acid catalyst, thus reducing considerably the amount of wastes generated.[32]

Besides their use as stoichiometric reagents, metal coordination or organometallic complexes have also been widely used as homogeneous catalysts, sometimes with extraordinary success. They are able to catalyze a large number of organic reactions, in many cases with extraordinary chemo-, regio- and enantioselectivities. The main advantage is the high activity and selectivity that can be achieved with these compounds, which is primarily due to the possibility to systematically change the organic ligands that coordinate to the central metal ion. In this way, the electronic and steric properties of the molecule can be finely modulated to obtain the desired catalytic properties. Additionally, the well defined structure of these metal complexes enables the study of their interaction with substrates by means of powerful quantum chemical calculations, which allows a precise knowledge on the reaction mechanism as well as to engage in predictive reactivity patterns. Homogeneous catalysts, however, are usually difficult to recover from the reaction medium and/or they decompose during the reaction.

Replacement of soluble metal coordination and organometallic complexes by heterogeneous catalysts can overcome these limitations, and in some cases, can even introduce additional valuable features, such as shape-selective properties. For instance, well-defined active sites ranging from protons to Lewis acids, and even redox sites, can be introduced in the frameworks of zeolites and mesoporous materials. When combining these well-defined framework or extra-framework active sites with the regular pore dimensions and topologies typical of zeolites, that can be selective towards different potential transition states (shape-selectivity), then highly active, selective, stable and recyclable catalysts can be obtained. Probably the most popular example of this class is titanium silicalite-1 (TS-1).[33] TS-1 in combination with hydrogen peroxide has allowed the industrial implementation of environmentally friendly technologies, such as phenol hydroxylation and cyclohexanone ammoximation.[33] However, owing to the relatively small pore diameter of TS-1 (5.5 Å), this zeolite catalyst cannot be used for reacting bulky substrates. Therefore, new titanium-containing large pore zeolites[34] and mesoporous materials[35] have been prepared to replace TS-1 when dealing with large substrates. Another four-valent metal, *viz.*, Sn, was successfully incorporated into the framework of pure silica zeolite beta and provided unique catalytic activity for the Baeyer–Villiger oxidation with hydrogen peroxide[36] and Meerwein–Ponndorf–Verley reductions[37] among others.[38] Later, zirconium,[39] niobium and tantalum[40] beta zeolites were also synthesized and successfully used in catalysis.

But the above heterogeneous catalysts have also their own limitations, which limit their performance and application. For instance, zeolites are suitable solid catalysts for gas phase reactions, but due to their limited available pore size, they generally undergo fast deactivation for liquid phase reactions. In addition, most of the efforts to develop zeolites with larger pores at the nanometre scale have met with failure, and only a handful of mesoporous zeolites have been prepared so far.[24,41–43] Another important limitation in heterogeneous catalysis is that fine modulation of the electronic and coordination properties of the metal active sites of the solid is usually much more difficult than in the case of the homogeneous metal coordination and organometallic complexes.

In an attempt to overcome these limitations of both homogeneous and heterogeneous catalysts, scientists have developed preparative methods to combine the well-controlled active sites of transition metal complexes with the adsorption and pore selectivity effects and recyclability of solid catalysts. Heterogeneization has been achieved by grafting or impregnation of the active metal coordination complexes on solid carriers, by intercalation within layered compounds, by introducing them in zeolite cavities by a "ship in a bottle" technique,[44] or by forming structured or non-structured mesoporous organic–inorganic hybrid systems.[45,46] In some cases, the objective of heterogeneization was simply to anchor the active species on a solid carrier to achieve well isolated, uniform single sites that will not interact between them and decompose. However, in other cases the heterogeneization went further in such a way that the solid also intervenes in the catalytic process, either by stabilizing transition states or by introducing additional active sites.[47,48] Unfortunately, these heterogeneization methods have no general application, since not all the homogeneous catalysts can be easily and successfully anchored to suitable supports. Sometimes all the necessary steps for immobilization make the synthesis procedure too complex and can severely increase the price of the catalyst, making it not competitive with respect to alternative processes.

In this context, porous crystalline MOFs can come into consideration, since these compounds can be considered as complementing and expanding the work of zeolites in heterogeneous catalysis, owing to the outstanding properties described in the previous section. In particular, it is very interesting that there is the possibility to prepare a large variety of MOFs with pore dimensions larger than those of classical zeolites, with a range of compositions and chemical properties, and with access to chiral catalysts. This could overcome the limitations of zeolites when used in liquid phase reactions or with large organic substrates, since wide pore MOFs could avoid diffusion control of the reaction. On the other hand, the modular construction of MOFs through the assembly of metals and organic ligands by coordination bonds allows for systematically changing the electronic and steric properties of the central metal ions, much like in soluble metal complexes. For instance, de Vos and co-workers have shown very recently that the Lewis acid character of a MOF can be electronically boosted by introducing side substituents in the organic linkers.[49] However, and in contrast with soluble metal complexes, MOFs will be easily recovered from the reaction medium by simple filtration, being solid materials, and this will allow recycling the material for further use. Thus, when the catalysis is based on metal activity at the nodes or at the ligands forming the walls, MOFs appear directly as solid counterparts of homogeneous catalysts. In summary, the rational design of MOFs and the large versatility in the engineering of their structures and compositions can certainly help in bridging the gap between homogeneous and heterogeneous catalysis.

7.2 Designing MOFs for Catalytic Applications

Porous MOFs contain three well differentiated parts: the metallic component, the organic linker and the porous system. As we will see in this section, it is possible to prepare MOFs in which the catalytic function is located at any of these three parts, or even containing simultaneously two or more types of catalytic sites. In some cases and for certain MOFs, the as-prepared material can be used directly as a catalyst, since it already contains the necessary active centers to catalyze the chemical reaction. This category includes those materials in which the metal sites can directly coordinate to the substrates and catalyze the reaction, as well as MOFs in which the organic ligands contain functional groups that can act as (organo) catalysts. But unfortunately, the MOFs that can be directly used as catalysts in the as-synthesized form represent only a tiny proportion of the whole family of MOFs. In most cases, it has been necessary to develop specific strategies to modify the starting material before it can be used in a catalytic reaction. In the following, we will review the main situations that we can find when facing the application of a MOF in catalysis, depending on whether the catalytic function is introduced at the metallic site, at the organic linker or inside the pore system. At this point, only the fundamentals and the conceptual design of the different preparative strategies are presented, while more detailed descriptions of the catalytic application of the materials are generally given in the following chapters of the book. The idea is to provide here in a single chapter a general overview of the various scenarios that are found when designing a MOF in view of its application in catalysis.

7.2.1 Catalysis at the Metallic Site

7.2.1.1 As-Synthesized Active MOFs and MOFs with Coordinatively Unsaturated Sites.

Some MOFs contain metal ions that can directly coordinate to the substrates to catalyze a chemical transformation, and these are what we refer to as *as-synthesized active MOFs*. Coordination of the substrate to the metal requires either an expansion of the coordination sphere of the metal ion, or a (reversible) displacement of one of the ligands forming the MOF originally coordinated to the metal site, as shown schematically in Figure 7.4. In either case, the crystalline network of the MOF has to be highly flexible to prevent the collapse of the structure as a consequence of the local distortions produced upon substrate coordination.

We have recently found that copper imidazolate, $[Cu(im)_2]$ (im = imidazolate)[50] and copper pyrimidinolate, $[Cu(2\text{-pymo})_2]$ (2-pymo = 2-hydroxypyrimidinolate),[51] both feature highly flexible networks that can readily

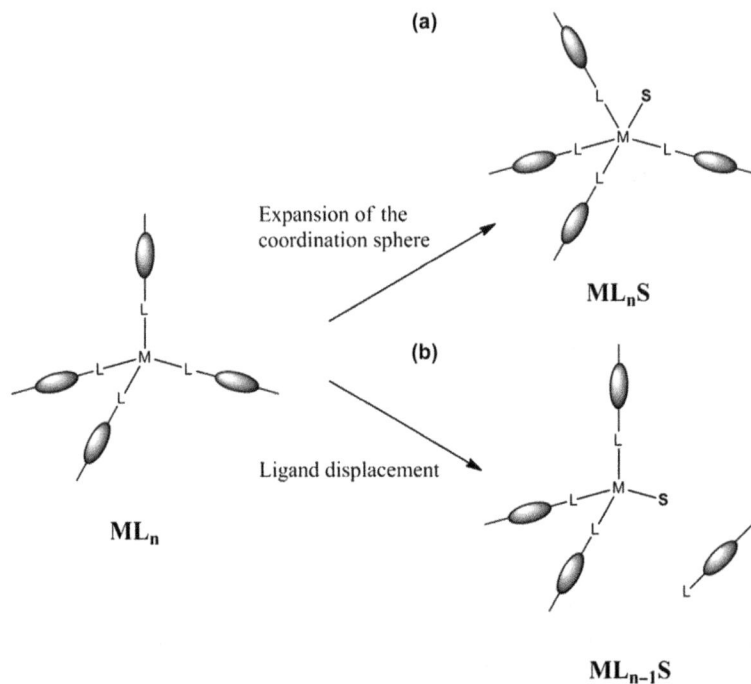

(a)

ML$_n$S

(b)

Expansion of the
coordination sphere

Ligand displacement

ML$_n$

ML$_{n-1}$S

Figure 7.4   Interaction of a substrate molecule, **S**, with a metal site, **M**, through (a) expansion of the coordination sphere around the metal ion; or (b) (reversible) displacement of one of the ligands.

accommodate changes in the coordination sphere of copper upon substrate binding while preserving the structure integrity. This property makes the two materials act as active catalysts for a number of reactions, including oxidation of activated alkanes,[52] 1,3-dipolar cycloaddition of azides to terminal alkynes,[53] or three-component coupling of aldehydes, alkynes and amines.[54] We have shown by DFT calculations on model clusters that both MOFs can coordinate a hydroperoxide molecule (HOOR) directly onto the copper center to form an adsorption complex of the type [Cu^{2+}-HOOR]. This coordination produced in both materials an expansion of the coordination sphere of the central Cu^{2+} ions from 4 to 5.[52] However, both MOFs revealed a different reactivity pattern on the subsequent dissociation of the adsorbed hydroperoxide into a hydroxyl radical (that remains adsorbed on the copper centre [Cu^{2+}-OH]) and a free alkyl radical, · OR. On one hand, Cu^{2+} ions in [Cu(im)$_2$] expanded their coordination sphere upon binding of a · OH radical, as in the case of the coordination to HOOR. On the contrary, the interaction of the · OH radical with the Cu^{2+} centre in [Cu(2-pymo)$_2$] implies the (reversible) de-coordination of one of the four 2-pymo ligands. Note that this ligand displacement would not necessary imply the collapse of the crystalline structure of the MOF. Actually, the Cu^{2+} centers would still remain connected to the framework through three out of the four initial 2-pymo ligands. Once the catalytic cycle is finished and the product desorbs from the active site, the 2-pymo ligand that has been displaced can coordinate again to the Cu^{2+} site to recover the initial catalytic centre.[52] A similar ligand displacement and re-coordination cycle has been demonstrated to occur in a series of zinc(II) benzoate coordination polymers during transesterification reactions.[55]

On the other hand, MOFs with coordinatively unsaturated sites (*cus*) are MOFs in which one of the coordination positions of the metal centers is occupied by a labile ligand, which can be removed without causing the collapse of the crystalline structure. In most cases, the labile ligands are solvent molecules that, when thermally removed, leave a free coordination position in the metal, which become available for adsorbed substrates. This type of compounds is schematically depicted in Figure 7.5. Relevant examples of this type of MOFs are the copper trimesate HKUST-1,[56] the chromium tereph-thalate MIL-101,[26] and related materials, in which the metal sites are coordinated to an apical water molecule that can be removed upon thermal activation under vacuum. A more detailed description of this type of compounds and how they can be modified is given in Chapter 3.

What is relevant here is that, upon creation of a coordination vacancy by removal of a labile ligand, the resulting metal center will be prone to accept electron density from any donor molecule that can be present in the medium; *i.e.*, the metal will behave as a Lewis acid center. As such, the resulting MOF can be used as a *Lewis acid catalyst* or even as *redox catalyst* when suitable oxidizing agents, such as O_2, H_2O_2 or hydroperoxides, are present in the reaction medium. There are several examples describing the use of this type of MOFs having *cus* as Lewis acid catalysts for the cyanosilylation of carbonyl compounds,[31,57,58] Mukaiyama-aldol condensation,[57] Friedel–Crafts benzi-lation,[59] isomerization of α-pinene oxide and conversion of citronellal into isopulegol,[49,60,61] as well as catalysts for the oxidation of alcohols,[62] sulfides,[63]

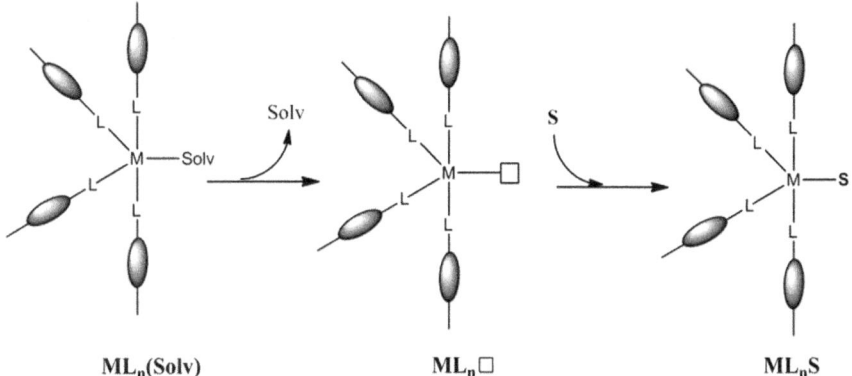

ML$_n$(Solv)   ML$_n$□   ML$_n$S

Figure 7.5 A coordinated labile solvent molecule (Solv) can be removed from the coordination sphere of a metal ion (**M**) leaving a coordination vacancy (□). The metal ion becomes thus prone to coordinate to substrate molecules (**S**).

olefins, paraffins,[64] or CO,[65] to name only a few examples. A more detailed description on the use of MOFs having *cus* as catalysts is given in Chapter 8.

7.2.1.2 MOFs Containing Metal Nodes with Semiconducting Properties

This group of materials is formed by MOFs featuring metal oxide nanoclusters at their nodes, whose electronic configuration correspond to that of a semiconductor, *i.e.*, with a band structure containing a completely filled valence band and an empty conduction band separated by a relatively small energy gap, E_g. It has been demonstrated that the materials that fulfill these requirements can actually have semiconducting properties.[66,67] Upon adequate optical, electronic or thermal excitation, electrons can be excited from the valence to the conduction band, thus generating an electron–hole pair. Examples of these materials are MOFs containing tetranuclear Zn_4O or hexanuclear $Zn_6O_4(OH)_4$ clusters. We have recently shown that both the electrons in the conduction band and the holes in the valence band are long lived species that decay to the ground state in the microsecond time scale.[66] Therefore, the lifetime of the charge separated state is long enough to allow it to interact with suitable electron donors or acceptors that can be present in the medium. We and others have demonstrated the utility of this type of compounds as photocatalysts,[66–69] *i.e.*, the energy needed to create charge-separated states was provided by UV light absorption (photons with energy higher than the band gap of the material). A more extended description of this type of compounds and their use as photocatalysts is given in Chapter 12.

7.2.1.3 Anchoring the Catalytic Active Species to the Metal Nodes

In Section 7.2.1.1, we have shown that it is possible to generate coordination vacancies on the metal ions of certain MOFs by the reversible thermal displacement of labile ligands. A smart strategy for introducing a catalytic center in the MOF consists of using these vacancies as anchoring points for introducing additional functionalities. The first example of the use of this strategy was reported by Hwang *et al.*[70] They prepared the chromium terephthalate MIL-101, containing Cr^{3+} sites in which one coordination position is occupied by a water molecule. After removing this water molecule by a thermal treatment under vacuum, the resulting vacancy was used for grafting ethylenediamine, in such a way that one of the $-NH_2$ groups is coordinated to the Cr^{3+} ion, while the second $-NH_2$ group remained free and pointing towards the center of the pores (see Figure 7.6a). The authors also demonstrated that these amino groups can be used as catalytic basic sites for Knoevenagel condensation.[70] Similarly, Banerjee *et al.* used the Cr^{3+} ions of MIL-101 as anchoring sites for introducing a proline derivative containing a pyridine group as anchoring point (Figure 7.6b).[22] The proline-containing

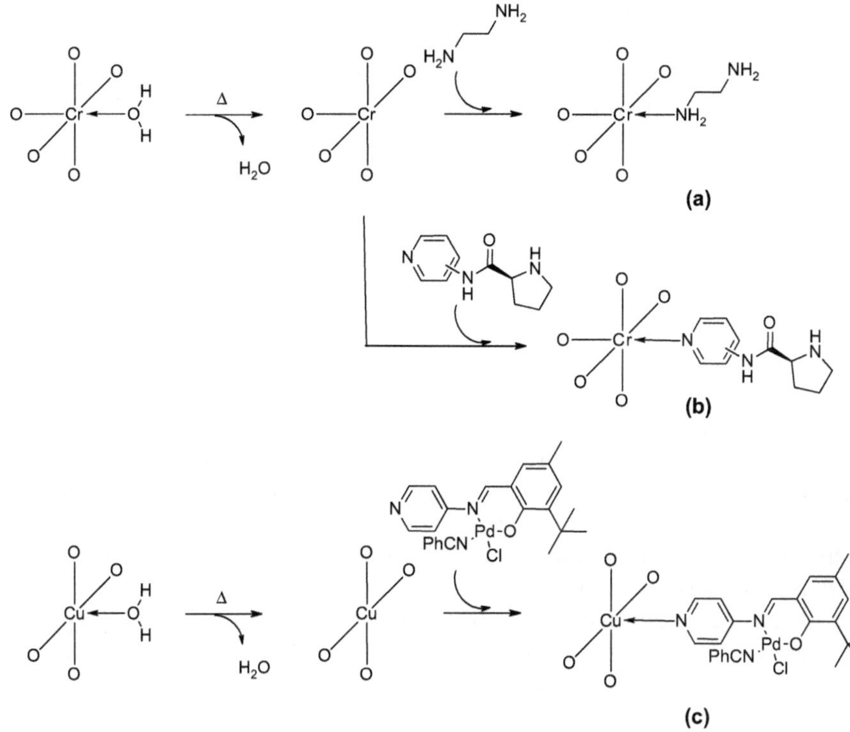

Figure 7.6 Some selected examples illustrating the anchoring of the catalytic species to the metal nodes of a MOF.

MOF thus prepared showed activity as a chiral organocatalyst for asymmetric aldol reactions between different aldehydes and ketones. But the technique can also be used to introduce not only organic functionality, but also a second metal site with potential catalytic activity. Very recently, Arnanz *et al.* have reported the preparation of a bimetallic Cu,Pd-MOF in which the Cu^{2+} *cus* sites at the framework nodes were used as anchoring points to coordinate a pyridine terminated Schiff base Pd^{2+} complex (Figure 7.6c).[71] The bifunctional catalyst was used to perform a tandem Sonogashira and 1,3-dipolar cyclo-addition from 2-iodobenzylbromide, sodium azide and alkynes, catalyzed by the Pd^{2+} and the Cu^{2+} centers, respectively. These are simply examples that serve to illustrate the concept and to get an impression of the potential of this preparative technique. Since it is actually a post-synthesis modification route, a more detailed description is given in Chapter 3.

7.2.2 Catalysis at the Organic Linker

7.2.2.1 *MOFs with Organic Functional Groups*

This is a group of MOFs containing functional groups at the organic linker that can catalyze a chemical reaction; *i.e.*, they behave as organocatalysts.

Therefore, and contrary to what occurs in the previous group, the catalytic function is located at the organic linker and not at the metal site. It is evident that the linkers that form this type of MOFs need to contain two different types of organic functional groups, as shown in Figure 7.7a: *coordinative* functional groups, G_1, that coordinate to the metal sites to hold the crystalline framework; and *reactive* functional groups, G_2, which are not coordinated to the metals and will be responsible for the catalytic properties of the material. A prototypic example of MOFs having two types of functional groups are those containing the ligand 2-aminoterephthalate, such as IRMOF-3, UiO-66-NH$_2$, M-MIL-53-NH$_2$ or M-MIL-101-NH$_2$ (M = Cr^{3+}, Fe^{3+}, Al^{3+}, V^{3+}). In all these compounds (and others), the ligand coordinate to the metal ions through the two carboxylate groups (*coordinative* groups) while the amino groups remain not coordinated and accessible (*reactive* groups), as shown in Figure 7.7b. It has been demonstrated that these amino groups can confer basic properties to the material.[72] Another historic example of ligand containing two types of functional groups that has been used for preparing a MOF is 1,3,5-benzene tricarboxylic acid tris[*N*-(4-pyridyl)amide].[73] This molecule coordinates to the metal ions through the pyridyl N atoms, while the amide groups remain free and can be used as active sites for base catalysis (see Figure 7.7c). A more recent example has been reported by Hupp and co-workers. These authors prepared the material Nu-601, containing 2D layers of Zn paddlewheel dimmers connected to the urea ligand depicted in Figure 7.7d and pillared with 4,4′-bipyridine. Nu-601 was found to be an active hydrogen-bond-donor

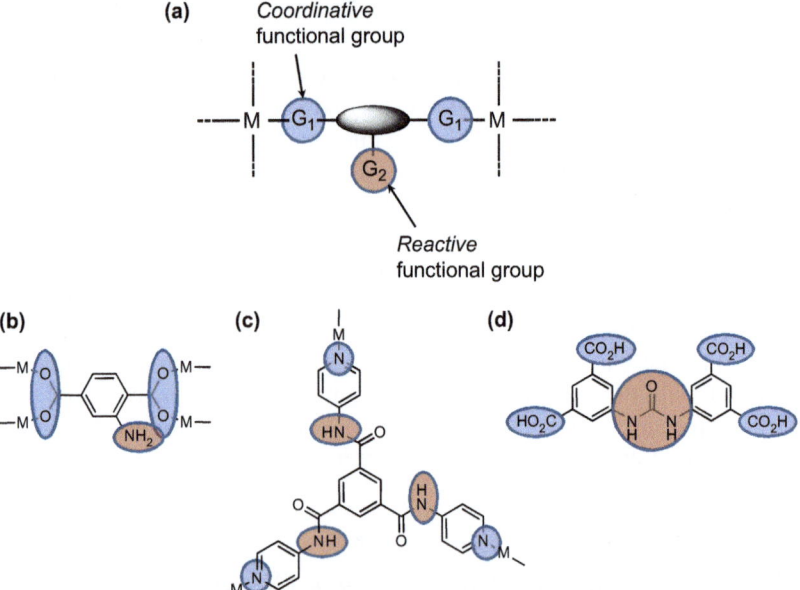

Figure 7.7   General structure and selected examples of ligands containing *coordinative* and *reactive* functional groups.

catalyst for Friedel – Crafts reactions between pyrroles and nitroalkenes.[74] A much more detailed list of compounds with functional organic groups can be found in Chapter 9.

However, it is not straightforward to prepare MOFs with reactive functional groups free and accessible to catalytic substrates, given the large tendency of the metal ions to coordinate to all the available functional groups of the ligand. Moreover, if we want to incorporate a certain functional group into the MOF framework directly during the synthesis, the ligand used must be soluble and the functional group must be resistant under the synthesis conditions; *i.e.*, it must be resistant to chemical transformation into another (inactive) functional group during synthesis. There are different approaches that can be adopted to control the introduction of a certain functional group into a MOF framework. Among them we can include:

(a) *Use of Protecting Groups.* Sometimes it is not possible to directly incorporate a functional group into the MOF structure, or it is not stable under the synthesis conditions and is chemically transformed into a new inactive moiety. For instance, it is possible to prepare homochiral MOFs containing proline as the organic ligand, but its organocatalytic activity is lost due to the protonation of the nitrogen atom.[75] This limitation has been solved by Lun *et al.* by using proline protected with thermolabile *tert*-butoxycarbonyl (Boc) groups. Once the material is prepared, the catalytically active proline form is recovered by a simple post-synthesis thermal treatment to 165 °C in DMF using microwave irradiation, as shown in Figure 7.8.[76] Meanwhile, the presence of the bulky Boc group in the ligand prevents interpenetration in the as-synthesized material, thus resulting in a higher accessible pore volume upon deprotection. While the catalytic performance of this MOF containing organocatalytic proline pendant groups was only modest concerning enantioselectivity,

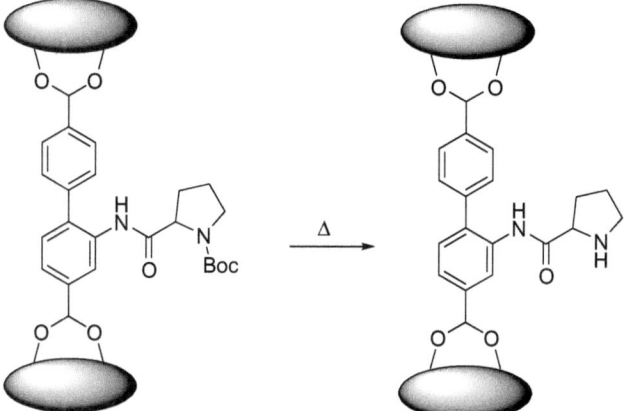

Figure 7.8 A thermolabile Boc protecting group can be used to preserve the protonated state of the proline moiety.

the authors presented a method that can be extended to the preparation of other interesting solids.

(b) *(Post-synthesis) Ligand Exchange.* A MOF can be viewed as a network of multidentate organic ligands connecting isolated metal ions or inorganic secondary building blocks, SBU (clusters of discrete size). From this point of view, Férey and co-workers proposed a *de novo* synthesis procedure starting from pre-formed trimmeric SBUs consisting of acetates of trivalent cations. When put in contact with solutions of dicarboxylate ligands, the authors showed that a direct exchange occurred between the monocarboxylate (acetate groups) and the dicarboxylates in which the parent trimmeric SBUs remained intact, thus leading to a three dimensional network.[77] Some years later, the same authors demonstrated that this strategy can also be used to prepare porous zirconium dicarboxylates exhibiting the UiO-66 architecture starting from pre-formed hexammeric zirconium methacrylate oxoclusters $[Zr_6O_4(OH)_4(OMc)_{12}]$ (OMc = methacrylate) by ligand exchange.[78] This strategy is shown in Figure 7.9.

Additionally, Cohen and co-workers have demonstrated that a post-synthesis ligand exchange strategy can also be used to transform an existing MOF with a ligand (L_1) into a new material, by simply contacting the solid with a solution containing a new isoreticular ligand (L_2), or even by directly mixing together two solid isoreticular MOFs, one with the ligand L_1 and the other with the ligand L_2, as shown in Figure 7.10.[79] In this way, the authors prepared mixed ligand MOFs with either the MIL-53(Al) or the MIL-68(In) structure containing both

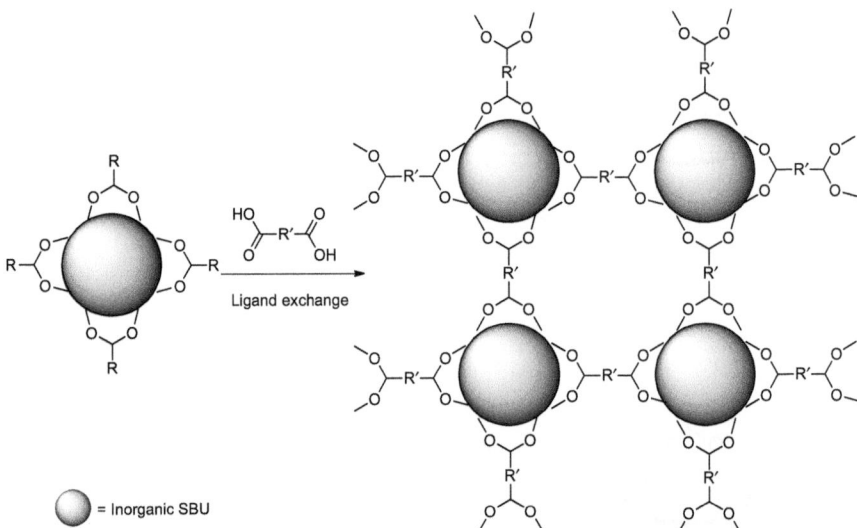

Figure 7.9   A MOF can be formed by ligand exchange of preformed inorganic SBU of different nuclearity.

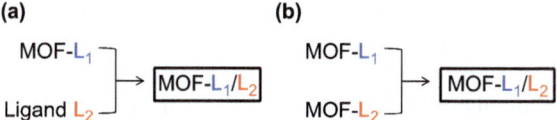

Figure 7.10 Mixed-ligand MOFs can be prepared through post-synthesis ligand exchange by contacting either (a) a MOF with a solution of another ligand; or (b) two pre-formed MOFs contacted in the solid state.

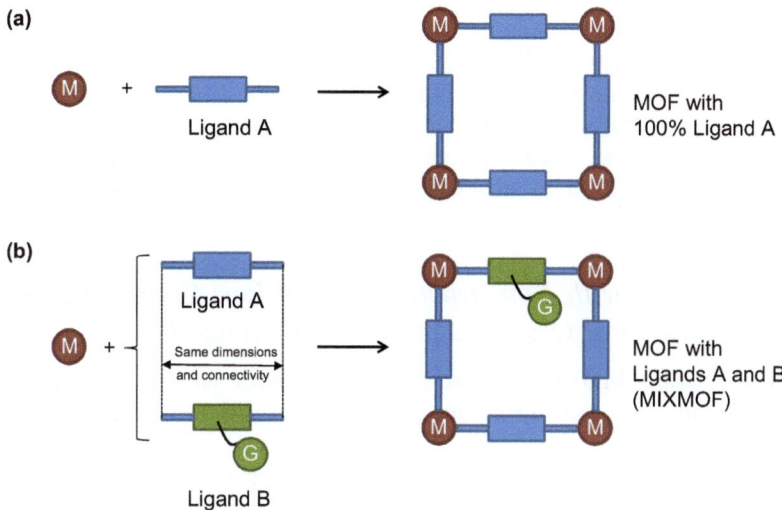

Figure 7.11 Synthesis of mixed-ligand MOFs (MIXMOFs) is a way to control the concentration of the functional group (G) by simply changing the ratio between both ligands used in the synthesis.

aminoterephthalate and bromoterephthalate ligands by contacting either pure MIL-53(Al)-NH$_2$ and MIL-53(Al)-Br solids, or MIL-68(In)-NH$_2$ and MIL-68(In)-Br solids, respectively. The methods however failed to prepare a mixed MIL-101(Cr) having simultaneously terephthalate and bromoterephthalate in the structure.

(c) *Use of Mixed-linkers (MIXMOFs).* In this approach, shown in Figure 7.11, two ligands of similar geometric properties (dimensions and connectivity) are used together in the synthesis of a MOF, with one of the ligands having an additional functional group (G) not involved in the coordination with the metals, which will become the active center in the final solid. In this way both ligands are randomly distributed in the framework forming a mixed structure. Using this strategy it is possible to control the concentration of the functional group by simply changing the ratio between both ligands used in the synthesis. This avoids preparing materials with a too high (stoichiometric) concentration of active sites which can lead to mutual deactivation or to a severe decrease of the available specific surface area, with the corresponding diffusion problems.

As an example to illustrate this strategy, Baiker and co-workers[80] have used the MIXMOF approach to prepare a series of isoreticular solids having the structure of MOF-5 and containing different proportions of terephthalate and aminoterephthalate ligands, with general formula $[Zn_4(O)(BDC)_{3-x}(ATA)_x]$ (BDC = terephthalate, ATA = aminoterephthalate, $x = 0 - 1.2$ corresponding to maximum 40% ATA incorporation). These mixed-ligand MOFs were used as catalysts to prepare propylene carbonate from propylene oxide and CO_2 using tetraalkylammonium halides as promoters. The authors found that the catalytic activity of the MIXMOF series increased with the content of ATA, although they did not include in their study a comparison with the activity of pure IRMOF-3, $[Zn_4(O)(ATA)_3]$. The amino group at the functionalized linker proved also to be beneficial for the immobilization of Pd species, which allowed the use of MIXMOFs in the oxidation of CO at high temperatures.[81] More recently, Huang *et al.* used a similar approach to prepare Pd nanoparticles supported and stabilized on mixed-ligand MIL-53(Al) as catalysts for the Heck coupling reaction.[82]

7.2.2.2   *MOFs with Non-structural Metal Ions Coordinated to Structural Organic Linkers (Metalloligands)*

In most existing MOFs, the structural metal sites have no coordination positions available to substrates, nor can these vacancies be created by removal of labile ligand molecules. On the contrary, the metal sites are completely blocked by tightly coordinative linkers forming the crystalline network, so they are not accessible for catalysis. A possible alternative that can be used to introduce metal active sites into a MOF is by binding the metal ion to a suitable organic molecule to form a metal coordination complex (a *metalloligand*). Then, this metalloligand is used as a linker to form the MOF. Therefore, the MOFs considered here contain two types of metal ions, as shown in Figure 7.12a: one of them (M_1), which belongs to the metalloligand, is responsible for the catalytic activity of the MOF, while a second type of metal ions, M_2, has only a structural role and is not directly involved in catalysis.

This strategy has some interesting advantages: we can finely control the oxidation state, electronic properties and coordination environment (including chiral environments), of the catalytic site through the preparation of suitable metalloligands. This could grant a means for controlling the catalytic properties only rivaled by transition metal molecular catalysts. Additionally, the immobilization of the active site on the pore walls of the MOF can decrease the mutual deactivation commonly encountered in molecular catalysts in solution, thus extending the catalyst lifetime.[83] In the following we will illustrate the use of this strategy through some selected examples. Kitagawa and co-workers[84,85] prepared a MOF containing $Cu(2,4\text{-pydca})_2$ metalloligands (2,4-pydca = pyridine-2,4-dicarboxylate, Figure 7.12b) coordinated to Zn^{2+} cations through one of the carboxylate groups to form a 3D structure. In this material, Zn^{2+} acted as a mere structural element, while Cu^{2+} ions were accessible for guest coordination. Similarly, the same group[86] prepared a series

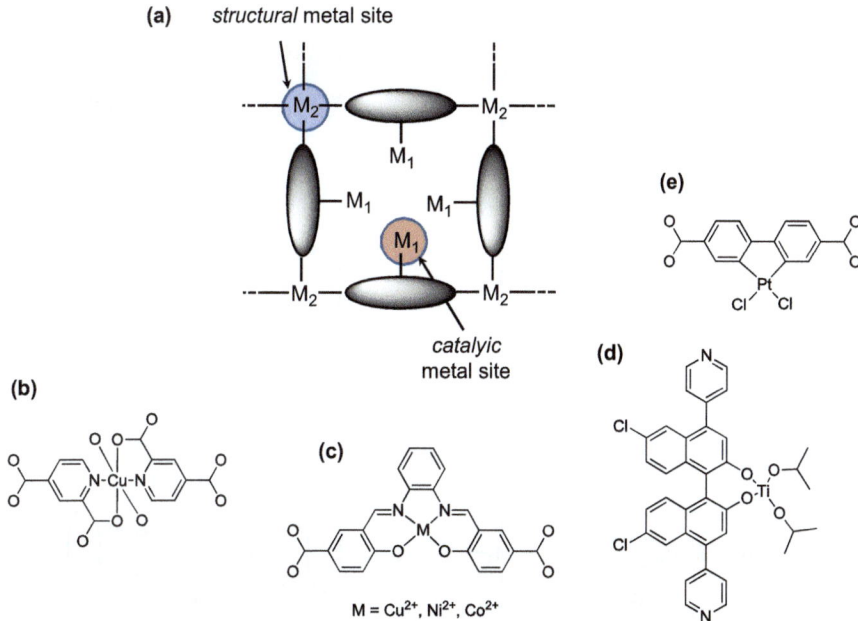

(a) *structural* metal site

catalyic metal site

(b)

(c)

M = Cu²⁺, Ni²⁺, Co²⁺

(d)

(e)

Figure 7.12 Some MOFs contain two types of metal ions: *catalytic* (M_1) and *structural* (M_2) metal sites. Selected examples of metalloligands used as building blocks for constructing MOFs are shown.

of materials containing metal Schiff base complexes, $M(H_2salphdc)$ ($M = Cu^{2+}$, Ni^{2+} or Co^{2+}, salphdc = N,N'-phenylenebis(salicylideneimine)dicarboxylate, Figure 7.12c), with Zn^{2+} cations at the nodes. Lin and co-workers[87] have prepared a homochiral MOF containing Cd^{2+} ions and the chiral ligand (R)-6,6'-dichloro-2,2'-dihydroxy-1,1'-binaphthyl-4,4'-bipyridine as the organic building unit. The ligand coordinates to Cd^{2+} through chlorine and the pyridine nitrogen, while the two hydroxyl groups of the binaphthyl moiety remain uncoordinated and pointing to the channels. Post-synthesis modification of this material by adding titanium isopropoxide yielded a titanium containing material, with titanium di-isopropoxide grafted to the walls of the MOF through the dihydroxy groups (Figure 7.12d). Szeto et al. prepared bimetallic materials containing Gd^{88} or Yb^{89} ions at the nodes, and Pt^{2+} ions four-coordinated by two Cl and by two N atoms of 2,2'-bipyridine-4,4'-dicarboxylate (Figure 7.12e), which could act as potential catalytic sites. More interesting examples are given in Chapter 9, along with a description of their application as heterogeneous catalysts.

7.2.2.3 MOFs with Chelating Linkers

In this strategy, a MOF is first prepared using as organic linker one having some functional group (G_2 in Figure 7.13) that does not coordinate to the metal ions (such as those described in Section 7.2.2.1). Using a covalent post-synthesis

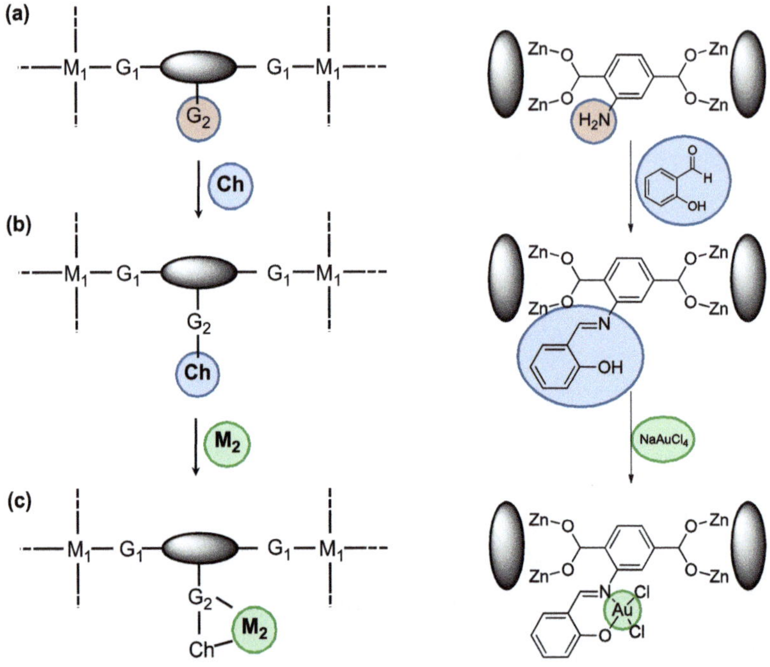

Figure 7.13 General procedure for the preparation of MOFs with chelating groups to introduce a second metal ion, and representative illustration of this method applied to the preparation of IRMOF-3-SI-Au.[90]

modification of the MOF, this functional non-coordinative group G_2 is used as anchoring point for introducing a new chelating ligand (Ch) that can be used to coordinate to a second metal ion, M_2, as shown in Figure 7.13. This can be seen as a combination of the two situations shown in Figures 7.7 and 7.12. In this way, we described the preparation of the Au(III)-containing MOF shown in the right part of Figure 7.13.[90] In this material, the amino groups of IRMOF-3 were modified post-synthesis by reacting with salicylaldehyde to form the corresponding imine, which was then used to coordinate and stabilize Au(III) cation. The resulting MOF containing Schiff base Au(III) complexes decorating the organic ligands was found to be an active catalyst for the selective hydrogenation of butadiene and for the synthesis of indoles through a three-component coupling of amines, aldehydes and alkynes followed by intramolecular cyclization.[90]

7.2.3   MOFs as Host Matrices or Nanometric Reaction Cavities

The realization of MOFs with unprecedentedly high values of surface areas and pore volumes both fascinated and challenged scientists to make and efficient use of such a huge pore space available. Obviously, applications for MOFs in gas storage (H_2, CO_2 and CH_4) did not take long to appear. But the porosity of MOFs also plays a prominent role in heterogeneous catalysis, contributing to a number of aspects.

It is evident that a high porosity of a heterogeneous catalyst facilitates the contact between the substrate of the reaction and the catalytic sites while decreasing mass transport (diffusion) limitations. Note that the catalytic performance of a solid catalyst with low porosity or with narrow pores with respect to the substrate dimensions can be severely decreased by diffusion control of the reaction rate.

As it will be shown and discussed in Chapter 11, MOFs with a homochiral pore system can be readily prepared and systematically tuned, thus endowing the solid with enantioselective properties and opening the door to applications in asymmetric catalysis and separations. This is in large contrast with other porous solids, such as zeolites, for which the creation of chiral pore systems is much more difficult and less controllable.

But besides the obvious advantages of a highly porous material mentioned above, the pore system of the solid can also be used as either a *host matrix* to introduce additional catalytic species, or as the physical space where a chemical reaction takes place (*nanometric reaction cavity*).

7.2.3.1 MOFs as Host Matrices

We have dedicated an entire chapter of this book (chapter 10) to describe the examples of encapsulated species inside MOFs, so only a brief description of the general strategy is given here.

As has been extensively described first for zeolites, zeotypes and mesoporous silicas, and more recently also for MOFs, the regular system of channels and cavities of these solids can be used to encapsulate various kinds of species, including metal or metal oxide nanoparticles or molecular catalysts. This process leads to the preparation of composite materials that can combine the properties of both the host and the guest, and can also present new synergic effects due to their mutual interaction. In this way, encapsulated species can benefit from aspects such as increased stability, higher dispersion, controlled size/shape, or suppression of self-deactivation, as compared to the same species in solution. On the other hand, the species adsorbed inside the pores of a material can also affect or modulate the activity of the host. A clear example of this is the well known antenna or sensitization effect of organic dyes inside the pores of a semi-conductor metal oxide, enhancing their electronic or optical properties.

In the more specific case of encapsulated species for catalytic applications of MOFs, we can find two limit situations (and of course, all the intermediate situations within):

(1) The MOF can simply act as an inert solid matrix in which the encapsulated species is dispersed. In these materials, none of the components forming the MOF is directly involved in the catalytic properties of the resulting material, and there is not any specific interaction between the MOF and the encapsulated species. Still, the MOF can affect the reactivity or the selectivity of the encapsulated species by introducing spatial constraints that can alter the distribution of reaction products.

(2) A more interesting situation is found when the MOF is not acting as a simple "spectator", but it can directly participate in the catalytic process contributing with additional functionalities not provided by the encapsulated moieties, such as acid or basic sites located at the metal nodes or the organic linkers. In this case, the use of MOFs as the host matrix leads to a multi-functional catalytic system, as will be described in the last chapter of this book along with some selected examples.

7.2.3.2 *MOFs as Nanometric Reaction Cavities*

Sometimes the confinement space of the pore system can largely influence the product selectivity, and this is more likely to occur when a substrate or product of the reaction and the host matrix in which it is contained have similar dimensions. One relevant example of performing a reaction inside the pores of a MOF is styrene polymerization.[91] Although this is a radical reaction catalyzed by AIBN (azobisisobutyronitrile), when it was carried out inside MOFs of general formula $[M_2(bdc)_2(teda)]$ (M: Zn^{2+} or Cu^{2+}, teda = triethylenediamine), the authors observed that the polymers recovered after dissolving the MOF exhibit a remarkably low polydispersity (1.66). In contrast, the analogous polymerization procedure performed in the absence of the MOF leads to a polydispersity of 4.68. Moreover, EPR spectroscopy during the polymerization of styrene inside the MOF showed an intense signal assigned to the propagating living radical. The signal did not disappear even after storing the sample for one week at 70 °C. This was attributed to the suppression of termination reaction and radical transfer inside the MOF channels.

The products formed during a photochemical reaction can also be modified when the process is conducted inside the pores of a MOF. One example is the reaction of *o*-methyl dibenzyl ketone inside the pores of the material $[Co_3(4,4'\text{-BPhDC})_3(4,4'\text{-bpy})]$ (BPhDC = biphenyl dicarboxylate; 4,4'-bpy = 4,4'-bipyridine). Asymmetric dibenzyl ketones can undergo homolytic cleavage in the α-position of the CO to give two differently substituted benzyl radicals that can recombine to give three possible diarylethanes. In solution, the product distribution arises from the random coupling of the two benzyl radicals, giving a 25, 50, 25% ratio of the three possible A–A, A–B and B–B diarylethanes. However, when diffusion is restricted due to confinement effects, the product distribution changes, favoring the asymmetric A–B diarylethane arising from recombination of the geminate radical pair. Accordingly, when the photolysis of *o*-methyl dibenzyl ketone was carried out inside $[Co_3(4,4'\text{-BPhDC})_3(4,4'\text{-bpy})]$, asymmetric *o*-tolyl phenyl ethane was formed with a 60% yield and 100% selectivity, accompanied with 40% of intramolecular hydrogen abstraction, as shown in Figure 7.14. The absence of ditolyl (A–A) and diphenyl ethane (B–B) adducts indicates that the MOF as host exerts a 100% cage effect allowing exclusively recombination of the geminate benzyl radicals. To put this value into context, photolysis of the same compound in NaX gives a 78% cage effect.[92]

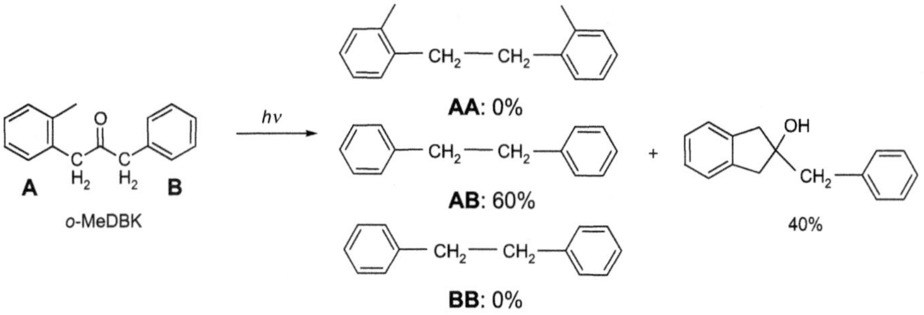

Figure 7.14 Product distribution in the photolysis of *o*-methyl dibenzyl ketone when the photolysis is carried out inside the pores of [Co$_3$(4,4'-BPhDC)$_3$(4,4'-bpy)].

Table 7.1 Selected catalytic results for the cyclopropanation reaction between the indicated olefins and diazo compounds catalyzed by Cu- and Au-containing MOFs.[a] Table elaborated with data taken from ref. 93.

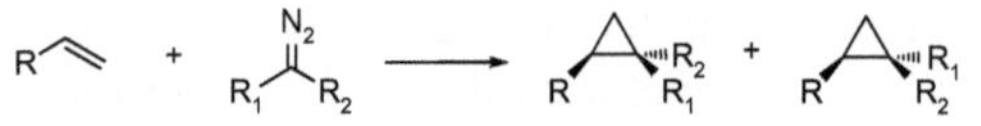

Catalyst	Olefin	Diazo compound	Yield(%)[b]	d.r. (%)[c]
HKUST-1[d]	Styrene	PhEDA	85	100
	α-methylstyrene	PhEDA	50	100
	β-methylstyrene	PhEDA	30	100
	1-octene	PhEDA	90	98
	cyclooctene	PhEDA	70	98
	DMHD	EDA	60	100
	cyclohexene	EDA	99	98
IRMOF-3-SI-Au[e]	Styrene	PhEDA	60	95
	DMHD	EDA	25	100
	cyclohexene	EDA	50	100

[a]Reaction conditions: Diazo compound (11 mmol), olefin (8 mmol) and catalyst (5 mol% metal), in 5 mL CH$_2$Cl$_2$, room temperature.
[b]Yield of the cyclopropanes; the remaining diazo compound was converted into autocoupling products.
[c]Diastereomeric selectivity: (*trans–cis*)/(*trans + cis*).
[d]HKUST-1 prepared according to the procedure reported in 56.
[e]IRMOF-3-SI-Au prepared according to the procedure reported in 90. PhEDA: 2-phenyldiazoacetate; EDA: diazoacetate; DMHD: 2,5-dimethyl-2,4-hexadiene.

We recently showed another example of how the pore system of a MOF can affect the selectivity to a certain product of a reaction. In particular, we showed that the diastereoselectivity of cyclopropanation reaction by transfer of carbene fragments from diazo compounds to olefins can be significantly enhanced by the pore system of Cu- and Au-containing MOFs (see Table 7.1).[93]

7.2.4   Post-Synthesis Modification of MOFs

So far, we have shown that it is possible to use different strategies to design a MOF to introduce suitable catalytic active species. In some cases, the as-prepared material can be used directly, or after simply removing a labile ligand to create a coordination vacancy. In other cases, the preparation of MOF catalysts involves a more elaborated synthesis effort, sometimes requiring the use of pre-formed metalloligands or "exotic" organic molecules. When none of these strategies or situations are possible, we have shown that the pores of MOFs can still be used to encapsulate active guests or as confinement spaces in which a chemical reaction takes place. It is evident that in a chapter dedicated to the design of catalytic active centers in MOFs, post-synthesis modification strategies cannot be missing at all.

It is usually found in the preparation of heterogeneous catalysts in general, and of MOFs in particular, that the as-prepared material still does not contain the desired chemical functionality that will be responsible for the catalytic activity. Thus, post-synthesis treatments are required to modify the properties of the solid and to introduce (or activate) the catalytic sites. These post-synthesis treatments can be of very different types, depending on the type of material and the reaction that has to be catalyzed. In the case of zeolites, zeotypes, and other metal or metal oxide catalysts, post-synthesis treatments consisting of thermal activation, ionic exchange, steaming or hydrogenation are routinely performed to pre-activate the catalytic centers. In the case of MOFs, post-synthesis modifications are also applied, and we have already described some of them in previous sections of this chapter. A more detailed and up-to-date description of the existing post-synthesis modification routes is given in chapter 3. Let us simply remember here that post-synthesis modification in MOFs can be used either for conditioning a "latent" catalytic site already existing in the MOF (like, for instance, the thermal removal of labile ligands to get access to the Lewis acid character of the metal sites), or to introduce new catalytic species by a large variety of techniques: grafting, encapsulation, metal or ligand exchange, anchoring, covalent modification, metal complexation, *etc*. Taken together, all these techniques introduce an overwhelming number of new strategies to fine tune the catalytic activity of a MOF, adding to the already vast synthetic possibilities. It is evident that, with all these tools in hand, the development of MOFs as highly efficient, specific and selective catalysts is just a matter of time and imagination of the scientists.

7.3   Conclusions

Throughout this chapter, we have shown the enormous potential that metal organic frameworks may have concerning their application as heterogeneous catalysts. Different ways have been outlined for introducing catalytic active sites at the metallic nodes, at the organic linkers or inside the pores of the MOFs, as well as the possibility to use them as the physical space where a chemical reaction takes place. The potential of using post-synthesis modification methods for modifying the properties of pre-formed MOFs, and especially the covalent modification of the organic linkers, has also been outlined.

The next chapters of this book will be devoted to describe, through specific examples, the existing scenarios in which the catalytic potential of MOFs has already been successfully demonstrated.

References

1. J.-R. Li, R. J. Kuppler and H. Zhou, *Chem. Soc. Rev.*, 2009, **38**, 1477–1504.
2. J.-R. Li, J. Sculley and H. Zhou, *Chem. Rev.*, 2012, **112**, 869–932.
3. L. J. Murray, M. Dinca and J. R. Long, *Chem. Soc. Rev.*, 2009, **38**, 1294–1314.
4. R. Getman, Y.-S. Bae, C. E. Wilmer and R. Q. Snurr, *Chem. Rev.*, 2012, **112**, 703–723.
5. K. Sumida, D. L. Rogow, J. A. Mason, T. M. McDonald, E. D. Bloch, Z. R. Herm, T.-H. Bae and J. R. Long, *Chem. Rev.*, 2012, **112**, 724–781.
6. M. P. Suh, H. J. Park, T. K. Prasad and D.-W. Lim, *Chem. Rev.*, 2012, **112**, 782–835.
7. M. D. Allendorf, C. A. Bauer, R. K. Bhakta and R. J. T. Houk, *Chem. Soc. Rev.*, 2009, **38**, 1330–1352.
8. J. Rocha, L. D. Carlos, F. A. Almeida Paz and D. Ananias, *Chem. Soc. Rev.*, 2011, **40**, 926–940.
9. Y. Cui, Y. Yue, G. Qian and B. Chen, *Chem. Rev.*, 2012, **112**, 1126–1162.
10. P. Horcajada, R. Gref, T. Baati, P. K. Allan, G. Maurin, P. Couvreur, G. Ferey, R. E. Morris and C. Serre, *Chem. Rev.*, 2012, **112**, 1232–1268.
11. L. E. Kreno, K. Leong, O. K. Farha, M. D. Allendorf, R. P. Van Duyne and J. T. Hupp, *Chem. Rev.*, 2012, **112**, 1105–1125.
12. M. Kurmoo, *Chem. Soc. Rev.*, 2009, **38**, 1353–1379.
13. H. Furukawa, N. Ko, Y. B. Go, N. Aratani, S. B. Choi, E. Choi, A. O. Yazaydin, R. Q. Snurr, M. O'Keeffe, J. Kim and O. M. Yaghi, *Science*, 2010, **329**, 424–428.
14. S. Kitagawa and M. Kondo, *Bull. Chem. Soc. Jpn.*, 1998, **71**, 1739–1753.
15. S. R. Batten, B. F. Hoskins and R. Robson, *J. Am. Chem. Soc.*, 1995, **117**, 5385–5386.
16. B. F. Hoskins and R. Robson, *J. Am. Chem. Soc.*, 1990, **112**, 1546–1554.
17. B. F. Hoskins, R. Robson and N. V. Y. Scarlett, *Angew. Chem., Int. Ed. Engl.*, 1995, **34**, 1203–1204.
18. M. Eddaoudi, J. Kim, N. Rosi, D. Vodak, J. Wachter, M. O'Keeffe and O. M. Yaghi, *Science*, 2002, **295**, 469–472.
19. J. H. Cavka, S. Jakobsen, U. Olsbye, N. Guillou, C. Lamberti, S. Bordiga and K. P. Lillerud, *J. Am. Chem. Soc.*, 2008, **130**, 13850–13851.
20. M. Meilikhov, K. Yusenko, D. Esken, S. Turner, G. Van Tendeloo and R. A. Fischer, *Eur. J. Inorg. Chem.*, 2010, 3701–3714.
21. J. Juan-Alcaniz, J. Gascon and F. Kapteijn, *J. Mater. Chem.*, 2012, **22**, 10102–10118.

22. M. Banerjee, S. Das, M. Yoon, H. J. Choi, M. H. Hyun, S. M. Park, G. Geo and K. Kim, *J. Am. Chem. Soc.*, 2009, **131**, 7524–7525.

23. L. Tang, L. Shi, C. Bonneau, J. Sun, H. Yue, A. Ojuva, B.-L. Lee, M. Kritikos, R. G. Bell, Z. Bacsik, J. Mink and X. Zou, *Nature Mater*, 2008, **7**, 381–385.

24. J. L. Sun, C. Bonneau, A. Cantin, A. Corma, M. J. Diaz-Cabanas, M. Moliner, D. L. Zhang, M. R. Li and X. D. Zou, *Nature*, 2009, **458**, 1154–U1190.

25. R. C. Rouse and D. R. Peacor, *Am. Mineral.*, 1986, **71**, 1494–1501.

26. G. Ferey, C. Mellot-Draznieks, C. Serre, F. Millange, J. Dutour, S. Surble and I. Margiolaki, *Science*, 2005, **309**, 2040–2042.

27. V. Colombo, S. Galli, H. J. Choi, G. D. Han, A. Maspero, G. Palmisano, N. Masciocchi and J. R. Long, *Chem. Sci.*, 2011, **2**, 1311–1319.

28. L. M. Huang, H. T. Wang, J. X. Chen, Z. B. Wang, J. Y. Sun, D. Y. Zhao and Y. S. Yan, *Microporous Mesoporous Mater.*, 2003, **58**, 105–114.

29. S. Hausdorf, J. Wagler, R. Mossig and F. O. R. L. Mertens, *J. Phys. Chem. A*, 2008, **112**, 7567–7576.

30. A. Dhakshinamoorthy, M. Alvaro and H. Garcia, *J. Catal.*, 2009, **267**, 1–4.

31. K. Schlichte, T. Kratzke and S. Kaskel, *Microporous Mesoporous Mater.*, 2004, **73**, 81–88.

32. M. Spagnol, L. Gilbert, H. Guillot and P. J. Tirel, *WO Patent Pat.* 9748665, 1997.

33. C. Perego, A. Carati, P. Ingallina, M. A. Mantegazza and G. Bellussi, *Appl. Catal. A: Gen.*, 2001, **221**, 63–72.

34. R. A. Sheldon, M. Wallau, I. W. C. E. Arends and U. Schuchardt, *Acc. Chem. Res.*, 1998, **31**, 485–493.

35. A. Corma, M. T. Navarro and J. Perez Pariente, *J. Chem. Soc., Chem. Commun.*, 1994, 147–148.

36. A. Corma, L. T. Nemeth, M. Renz and S. Valencia, *Nature*, 2001, **412**, 423–425.

37. A. Corma, M. E. Domine and S. Valencia, *J. Catal.*, 2004, **215**, 294–304.

38. A. Corma and M. Renz, *Chem. Commun.*, 2004, 550–551.

39. Y. Zhu, G. Chuah and S. Jaenicke, *J. Catal.*, 2004, **227**, 1–10.

40. A. Corma, F. X. Llabrés i Xamena, C. Prestipino, M. Renz and S. Valencia, *J. Phys. Chem. C*, 2009, **113**, 11306–11315.

41. A. Corma, M. J. Diaz-Cabanas, J. L. Jorda, C. Martinez and M. Moliner, *Nature*, 2006, **443**, 842–845.

42. M. Estermann, L. B. McCusker, C. Baerlocher, A. Merrouche and H. Kessler, *Nature*, 1991, **352**, 320–323.

43. K. G. Strohmaier and D. E. W. Vaughan, *J. Am. Chem. Soc.*, 2003, **125**, 16035–16039.

44. M. Alvaro, E. Carbonell, M. Esplá and H. Garcia, *Appl. Catal. B: Environm.*, 2005, **57**, 37–42.

45. U. Diaz, T. Garcia, A. Velty and A. Corma, *J. Mater. Chem.*, 2009, **19**, 5970–5979.

46. U. Diaz, J. A. Vidal-Moya and A. Corma, *Microporous Mesoporous Mater.*, 2006, **93**, 180–189.
47. A. Comas-Vives, C. Gonzalez-Arellano, M. Boronat, A. Corma, M. Iglesias, F. Sanchez and G. Ujaque, *J. Catal.*, 2008, **254**, 226–237.
48. A. Corma, *Catal. Rev. Sci. Eng.*, 2004, **46**, 369–417.
49. F. Vermoortele, M. Vandichel, B. Van de Voorde, R. Ameloot, M. Waroquier, V. Van Speybroeck and D. E. de Vos, *Angew. Chem., Int. Ed.*, 2012, **51**, 4887–4890.
50. N. Masciocchi, S. Bruni, E. Cariati, F. Cariati, S. Galli and A. Sironi, *Inorg. Chem.*, 2001, **40**, 5897–5905.
51. L. C. Tabares, J. A. R. Navarro and J. M. Salas, *J. Am. Chem. Soc.*, 2001, **123**, 383–387.
52. I. Luz, A. León and M. Boronat, F. X. Llabrés i Xamena and A. Corma, *Catal. Sci. Technol.*, 2013, **3**, 371–379.
53. I. Luz, F. X. Llabrés i Xamena and A. Corma, *J. Catal.*, 2010, **276**, 134–140.
54. I. Luz, F. X. Llabrés i Xamena and A. Corma, *J. Catal.*, 2012, **285**, 285–291.
55. H. Kwak, S. H. Lee, S. H. Kim, Y. M. Lee, B. K. Park, Y. J. Lee, J. Y. Jun, C. Kim, S.-J. Kim and Y. Kim, *Polyhedron*, 2009, **28**, 553–561.
56. S. S. Y. Chui, S. M. F. Lo, J. P. H. Charmant, A. G. Orpen and I. D. Williams, *Science*, 1999, **283**, 1148–1150.
57. S. Horike, M. Dinca, K. Tamaki and J. R. Long, *J. Am. Chem. Soc.*, 2008, **130**, 5854–5855.
58. A. Henschel, K. Gedrich, R. Kraehnert and S. Kaskel, *Chem. Commun.*, 2008, 4192–4194.
59. P. Horcajada, S. Surble, C. Serre, D. Y. Hong, Y. K. Seo, J. S. Chang, J. M. Greneche, I. Margiolaki and G. Ferey, *Chem. Commun.*, 2007, 2820–2822.
60. L. Alaerts, E. Seguin, H. Poelman, F. Thibault-Starzyk, P. A. Jacobs and D. E. De Vos, *Chem. Eur. J.*, 2006, **12**, 7353–7363.
61. F. G. Cirujano, F. X. Llabrés i Xamena and A. Corma, *Dalton Trans.*, 2012, **41**, 4249–4254.
62. C. N. Kato, M. Hasegawa, T. Sato, A. Yoshizawa, T. Inoue and W. Mori, *J. Catal.*, 2005, **230**, 226–236.
63. Y. K. Hwang, D. Y. Hong, J. S. Chang, H. Seo, M. Yoon, J. Kim, S. H. Jhung, C. Serre and G. Ferey, *Appl. Catal. A: Gen.*, 2009, **358**, 249–253.
64. J. Kim, S. Bhattacharjee, K.-E. Jeong, S.-Y. Jeong and W.-S. Ahn, *Chem. Commun.*, 2009, 3904–3906.
65. R. Q. Zou, H. Sakurai, S. Han, R. Q. Zhong and Q. Xu, *J. Am. Chem. Soc.*, 2007, **129**, 8402.
66. M. Alvaro, E. Carbonell, B. Ferrer, F. X. Llabrés i Xamena and H. Garcia, *Chem. Eur. J.*, 2007, **13**, 5106–5112.
67. T. Tachikawa, J. R. Choi, M. Fujitsuka and T. Majima, *J. Phys. Chem. C*, 2008, **112**, 14090–14101.

68. J. Gascon, M. D. Hernandez-Alonso, A. R. Almeida, G. P. M. van Klink, F. Kapteijn and G. Mul, *Chem Sus Chem*, 2008, **1**, 981–983.
69. C. Gomes Silva, I. Luz, F. X. Llabrés i Xamena, A. Corma and H. Garcia, *Chem. Eur. J.*, 2010, **16**, 11133–11138.
70. Y. K. Hwang, D. Y. Hong, J. S. Chang, S. H. Jhung, Y. K. Seo, J. Kim, A. Vimont, M. Daturi, C. Serre and G. Ferey, *Angew. Chem., Int. Ed.*, 2008, **47**, 4144–4148.
71. A. Arnanz, M. Pintado-Sierra, A. Corma, M. Iglesias and F. Sanchez, *Adv. Synth. Catal.*, 2012, **254**, 1347–1355.
72. J. Gascon, U. Aktay, M. D. Hernandez-Alonso, G. P. M. van Klink and F. Kapteijn, *J. Catal.*, 2009, **261**, 75–87.
73. S. Hasegawa, S. Horike, R. Matsuda, S. Furukawa, K. Mochizuki, Y. Kinoshita and S. Kitagawa, *J. Am. Chem. Soc.*, 2007, **129**, 2607–2614.
74. J. M. Roberts, B. M. Fini, A. A. Sarjeant, O. K. Farha, J. T. Hupp and K. A. Scheidt, *J. Am. Chem. Soc.*, 2012, 134.
75. M. J. Ingleson, J. Bacsa and M. J. Rosseinsky, 2007, 3036–3038.
76. D. J. Lun, G. I. N. Waterhouse and S. G. Telfer, *J. Am. Chem. Soc.*, 2011, **133**, 5806–5809.
77. C. Serre, F. Millange, S. Surble and G. Ferey, *Angew. Chem., Int. Ed.*, 2004, **43**, 6286–6289.
78. V. Guillerm, S. Gross, C. Serre, T. Devic, M. Bauer and G. Ferey, *Chem. Commun.*, **46**, 767–769.
79. M. Kim, J. F. Cahill, H. Fei, K. A. Prather and S. M. Cohen, *J. Am. Chem. Soc.*, 2012, **134**, 18082–18088.
80. W. Kleist, F. Jutz, M. Maciejewski and A. Baiker, *Eur. J. Inorg. Chem.*, 2009, 3552–3561.
81. W. Kleist, M. Maciejewski and A. Baiker, *Thermochim. Acta*, 2010, **499**, 71–78.
82. Y. Huang, S. Gao, T. Liu, J. Lu, X. Lin, H. Li and R. Cao, *Chem Plus Chem*, 2012, **77**, 106–112.
83. S. H. Cho, B. Q. Ma, S. T. Nguyen, J. T. Hupp and T. E. Albrecht-Schmitt, *Chem. Commun.*, 2006, 2563–2565.
84. S. Kitagawa, S. Noro and T. Nakamura, *Chem. Commun.*, 2006, 701–707.
85. S. Noro, S. Kitagawa, M. Yamashita and T. Wada, *Chem. Commun.*, 2002, 222–223.
86. R. Kitaura, G. Onoyama, H. Sakamoto, R. Matsuda, S. Noro and S. Kitagawa, *Angew. Chem., Int. Ed.*, 2004, **43**, 2684–2687.
87. C. D. Wu, A. Hu, L. Zhang and W. B. Lin, *J. Am. Chem. Soc.*, 2005, **127**, 8940–8941.
88. K. C. Szeto, C. Prestipino, C. Lamberti, A. Zecchina, S. Bordiga, M. Bjorgen, M. Tilset and K. P. Lillerud, *Chem. Mater.*, 2007, **19**, 211–220.

89. K. C. Szeto, K. P. Lillerud, M. Tilset, M. Bjorgen, C. Prestipino, A. Zecchina, C. Lamberti and S. Bordiga, *J. Phys. Chem. B*, 2006, **110**, 21509–21520.

90. X. Zhang, F. X. Llabrés i Xamena and A. Corma, *J. Catal.*, 2009, **265**, 155–160.

91. T. Uemura, K. Kitagawa, S. Horike, T. Kawamura, S. Kitagawa, M. Mizuno and K. Endo, *Chem. Commun.*, 2005, 5968–5970.

92. N. J. Turro, *Acc. Chem. Res.*, 2000, **33**, 637.

93. A. Corma, M. Iglesias, F. X. Llabrés i Xamena and F. Sanchez, *Chem.–Eur. J.*, 2010, **16**, 9789–9795.

CHAPTER 8

Catalysis at the Metallic Nodes of MOFs

FREDERIK VERMOORTELE, PIETERJAN VALVEKENS AND DIRK DE VOS*

Centre for Surface Chemistry and Catalysis, University of Leuven, Kasteelpark Arenberg 23, box 2461, 3001 Leuven, Belgium, *Email: Dirk.DeVos@biw.kuleuven.be

8.1   Introduction

Metal organic frameworks (MOFs) are by definition formed from the coordinative polymerization of metal ions or clusters and polyfunctional linker molecules. Because of their high intrinsic metal loading, MOFs have been intensively studied as catalysts in various reactions over the past few years. A large part of the catalysis research has focused on using the large surface area of these MOFs to incorporate other catalytically active species. For instance, these active species can be encapsulated in the pores or cages; the MOFs can serve as a host for inorganic clusters like Keggin[1] anions or metal nanoparticles (see Chapter 10); catalytic sites can be incorporated in the organic struts linking the metal nodes (Chapter 9),[2] or the organic linkers can be pre- or postsynthetically modified with catalytic metal species (Chapter 3). The most intriguing class of MOF catalysts however, is that in which the structural nodes themselves act as the catalytic centres. In this latter case, catalysis at the metal ions or metal clusters may undergo strong influences of the stereoelectronic properties of the organic linker. The metal nodes can have different catalytic roles depending on the nature of the metal. By accepting electron pairs from

RSC Catalysis Series No. 12
Metal Organic Frameworks as Heterogeneous Catalysts
Edited by Francesc X. Llabrés i Xamena and Jorge Gascon
© The Royal Society of Chemistry 2013
Published by the Royal Society of Chemistry, www.rsc.org

reactant molecules, open metal sites in MOFs can act as Lewis acids. When the MOFs are used in oxidation or hydrogenation reactions, there can be an additional requirement for the framework to accommodate metal ions with changing coordinative demands or even oxidation states. Especially when the metal ions of the MOF are alkaline earths, the MOF can also be used as a base catalyst, where the metal–ligand ensemble abstracts a proton from the reactant molecules.

In general, two groups of materials can be distinguished. First, in a limited number of MOFs, free coordination sites ('coordinatively unsaturated sites', or *cus*) are available in the structure as determined by crystallography and *in situ* or *ex situ* spectroscopy.[3] Simple solvent removal by evacuation at increased temperature, or solvent displacement by incoming reactants may suffice to initiate the catalytic reaction on the metal site. It is, however, often unclear how sterically crowded the site is, whether all the open metal sites can take part in catalysis, or how the site's activity can be further increased. Secondly, an increasing number of MOFs has been reported to be catalytically active in a spectrum of reactions, even if the regular crystallographic structure does not contain any clearly defined active site. Hupp and co-authors have referred to this in the past as 'opportunistic catalysis'.[4] In some cases, this may indeed imply that a poorly defined material has been formed by disintegration of an intrinsically labile MOF (*e.g.* Zn oxide impurities in a water-labile MOF-5). In other cases, the MOF material appears to be highly stable and recyclable, and increasing insight is gathered in how to create such active sites and how to modulate their properties.

This chapter will give an overview of some research on MOF catalysts in which the metal ions act as the active sites; several examples will be discussed in more detail. This chapter will be structured according to the reaction type, starting with base catalysed reactions, followed by Lewis acid reactions and redox reactions such as oxidations in the vapour or liquid phase, or hydrogenations.

8.2 MOFs as Base Catalysts

Until now, the development of catalysts employing the metal nodes of MOFs as the active site has prevalently focused on acid or redox catalyzed systems. For basic MOF catalysts, research is mainly limited to the use of amino-modified linker molecules or grafted amine sites on the metal sites.[5–6] In recent years, some attempts have been made to synthesize basic MOF catalysts even without such linkers.[7–10] Koner and co-workers have synthesised various MOFs containing Mg^{2+} and Ba^{2+} which are claimed to be active basic catalysts.

The aldol condensation reaction of different ketones or aldehydes can be catalysed either using a Lewis acid catalyst or using a basic catalyst. This reaction was used as a test for MOFs built up from Mg^{2+} or Ba^{2+} and pyrazole-3,5-dicarboxylic acid (H_2PyDC) or pyridine-2,5-dicarboxylic acid (H_2PDC). The authors have reported isolated yields up to 90% after 6 h reaction at 10 °C for the facile condensation of 4-nitrobenzaldehyde and acetone using the dehydrated [Mg(PDC)(H$_2$O)]. Similar results were obtained using the Ba(PDC) catalyst. A prerequisite for the successful conversion of the reactants, however, is the use of stoichiometric amounts of diethylamine (ED)[7–9], which raises questions concerning the basic nature of the framework and its true role in catalysis. The material consisting of Ba^{2+} and pyrazole-3,5-dicarboxylate, [Ba(PyDC)] was observed to catalyze the aldol condensation even in the absence of diethylamine.[10] While this is a promising result, the reaction in this case is likely limited to the outer surface of the material due to its limited porosity.

Although the development of basic MOF catalysts employing metal nodes as the active site evolves at a slower pace than that of acid or redox catalysis, the abovementioned steps towards such catalysts clearly indicate that this topic holds ample promise. New synthetic efforts should be made to make other MOFs bearing these metal ions while having a well accessible pore system.

8.3   MOFs as Oxidation Catalysts

Before discussing specific cases, it should be made clear that transition metals can facilitate liquid phase oxidations *via* several mechanism types. Early transition metals in their highest oxidation state, like Ti^{4+} or V^{5+}, activate peroxides because of their Lewis acidity. As their oxidation state does not vary in this process, the main event is a reversible expansion of the coordination sphere. By contrast, homogeneous oxidation catalysts like Fe, Cr, Co *etc.* undergo one- or two-electron redox changes in a typical cycle. If a similar mechanism is considered for a MOF-embedded metal ion, the structure should not only allow expansion of the metal ion's coordination sphere, but also a change of the valence. Clearly, if one of the states of the metal is held less efficiently in the lattice, the danger of metal leaching is imminent.[11,12]

8.3.1   Cu and Co-containing MOFs

As an example of redox-active MOFs with Co^{2+}/Co^{3+} ions embedded in the framework, MOFs containing the bicipital 1,4-bis[(3,5-dimethyl)pyrazol-4-yl]benzene linker have been studied. Two structures with distinct behaviour were considered. MFU-1 contains nodes consisting of 4 Co^{2+} tetrahedra linked by a central oxo anion; these nodes are connected in a pseudo-octahedral fashion by the bis-pyrazolates, leading to a structure akin to that of MOF-5.[13] In MFU-2, chains of tetrahedral Co^{2+} ions are connected by the bis-pyrazolate linkers.[14] In both structures, the strong affinity between the soft Co^{2+} ions and the soft Lewis base atoms of the heterocyclic ligands results in a remarkable stability to water. However, the stability of the materials in oxidations using an alkyl hydroperoxide like *tert*-butylhydroperoxide (t-BuOOH or TBHP) is not guaranteed. MFU-1 was structurally stable and acted as a truly heterogeneous catalyst in hydrocarbon oxidation with TBHP, while MFU-2 was degraded and exhibited metal leaching. This can be readily related to the stronger nucleo-philicity of alkyl hydroperoxides in comparison with water and shows that framework stability is a classical, but critical pitfall. Effects of crystal size were studied using MFU-1 materials prepared solvothermally or *via* microwave synthesis; the limited dependence of the activity on crystal size suggested that sites inside the crystals have a considerable contribution to the activity. A classical homogeneous HaberWeiss mechanism for Co-induced free radical chain oxidation implies that both Co^{2+} and Co^{3+} would react with hydroperoxides, resulting in one-electron oxidations or reductions respectively. The tBuOO$^\bullet$ radicals formed in this process can then lead to formation of allylic peroxides (Scheme 8.1).

In a first approximation, the fractions of Co present in the trivalent or divalent state would reflect the relative reaction rates of, respectively, Co^{2+} and Co^{3+} with the peroxide. In a heterogenized version of this mechanism, the challenge is then to keep the metal ions well fixed in the lattice in both oxidation states. In MFU-1, expansion of the 4-fold coordination around Co^{2+} only proceeds with difficulty, due to steric crowding around the tetrahedral Co sites and symmetry constraints imposed by the tetranuclear cluster (Figure 8.1). This likely decreases the reactivity of the Co^{2+}. XANES data of MFU-1 before and after exposure to the peroxide show that, at most, 8% of the Co is in the trivalent state. This suggests that the putative five-coordinate Co^{3+}–OOR intermediate has a strong tendency to revert to the divalent 4-coordinate state, thus keeping the Co inside the lattice. Summarizing, strict control over the coordination chemistry of the Co site allows a compromise between activity and stability towards leaching to be found. Similar conclusions were obtained for the use of Co-substituted aluminophosphates like CoAPO-11 in autoxidation of cycloalkanes;[15] in the latter case, starting from tetrahedral Co^{2+} surrounded by phosphates, the coordination number and the Co valence can be increased, but the Co eventually stays well associated with the lattice, allowing a truly heterogeneous catalysis.

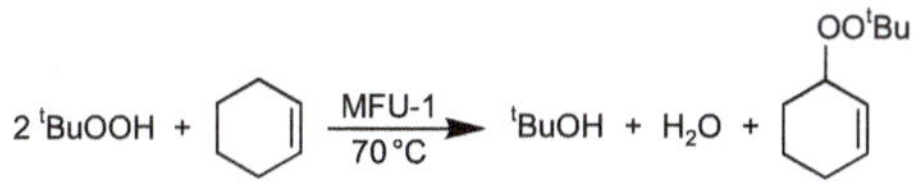

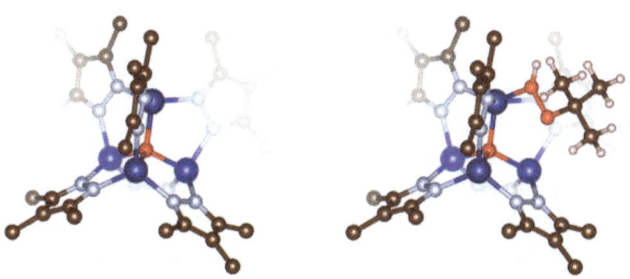

Scheme 8.1

Figure 8.1 Active site of MFU-1 before (left) and after coordination of TBHP (right).

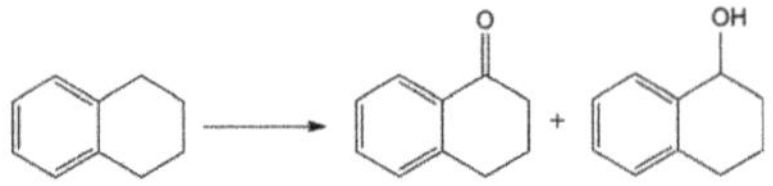

Scheme 8.2

In an even more attractive system, [Cu(2-PYMO)$_2$] and [Co(BIM)$_2$] were studied in the liquid-phase oxidation of alkanes using air as the oxidant.[16] The materials are built up from Cu^{2+} or Co^{2+} metal ions linked together through the N-atoms of pyrimidinolate (PYMO) or benzimidazolate (BIM). The oxidation of tetralin to tetralone was used as the test reaction (Scheme 8.2). A substrate-to-metal ratio of 2000 was employed. The two MOFs were found to have different characteristics in this reaction, depending on whether ketone or hydroperoxide formation is the aim. While on Cu(2-PYMO)$_2$ the ratio of tetralone *vs.* tetralol was not that high (2.7), this material was found to be very efficient and selective in producing the tetralinhydroperoxide in the initial stages of the reaction. [Co(BIM)$_2$], on the other hand, showed high tetralone *vs.* tetralol ratios (6.8–11.8), but was very inefficient in producing the hydroperoxide. With these results in mind, the authors used a physical mixture of the two MOFs to obtain a maximum performance of the catalytic system. Eventually it was found that an optimum composition of [Cu(2-PYMO)$_2$] *vs.* [Co(BIM)$_2$] of 90/10 gave the best performance.[16]

Very recently Asefa and Li *et al.* have reported on the catalytic activity of a layered [Co(OBA)$_2 \cdot 2H_2O$] material (H$_2$OBA = 4,4′-oxybis(benzoic acid)) as a highly active catalyst in olefin epoxidation reactions.[17] The 2D layered compound has an advantage over 3D-MOFs, because all of the metal sites are easily accessible for reactants. The material was used in the solvent-free epoxidation of aromatic olefins. The olefin was mixed with a double excess of anhydrous TBHP and brought into contact with a trace amount of catalytic MOF. Yields up to 92% were obtained for the oxidation of styrene to styrene oxide at 75 °C. Unfortunately, some leaching of Co^{2+} was evidenced during the analysis of the reaction mixture. Moreover, the possible formation of oligomeric byproducts was not assessed in detail, while it is well known that in styrene epoxidation by radicals such processes can interfere with the mass balance.

For open metal sites in MOFs to be able to catalyze oxidation reactions, the framework must typically be able to accommodate the modified coordinative demand of the metal ions undergoing a change of the oxidation state. This is, however, not always the case, as shown for instance by the oxidation catalyzed by the Cu-*trans*-1,4-cyclohexanedicarboxylate MOF.[18–19] Here, redox changes are situated in the ligated species, rather than in the framework metal atoms. Upon oxidation of a flexible Cu-*trans*-1,4-cyclohexanedicarboxylate MOF with H$_2$O$_2$, the Cu-paddlewheels are disconnected, rotated and again linked to each other by peroxo bridges. This material is a heterogeneous and reusable catalyst for oxidation of various aliphatic and aromatic alcohols with H$_2$O$_2$ with selectivities above 99%. A similar type of peroxo species might well be formed based on the Cu-paddlewheel sites of [Cu$_3$(BTC)$_2$], which was used for the TBHP assisted oxidation of benzylic compounds.[20]

Another MOF that shows the intrinsic potential of Cu-paddlewheel clusters to catalyze oxidations is the [Cu(5-MIPT)(H$_2$O)](H$_2$O)$_2$ MOF constructed from Cu^{2+} and 5-methylisophthalate (5-MIPT) ligands reported by Zou *et al.*[21] Each copper ion is coordinated to an apical water molecule. These coordinated water molecules, together with physisorbed water, can be thermally removed to open up the copper sites located in the channel walls. The catalytic potential of the MOF was demonstrated in the oxidation of CO to CO$_2$ using air. The activities are similar to or higher than those of the used CuO reference catalysts. Full conversion of CO was obtained at 200 °C. The porous lattice exhibits a stable activity and remains intact after catalytic reaction; no CuO formation was observed.

Recently, Zhang *et al.* synthesized a new Cu-MOF exhibiting high activity in CO oxidation.[22] [Cu$_3$(μ_3-OH)(PyCA)$_3$(H$_3$O)].2C$_2$H$_5$OH.4H$_2$O was synthesized starting from 1*H*-pyrazole-4-carboxylic acid (PyCA) and Cu(NO$_3$)$_2$ in DMF. Unlike the aforementioned Cu-paddlewheel type structures, this material is built up from trinuclear triangular arrangements of Cu^{2+}. The trinuclear unit consists of a planar Cu$_3$(μ_3-OH) core which is supported by bridging pyrazolyl groups. The carboxylate groups complete the square planar coordination of the Cu^{2+} ions. The framework itself is anionic, with water molecules (in the form of H$_3$O$^+$) residing in the channels. These can, however, be removed upon mild thermal treatment. The material was found to be active in the CO oxidation, with full oxidation being attained at 230 °C. The lower activity of [Cu$_3$(μ_3-OH)(PyCA)$_3$] in comparison to [Cu(5-MIPT)] was explained by the observation that only a fraction of the unsaturated Cu^{2+} sites were accessible for CO molecules.[22]

Other types of MOF Cu-sites have been explored as well for oxidation catalysis. [Cu(H$_2$BTEC)(BPY)] was used for the cyclohexene and styrene epoxidation with TBHP with respective yields of 65% and 24% after 24 h of reaction at 75 °C.[23] The efficiency of the catalyst is illustrated by turnover frequencies as high as 79 h^{-1} for cyclohexene. The oxidation cycle starts by coordination of the Cu^{2+} center to TBHP, resulting in an expansion of the Cu^{2+} coordination sphere. After a nucleophilic attack of the olefin substrate on this species, a concerted oxygen transfer takes place, leading to the departure of *tert*-butanol. Finally, the rupture of the Cu–O bond produces the epoxide and the regenerated catalyst.

Baiker *et al.* reported the highly selective allylic oxidation of cyclohexene to cyclohexylhydroperoxide using [Cu(BPY)(H$_2$O)$_2$(BF$_4$)$_2$(BPY)] with molecular oxygen.[24] Selectivities of up to 90% were obtained after reaction at 45 °C without solvent. The active Cu(II) sites are located at the surface of the framework and both BPY and water are involved in the active complex. Removal of the structural water at 100 °C under vacuum opens the pores of this catalyst but suppresses the oxidation activity due to the direct co-ordination of all BPY ligands to Cu^{2+} ions. The activity, however, is rather low with a TOF of around 2 h^{-1}. The same reaction was recently used with

$[Cu_2(OH)(BTC)(H_2O)]_n \cdot 2nH_2O$ and $[Co_2(DOBDC)(H_2O)_2] \cdot 8H_2O$ (CPO-27(Co), also known as MOF-74(Co)) in the solventless aerobic oxidation of cyclohexene.[25] It is observed that both the Cu^{2+}- and Co^{2+}-MOF (M-MOF) can oxidize cyclohexene to give 2-cyclohexen-1-ol and 2-cyclohexen-1-one as the main products. The authors propose that O_2 is activated on the open metal sites of the MOFs, as was also reported for homogeneous Cu^{2+}- and Co^{2+}-containing complexes. The superoxide adduct of the M-MOF can abstract a hydrogen atom from cyclohexene, leading to a M-OOH species. The cyclohexenyl radicals then initiate a classical free radical autoxidation, with formation of cyclohexylhydroperoxide and its ketone and alcohol decomposition products. A hot filtration experiment proved that these MOF catalysts are stable and recyclable under the reaction conditions.[25]

A nice example of how an intrinsically inactive material can be made active is shown in the work of Baiker and coworkers.[26] They modified $[Cu_3(BTC)_2]$ by partial substitution of BTC linkers by pyridine-3,5-dicarboxylate (P'DC). The incorporation of the P'DC leads to a defined number of defect sites at the Cu-paddlewheel clusters. Up to 50% of the BTC linkers could be replaced by P'DC without changing the structure of the resulting materials, as was confirmed by XRD. XAS revealed that the incorporation of the pyridine moiety into the crystal lattice leads to a local defect site at some of the dimeric Cu units, where normally a bridging carboxylate group would be coordinated. These structural defects cause a modified chemical and electronic environment of the metal centers, which consequently also affects their accessibility and activity in catalytic applications. $[Cu_3(BTC)_2]$ and substituted $[Cu_3(BTC-P'DC)_2]$ materials were found to catalyze the direct hydroxylation of aromatic compounds at low temperatures. In contrast to homogeneously dissolved Cu salts, a surprisingly high selectivity toward *ortho-* and *para-*cresol was observed in the reaction of toluene and hydrogen peroxide. The product selectivity was significantly different for pure $[Cu_3(BTC)_2]$ and the mixed-linker materials. When acetonitrile was used as the solvent, a maximum turnover frequency of up to 2.37 h^{-1} could be reached, twice as high as the TOFs that were previously reported for copper complexes encapsulated in Na-Y zeolites. The toluene concentration had a dramatic and opposite influence on the conversion and the product selectivity. Lower substrate concentrations led to an increased conversion, but also to an increased selectivity toward the undesired side products benzaldehyde and methylbenzoquinone. High selectivity toward the cresol products was mainly observed under solvent-free conditions. Using optimized reaction conditions, the direct hydroxylation of other aromatic substrates including benzene, xylenes, and aryl halides became feasible.[26]

8.3.2 Fe-containing MOFs

Several reports have also been published on the use of Fe-MOFs as oxidation catalysts. Dhakshinamoorthy *et al.* found Basolite® F300 (Fe-BTC) to be a reusable redox catalyst for the selective oxidation of thiols to disulfides.[27] Using acetonitrile as a solvent, a full conversion was obtained after 1 h when using thiophenol as substrate. Other substrates were also tested and it was observed that the material decomposes when thiobenzoic acid and pentanedithiol were used. Taking into account that MIL-100(Fe), the more crystalline analogue of F300, exhibits a reversible redox behaviour,[28] this could also be the case in Fe-BTC. The same groups found MIL-100(Fe) to be an active and selective catalyst for the oxidation of diphenylmethane (DPM) and triphenylmethane (TPM) using TBHP (5.5 M in decane)[29] as well as for the aerobic oxidation of thiophenol. The catalyst is stable, can be recycled and no deactivation was observed. The oxidations studied correspond to a heterogeneous catalytic process, where the reaction occurs within the pores of the MOF. To demonstrate this, the reaction was performed with MIL-100(Fe) of different crystallite sizes. When using small crystallites, the aerobic oxidation of thiophenol and the oxidation of DPM with TBHP were found not to be controlled by intracrystalline diffusion, while the oxidation of TPM occurs mainly at the external surface of the catalyst. Thus, in this case, the catalyst activity can be improved by preparing a MOF catalyst with smaller crystals.

In a related study, Fe-BTC and MIL-100(Fe) were compared in diphenylmethane and thiophenol oxidation.[30] It was found that MIL-100(Fe) is the best catalyst for these oxidations. This was ascribed to the higher purity of MIL-100(Fe), which may facilitate the Fe^{3+}/Fe^{2+} redox switching.

Very recently, the mesoporous Cr- and Fe-MIL-101 materials were reported as good catalysts for cyclohexane conversion.[31] MIL-101(Cr) had the highest conversions and selectivities when TBHP was used as oxidant, producing a mixture of cyclohexanol and cyclohexanone with 92% selectivity at 25% conversion (TOF = 7.2 h^{-1}). For MIL-101(Fe), the presence of O_2 was essential for good conversions. The selectivity of the Fe material towards the ketone and alcohol, however, was not that high with only 49% selectivity at 38% conversion. Both MOFs could be reused without loss of activity.

8.3.3 Oxidations with V-MOFs

Vanadium catalysts, in combination with peroxides, are commonly used for the epoxidation of allylic alcohols or simple olefins. The immobilization of homogeneous vanadium species however is often troublesome due to the different possible forms of V (such as the cationic form VO^{2+}, the anionic form VO_4^{3-} or neutral species like vanadyl acetylacetonate), and their ease of interconversion upon interaction with various oxidants (TBHP and H_2O_2) under various conditions (dry or aqueous, and various values of pH).[12]

An example of a V-MOF successfully used as a catalyst in the epoxidation of olefins is MIL-47[32–34] or its naphthalenedicarboxylate analogue COMOC-3.[35] MIL-47 has a three-dimensional framework, built up from linear chains of vanadium oxide connected by the terephthalate linkers forming one-dimensional rhombic pores. Each V^{4+}-center in this chain is coordinated to four oxygen atoms from four carboxylate groups, and to two oxygen atoms on the O-V-O axis, thus forming a saturated octahedral coordination sphere. When using TBHP in water as the oxidant in the oxidation of cyclohexene

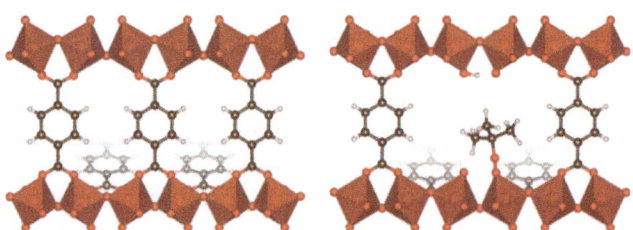

Scheme 8.3

Figure 8.2 Structure of MIL-47 with saturated V coordination spheres (left), and formation of the active site after decoordination of a linker molecule and coordination of *t*-BuOOH (right).

(Scheme 8.3), MIL-47 is prone to leaching of the metal ion, up to 13% after 1 h, and catalytic activities comparable to that of homogeneous vanadyl acetylacetonate (VO(acac)$_2$) have been measured. Catalyst stability improves when switching to TBHP in decane as the oxidant and the amount of vanadium leaching is negligible (<3%) after 7 h of reaction. In these conditions the catalytic activity is lower but still present.

The lack of *cus* in the framework, and the fact that V-catalyzed heterolytic epoxidations usually proceed *via* a purely Lewis acid mechanism with V in the +5 state rather than in the +4 state,[12,36,37] raise questions concerning the nature of the active site in MIL-47. The authors propose that the active site is created by the cleavage of a terephthalate-vanadium bond and the replacement of the ligand by a hydroxyl group. As the formation of such a hydroxyl group will proceed more swiftly in a water-based medium, the initial rate is expected to be higher when using TBHP in water rather than TBHP in decane. In the latter case, TBHP is thought to be responsible for the activation of the catalyst (Figure 8.2). As this activation proceeds more gradually in the absence of water, the framework structure remains better conserved during the reaction when using TBHP in decane.

To further investigate the nature of this active site and the different possible reaction mechanisms, an *ab initio* computational study was performed. Although several catalytic pathways may co-exist and compete with one another, two main pathways were investigated in greater detail. In the first pathway, a V^{4+}-species is responsible for the direct epoxidation of the olefin. In the second pathway, indicated as the radical pathway, homolytic cleavage of the peroxy bond results in the oxidation of V^{4+} to V^{5+}. This complex can be activated with TBHP and can subsequently epoxidize cyclohexene. Both reaction routes are completed by a regeneration step that closes the cycle. EPR and NMR measurements indicate that approximately 20% of the V^{4+} sites are oxidized to V^{5+} in the first minutes of the catalytic reaction, confirming the coexistence of the different reaction pathways.

Summarizing, careful adaptation of the reaction conditions to the catalyst stability allows for a balance of the catalyst activation, *via* temporary deco-ordination of linker molecules, and loss of the catalyst structure. Hence, the active V-centres are efficiently immobilized in the metal organic framework. Previous attempts to immobilize V epoxidation catalysts, *e.g.* on microporous and mesoporous molecular sieves such as V-AlPO-5[12,38] and V-HMS[39] respectively, were only partly successful, and V leaching was more often encountered than not. Many factors can affect the vanadium leaching from such materials during liquid phase epoxidations, such as the nature of the vanadium source, the solvent, the substrate and the oxidant.

MIL-47 and the related compound MOF-48 were also shown to catalyze the conversion of methane to acetic acid.[40] MOF-48 is synthesized using 2,5-dimethylterephthalic acid rather than terephthalic acid. Reactions were carried out in a mixture of potassium peroxydisulfate in trifluoroacetic acid (Scheme 8.4). Hot filtration tests confirm that catalysis proceeds heterogen-eously. The overall performance of the catalysts is reported to rival those of homogeneous vanadium catalysts and to exceed those of other heterogeneous catalysts such as Pt/alumina and sulfonated zirconia. Although the reaction mechanism was not addressed, the oxidation of a fraction of the V-sites in the MOFs may again be key in the mechanism of the MOF-based oxidation of methane to acetic acid.[37]

8.4 MOFs as Reduction Catalysts

While MOF-supported metal nanoparticles can readily be used for a variety of hydrogenations,[41–44] we here discuss some examples of MOF structures in which the lattice ions themselves mediate selective reduction reactions, either by hydrogen or by an alternative reducing agent. A first example is the 3D indium terephthalate MOF reported by Gómez-Lor *et al.*[45] This structure, built up from hexagonal sheets of $[In_2(OH)_3]^{3+}$ connected by terephthalate molecules, displays activity in the reduction of nitroaromatic compounds using H_2 as a reducing agent (Scheme 8.5). It is well known that zero-valent In is a useful *reagent* for selective reduction of imines, heterocycles and nitro compounds;[46]

$$2\ CH_4 + 2\ H_2O + 4\ K_2S_2O_8 \xrightarrow[\text{TFA, 80\,°C, 20 h}]{\text{MIL-47}} CH_3COOH + 8KHSO_4$$

$$CH_4 + CO + H_2O + K_2S_2O_8 \xrightarrow[\text{TFA, 80\,°C, 20 h}]{\text{MIL-47}} CH_3COOH + 2KHSO_4$$

Scheme 8.4

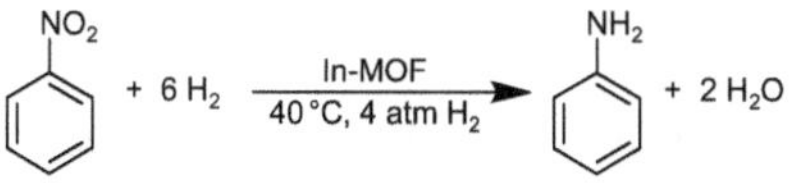

Scheme 8.5

however, there are no reports on cationic In^{3+} being capable of catalytic hydrogenations using H_2. In the In-MOF, the indium seems unable to vary its oxidation state. The authors suggest that coordinatively unsaturated sites, *e.g.* at the crystal edges, are involved in the mechanism. As In^{3+} cannot undergo an oxidative addition of H_2, the mechanism must involve a heterolytic hydrogen cleavage, as is common *e.g.* on Ru^{3+}, resulting in a In^{3+}-bound hydride ligand. The catalytic activity is remarkable, with 100% yield and average turnover frequencies $>100\,h^{-1}$, for instance in the reduction of 1-nitro-2-methylnaphthalene at 313 K and 4 bar H_2. As the active sites are exclusively located at the outer surface, a decrease in particle size should increase the number of active sites and the catalytic activity.

Another MOF with pronounced catalytic activity in reduction reactions is $Al_2(BDC)_3$,[47] which was employed as the commercial material Basolite® A100, and likely has a similar structure as MIL-53.[48] Although this structure is lacking *cus*, the activity of this MOF for the chemoselective reduction of carbon–carbon multiple bonds with hydrazine surpasses that of other *cus* containing structures such as Fe-BTC (Basolite® F300) or $Cu_3(BTC)_2$ (Basolite® C300). As the reaction rate was unaffected by the presence of electron-donating or electron-withdrawing groups in the reactants, the intermediacy of carbocations or the development of a partially positively charged transition state in the rate determining step was ruled out. The authors propose a mechanism in which a coordinatively unsaturated Lewis acid Al^{3+} participates, but do not prove how this site should be formed. The possibility of a reversible decoordination of a linker molecule is mentioned; defects or outer surface sites are alternative plausible explanations for the catalytic activity presented by Basolite® A100.

In 2006, Navarro *et al.* reported the olefin hydrogenation using a MOF built up from Pd and 2-hydroxypyrimidine, called $Pd(2\text{-}PYMO)_2$.[49] The structure consists of sodalite type cages which are connected with 2 different hexagonal windows with respective free openings of 0.48 nm and 0.88 nm. The material was also found to be active in other types of reactions, such as oxidations and Suzuki coupling reactions.[50] The authors describe the hydrogenation of 1-octene under mild conditions, with full conversion of the reactant and a 59% yield to octane after 40 minutes, 2-octene being formed as a side product. Shape selectivity was observed, as a more bulky olefin like cyclododecene was not hydrogenated even after 5 hours of reaction. Opelt and co-workers investigated the stability and reusability of this MOF under the given reaction conditions.[51] It was observed that during the first 4 hours of reaction, the material was indeed size-selective for 1-octene in comparison to cyclododecene. After 4 h, the material started to get hydrogenated with the liberation of the uncoordinated linker molecules and Pd^0; at the same time, it started converting cyclododecene indicating that longer exposure times to hydrogen lead to a collapse of the framework. This was confirmed by X-ray powder patterns in which Pd^0 appeared. In a follow up study, a more detailed study on the nature of the active site was performed.[52] The changes at the metal node during the hydrogenation using *in situ* X-ray absorption spectroscopy (XAS) and IR spectroscopy were followed. With the help of these techniques, the authors showed that in fact in the first hours, no Pd^0 was formed, indicating that the Pd^{2+} is the active site in the 1-octene hydrogenation. Moreover, it was shown that when performing the solventless hydrogenation of 1-octene, no Pd^0 formation could be detected within 24 hours of reaction. This hints that the local concentration of the reactant at the active site is crucial to avoid reduction of the Pd^{2+} form of the MOF.

A final example of a MOF active in reduction reactions is $Ca(HFIPBB)(H_2HFIPBB)_{0.5}(H_2O)$.[53] This material has 2 types of cages, one with an accessible window of 0.58 nm × 0.86 nm, and a microporous cavity with a pore diameter of 0.33 nm. The material was used in the hydrogenation of styrene in toluene under relative mild conditions (373 K and 5.05 bar H_2). The material is highly active reaching full conversion at 100% selectivity, with a TOF of 254 h^{-1}. The authors propose a hydrogenation mechanism in which the calcium coordination number is increased from 7 to 8 to allow substrate coordination and simultaneous hydride formation, although no experiments were performed to support this mechanism.

8.5 MOFs as Catalysts for Electrophilic Aromatic Substitutions

The MIL-100(Fe) structure, with its large number of available *cus*, was early on reported to display activity in Friedel–Crafts benzylation reactions using benzyl chloride as the alkylating agent.[54] Recently also the acylation of *p*-xylene with benzoyl chloride using the same MOF was reported.[55]

However, also materials like MOF-5 (IRMOF-1) and IRMOF-8 have been shown to be active catalysts for electrophilic aromatic substitutions.[56] These MOFs, however, do not contain *cus*; their structure, in which Zn_4O subunits are linked by octahedral arrangements of terephthalate or 2,6-naphtalenedi-carboxylate respectively, is coordinatively saturated. IR and ^1H NMR studies suggest that non-structural hydroxyl moieties, present as defects of the structure or as non-XRD detectable MOF-69c microcrystallites, may provide the Brönsted acidity required to catalyze these reactions.[56]

As structural defects are thought to be responsible for the catalytic activity of MOF-5, efforts have been made to increase the catalytic activity *via* the deliberate introduction of such defects.[57] First, the use of a fast precipitation procedure instead of the conventional solvothermal synthesis was assessed as a method for the introduction of defects. Catalytic testing shows that this led to a noticeable increase in catalytic activity while maintaining the high *para*-selectivity of the process. In order to further increase the number of defects, monodentate linkers were added to the reaction mixtures, such as 2-methylbenzoate (Figure 8.3). It was believed that this strategy would yield defects at the nodes that are adjacent to the monodentate linkers. Catalytic testing, however, shows that the introduction of these monodentate linkers does not significantly alter the catalytic performance. This limited effect of the

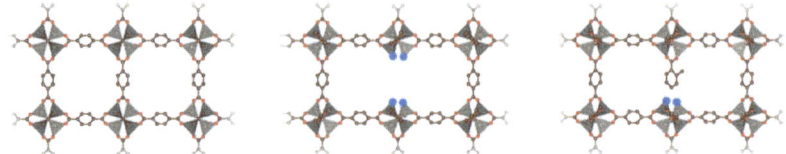

Figure 8.3 Structure of MOF-5 without defects (left), with defects induced by fast precipitation (middle) and with defects induced using monodentate carboxylates (right).

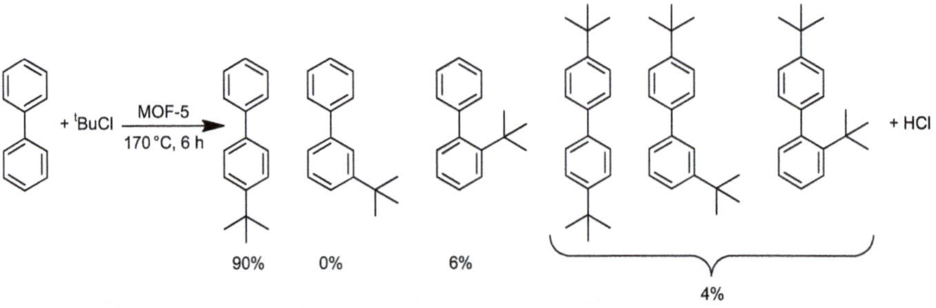

Scheme 8.6

newly formed defect sites may be due to steric hindrance caused by the proximity of the monodentate carboxylate.

The catalytic performance of these materials was compared with that of H-Beta zeolite in the *tert*-butylation of biphenyl. This shows that although the substrate conversion is higher when using H-Beta, the *para*-selectivity of the products is higher and less di-alkylated products are formed on MOF-5 and IRMOF-8 (Scheme 8.6). The selectivity of H-Beta can, however, be increased *via* dealumination procedures. As the pore windows of MOF-5 and IRMOF-8 are larger than those of H-Beta, the shape selective properties may result from a classical transition state shape selectivity mechanism. It is suggested that the substrate is adsorbed in a specific manner to allow the formation of the transition state for the *para*-oriented product. Once alkylated, the product cannot be activated in the same manner because of steric hindrance and double alkylation cannot proceed.

While the activity of MOF-5 depends on defect formation,[56,57] many MOF structures bear structural hydroxyl groups, like the MIL-53 family of materials (M(OH)(DBC), M = Al, Cr, Fe, Ga or In and BDC = terephthalate).[58] Strikingly, in the *tert*-butylation of biphenyl, MIL-53(Al) is fully inactive, while MIL-53(Ga), also called IM-19, displays a strong activity.[59] This demonstrates that the catalytic performance of these materials is not only governed by the presence of the μ_2-OH-groups, but also by the choice of the metal. *In-situ* FTIR studies indicate a significantly higher acidity for MIL-53(Ga) than for MIL-53(Al). An *ab initio* molecular modelling study,

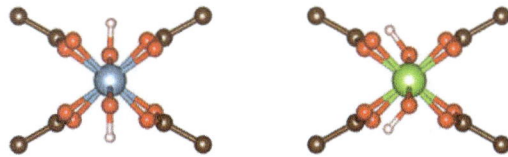

Figure 8.4   Structural hydroxyls on MIL-53(Al) (left) and MIL-53(Ga) (right).

however, suggests that solely the mild acidity of the Ga-μ_2-OH-Ga species does not explain the high catalytic activity of MIL-53(Ga). DFT calculations show that the structural OH groups in MIL-53(Ga) must be tilted (Figure 8.4). This induces a much stronger stabilization of the positively charged intermediates than in the non-polar MIL-53(Al), increasing the reaction rate. Hence, subtle differences in catalyst structure due to a change in metal ion may lead to large differences in catalytic activity.

MIL-53(Ga) displays a higher activity and *para*-selectivity in the alkylation of biphenyl with *t*-BuCl than H-Beta and H-Mordenite zeolites. When compared to homogeneous $AlCl_3$, the activity of MIL-53(Ga) is lower but the *para*-selectivity is greatly increased.

8.6   MOFs in other Lewis Acid Catalyzed Reactions

Many other Lewis acid catalysed reactions have already been used to prove the catalytic potential of MOFs. One of the most used reactions for this purpose is the cyanosilylation reaction. This reaction is often employed because it requires mild Lewis acid sites and is performed under moderate conditions. Since the first report on cyanosilylation of carbonyl compounds using a Cd-bipyridine coordination polymer as catalyst,[60] many more MOFs containing mostly divalent metals such as Cu^{2+}, Zn^{2+}, Mn^{2+}, *etc.* were reported to be able to catalyse this reaction.[61–69] One of the most active MOFs was found to be MIL-101(Cr).[70] This material is built up from a trimeric building unit in which two out of three chromium ions are ligated to a water molecule which can be removed by thermal treatment. This high number of open metal sites,[71] combined with its high stability make it an interesting material to look at for catalysis. In the cyanosilylation of benzaldehyde with trimethylsilylcyanide (scheme 8.7) MIL-101(Cr) gave a yield of 98.5% and a TOF of >150h^{-1} (after 1 h). The material can be reused multiple times, with only a small reduction of the activity.

Epoxide methanolysis is another Lewis acid catalysed reaction, sometimes used in MOF catalysis. It was shown for the well known $[Cu_3(BTC)_2]$ that reducing the crystal size of these MOFs has a pronounced influence on the methanolysis of styrene oxide.[72] The Lewis acid properties of this MOF were already exemplified using several reactions, such as the organic isomerizations of α-pinene oxide, (+)-citronellal and the ethylene ketal of bromo-propiophenone,[73] cyanosilylation reactions,[61] Friedländer condensation,[74] alkene cyclopropanation,[75] CO oxidation,[76] *etc.* When using the nanosized

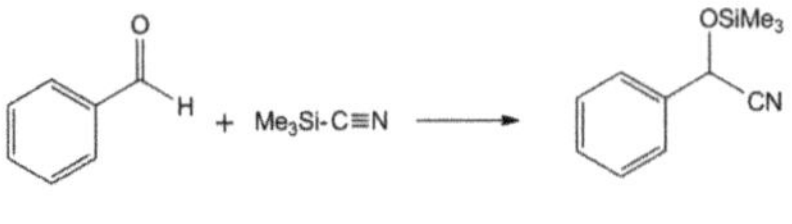

Scheme 8.7

[Cu$_3$(BTC)$_2$] particles synthesized *via* a freeze-drying step, a 90% yield of the corresponding 2-methoxy-2-phenylethanol was obtained, whereas the normal hydrothermally synthesized [Cu$_3$(BTC)$_2$] gave only 2% conversion in the same time.[72] As both reactants should normally be able to enter the pores of this material, the results of this study with smaller MOF particles raise the question of whether the catalytic activity of [Cu$_3$(BTC)$_2$] in conversion of large substrates may largely be due to open metal sites accessible on the outer surfaces of the crystal.

In another example, we found very recently that the mild Lewis acidity of the Bi^{3+} ions in CAU-7 is perfect for catalysing the hydroxymethylation of 2-methylfurfural. CAU-7 is one of the first Bi-containing MOFs, built up from Bi^{3+} ions and 1,3,5-benzenetrisbenzoate (BTB) ligands.[77] The Bi^{3+} ions are nine-coordinated by oxygen atoms of the BTB linkers; forming face-sharing BiO$_9$ polydehdra. The hydroxymethylation reaction is very sensitive for excessively strong acidity, leading to condensation and polymerization of the 2-methylfuran. It was shown that CAU-7, unlike other MOFs like UiO-66 or MIL-53(Ga), gave a high yield to the 5-methylfurfurylalcohol at high conversions, largely avoiding consecutive condensation reactions. Acetonitrile chemisorption indicated that CAU-7 possesses mild Lewis acid sites, probably due to some missing linkers in the structure. Moreover, the hydrophobic nature of the material facilitates the fast desorption of the reaction product, lowering the chance of consecutive condensation reactions.

A remarkable example of metal node catalysis is the use of ZIF-8 as a transesterification catalyst. In principle, the zeolitic imidazolate framework ZIF-8, synthesized from 2-methylimidazole and a zinc precursor, is equally a structure with fully coordinated Zn^{2+}, without *cus*. Nevertheless, the material has proven to be an efficient catalyst for several reactions, such as the conversion of CO$_2$ with epichlorohydrine to chloropropene carbonate[78] or the Knoevenagel reaction.[79] The nature of the active sites in these catalysts has been studied to a great extent for their use as a transesterification catalyst.[80] By combining FTIR-monitored carbon monoxide adsorption and density functional theory (DFT) calculations, three hypotheses for the formation of the active sites were brought forward:

1. The saturated node is able to temporarily increase its coordination number, thus acting as a Lewis acid;
2. Temporary de-coordination of one of the node-to-linker bonds generates a transient species able to activate reactants.
3. The active site may be located at the external surface or at structural defects due to the expected presence of dangling bonds.

DFT was used to calculate the IR-frequency shifts of CO when adsorbed on different types of possible active sites in the bulk and at the surface of ZIF-8. Comparison of these shifts with those recorded experimentally suggests that the acido-basic sites are located at the external surface or at structural defects rather than in the micropores of ZIF-8. The nature of the sites at the external surface of the catalyst has proven to be very diverse: OH and NH-groups, hydrogenocarbonates, low-coordinated Zn atoms and free N^- moieties belonging to linkers, giving rise to strong Lewis acid sites (*e.g.* Zn^{II}-species), as well as Brønsted acid sites (NH groups) together with basic sites (OH groups and N^- moieties).

Another zeolitic imidazolate framework with reported catalytic activity in the Knoevenagel reaction is ZIF-9,[81] synthesized from benzimidazole and a cobalt source. As the pore windows of ZIF-9 are too small to accommodate the aromatic reactant, catalysis is limited to the outer surface of the material. The nature of the sites at the outer surface is thought to be similar to those found for ZIF-8.

A final example of a MOF with unexpected catalytic activity at the metal sites, is the UiO-66 type MOF, containing $[Zr_6O_4(OH)_4]^{12+}$ clusters which, according to the structure model, are surrounded by twelve bidentate linkers such as terephthalate and its substituted variants. In the perfectly crystalline material, the Zr has its maximal coordination number of 8.[82] Remarkably, the $Zr_6O_4(OH)_4$ cluster can be reversibly dehydroxylated to $[Zr_6O_6]^{12+}$ at temperatures between 373 and 523 K, depending on the substituents on the 1,4-benzenedicarboxylate linker. This decreases the coordination number from 8 to 7, which still leaves the Zr in a highly coordinated state, with little room for incoming reactants. Nevertheless, catalytic experiments prove that UiO-66 type MOFs display an activity as expected for Lewis acids, for instance in the Oppenauer oxidation of alcohols,[83] in the cross-aldol condensation of heptanal and benzaldehyde,[84] or in the cyclization of citronellal to isopulegol,[83] which is an ene type cyclization. Detailed physicochemical characterization, in particular *via* thermogravimetric analysis, reveals that the number of linkers surrounding the Zr clusters is significantly lower than the twelve that are theoretically expected, even in materials that appear as perfectly crystalline in the diffractograms.[85] Thus, in this particular structure, linker deficiencies of up to 25% create a large number of active sites in the material. Calculations confirm that removal of one or more linkers from Zr is necessary before reactants can access the active sites.

An additional remarkable property of the Zr active sites is that their Lewis acidity can be modulated by the electronic properties of the benzenedicarboxylate linkers.[83] Thus, for a nitrosubstituted UiO-66, the initial rate in the citronellal cyclization is at least 40 times higher than for the non-substituted material. Enhancing the electron withdrawing character of the organic part of the MOF therefore results in a stronger Lewis acidity. Calculations showed that in the nitro-substituted material, both the free energies of the adsorbed state and of the transition state are significantly lowered.

8.7   Conclusions

In the traditional catalysis community, the topic of MOF catalysis is looked at with critical or sometimes even sceptical attitudes. The examples discussed in the foregoing sections prove that these opinions are far from justified, and are still rooted in the naive vision that MOFs should outperform and eventually displace well established catalyst classes like zeolites or oxides. Instead, the heterogeneous catalysis community should regard MOFs as a golden opportunity to reclaim a large field of applications from homogeneous catalysis, where until now dissolved metal salts and complexes have been used. There is now sufficient knowledge to deal with issues like leaching and lattice stability; and many of the catalytically interesting MOFs are based on affordable and commercial linker molecules. The tunability of the MOF catalysts and their smooth recyclability are sufficient assets such that, within a reasonable timeframe, industry will start to apply these materials, at least if the MOF community dares to demonstrate the full potential of these fascinating catalysts.

List of Abbreviations

2-PYMO	2-hydroxypyrimidine
5-MIPT	5-methylisophthalate
BDC	1,4-benzenedicarboxylate
BPY	4,4′-bipyridine
BTB	1,3,5-benzenetrisbenzoate
BTC	1,3,5-benzenetricarboxylate
BTEC	1,2,4,5-benzenetetracarboxylate
cus	coordinatively unsaturated sites
DMF	dimethylformamide
DMSO	dimethyl sulfoxide
DOBDC	2,5-dihydroxyterephthalic acid
ED	ethylenediamine
H_2HFIPBB	4,4′-(hexafluoroisopropylidene)bis(benzoic)acid
MOF	metal organic framework
OBA	4,4′-oxybis(benzoate)
PDC	pyridine-2,5-dicarboxylate
P′DC	pyridine-3,5-dicarboxylate
PyCA	1*H*-pyrazole-4-carboxylic acid
PyDC	pyrazole-2,5-dicarboxylate
SBU	secondary building unit
TBHP	*tert*-butylhydroperoxide
TOF	turnover frequency

Acknowledgements

Our efforts in MOF catalysis were supported by FWO Vlaanderen, IAP Belspo and KULeuven CASAS funding.

References

1. L. H. Wee, S. R. Bajpe, N. Janssens, I. Hermans, K. Houthoofd, C. E. A. Kirschhock and J. A. Martens, *Chem. Commun.*, 2010, **46**, 8186–8188.
2. D. Feng, Z.-Y. Gu, J.-R. Li, H.-L. Jiang, Z. Wei and H.-C. Zhou, *Angew. Chem., Int. Ed.*, 2012, **51**, 10307–10310.
3. C. Volkringer, H. Leclerc, J.-C. Lavalley, T. Loiseau, G. Férey, M. Daturi and A. Vimont, *J. Phys. Chem. C*, 2012, **116**, 5710–5719.
4. J. Lee, O. K. Farha, J. Roberts, K. A. Scheidt, S. T. Nguyen and J. T. Hupp, *Chem. Soc. Rev.*, 2009, **38**, 1450–1459.
5. J. Gascon, U. Aktay, M. D. Hernandez-Alonso, G. P. M. v. Klink and F. Kapteijn, *J. Catal.*, 2009, **261**, 75–87.
6. Y. K. Hwang, D. Y. Hong, J. S. Chang, S. H. Jhung, Y. K. Seo, J. Kim, A. Vimont, M. Daturi, C. Serre and G. Ferey, *Angew. Chem., Int. Ed.*, 2008, **47**, 4144.
7. R. Sen, D. Saha and S. Koner, *Chem.–Eur. J.*, 2012, **18**, 5979–5986.
8. D. Saha, T. Maity, R. Sen and S. Koner, *Polyhedron*, 2012, **43**, 63–70.
9. D. Saha, R. Sen, T. Maity and S. Koner, *Dalton Trans.*, 2012, **41**, 7399–7408.
10. T. Maity, D. Saha, S. Das and S. Koner, *Eur. J. Inorg. Chem.*, 2012, **30**, 4914–4920.
11. R. A. Sheldon and J. K. Kochi, Metal-catalyzed oxidations of organic compounds: mechanistic principles and synthetic methodology including biochemical processes, Academic Press, New York, 1981.
12. D. E. De Vos, B. F. Sels and P. A. Jacobs, in *Advances in Catalysis*, Academic Press, 2001, vol. 46, pp. 1–87.
13. M. Tonigold, Y. Lu, B. Bredenkötter, B. Rieger, S. Bahnmüller, J. Hitzbleck, G. Langstein and D. Volkmer, *Angew. Chem., Int. Ed.*, 2009, **48**, 7546–7550.
14. M. Tonigold, Y. Lu, A. Mavrandonakis, A. Puls, R. Staudt, J. Möllmer, J. Sauer and D. Volkmer, *Chem. Eur. J.*, 2011, **17**, 8671–8695.
15. D. L. Vanoppen, D. E. De Vos, M. J. Genet, P. G. Rouxhet and P. A. Jacobs, *Angew. Chem., Int. Ed.*, 1995, **34**, 560–563.
16. F. X. Llabrés i Xamena, O. Casanova, R. Galiasso Tailleur, H. Garcia and A. Corma, *J. Catal.*, 2008, **255**, 220.
17. J. Zhang, A. V. Biradar, S. Pramanik, T. J. Emge, T. Asefa and J. Li, *Chem. Commun.*, 2012, **48**, 6541–6543.
18. C. Kato, M. Hasegawa, T. Sato, A. Yoshizawa, T. Inoue and W. Mori, *J. Catal.*, 2005, **230**, 226.
19. C. Kato and W. Mori, *C. R. Chimie*, 2007, **10**, 284.
20. A. Dhakshinamoorthy, M. Alvaro and H. Garcia, *J. Catal.*, 2009, **267**, 1.
21. R. Zou, H. Sakurai, S. Han, R. Zhong and Q. Xu, *J. Am. Chem. Soc.*, 2007, **129**, 8402.
22. S. Su, Y. Zhang, M. Zhu, X. Song, S. Wang, S. Zhao, S. Song, X. Yanga and H. Zhang, *Chem. Commun.*, 2012, **48**, 11118–11120.

23. K. Brown, S. Zolezzi, P. Aguirre, D. Venegas-Yazigi, V. Paredes-Garcia, R. Baggio, M. Novak and E. Spodine, *Dalton Trans.*, 2009, 1422.
24. D. Jiang, T. Mallat, D. M. Meier, A. Urakawa and A. Baiker, *J. Catal.*, 2010, **270**, 26.
25. Y. Fu, D. Sun, M. Qin, R. Huang and Z. Li, *RSC Adv.*, 2012, **2**, 3309–3314.
26. S. Marx, W. Kleist and A. Baiker, *J. Catal.*, 2011, **281**, 76–87.
27. A. Dhakshinamoorthy, M. Alvaro and H. Garcia, *Chem. Commun.*, 2010, **46**, 6476–6478.
28. J. Yoon, Y.-K Seo, Y. Hwang, J.-S. Chang, H. Leclerc, S. Wuttke, P. Bazin, A. Vimont, M. Daturi, E. Bloch, P. L. Llewellyn, C. Serre, P. Horcajada, J.-M. Grenèche, A. E. Rodrigues and G. Férey, *Angew. Chem., Int. Ed.*, 2010, **49**, 5949–5952.
29. A. Dhakshinamoorthy, M. Alvaro, Y. K. Hwang, Y.-K. Seo, A. Corma and H. Garcia, *Dalton Trans.*, 2011, **40**, 10719–10724.
30. A. Dhakshinamoorthy, M. Alvaro, P. Horcajada, E. Gibson, M. Vishnuvarthan, A. Vimont, J.-M. Grenèche, C. Serre, M. Daturi and H. Garcia, *ACS Catal.*, 2012, **2**, 2060–2065.
31. N. V. Maksimchuk, K. A. Kovalenko, V. P. Fedin and O. A. Kholdeeva, *Chem. Commun.*, 2012, **48**, 6812–6814.
32. K. Barthelet, J. Marrot, D. Riou and G. Férey, *Angew. Chem., Int. Ed.*, 2002, **41**, 281–284.
33. K. Leus, I. Muylaert, M. Vandichel, G. B. Marin, M. Waroquier, V. Van Speybroeck and P. Van Der Voort, *Chem. Commun.*, 2010, **46**, 5085–5087.
34. K. Leus, M. Vandichel, Y.-Y. Liu, I. Muylaert, J. Musschoot, S. Pyl, H. Vrielinck, F. Callens, G. B. Marin, C. Detavernier, P. V. Wiper, Y. Z. Khimyak, M. Waroquier, V. Van Speybroeck and P. Van Der Voort, *J. Catal.*, 2012, **285**, 196–207.
35. Y.-Y. Liu, K. Leus, M. Grzywa, D. Weinberger, K. Strubbe, H. Vrielinck, R. Van Deun, D. Volkmer, V. Van Speybroeck and P. Van Der Voort, *Eur. J. Inorg. Chem.*, 2012, **16**, 2819–2827.
36. M. L. Kuznetsov and J. C. Pessoa, *Dalton Trans.*, 2009, 5460–5468.
37. M. V. Kirillova, M. L. Kuznetsov, P. M. Reis, J. A. L. da Silva, J. J. R. Fraústo da Silva and A. J. L. Pombeiro, *J. Am. Chem. Soc.*, 2007, **129**, 10531–10545.
38. M. S. Rigutto and H. van Bekkum, *J. Mol. Catal.*, 1993, **81**, 77–98.
39. J. Sudhakar Reddy, P. Liu and A. Sayari, *Appl. Catal., A*, 1996, **148**, 7–21.
40. A. Phan, A. U. Czaja, F. Gándara, C. B. Knobler and O. M. Yaghi, *Inorg. Chem.*, 2011, **50**, 7388–7390.
41. Y. Pan, D. Ma, H. Liu, H. Wu, D. He and Y. Li, *J. Mater. Chem.*, 2012, **22**, 10834–10839.
42. H. Zhao, H. Song and L. Chou, *Inorg. Chem. Commun.*, 2012, **15**, 261–265.
43. H. Liu, Y. Li, R. Luque and H. Jiang, *Adv. Synth. Catal.*, 2011, **353**, 3107–3113.
44. A. Henschel, K. Gedrich, R. Kraehnert and S. Kaskel, *Chem. Commun.*, 2008, 4192–4194.

45. B. Gomez-Lor, E. Gutiérrez-Puebla, M. Iglesias, M. A. Monge, C. Ruiz-Valero and N. Snejko, *Inorg. Chem.*, 2002, **41**, 2429–2432.
46. M. R. Pitts, J. R. Harrison and C. J. Moody, *J. Chem. Soc., Perkin Trans.*, 2001, **1**, 955–977.
47. A. Dhakshinamoorthy, M. Alvaro and H. Garcia, *Adv. Synth. Catal.*, 2009, **351**, 2271–2276.
48. T. Loiseau, C. Serre, C. Huguenard, G. Fink, F. Taulelle, M. Henry, T. Bataille and G. Férey, *Chem.–Eur. J.*, 2004, **10**, 1373–1382.
49. J. A. R . Navarro, E. Barea, J. M. Salas, N. Masciocchi, S. Galli, A. Sironi, C. O. Ania and J. B. Parra, *Inorg. Chem.*, 2006, **45**, 2397.
50. F. X. Llabrés i Xamena, A. Abad, A. Corma and H. Garcia, *J. Catal.*, 2007, **250**, 294.
51. S. Opelt, V. Krug, J. Sonntag, M. Hunger and E. Klemm, *Microporous Mesoporous Mater.*, 2012, **147**, 327–333.
52. S. Schuster, E. Klemm and M. Bauer, *Chem. Eur. J.*, 2012, **18**, 15831–15837.
53. A. E. Platero Prats, V. A. de la Peña-O'Shea, M. Iglesias, N. Snejko, Á. Monge and E. Gutiérrez-Puebla, *ChemCatChem*, 2010, **2**, 147–149.
54. P. Horcajada, S. Surble, C. Serre, D.-Y. Hong, Y.-K. Seo, J.-S. Chang, J.-M. Greneche, I. Margiolaki and G. Ferey, *Chem. Comm.*, 2007, 2820–2822.
55. L. Kurfiřtová, Y.-K. Seo, Y. K. Hwang and J.-S. Chang, J. Čejka, *Catal. Today*, 2012, **179**, 85–90.
56. U. Ravon, M. E. Domine, C. Gaudillere, A. Desmartin-Chomel and D. Farrusseng, *New J. Chem.*, 2008, **32**, 937–940.
57. U. Ravon, M. Savonnet, S. Aguado, M. E. Domine, E. Janneau and D. Farrusseng, *Microporous Mesoporous Mater.*, 2010, **129**, 319–329.
58. C. Serre, F. Millange, C. Thouvenot, M. Noguès, G. Marsolier, D. Louër and G. Férey, *J. Am. Chem. Soc.*, 2002, **124**, 13519–13526.
59. U. Ravon, G. Chaplais, C. Chizallet, B. Seyyedi, F. Bonino, S. Bordiga, N. Bats and D. Farrusseng, *ChemCatChem*, 2010, **2**, 1235–1238.
60. M. Fujita, Y. J. Kwon, S. Washizu and K. J. Ogura, *J. Am. Chem. Soc.*, 1994, **116**, 1151.
61. K. Schlichte, T. Kratzke and S. Kaskel, *Microporous Mesoporous Mater.*, 2004, **73**, 81.
62. S. Horike, M. Dinca, K. Tamaki and J. R. Long, *J. Am. Chem. Soc.*, 2008, **130**, 5854.
63. P. Phuengphai, S. Youngme, P. Gamez and J. Reedijk, *Dalton Trans.*, 2010, **39**, 7936–7942.
64. T. Ladrak, S. Smulders, O. Roubeau, S. J. Teat, P. Gamez and J. Reedijk, *Eur. J. Inorg. Chem.*, 2010, **24**, 3804–3812.
65. A. E. Platero-Prats, M. Iglesias, N. Snejko, A. N. Monge and E. Gutiérrez-Puebla, *Cryst. Growth Des.*, 2011, **11**, 1750–1758.
66. D. Sarma and S. Natarajan, *Indian J. Chem., Sect. A: Inorg., Bio-inorg., Phys., Theor. Anal. Chem.*, 2011, **50**, 1281–1289.

67. D. Sarma, K. V. Ramanujachary, N. Stock and S. Natarajan, *Cryst. Growth. Des.*, 2011, **11**(4), 1357–1369.
68. R. K. Das, A. Aijaz, M. K. Sharma, P. Lama and P. K. Bharadwai, *Chem.–Eur. J.*, 2012, **18**, 6866–6872.
69. A. Mallick, E.-M. Schön, T. Panda, K. Sreenivas, D. Díaz Díaz and R. Banerjee, *J. Mater. Chem.*, 2012, **22**, 14951–14963.
70. A. Henschel, K. Gedrich, R. Kraehnert and S. Kaskel, *Chem. Commun.*, 2008, 4192.
71. G. Ferey, C. Mellot-Draznieks, C. Serre, F. Millange, J. Dutour, S. Surble and I. Margiolaki, *Science*, 2005, **309**, 2040.
72. L. H. Wee, M. R. Lohe, N. Janssens, S. Kaskel and J. A. Martens, *J. Mater. Chem.*, 2012, **22**, 13742–13746.
73. L. Alaerts, E. Seguin, H. Poelman, F. Thibault-Starzyk, P. A. Jacobs and D. E. De Vos, *Chem.Eur. J.*, 2006, **12**, 7353.
74. E. Perez-Mayoral and J. Ceijka, *ChemCatChem*, 2011, **3**, 157–159.
75. A. Corma, M. Iglesias, F. X. Llabrés i Xamena and F. Sanchez, *Chem. Eur. J.*, 2010, **16**, 9789–9795.
76. J.-Y. Ye and C.-J. Liu, *Chem.Commun.*, 2011, **47**, 2167–2169.
77. M. Feyand, E. Mugnaioli, F. Vermoortele, B. Bueken, J. M. Dieterich, T. Reimer, U. Kolb, D. De Vos and N. Stock, *Angew. Chem., Int. Ed.*, 2012, **51**, 10373–10376.
78. C. M. Miralda, E. E. Macias, M. Zhu, P. Ratnasamy and M. A. Carreon, *ACS Catal.*, 2011, **2**, 180–183.
79. U. P. N. Tran, K. K. A. Le and N. T. S. Phan, *ACS Catal.*, 2011, **1**, 120–127.
80. C. Chizallet, S. Lazare and D. Bazer-Bachi, F. Bonnier, V. Lecocq, E. Soyer, A. A. Quoineaud and N. Bats, *J. Am. Chem. Soc.*, 2010, **132**, 12365–12377.
81. L. T. L. Nguyen, K. K. A. Le, H. X. Truong and N. T. S. Phan, *Catal. Sci. Technol.*, 2012, **2**, 521–528.
82. J. H. Cavka, S. Jakobsen, U. Olsbye, N. Guillou, C. Lamberti, S. Bordiga and K. P. Lillerud, *J. Am. Chem. Soc.*, 2008, **130**, 13850–13851.
83. F. Vermoortele and M. Vandichel, B. Van de Voorde, R. Ameloot, M. Waroquier, V. Van Speybroeck and D. E. De Vos, *Angew. Chem., Int. Ed.*, 2012, **51**, 4887–4890.
84. F. Vermoortele, R. Ameloot, A. Vimont, C. Serre and D. De Vos, *Chem. Commun.*, 2011, **47**, 1521–1523.
85. L. Valenzano, B. Civalleri, S. Chavan, S. Bordiga, M. H. Nilsen, S. Jakobsen, K. P. Lillerud and C. Lamberti, *Chem. Mater.*, 2011, **23**, 1700–1718.

CHAPTER 9

Catalysis at the Organic Ligands

JOSEPH E. MONDLOCH,*[a] OMAR K. FARHA*[a] AND
JOSEPH T. HUPP*[a,b]

[a] Department of Chemistry, Northwestern University, 2145 Sheridan Road,
Evanston, IL, 60208, U.S.A.; [b] Argonne National Laboratory, Argonne, IL,
60439, U.S.A.,
*Email: joseph.mondloch@northwestern.edu; o-farha@northwestern.edu;
j-hupp@northwestern.edu

9.1 Introduction

Metal organic frameworks (MOFs) are an intriguing class of porous, crystalline
materials composed of inorganic nodes (metal ions or clusters) and organic
linkers (structural ligands). Many MOFs also contain non-structural ligands.
By combining these components, one can imagine accessing an almost limitless
number of rationally designed and therefore functional materials—including
materials functional for catalysis. Indeed, heterogeneous catalysis was among
the first proposed applications of MOF chemistry,[1] as well as one of the first
experimentally demonstrated.[2] Two decades later the concept of catalysis
by MOFs has been both rigorously and broadly demonstrated.[3–6] As illustrated
in Scheme 9.1, examples encompass: (i) catalysis occurring at metal nodes
(see Chapter 8 herein); (ii) catalysis driven by encapsulated species (see
Chapter 10 herein); (iii) catalysis involving encapsulants and frameworks
working in tandem;[7,8] (iv) catalysis accomplished solely by metal-free organic
linkers; (v) catalysis facilitated by non-structural metal ions that are coordinated and thereby made available by organic linkers (e.g., salens, BINOLs,

RSC Catalysis Series No. 12
Metal Organic Frameworks as Heterogeneous Catalysts
Edited by Francesc X. Llabrés i Xamena and Jorge Gascon
© The Royal Society of Chemistry 2013
Published by the Royal Society of Chemistry, www.rsc.org

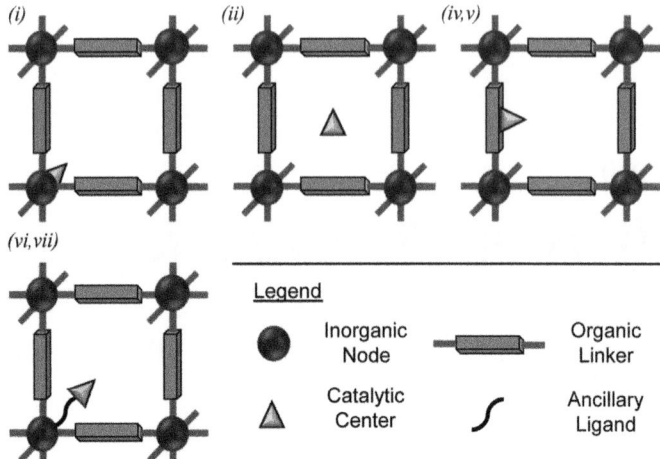

Scheme 9.1   Illustrative examples of many of the strategies utilized to create MOFs with catalytically active moieties.

porphyrins); (vi) catalysis based on ancillary organic ligands and (vii) catalysis based on non-structural metal ions that are coordinated by ancillary organic ligands (*e.g.*, tethered catecholates). Considered here are catalysts that make use of approaches (iv), (v), (vi) and (vii)—approaches where catalysis occurs either at a structural or ancillary organic ligand (*i.e.*, at the "*organic ligands*").

Catalysis at the organic ligands often takes the form of "heterogenized" homogeneous catalysts.[9] So-called single-site catalysts[9] have garnered great attention within the broader catalysis community given their potential to combine many useful attributes from both homogenous (*e.g.*, selectivity and rational design) and heterogeneous (*e.g.*, ease of separation and recyclability) catalysis. Given their atomically defined and readily modified structures, MOFs have the potential to be a superior, next generation of single-site catalysts.[10] Some of the more interesting examples at the organic ligands include: (i) MOFs that exhibit increased activities and lifetimes *vs.* their homogeneous counterparts; (ii) MOFs that are enantioselective catalysts; (iii) MOFs exhibiting enzyme-like enantio-selective behavior; (iv) MOFs containing multiple distinct catalytic sites at the organic linkers; (v) MOFs that can be utilized for sequential or concurrent tandem catalysis (CTC); (vi) MOFs that exhibit multifunctional behavior (*e.g.*, sorption followed by catalysis) and (vii) MOFs that are involved in pathways relevant to the production of solar fuels (*e.g.*, H_2O oxidation). Given the proliferation of studies regarding catalysis at the organic ligands (Table 9.1), throughout this book chapter we have been selective in highlighting examples that we are particularly familiar with or that are illustrative of unusual catalysis and/or are conceptually interesting. However, first we highlight a few general considerations that the reader should be cognizant of while doing catalysis with MOFs.

Table 9.1 A catalogue of known MOFs containing catalysts at the organic ligands and the reactions they catalyzed.

Entry	MOF Material	Reaction(s) Catalyzed	Ref.
De Novo Incorporation of Metal-Based Catalytic Sites			
1	[(Mn(L1)Mn$_{1.5}$)(DMF)]	Oxidation of alkenes and alkanes	11
2	[Zn$_2$(bpdc)$_2$L2]	Asymmetric epoxidation of olefins	12
3	[Zn$_2$(Zn–L3)(tcpb)$_{0.5}$]	Intermolecular acyl transfer	13
4	[Zn$_4$(μ_4–O)(L4)$_3$ or L5 or L6]	Asymmetric epoxidation of olefins	14
5	[Zn$_4$(μ_4–O)((Ru$^{III\ or\ II}$(L7)(py)$_2$Cl)$_3$]	Asymmetric cyclopropanation of olefins	15
6	[Zn$_2$(Zn–L8)(Mn–L3)]	Asymmetric epoxidation of olefins	16
7	[Zn$_2$(tcpb)$_{0.5}$L2]	Asymmetric epoxidation of olefins	17
8	[Zn$_4$(μ_4–O)(L11)$_3$]	Asymmetric epoxidation of olefins and epoxide ring opening	18
9	[Zr$_6$(O)$_4$(OH)$_4$(bdc)$_{6-x}$(L24a,b,c)$_x$]	Water oxidation	19
10	[Zr$_3$(OH)$_8$(M–L8)] (M = Mn, Co, Ni, Cu & Zn)	Alcohol oxidation	20
11	[Co$_2$(μ–H$_2$O)(H$_2$O)$_4$][Co–dcdbp]	Alkene epoxidation	21
12	[Mn$_5$Cl$_2$(MnCl–ocpp–(DMF)$_4$(H$_2$O)$_4$], [Mn$_5$Cl$_2$(Ni–ocpp)(H$_2$O)$_8$], [Cd$_5$Cl$_2$(MnCl–ocpp)–(H$_2$O)$_6$]	Alkylbenzene oxidations	22
13	[Cd(Ni–L7)(DMF)$_4$] and [Cd(Co–L7)(DMF)$_4$–(OAc)$_4$]	Hydrolytic kinetic resolution of epoxides	23
14	[Ru(L7a)(py)$_2$]Cl	Asymmetric cylopropanation of olefins	24
Post-Synthesis Modification to Include Metal-Based Catalytic Sites			
15	Ti(O^{i}Pr)$_2$[Cd$_3$Cl$_6$L10]	Asymmetric alkylation of aldehydes	25
16	Ti(O^{i}Pr)$_2$[Cd$_3$L10(NO$_3$)$_6$]	Asymmetric alkylation of aldehydes	26
17	[(Zn$_4$O(bdc–NH$_2$)$_{2.6}$(V–sal)$_{0.4}$	Alkene oxidation	27
18	Zn$_4$O(bdc–NH$_2$)$_{1-x}$(btb)$_{4/3}$(Fe$_{0.5x}$–AMsal)$_x$]	Carbon–carbon coupling	28
19	Zn$_4$O(sita–AuCl$_2$)$_x$(bdc–NH$_2$)$_{3-x}$]	Hydrogenation of dienes and domino coupling	29
20	Zn$_4$O(bdc–NH$_2$)$_{1-x}$(btb)$_{4/3}$(M–AMsal)$_x$] (M = In, Fe) and [Zn$_4$O(bdc–NH$_2$)$_{1-x}$(btb)$_{4/3}$(M–AMpz)$_x$] (M = In, Cu)	Epoxide ring opening	30
21	[Zn$_4$O(bdc)$_{3-x}$(Pd–abdc)$_x$]$_n$	Carbon monoxide oxidation	31
22	Ti(O^{i}Pr)$_2$[Zn$_2$(L12c)–(DMF)(H$_2$O)]	Asymmetric alkylation of aldehydes	32
23	Ti(O^{i}Pr)$_2$[(L12a–d)Cu$_2$(DEF)$_2$]	Asymmetric alkylation of aldehydes	33
24	[Zn$_4$O(bdc–NH$_2$)$_{2.71}$(Mn(acac)$_2$)$_{0.21}$]	Epoxidation of alkenes	34
25	[Cu$_2$(L13)$_{0.24}$(Pd–L13)$_{0.76}$(MeOH)$_2$]	Carbon–carbon coupling	35

Table 9.1 (Continued)

Entry	MOF Material	Reaction(s) Catalyzed	Ref.
26	Ti(O^{i}Pr)$_2$/Zn^{II}[Cu$_2$((S)-L14)](H$_2$O$_2$)	Asymmetric hetero-Diels–Alder	36
27	[Zn$_2$(tcpb)$_{0.5}$(Mn^{II}-L15)]	Asymmetric epoxidation of olefins	37
28	[Zn$_2$(Zn-L8)(Zn$_{1-x}$M$_x$-L3)] (M = 2H$^+$, Al^{III} Sn^{IV})	Epoxide ring opening	38
De Novo Incorporation of Organocatalytic Sites			
29	[Cu$_2$(pzdc)$_2$(pyrazine)]	Anionic polymerization of acetylenes	39
30	[Cd(4-btapa)$_2$(NO$_3$)$_2$]	Knoevenagel condensation	40
31	[Cu$_2$[L16]Cl$_2$]	Biginelli reaction and asymmetric 1,2-additions to ketones	41
32	[Zn$_4$O(bdc–NH$_2$)$_3$] and [Al(OH)(bdc–NH$_2$)]	Knoevenagel condensation	42
33	[Zn$_2$(tpt)$_2$(bdc–NH$_2$)I$_2$]	Knoevenagel condensation	43
34	[Zn(2-methylimidazole)$_2$]	Knoevenagel condensation	44
35	[Cr$_3$(H$_2$O)$_3$O(bdc–SO$_3$H)$_2$(bdc–SO$_3$)]	Cellulose hydrolysis	45
36	[Al$_3$O(DMF)(bdc–NH$_2$)$_3$]	Knoevenagel condensation	46
37	[Cu$_2$(pdai)(H$_2$O)]	^{a}Tandem deacetilization–Knoevenagel condensation	47
38	[Zn$_2$(bipy)$_2$L22]	Friedel–Crafts reaction	48
39	[(R-L25)Cu$_2$(H$_2$O)$_2$] and [((R)-L26)Cu$_2$(H$_2$O)$_2$]	Asymmetric Friedel–Crafts reaction	49
Post-Synthesis Modification to Include Organocatalytic Sites			
40	[Zn$_3$(μ-O)(L17–H)$_6$(H$^+$)$_2$]	Transesterfication	50
41	[Cu(L- or D-asp)bpe$_{0.5}$(HCl)]	Asymmetric methanolysis of epoxides	51
42	[Al(OH)(bdc–NH$_2$)$_{0.57}$(AMmal)$_{0.43}$]	Methanolysis of epoxides	52
43	[Zn$_4$O(bpdc–NH–pro)$_3$]	Asymmetric aldol reaction	53
44	[Cr$_3$(F,OH)–(H$_2$O)$_2$O(bdc)$_{3-x}$(bdc–SO$_4$H)$_x$] and [Al(OH)(bdc)$_{1-x}$(bdc–SO$_4$H)$_x$]	Esterification	54
45	[CoCl$_6$(L27)$_2$]	Addition of alchols to (E)-hex-4-ene-3-one	55
46	[Zn$_2$(bipy)(dpyi–L23)]	Asymmetric aldol reactions	56
Post-Synthesis Incorporation of Ancillary Catalytic Sites			
47	[ED– or DTA– or APS–Cr$_3$(F,OH)–(H$_2$O)$_2$O(bdc)$_3$]	Knoevenagel condensation	57
48	[Cr$_3$(F,OH)–(L18 or L19)$_{1.8}$O(bdc)$_3$]	Asymmetric aldol reactions	58
49	[Cd$_3$(bib)$_2$L20]	Asymmetric aldol reactions	59
50	[Cr$_3$(F,OH)O(bdc)$_3$(VO-dop)$_{0.13}$]	Oxidation of thioanisole	60

^{a}Tandem catalysis occurred as a result of coupling the node and linker.

9.2 General Considerations

Two broad strategies exist for incorporating catalysts at the organic ligands in MOFs (Scheme 9.2). In the *de novo* approach all the components of the active catalyst are incorporated during the initial MOF synthesis. In some instances the linker contains the catalytically active site prior to the MOF synthesis, while in other instances the catalytically active site is incorporated *in situ* (*e.g.*, spontaneous metallation in porphyrin-based MOFs, *vide infra*). In contrast, post-synthesis modification refers to the insertion of catalytically active moieties after the MOF structure is formed. Each strategy has advantages and disadvantages, some of which will become apparent throughout this book chapter.

Synthesis also governs catenation—catenation results when two or more identical networks become interwoven or interpenetrated and appears more frequently as the organic linker is lengthened. Catenation can affect key catalytic properties (such as activity, selectivity and lifetime) in MOFs, *vide infra*. Some known strategies for suppressing catenation include controlling the reactant concentration,[61] temperature,[62] templating organic molecules,[14] and through ligand-design.[63]

Permanent micro- and/or mesoporosity along with internal surface areas capable of accommodating substrates are prerequisites for gas-phase heterogeneous catalysis. It is therefore necessary to remove guest solvent molecules (incorporated during synthesis) from the pores of the MOF. This process is often termed "activation". While simple heating under vacuum can result in activated MOFs, often it does not. Methods such as low-boiling point solvent exchange (followed by activation),[61] freeze-drying[64] and supercritical carbon dioxide activation[65] have proven useful in obtaining more thoroughly activated MOFs. Supercritical carbon dioxide activation has also helped in the general handling of large MOF crystals that often fragment prior to, or during,

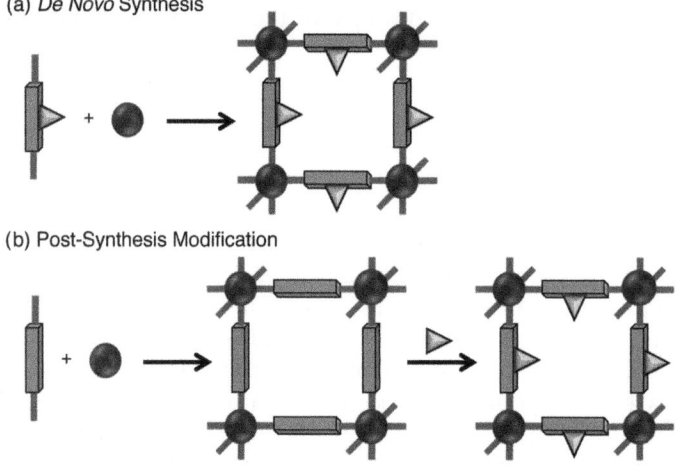

Scheme 9.2 An illustration of the (a) *de novo* and (b) post-synthesis modification strategies for the preparation of MOFs with catalysts at the organic ligands.

catalysis.[17] In contrast, permanent micro- or mesoporosity may *not* be essential for condensed phase catalysis. Solution-based porosity can be implied from X-ray diffraction (XRD) measurements, measured *via* thermal gravimetric analysis (TGA) of solution-filled MOFs[12] or by dye-uptake assay.[33]

Finally, several alternative hypotheses should be considered to ensure that the observed catalysis is consistent with the proposed heterogeneous MOF entity. In condensed phase catalytic reactions, the organic linkers (or closely related species) may dissociate from the MOF and homogeneous catalysis may ensue. Filtration/separation of the putative homogeneous complex from the MOF, followed by subsequent monitoring of the filtrate for catalytic activity is a useful control experiment that can rule out the putative homogenous phase. Catalysis could also be occurring at the surface of the MOF—coordinatively unsaturated metal nodes and terminal ligands can act as Lewis acids and bases respectively, while defect sites can introduce ill-defined and non-desirable catalytic sites. Competitive size selective substrate experiments can constitute "good circumstantial evidence for heterogeneous reactivity".[3] In addition, size selective poisoning,[11] deliberate selective demetallation of the crystal surface,[17] and grinding the MOF crystals can change the ratio of surface to interior catalytic sites and be of further use in delineating surface *vs.* interior catalysis. While not always rigorously demonstrated within the literature, appropriate control reactions ensure that the observed catalysis is occurring at the MOF.

9.3   Catalysis at the Organic Ligands

Table 9.1 contains fifty-one papers regarding catalysis at the organic ligands. We have organized the illustrative examples that follow by the type of linker utilized. Figure 9.1 shows the various organic ligands that are discussed throughout this book chapter while Figure 9.2 illustrates how those linkers are incorporated (either as an active catalytic component or structurally) to form MOFs. The abbreviations used throughout the rest of this chapter are catalogued at the end of the chapter.

9.3.1   Salen-based Linkers

Salen ligands have been rigorously utilized in coordination chemistry and homogeneous catalysis.[66] One classic example is the Jacobsen–Katsuki catalyst which is useful for transforming prochiral alkenes into chiral epoxides[67] and has found utility in natural product synthesis.[68] There has therefore been significant interest in incorporating metallosalens into MOFs and subsequently studying their catalytic properties.[12,14,15,17,18,23,24,37] In 2006, Cho *et al.* used a *de novo* approach to directly incorporate a Jacobsen–Katsuki-like Mn-salen ligand (L2), along with 4,4′-diphenylcarboxlic acid (bpdc), into a robust Zn-based pillared-paddlewheel MOF ([Zn$_2$(bpdc)$_2$L2]).[12] [Zn$_2$(bpdc)$_2$L2] is two-fold catenated and contains medium sized pores ($a \times b = 15.7$ Å$\times 6.2$ Å) that are substrate accessible.

Cho *et al.* utilized [Zn$_2$(bpdc)$_2$L2] for the asymmetric epoxidation of 2,2-dimethyl-2H-chromene with 2-(tert-butylsulfonyl)iodosylbenzene as the

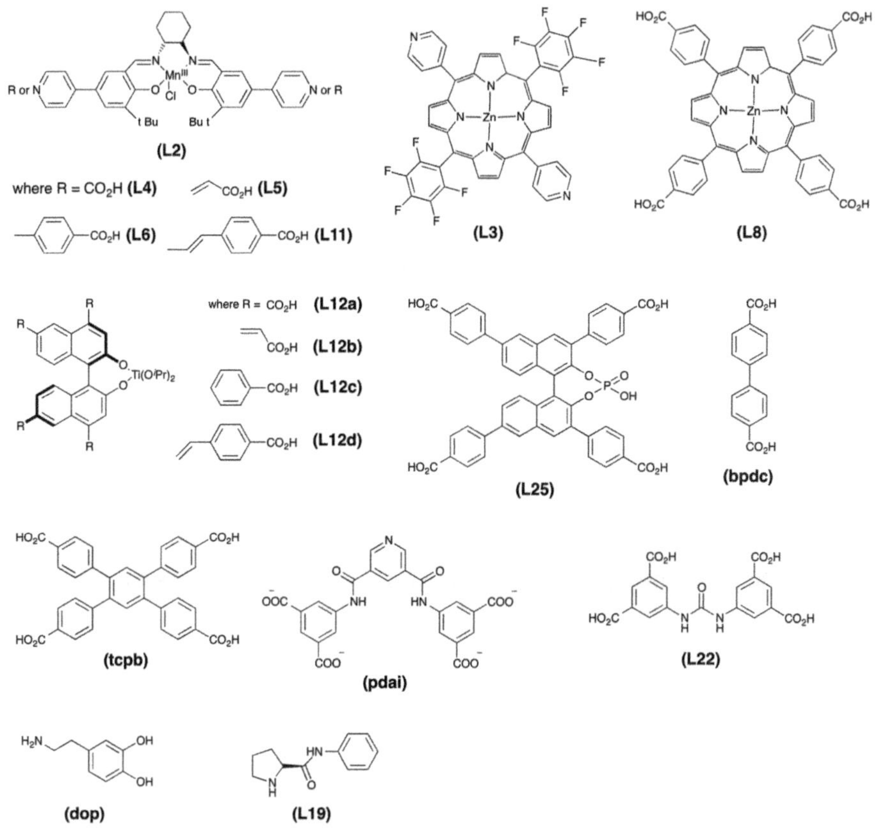

Figure 9.1 Structures of various organic ligands discussed within this book chapter. Their full names are given in at the end of the chapter.

oxidant. The epoxide was obtained in 71% yield and 82% *ee*. [Zn$_2$(bpdc)$_2$L2] could be separated from the reaction mixture, redispersed, and reused up to three times with minimal loss in both activity and selectivity—a result which demonstrated the heterogeneous nature of the catalyst. The total turnover numbers, for both [Zn$_2$(bpdc)$_2$L2] and its homogeneous analog L2, are shown in Figure 9.3. Examination of the data reveals that [Zn$_2$(bpdc)$_2$L2] significantly outperforms L2 at longer reaction times (*i.e.*, > ~0.1 h). Bimolecular catalyst deactivation pathways—which are common in homogeneous systems[12]—are eliminated upon framework immobilization, highlighting an important advantage of MOF-based catalysts over their homogeneous counterparts. [Zn$_2$(bpdc)$_2$L2] also showed similar *ee vs.* the homogeneous catalyst indicating some flexibility in the framework.

In a subsequent study, Shultz *et al.* utilized the tetratopic ligand, tetrakis(4-caroxyphenyl)-benzene (tcpb), in combination with L2 and Zn(NO$_3$)$_2$ · 6H$_2$O to synthesize the MOF [Zn$_2$(tcpb)$_{0.5}$L2].[17] The importance of the larger (*vs.* bpdc), sterically demanding, tcpb ligand is three-fold: (i) it

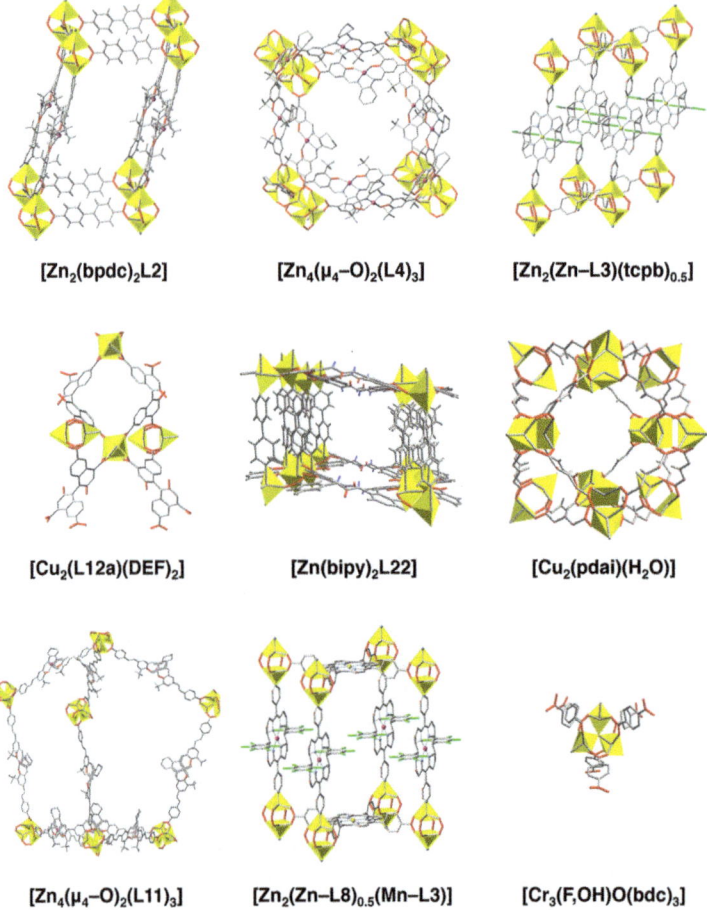

[Zn₂(bpdc)₂L2]   **[Zn₄(µ₄-O)₂(L4)₃]**   **[Zn₂(Zn-L3)(tcpb)₀.₅]**

[Cu₂(L12a)(DEF)₂]   **[Zn(bipy)₂L22]**   **[Cu₂(pdai)(H₂O)]**

[Zn₄(µ₄-O)₂(L11)₃]   **[Zn₂(Zn-L8)₀.₅(Mn-L3)]**   **[Cr₃(F,OH)O(bdc)₃]**

Figure 9.2 Structures of many of the MOFs discussed within this chapter. Solvent molecules and interpenetrated networks have been omitted for clarity.

suppresses catenation, (ii) it leads to larger ($a \times b = 22.4\,\text{Å} \times 11.7\,\text{Å}$), more open substrate accessible channels and (iii) it increases the crystallite size to $\sim 1\,\text{mm} \times 0.3\,\text{mm}$ plates (*vs.* $\sim 0.1\,\text{mm} \times 0.1\,\text{mm}$ plates for [Zn₂(bpdc)₂L2]). The non-catenated [Zn₂(*tcpb*)₀.₅L2] structure yields $\sim 75\%$ more turnovers for the epoxidation of 2-dimethyl-2H-chromene with 2-(tert-butylsulfonyl)iodosylbenzene as the oxidant *vs.* the catenated [Zn₂(*bpdc*)₂L2] structure. The general result that catenation suppresses turnover was also observed by Song *et al.* in a similar salen-based system.[14] The increase in turnover is readily explained by the fact that all the catalytically active Mn-salen sites are available for catalysis in the non-catenated [Zn₂(*tcpb*)₀.₅L2] structure (while only half were available in the [Zn₂(*bpdc*)₂L2] structure). This example demonstrates one of the most attractive features of MOFs for catalysis—their structures, and therefore resultant catalytic properties, are readily tunable by synthetic modification!

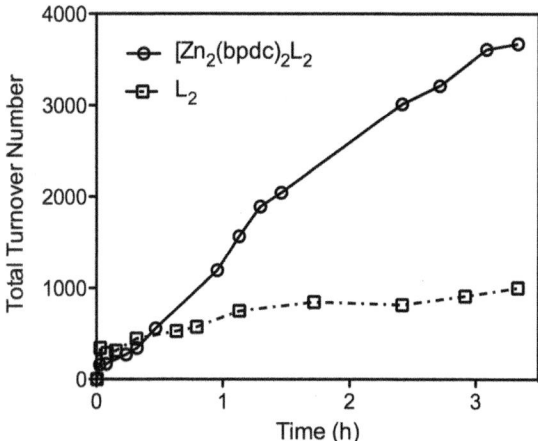

Figure 9.3 Total turnover number *vs.* time profiles for the asymmetric epoxidation of 2,2-dimethyl-2H-chromene with [Zn2(bpdc)2L2] and L2 as catalysts. Adapted from ref. 12 with permission from The Royal Society of Chemistry.

Song. recently demonstrated that a salen-derived (L11) MOF is capable of *sequentially* catalyzing a series of chemical reactions.[18] In particular, they directly incorporated a Mn-based metallosalen ligand (L11) into the MOF [Zn4(μ_4-O)2(L11)3] which was competent for the epoxidation of alkenes and the subsequent ring opening reaction, albeit with a solvent change. The epoxidation reaction occurred at the organic linker (*i.e.*, @ Mn^{III} salen sites), while the ring-opening reaction occurred predominantly at Zn^{II} defect sites within the Zn4(μ-O)4 secondary building units (Figure 9.4). The enantioselectivity from the epoxidation reaction was conserved for the ring opening reaction. The study by Song *et al.* demonstrates that immobilization of multiple catalytic units within a single MOF structure may lead to multifunctional materials that are attractive for sequential or even CTC strategies (*vide infra*).

Given that many metallosalen complexes are known homogenous catalysts (*e.g.*, Co and Cr),[69] a general synthetic route to a homologous series of metallosalen-based MOFs is highly desirable. In principle, the appropriate metallosalen complexes could be directly incorporated into known MOF structures such as [Zn2(tcpb)L2]. Unfortunately, such syntheses are not always straightforward. Schultz *et al.* demonstrated a post-synthesis modification strategy for the *de-* and *re-*metallation of [Zn2(tcpb)L2].[37] Demetallation was accomplished by soaking the [Zn2(tcpb)L2] crystals in MeOH and H_2O_2 overnight. Subsequent remetallation occurred after soaking the demetallated crystals in solutions containing M^{II} ions (M = Cr, Co, Mn, Ni, Cu and Zn) for 24 h, as depicted in Figure 9.5. Remetallation could be observed visually and was confirmed *via* ICP-OES. Most importantly, this post-synthesis metallation strategy allows incorporation of metals, such as Co, that have proven difficult to incorporate into salen-based MOF structures (due to self-association),[37] and should open the door for new opportunities in catalysis.

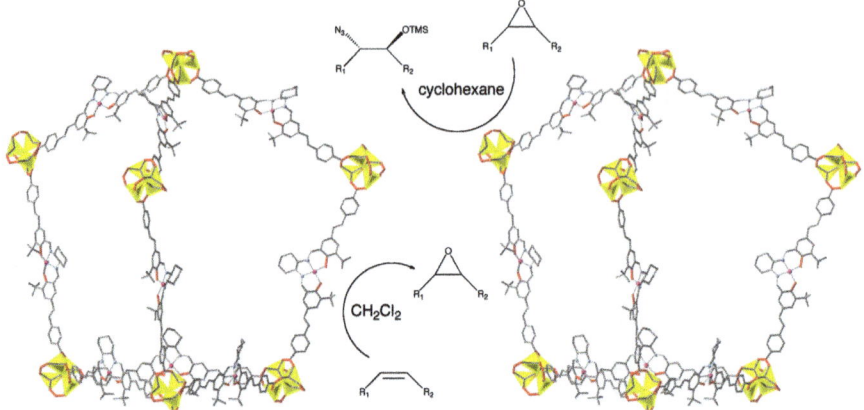

Figure 9.4 An illustration of sequential asymmetric catalysis as described by Song
et al.
Adapted from ref. 18 with permission from The Royal Society of
Chemistry.

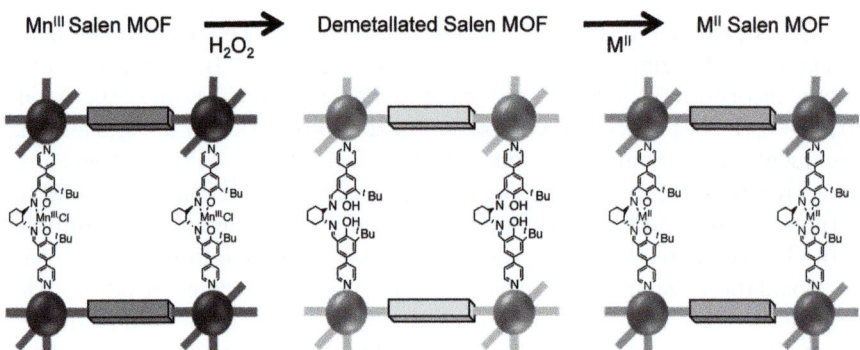

Figure 9.5 An illustration of the synthetic strategy employed by Shultz *et al.* for the
demetallation, and subsequent remetallation, of [Zn₂(tcpb)₀.₅L2].

9.3.2 Metalloporphyrin-based Linkers

Metalloporhyrins are well-known molecular catalysts[70] that can function as
mimics of metallo-enzymes.[71] Hence, MOFs incorporating catalytically active
metalloporphyrins have received considerable interest.[11,13,16,21–22,38] Suslick
et al. were able to synthesize a Mn^{III} tetra(*p*-carboxyphenyl) porphyrin
containing MOF, [(Mn(L1)Mn₁.₅)(DMF)].[11] [(Mn(L1)Mn₁.₅)(DMF)] is an
active catalyst for the hydroxylation of linear and cyclic alkanes and alkenes.
While it was ultimately concluded that catalysis occurs on the exterior of the
[(Mn(L1)Mn₁.₅)(DMF)] crystals, Suslick *et al.*'s studies demonstrate several
important strategies for delineating interior *vs.* exterior MOF-based catalysis.
Two lines of evidence were consistent with catalysis occurring on the exterior of
the MOF crystals: (i) no shape selectivity was observed for the hydroxylation

reactions, despite the fact that [(Mn(L1)Mn$_{1.5}$)(DMF)] preferentially adsorbs "small, slender, hydrophilic guests"[11] and (ii) large, bulky bases capable of binding to surface Mn-porphyrin sites but too large to fit within the pores halted catalysis (*e.g.*, 3,5-di-bromo-pyridine). Surprisingly, selective poisoning studies are nearly non-existent in MOF-based catalysis despite their utility.

With the goal of synthesizing catalytically active and substrate accessible metalloporphyrin sites within MOFs, Schultz *et al.* combined Zn(NO$_3$)$_2$· 6H$_2$O, 5,15-dipyridyl-10,20-bis(pentafluorophenyl)-porphyrin (L3) and 1,2,4,5-tetrakis(4-carboxyphenyl)benzene (tcbp).[13] The free-base porphyrin was spontaneously metallated with Zn^{II} ions during the synthesis to give the MOF, [Zn$_2$(Zn–L3)(tcpb)$_{0.5}$]. Large substrate accessible pores were confirmed *via* ^{1}H NMR of solvent/substrate soaked crystals. [Zn$_2$(Zn–L3)(tcpb)$_{0.5}$] was ~2400-fold more active for the acyl transfer between 3-pyridylcarbinol (3-PC) and *N*-acetylimidazole (NAI) *vs.* the uncatalyzed background reaction.

Three hypotheses were considered to explain the observed rate enhancement: (i) stabilization of the transition state *via* Lewis acid activation (a charge stabilization effect, Figure 9.6a), (ii) stabilization of the transition state *via* substrate alignment (Figure 9.6b), and (iii) preconcentration of the substrates within the MOF cavities (a higher local concentration leading to a higher rate), Figure 9.6c. Preferential substrate alignment was ruled out by examining substrate dependencies on the acyl transfer. 2- and 4-PC—substrates known to have different reactivity in similar soluble supramolecular systems[72]—showed no significant change in the observed rate. Lewis acid activation was ruled out by comparison to a homogeneous Zn-porphyrin analog; only a two-fold rate enhancement *vs.* the uncatalyzed reaction was observed. In short, the results are consistent with and supportive of *the rate enhancement being largely due to a preconcentration effect*. Of relevance here, similar rate enhancements were recently observed in two separate and different MOF-based catalytic systems.[21,23] In those systems, the rate enhancements were attributed to substrate alignment ("cooperativity") between two proximal catalytic sites.

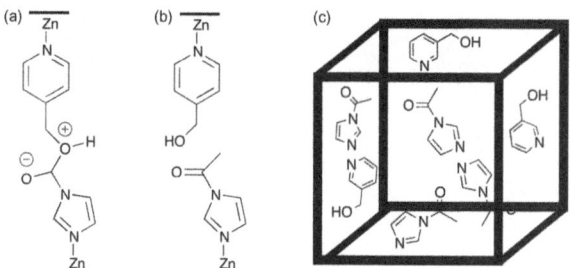

Figure 9.6 An illustration of three hypotheses that could explain the ~2400-fold increase in activity for the acyl transfer between 3-PC and NAI by [Zn$_2$(Zn–L3)(tcpb)$_{0.5}$]: (a) stabilization of the transition state *via* Lewis acid activation; (b) substrate alignment and (c) preconcentration within the MOF pores.

MOFs containing multiple accessible metal sites are "attractive candidates for applications that require multi-functional performance."[38] Two recent reports have demonstrated a series of strategies for incorporating multiple metalloporphyrin ligands into a single catalytic MOF structure.[16,38] Farha *et al.* demonstrated a *de novo* approach by replacing tcpb in [Zn$_2$(Zn–L3)(tcpb)$_{0.5}$], with a second meso-tetrakis(4-caroxyphenyl)porphyrinato (L8) ligand.[16] The resulting MOF, [Zn$_2$(Zn–L8)(Mn–L3)], contains two distinct porphyrin linkers and therefore can contain pairs of distinct metals (*e.g.*, Zn^{II}, Pd^{II}, Fe^{III}, Mn^{III} and Al^{III} were incorporated in various combinations). Subsequently, Takaishi *et al.* utilized a solvent-assisted linker exchange (SALE)[73–76] strategy, Figure 9.7, to exchange the Zn-dipyridal pillars (Zn–L3) for dipyridal pillars containing 2H$^+$ (*i.e.*, free base), Al^{III}, or Sn^{IV}. In general it is difficult to incorporate free-base porphyrins into MOFs given their tendency to spontaneously metallate during the MOF synthesis—for example, free-base cannot be incorporated *de novo* into [Zn$_2$(Zn–L8)(Mn–L3)]. During SALE, however, no free metal ions are present in solution and spontaneous metallation does not occur. The free-base SALE porphyrin, [Zn$_2$(Zn–L8)(2H$^+$–L3)], could also be remetallated with Co^{II}. Both the *de novo* material ([Zn$_2$(Zn–L8)(Mn–L3)]), and the SALE material ([Zn$_2$(Zn–L8)(Al^{III}–L3)]) were demonstrated to be active catalysts. Demonstrating that multiple metals/ porphyrin combinations can be incorporated within a single MOF is exciting and may open new avenues *en route* to multi-functional materials—including materials that could be utilized in sequential and possibly CTC. In addition, SALE may be amenable for creating core–shell type MOF architectures that prove to be multi-functional and contain additional unique properties that are not accessible through traditional synthetic strategies.

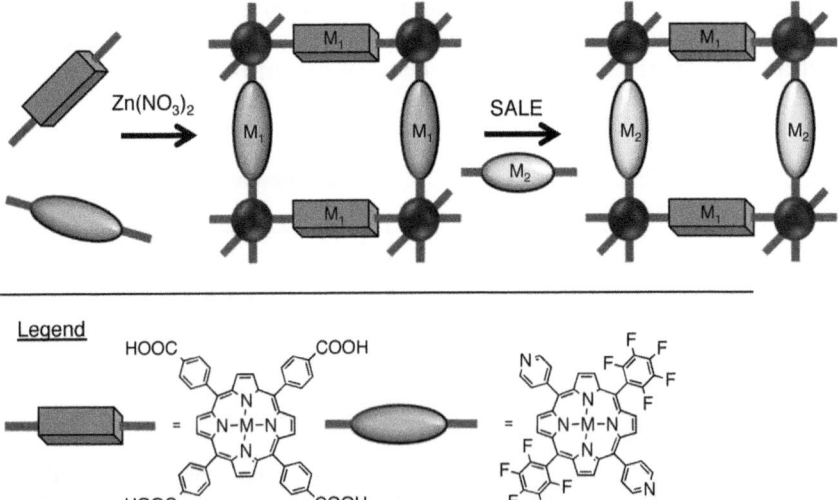

Figure 9.7   An illustration of the SALE strategy utilized by Takaishi *et al.* to incorporate multiple metallated porphyrins within a single MOF.

9.3.3 Binapthyl-based Linkers

Binapthyl ligands have served as extensive platforms in homogeneous enantioselective catalysis.[77] They are readily modifiable and accommodate a variety of catalytically active centers. A series of isoreticular MOFs with binapthyl ligands of varying size (L12a–d) were linked to carboxylate containing Cu paddle-wheel secondary building units by Ma et al.[33] This series was post-synthetically functionalized with Ti(O^{i}Pr)$_4$, and the resultant MOFs (Ti(O^{i}Pr)$_2$[(L12a–d)Cu$_2$(DEF)$_2$]) have channel sizes varying from 1.3×1.1 nm^2 to 3.2×2.4 nm^2. Ti(O^{i}Pr)$_2$[(L12a–d)Cu$_2$(DEF)$_2$] was active (and in some instances selective) for the addition of ZnEt$_2$ to a variety of aromatic aldehydes. Ma et al. were able to demonstrate that the *ee* decreases with the percentage of (calculated) void space within the MOF pores, suggesting that large open channels are necessary for efficient enantioselective reactions to proceed. This was experimentally verified by correlating the MOF void space, *ee* and the amount of solvent present by dye uptake (or lost via TGA). Other studies have shown that *ee* is also dependent on the MOF crystallite size,[12,17] hence multiple factors need to be considered when designing enantioselective MOF catalysts.

Ma et al. also prepared a very similar MOF by linking the dicarboxylate moieties of L12c to Zn$_2$ nodes.[32] Large 1.5×2.0 nm^2 channels result, despite the fact that the frameworks are two-fold catenated. Treatment of [(L12c)Zn$_2$(DMF)(H$_2$O)] with Ti(O^{i}Pr)$_4$ yielded an active catalyst for the addition of ZnEt$_2$ to aromatic aldehydes. However, the observed *ee*'s were <30% over the four substrates examined. These results are in stark contrast to the higher ee typically observed for similar Ti(O^{i}Pr)$_4$ treated MOFs.[33] Examination of the Ti(O^{i}Pr)$_4$ treated [(L12c)Zn$_2$(DMF)(H$_2$O)] single crystal structure revealed that an *inter*molecular Ti(binapthyl)$_2$ complex was formed rather than the desired *intra*molecular product (Figure 9.8). This is a result of the catenated MOF structure where two binapthyls are in close enough proximity to allow the *inter*moleclar cross-linking. The results suggest that care

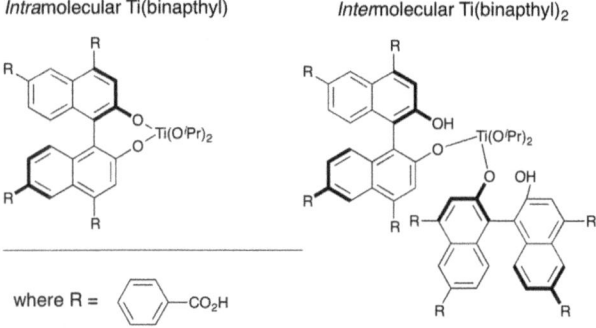

Figure 9.8 The *intra*molecular and *inter*molecular Ti(binaphthyl)$_2$ complexes observed by Ma et al. The *inter*molecular linking between two binpathyl linkers leads to a drastic decrease in the observed *ee* when compared to the *intra*molecular analog.

must be taken during post-synthesis treatments of catenated MOF structures. We point out that insights such as these are difficult to obtain in almost all other areas of heterogeneous catalysis (zeolites being the well-known exception) given that atomic resolution single crystal X-ray diffraction studies are simply not available.

9.3.4 Organocatalysis from Linkers

Broadly defined "organocatalysis" encompasses catalytic reactions occurring in the presence of organic molecules that do not contain metal atom(s).[78] One relevant example is the so-called hydrogen-bond-donating (HBD) catalysts— "HBD catalysis is a biomimetic inspired alternative to Lewis acid activation."[48] Roberts *et al.* recently rationalized that HBD catalysts may benefit from immobilization within MOFs given their propensity to deactivate through self-associated dimerization and oligomerization pathways.[48] They connected the thiourea HBD ligand L22, along with bipy pillaring struts, to Zn paddlewheel SBUs to yield [$Zn_2(bipy)_2L22$]. [$Zn_2(bipy)_2L22$] was more active than a similar homogeneous HBD catalyst for the Friedel–Crafts reaction between *N*-methylpyrrole and (*E*)-1-nitroprop-1-ene. The solvent played a crucial role in the observed catalysis. No reaction was observed in the nonpolar solvent toluene, while 98% conversion was observed in the more polar mixture of $MeNO_2$/THF, a mixture that likely facilitates H-bonding exchange.[48] Clearly immobilization of organocatalysts within MOFs is an attractive strategy *en route* to improved heterogenized catalysts.

Much effort has been put forth in designing MOFs for asymmetric catalysis.[4,6] In most instances chirality is intrinsic, that is it is derived from the catalytically active site.[49] Zheng *et al.* demonstrated, however, that chirality can be induced by the MOF pore rather than the active site.[49] They used a *de novo* approach to incorporate the chiral P-based BINOL ligand (L25), along with Cu paddlewheel SBUs, to synthesize the MOF [((*R*)-L25)$Cu_2(H_2O)_2$]. Friedel–Craft reactions between indole and a series of (*E*)-*N*-sulfonimides in the presence of the homogeneous ligand (L25) resulted exclusively in the formation of *S* enantiomers while the *R* enantiomers were obtained in the presence of [((*R*)-L25)$Cu_2(H_2O)_2$]. Circular dichroism confirmed the products were mirror images of each other and theoretical calculations suggest that the switch in chirality results from substrate interactions with the MOF walls. The result is striking, and resembles the enantioselectivity often observed in enzymes.[49] Clearly the idea that MOFs can be designed analogues of enzymes[3] is starting to be realized.

Concurrent tandem catalysis (CTC) "involves the cooperative action of two or more catalytic cycles in a single reactor."[79] While attractive, execution of CTC can be difficult given that the substrates, intermediates, products and multiple catalysts all must be compatible under the relevant reaction conditions—furthermore, the kinetics of each catalytic step must be optimized for efficient catalysis to proceed. MOFs offer a platform to immobilize multiple catalysts and eliminate catalyst–catalyst deactivation pathways. Park *et al.*

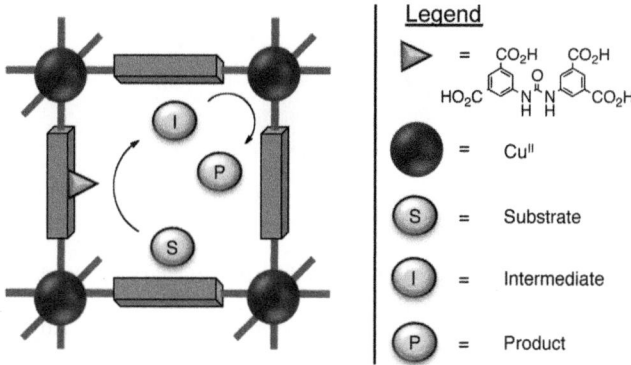

Figure 9.9 An illustration of CTC between an organic linker and a metal node within a MOF, as demonstrated by Park *et al.*

recently used a *de novo* approach to link the organocatalyst pdai to Cu paddlewheel SBUs.[47] The resultant MOF, [Cu$_2$(pdai)(H$_2$O)], was active for deacetilization/Knoevenagel CTC. Appropriate control reactions demonstrated that the coordinatively unsaturated Cu^{II} nodes were responsible for deacetilization, while pdai was responsible for the Knoevenagel condensation. Park *et al.*'s study offers the first demonstration of CTC within a MOF. MOFs containing multiple catalytic units at structural linkers may also be highly desirable for CTC (Figure 9.9).

9.3.5 Catalysis by Ancillary Ligands

Banerjee *et al.* post-synthetically modified the coordinatively unsaturated Cr metal centers of MIL-101(Cr) by treating them with the ancillary ligands L18 and L19.[58] This treatment turns the achiral MIL-101(Cr) into a chiral MOF, [Cr$_3$(F)-(L18 or L19)$_{1.8}$(H$_2$O)$_{0.2}$O(bdc)$_3$]. The resultant MOF is active for a variety of asymmetric aldol reactions between aldehydes and ketones. Interestingly, the heterogeneous catalysts, [Cr$_3$(F)-(L18 or L19)$_{1.8}$(H$_2$O)$_{0.2}$O(bdc)$_3$], were more selective (*e.g.*, 69 *vs.* 29 % *ee*) than their soluble homogeneous counterparts under identical conditions, which may be a result of restricted substrate movement within the MOF cavities. A similar enhancement in *ee* was also observed in a different system by Dang *et al.*,[59] suggesting that the idea of "reactivity defining microenvironments"[3] is achievable in MOF-based catalysis.

Nguyen *et al.* recently employed a similar ancillary functionalization strategy to modify MIL-101 with dopamine, Figure 9.10.[60] The amine moiety of dopamine binds to the dehydrated and coordinatively unsaturated Cr^{III} nodes, while the catechol moiety remained open for further modification. The catechol moiety was subsequently metallated with VO(acac)$_2$, albeit on the surface and outermost cavities of the MIL-101 only. Regardless, the [Cr$_3$(F,OH)O(bdc)$_3$(VO-dop)$_{0.13}$] MOF was active for the oxidation of

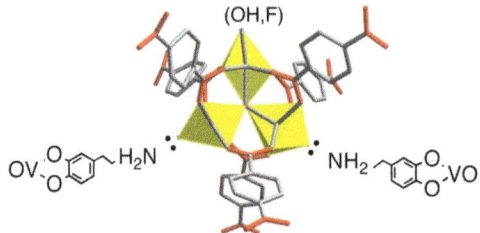

Figure 9.10   An illustration of –VO tethered to the ancillary ligand dopamine at the coordinatively unsaturated Cr^{III} sites in MIL-101.
                   Adapted from ref. 60 with permission from the Royal Society of Chemistry.

thioanisole. Such a strategy is amenable for tethering additional metal sites on coordinatively unsaturated metal nodes and may also be amenable for producing unique core–shell MOFs.

9.4 Summary

Catalysis "at the organic ligands" (of crystalline MOFs) is an emerging sub-field of heterogeneous catalysis. Examples include both traditional metal-based catalysts as well as organocatalysts that can be synthesized *de novo* or post-synthetically. MOFs offer a heterogeneous platform that allows separation from the catalytic reaction mixture, increased catalytic lifetimes and improvements in selectivity. It is clear that many of the attractive visions for MOF catalysts—including, "multi-catalyst architectures", "metal coordination environments that can be achieved in no other ways" and "reactivity-defining microenvironments"[3]—are starting to be realized. We believe therefore that MOFs are poised to become an improved, next generation of single-site heterogeneous catalysts.

Abbreviations List

abdc	2-aminobenzen-1,4-dicarboxylate
acac	acetlyacetonato
AMmal	4-((2,5-bis(methoxy-carbonyl)phenyl)-amino)-4-oxobut-2-enoic acid
AMpz	2,3-pyrazinedicarboxylic anhydride
AMsal	3-hydroxyphthalic anhydride
aps	3-aminopropyltrialkoxysilane
asp	aspartate
bdc	1,4-benzenedicarboxylate
bpdc	biphenyldicarboxylate
bipy	bipyridine
bpe	1,2-bis(4-pyridyl)ethylene
4-btapa	1,3,5-benzene tricarboxylic acid tris N-(4-pyridyl)amide
btb	benzene,-1,3,5-tribenzoate

dcdbp	5,15-bis(3,5-dicarboxyphenyl)-10,20-bis(2,6-dibromophenyl)porphyrin
deta	diethylenetriamine
dpyi	dimethyl-5-(prop-2-ynyloxy)isophthalic acid
DMF	dimethylformamide
dop	dopamine
ed	ethylenediamine
OAc	acetic acid
O^{i}Pr	isopropoxide
ocpp	5,10,15,20-tetrakis(3,5-biscarboxylphenyl)porphyrin
pdai	5,5'-((pyridine-3,5-dicarbonyl)bis(azanediyl))diisophosphthalate
pro	proline
py	pyridine
pzdc	pyrazine-2,3-dicarboxylate
sal	salicylidene moiety
sita	2-salicylideneimine terephthalate
tpt	tris(4-pyridyl)triazine
tcpb	1,2,4,5-tetrakis(4-carboxyphenyl)benzene
L1	tetra(p-carboxyphenyl)porphyrins
L2	(*R,R*)-(−)-1,2-cyclohexanediamino-N,N'-bis(3-*tert*-butyl-5-(4-pyridyl)salicylidene)Mn^{III}Cl
L3	(5,15-dipyridyl-10,20-(pentafluorophenyl)porphyrin
L4	(*R,R*)-(−)-N,N'-bis(3-carboxyl-5-*tert*-butylsalicylidene)-1,2-cyclo-hexanediamino-Mn^{III}Cl
L5	(2*E*,2'*E*)-3,3'-(5,5'-(1*E*,1'*E*)-(1*R*,2*R*)-cyclohexane-1,2-diylbis(azan-1-yl-1-ylidene)bis(meth-an-1-yl-1-ylidene)bis(3-*tert*-butyl-4-hydroxy-5,1-phenylene)diacrylic acid)Mn^{III}Cl
L6	5',5''-(1E,1'E)-(1R,2R)-cyclohexane-1,2-diylbis(azan-1-yl-1-ylidene)bis(methan-1-yl-1-ylidene)bis(3'-*tert*-butyl-4'hydroxybiphenyl-4-carboxylic acid)Mn^{III}Cl
L7	(*R,R*)-(−)-1,2-cyclohexane diamino-N,N'-bis(3-tert-butyl-salicylidene)Ru^{II}
L7a	(*R,R*)-(−)-N,N'-bis(3-acrylate-5-tert-butyl-salicylidene)-1,2-cycloheanediamine
L8	meso-tetrakis(4-carboxyphenyl) porphyrin tetracarboxylate (metal = Zn, Mn, Co, Ni, Cu)
L10	(*R*)-6,6'-dichloro-2,2'-dihydroxy-1,1'-binapthyl-4,4'-bipyridine
L11	4,4'-(1*E*,1'*E*)-2,2'-(5,5'-(1*E*,1'*E*)-(1*R*,2*R*)-cyclohexane-1,2-diylbis(azan-1-yl-1-yli-dene)-bis(methan-1-yl-1-ylidene)bis(3-tert-butyl-4-hydroxy-5,1-phenylene))bis(ethene-2,1-diyl) dibenzoic acid Mn^{III}Cl
L12a	(*R*)-2,2'-diethoxy-1,1'-binapthyl-4,4',6,6'-tetracarboxylic acid
L13	double azolium *N*-heterocyclic carbene
L14	(*S*)-2,2'-dihydroxy-6,6'-dimethyl-(1,1'-biphenyl)-4,4'-dicarboxylic acid

L15	(*R,R*)-(−)-1,2-cyclohexanediamino-*N,N'*-bis(3-*tert*-butyl-5-(4-pyridyl)salicylidene)
L16	(*S*)-3-hydroxy-2-(pyridin-4-ylmethylamino)propanoic acid
L17	D-tartaric acid
L18	(S)-*N*-(pyridin-3-yl)-pyrrolidine-2-carboxamide
L19	(S)-*N*-(pyridin-4-yl)-pyrrolidine-2-carboxamide
L20	L-*N-tert*-butoxy-carbonyl-2-(imidazole)-1-pyrrolidine
L22	5,5'-carbonylbis(azanediyl)diisophthalic acid
L23	L-or D-2-azidomethylpyrrolidine
L24a	chloro(η^5-pentamethylcyclopentadienyl)(2-(4-carboxyl)phenyl-(5-carboxyl)pyridine-C^2,N')iridium(III)
L24b	chloro(η^5-pentamethylcyclopentadienyl)(5,5'-dicarboxyl-2,2'-bipyridine)iridum(III) chloride
L24c	diaqua-bis(2-(4-carboxyl)phenyl-(5-carboxyl)pyridine-C^2,N')iridium(III) triflate
L25	(*R*)-3,3',6,6'-tetrakis(4-benzoic acid)-1,1'-binapthyl phosphate
L26	(*R*)-4,4',6,6'-tetrakis(4-benzoic acid)-1,1'-binapthyl phosphate
L27	Trimethylimidazole-2,4,6-triethyl benzene

Acknowledgements

JEM was supported in part by the Department of Energy (DOE) Office of Energy Efficiency and Renewable Energy (EERE) Postdoctoral Research Awards under the EERE Fuel Cell Technologies Program administered by the Oak Ridge Institute for Science and Education (ORISE) for the DOE. ORISE is managed by Oak Ridge Associated Universities (ORAU) under DOE contract number DE-AC05-06OR23100. All opinions expressed in this paper are the author's and do not necessarily reflect the policies and views of DOE, ORAU, or ORISE.

References

1. B. F. Hoskins and R. Robson, *J. Am. Chem. Soc.*, 1990, **112**, 1546–1554.
2. M. Fujita, Y. J. Kwon, S. Washizu and K. Ogura, *J. Am. Chem. Soc.*, 1994, **116**, 1151–1152.
3. J. Lee, O. Farha, J. Roberts, K. Scheidt, S. T. Nguyen and J. T. Hupp, *Chem. Soc. Rev.*, 2009, **38**, 1450.
4. L. Ma, C. Abney and W. Lin, *Chem. Soc. Rev.*, 2009, **38**, 1248.
5. A. Corma and H. Garcia, and F. X. Llabres i Xamena, *Chem. Rev.*, 2010, **110**, 4606–4655.
6. M. Yoon, R. Srirambalaji and K. Kim, *Chem. Rev.*, 2011, **112**, 1196–1231.
7. G. Lu, S. Li, Z. Guo, O. K. Farha, B. G. Hauser, X. Qi, Y. Wang, X. Wang, S. Han, X. Liu, J. S. DuChene, H. Zhang, Q. Zhang, X. Chen, J. Ma, S. C. J. Loo, W. D. Wei, Y. Yang, J. T. Hupp and F. Huo, *Nat. Chem.*, 2012, **4**, 310–316.

8. C. Wang, K. E. deKrafft and W. Lin, *J. Am. Chem. Soc.*, 2012, **134**, 7211–7214.
9. *Modern Surface Organometallic Chemistry*, ed. J.-M. Basset, R. Psaro, D. Roberto and R. Ugo, Wiley-VCH, Weinheim, 2009.
10. D. Gajan and C. Coperet, *New J. Chem.*, 2011, **35**, 2403–2408.
11. K. S. Suslick, P. Bhyrappa, J. H. Chou, M. E. Kosal, S. Nakagaki, D. W. Smithenry and S. R. Wilson, *Acc. Chem. Res.*, 2005, **38**, 283–291.
12. S. Cho, B. Ma, S. T. Nguyen, J. T. Hupp and T. Albrecht-Schmitt, *Chem. Commun.*, 2006, 2563.
13. A. Shultz, O. Farha, J. T. Hupp and S. T. Nguyen, *J. Am. Chem. Soc.*, 2009, **131**, 4204–4205.
14. F. Song, C. Wang, J. Falkowski, L. Ma and W. Lin, *J. Am. Chem. Soc.*, 2010, **132**, 15390–15398.
15. J. Falkowski, C. Wang, S. Liu and W. Lin, *Angew. Chem., Int. Ed.*, 2011, **50**, 8674–8678.
16. O. K. Farha, A. M. Shultz, A. A. Sarjeant, S. T. Nguyen and J. T. Hupp, *J. Am. Chem. Soc.*, 2011, **133**, 5652–5655.
17. A. Shultz, O. Farha, D. Adhikari, A. A. Sarjeant, J. T. Hupp and S. T. Nguyen, *Inorg. Chem.*, 2011, **50**, 3174–3176.
18. F. Song, C. Wang and W. Lin, *Chem. Commun.*, 2011, **47**, 8256–8258.
19. C. Wang and Z. Xie, K. E. deKrafft and W. Lin, *J. Am. Chem. Soc.*, 2011, **133**, 13445–13454.
20. D. Feng, Z.-Y. Gu, J.-R. Li, H.-L. Jiang, Z. Wei and H.-C. Zhou, *Angew. Chem. Int. Ed*, 2012, **51**, 10307–10310.
21. L. Meng, Q. Cheng, C. Kim, W.-L. Gao, L. Wojtas, Y.-S. Chen, M. J. Zaworotko, X. P. Zhang and S. Ma, *Angew. Chem., Int. Ed.*, 2012, **51**, 10082–10085.
22. X.-L. Yang, M.-H. Xie, C. Zou, Y. He, B. Chen, M. O'Keefe and C.-D. Wu, *J. Am. Chem. Soc.*, 2012, **134**, 10638–10645.
23. C. Zhu, G. Yuan, X. Chen, Z. Yang and Y. Cui, *J. Am. Chem. Soc.*, 2012, **134**, 8058–8061.
24. J. M. Falkowski, S. Liu, C. Wang and W. Lin, *Chem. Commun.*, 2012, **48**, 6508–6510.
25. C. D. Wu, A. Hu, L. Zhang and W. B. Lin, *J. Am. Chem. Soc.*, 2005, **127**, 8940–8941.
26. C. Wu and W. Lin, *Angew. Chem., Int. Ed.*, 2007, **46**, 1075–1078.
27. M. Ingleson, J. P. Barrio, J. Guilbaud, Y. Khimyak and M. Rosseinsky, *Chem. Commun.*, 2008, 2680–2682.
28. K. K. Tanabe and S. Cohen, *Angew. Chem. Int. Ed.*, 2009, **48**, 7424–7427.
29. X. Zhang, F. X. Llabrés i Xamena and A. Corma, *J. Catal.*, 2009, **265**, 155–160.
30. K. Tanabe and S. Cohen, *Inorg. Chem.*, 2010, **49**, 6766–6774.
31. W. Kleist, M. Maciejewski and A. Baiker, *Thermochim. Acta*, 2010, **499**, 71–78.
32. L. Ma, C. Wu, M. Wanderley and W. Lin, *Angew. Chem., Int. Ed.*, 2010, **49**, 8244–8248.

33. L. Ma, J. Falkowski, C. Abney and W. Lin, *Nat. Chem.*, 2010, **2**, 838–846.
34. S. Bhattacharjee, D. Yang and W. Ahn, *Chem. Commun.*, 2011, **47**, 3637–3639.
35. G. Kong, X. Xu, C. Zou and C. Wu, *Chem. Commun.*, 2011, **47**, 11005–11007.
36. K. Jeong, Y. Go, S. Shin, S. Lee, J. Kim, O. Yaghi and N. Jeong, *Chem. Sci.*, 2011, **2**, 877–882.
37. A. M. Shultz, A. A. Sarjeant, O. K. Farha, J. T. Hupp and S. T. Nguyen, *J. Am. Chem. Soc.*, 2011, **133**, 13252–13255.
38. S. Takaishi, E. J. DeMarco, M. J. Pellin, O. K. Farha and J. T. Hupp, *Chem. Sci.*, 2012, In Press.
39. T. Uemura, R. Kitaura, Y. Ohta, M. Nagaoka and S. Kitagawa, *Angew. Chem. Int. Ed.*, 2006, **45**, 4112–4116.
40. S. Hasegawa, S. Horike, R. Matsuda, S. Furukawa, K. Mochizuki, Y. Kinoshita and S. Kitagawa, *J. Am. Chem. Soc.*, 2007, **129**, 2607–2614.
41. M. Wang, M. Xie, C. Wu and Y. Wang, *Chem. Commun.*, 2009, 2396–2398.
42. J. Gascon, U. Aktay, M. D. Hernandez-Alonso, G. P. M. van Klink and F. Kapteijn, *J. Catal.*, 2009, **261**, 75–87.
43. Y. Tan, Z. Fu and J. Zhang, *INOCHE*, 2011, **14**, 1966–1970.
44. U. P. N. Tran, K. K. A. Le and N. T. S. Phan, *ACS Catal.*, 2011, **1**, 120–127.
45. G. Akiyama, R. Matsuda, H. Sato, M. Takata and S. Kitagawa, *Adv Mater*, 2011, **23**, 3294.
46. P. Serra-Crespo, E. V. Ramos-Fernandez, J. Gascon and F. Kapteijn, *Chem. Mater.*, 2011, **23**, 2565–2572.
47. J. Park, J.-R. Li, Y.-P. Chen, J. Yu, A. A. Yakovenko, Z. U. Wang, L.-B. Sun, P. B. Balbuena and H.-C. Zhou, *Chem. Commun.*, 2012, **48**, 9995–9997.
48. J. M. Roberts, B. M. Fini, A. A. Sarjeant, O. K. Farha, J. T. Hupp and K. A. Scheidt, *J. Am. Chem. Soc.*, 2012, **134**, 3334–3337.
49. M. Zheng, Y. Liu, C. Wang, S. Liu and W. Lin, *Chem. Sci.*, 2012, **3**, 2623–2627.
50. J. S. Seo, D. Whang, H. Lee, S. I. Jun, J. Oh, Y. J. Jeon and K. Kim, *Nature*, 2000, **404**, 982–986.
51. M. Ingleson, J. Barrio, J. Bacsa, C. Dickinson, H. Park and M. Rosseinsky, *Chem. Commun.*, 2008, 1287–1289.
52. S. Garibay, Z. Wang and S. Cohen, *Inorg. Chem.*, 2010, **49**, 8086–8091.
53. D. J. Lun, G. I. N. Waterhouse and S. G. Telfer, *J. Am. Chem. Soc.*, 2011, **133**, 5806–5809.
54. M. G. Goesten, J. Juan-Alcaniz, E. V. Ramos-Fernandez, K. Gupta, E. Stavitski, H. van Bekkum, J. Gascon and F. Kapteijn, *J. Catal.*, 2011, **281**, 177–187.
55. M. B. Lalonde, O. K. Farha, K. A. Scheidt and J. T. Hupp, *ACS Catal.*, 2012, **2**, 1550–1554.
56. W. Zhu, C. He, P. Wu, X. Wu and C. Duan, *Dalton Trans.*, 2012, **41**, 3072–3077.

57. Y. K. Hwang, D. Hong, J. Chang, S. H. Jhung, Y. Seo, J. Kim, A. Vimont, M. Daturi, C. Serre and G. Ferey, *Angew. Chem., Int. Ed.*, 2008, **47**, 4144–4148.

58. M. Banerjee, S. Das, M. Yoon, H. J. Choi, M. H. Hyun, S. M. Park, G. Seo and K. Kim, *J. Am. Chem. Soc.*, 2009, **131**, 7524.

59. D. Dang, P. Wu, C. He, Z. Xie and C. Duan, *J. Am. Chem. Soc.*, 2010, **132**, 14321–14323.

60. H. G. T. Nguyen, M. H. Weston, O. K. Farha, J. T. Hupp and S. T. Nguyen, *CrystEngComm*, 2012, **14**, 4115–4118.

61. M. Eddaoudi, J. Kim, N. Rosi, D. Vodak, J. Wachter, M. O'Keeffe and O. M. Yaghi, *Science*, 2002, **295**, 469–472.

62. J. Zhang, L. Wojtas, R. W. Larsen, M. Eddaoudi and M. J. Zaworotko, *J. Am. Chem. Soc.*, 2009, **131**, 17040.

63. O. K. Farha and J. T. Hupp, *Acc. Chem. Res.*, 2010, **43**, 1166–1175.

64. L. Ma, A. Jin, Z. Xie and W. Lin, *Angew. Chem., Int. Ed.*, 2009, **48**, 9905–9908.

65. A. P. Nelson, O. K. Farha, K. L. Mulfort and J. T. Hupp, *J. Am. Chem. Soc.*, 2009, **131**, 458.

66. C. Baleizao and H. Garcia, *Chem. Rev.*, 2006, **106**, 3987–4043.

67. W. Zhang, J. L. Loebach, S. R. Wilson and E. N. Jacobsen, *J. Am. Chem. Soc.*, 1990, **112**, 2801–2803.

68. L. Deng and E. N. Jacobsen, *J. Org. Chem.*, 1992, **57**, 4320–4323.

69. E. N. Jacobsen, *Acc. Chem. Res.*, 2000, **33**, 421–431.

70. B. Meunier, *Chem. Rev.*, 1992, **92**, 1411–1456.

71. S. J. Lippard and J. M. Berg, *Principles of Bioinorganic Chemistry*, University Science Books, Mill Valley, 1994.

72. C. G. Oliveri, N. C. Gianneschi, S. T. Nguyen, C. A. Mirkin, C. L. Stern, Z. Wawrzak and M. Pink, *J. Am. Chem. Soc.*, 2006, **128**, 16286–16296.

73. B. J. Brunett, P. M. Barron, H. Chunhua and W. Choe, *J. Am. Chem. Soc.*, 2011, **133**, 9984–9987.

74. C. Y. Lee, O. K. Farha, B. J. Hong, A. A. Sarjeant, S. T. Nguyen and J. T. Hupp, *J. Am. Chem. Soc.*, 2011, 15858–15861.

75. M. Kim, J. F. Cahill, Y. Su, K. A. Prather and S. M. Cohen, *Chem. Sci.*, 2012, **3**, 126–130.

76. O. Karagiaridi, W. Bury, A. A. Sarjeant, C. L. Stern, O. K. Farha and J. T. Hupp, *Chem. Sci.*, 2012, **3**, 3256–3260.

77. Y. Chen, S. Yekta and A. K. Yudin, *Chem. Rev.*, 2003, **103**, 3155–3211.

78. P. I. Dalko and L. Moisan, *Angew. Chem., Int. Ed*, 2004, **43**, 5138–5175.

79. J.-C. Wasilke, S. J. Obrey, R. T. Baker and G. C. Bazan, *Chem. Rev.*, 2005, **105**, 1001–1020.

CHAPTER 10

MOFs as Nano-reactors

JANA JUAN-ALCAÑIZ,*
ENRIQUE V. RAMOS-FERNANDEZ,
FREEK KAPTEIJN AND JORGE GASCON

Catalysis Engineering/ChemE, TUDelft, Julianalaan 136, 2628BL, Delft,
The Netherlands
*Email: janajuanalcaniz@gmail.com

10.1 What is a Nanoreactor?

A nanoreactor is a nanosized container used for accommodating chemical reactions. However, a nanoreactor is not simply a holding vessel. The limited reaction space inside a nanoreactor is a critical part of the chemical process, due to its strong effect on movement and interactions among the molecules inside. Examples of nanoreactors have been widely utilized in nature, in the form of organized biological microphases where complex biochemical reactions take place. Nanoreactors like the nucleus, mitochondria, Golgi apparatus, lysosomes, and the pores of channel proteins, deliver specific concentrations and arrangements of molecules, with profound consequences on the chemical processes carried out inside.

The kinetics and mechanisms of chemical reactions in small-scale confined geometries are different compared with the same reactions in bulk solutions. Several consequences on the chemical processes developed from the following inherent properties of such small confined spaces:[1]

1. Small volume: a discrete number of molecules creates large fluctuations in the number of reagents per nanoreactor leading to very different kinetics,

RSC Catalysis Series No. 12
Metal Organic Frameworks as Heterogeneous Catalysts
Edited by Francesc X. Llabrés i Xamena and Jorge Gascon
© The Royal Society of Chemistry 2013
Published by the Royal Society of Chemistry, www.rsc.org

and sometimes pathways, within a set of nanoreactors. The average behaviour of the ensemble is not the same as would be for bulk measurements. Small spaces increase diffusion and heat transfer with respect to bulk reaction, due to the short pathway that molecules need to follow to meet each other. In addition, the restricted space of a nano-reactor might induce segregation or phase separation of solvents and reactants inside. For instance, a given solvent might displace the absorbed reactants, leading to a different reaction mechanism.

2. Large surface-to-volume ratio: frequency and type of interactions between molecules inside the space may be influenced by the properties of the wall. Since surface effects cannot be neglected, this influence may result in alteration of mechanisms (intermediates) and reaction rates. The way reactants interact with the wall changes reactivity, guiding formation and evolution of the reaction transition state. However, if the nanoreactor space is restrictive, then the reactants may not be able to align themselves adequately to achieve specific transition state conformations, or to relax fully once it is formed, changing the product selectivity. If the concentration of reactants is high inside the nanoreactor, the reaction rate, in the absence of diffusion limitations, increases, since the existence of the wall shortens the mean free path.

The dominant effects will be determined by the dimension of the confined space, the number of molecules inside, the interaction between the reactants and the wall, and the presence of a catalytic functionality. Nanoreactors can be prepared with different sizes and shapes, leading not only to the fine tuning of the above mentioned properties, but also towards shaping the production in the confined space. The nanoreactor space can be used to form nanomaterials (like nanoparticles), localized chemical and/or physico-chemical environments, and host–guest species to conduct chemical reactions. In the case of immobilized catalysts, the confined space will modify the concentration of the reactant close to the active site and the catalyst–reactant interaction, influencing directly the mechanism and reaction rate.

As a relatively new concept in science and engineering, nanoreactors first emerged in the late 1990s, when several early reviews pointed towards their potential in chemical transformations. During the last few years, the interest in nanoreactors has increased.[2,3] Several reviews highlighted the synthesis and general characterization of specific categories of nanoreactors.[3–5] There are essentially three types of, widely accepted, potential nanoreactors:[1]

1. Biological macromolecules such as proteins and nucleic acids, are the smallest organic nanoreactor structures composed of one or a few large molecules forming hollow spaces into which, at least, one other molecule can fit. The caging nanoreactor or molecular basket may or may not participate in the transformation of the molecules entrapped, but its presence influence the outcome. It has a broad application in biology and biomedical transformation.

2. Self-assembly of molecules like micelles, vesicles or emulsions, are composed essentially of polymers and are particularly rich in terms of structural variety.[5,6] These types of nanoreactors are mainly used for templating the synthesis of other nanostructures as well as forming chemical reservoirs for drugs, chromophores, and other reagents.

3. Natural or synthetic nanoporous solids like zeolites, nanotubes, and lately, metal organic frameworks. In general, inorganic nanoreactor structures have been of interest for industrial applications at high-temperature or high-pressure, due to elevated chemical and mechanical resistance of the inorganic lattice. Their porous structures are considered to be an interconnected network of nanoreactors. This group is composed of (i) inorganic solids such as zeolites, mesoporous silicas and metal oxide frameworks; (ii) organic nanostructured solids such as covalent organic frameworks or microporous polymers; and (iii) hybrid solids such as coordination polymers or metal organic frameworks. These solids are mainly used in catalysis, synthesis of nanomaterials and encapsulation of actives species.[7]

This chapter is a literature review focussed on understanding the role of metal organic frameworks as nanoreactors. The easy tunability of these hybrid porous crystalline structures, offers a plethora of nanosize confinement possibilities. Encapsulation of active species in such spaces has been widely studied during the last few years. Nanoparticles, macromolecules, enzymes or polyoxometalates are some of the encapsulated species that will be further discussed.

10.1.1   Nanospace for Encapsulation

The concept of encapsulation is by no means young. It was originally conceived as a process where a continuous thin coating is formed around solid particles, liquid droplets, or gas bubbles, fully contained within the capsule wall. It rapidly became a topic of interest in a wide range of scientific and industrial areas varying from pharmaceutics[8,9] to agriculture[10] and from pesticides[11] to enzymes.[12] Encapsulation technology has been used in the food industry for more than 60 years as a way to provide liquid and solid ingredients with an effective barrier from environmental and/or chemical interactions until release is desired.[13,14] The concept of encapsulation was later expanded to inorganic supports like zeolites.[15] The first examples using zeolites as scaffolds for encapsulation, dates from the late 1970s and opened the door to the encapsulation of gaseous species on small pore zeolites and clathrates,[16–18] and to the encapsulation of more attractive moieties like large homogenous catalysts, molecular wires or semi-conductors.[19–25] It is worth clarifying that when dealing with encapsulated species in a porous material, the application scope differs from that of other approaches such as self-assembly encapsulation. In the latter, protection and eventual release of the encapsulated moieties is envisaged. In contrast, when active species are encapsulated in the pores of a support, the active moieties cannot leave the support but their main properties (catalytic, opto-electronic . . .) can be used and even enhanced by confinement effects in the resulting composite.

To satisfy the condition of encapsulation in a porous matrix, the entrapped species must be larger than the pores of the support material. As the active species are larger than the pores of the support, techniques such as impregnation cannot be used. Therefore, two different synthetic approaches can be followed (see Figure 10.1):

1. Assembling the active species within the pores of the support, known as "*ship in bottle*" approach.
2. Assembling the support around the active species, also known as "*bottle around ship*" or "*templated synthesis*" approach.

The route used for the encapsulation is dictated by the chemistry of the support and the system to be encapsulated. If the targeted species are to be assembled within the pores of the support, then the support needs to be stable under the reaction conditions used. If the support is assembled around the active species, then the active species needs to be stable to the synthesis conditions of that support. Hence, if the active species can be made easily in a small number of steps then assembly within the pores is preferred. If the active species are difficult to synthesize but are intrinsically stable then the assembly of the support around the active species is preferred.

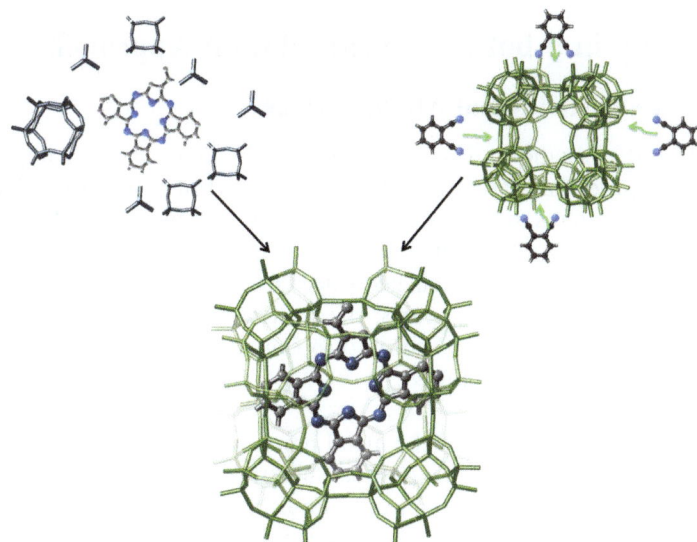

Figure 10.1 Different strategies for the encapsulation of an active moiety in a porous host: "*bottle around ship*" or "*templated synthesis*" approach (top left) and "*ship in bottle*" approach (top right). The composite after encapsulation by either method is depicted in the bottom.
Adapted from ref. 26.

In view of the different methodologies towards encapsulation, MOFs offer, in principle, several advantages over zeolites. Indeed, several reviews have been summarized the work done on encapsulation of guest species in MOFs scaffolds.[26,27] The almost unlimited topological richness of MOFs opens the door to the encapsulation of a much larger variety of moieties. Furthermore, MOFs are synthesized in most cases under milder conditions than zeolites and usually in the presence of organic solvents, allowing the direct encapsulation of fragile organic molecules. In addition, in the case of MOFs, no structure directing agent other than the synthesis solvent is needed and therefore post-synthetic calcination steps are not required.

The porosity of MOFs is, in general, much larger than that of their inorganic counterpart, zeolites, justifying the designation 'framework' and challenging the scientific community to make an effective use of such an empty space (confined space). Indeed, the presence of confined empty spaces inside MOFs offers unprecedented possibilities for "interior architecture" and for exploring the concept of nanoreactors. Following the architecture similarity, when decorating a house one can hold furniture on the walls (grafting of active, small, molecules via covalent bonding or electrostatic interactions) or one can build bigger furniture within the room. When functional sites are built inside a microporous material in such a way that they cannot leach due to geometrical constraints, the active sites are said to be encapsulated. This chapter deals with the "interior architecture" of metal organic frameworks and how chemical transformations can be performed in such small volume.

10.2   "Ship in a bottle" Encapsulation Approach

10.2.1   Nanoparticles and Metal Oxides

The use of MOFs as nano-molds for hosting functional inorganic nanoparticles has attracted quite some attention in recent years. For a detailed review on Metals@MOFs systems, we recommend the recent reviews by Fischer and co-workers and Garcia et al.[28,29] In this work we will mostly focus on the most recent and relevant advances published since 2010 in the field of catalysis.

As it is the case for other supports, the strategy for the assembly of nano-clusters always starts by the deposition of precursors in the MOF pore space. Several approaches can be followed:

- Impregnation: a solution of the metal precursor (usually in the form of a salt) is contacted with the porous substrate.[30] When the amount of impregnation solution is equal to the pore volume of the MOF the method is called "incipient wetness". One of the main drawbacks of this method is the necessity of a subsequent reduction of the metal precursor to form the corresponding nano-particles. In order to avoid high temperatures and possible MOF decomposition, chemicals like hydrazine or $NaBH_4$ can be used as reducing agents. Special attention should be put on the type of precursor, for instance nitrates or chlorines must be washed out (before or after reduction) in order to avoid future poisoning of the active phase.

– Solvent free gas phase loading: chemical vapour deposition precursors can be incorporated into the pores or channels of MOFs by a sublimation process.[31] Since the method is solvent free, it allows for high metal loadings. The decomposition of the precursor can be achieved by a thermal or chemical treatment, UV irradiation of the sample, or a reactive gas atmosphere. A variation of this method is the so-called solid grinding.[32] In this case, the metal precursors are firstly carefully ground together with the MOF and later infiltrated into the MOF porosity and reduced by a mild temperature treatment in a reducing atmosphere.

Depending upon the method and the conditions applied for the loading and decomposition of precursors, nano-particle encapsulation might only be partially successful. Three possible case-scenarios have been defined when classifying metal nano-particle distribution in MOFs (see Figure 10.2):[28]

– In class A most of the nano-particles are preferentially deposited at the outer surface of a MOF crystal, resulting in a wide particle size distribution and a poor metal dispersion, especially in the case of high loadings.
– In class B most of the nano-particles are deposited in the porosity of the MOF but they display a rather broad particle size distribution that might arise from partial destruction of the MOF skeleton.
– In class C, nano-particles with a homogeneous particle size distribution close to that of the MOF pores or cavities are evenly dispersed throughout the porous host.

Both systems A and B cannot be considered as nanoreactors, since the active species are not located in the confined spaces (A) or the confined spaces have been partially damaged (B), leading to a broad particle size distribution.

An important point addressed by Fischer and co-workers[28] and clearly exemplified by Allendorf *et al.*[33] deals with artefacts during characterization.

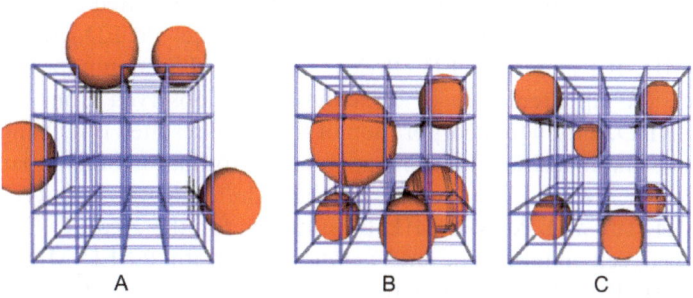

Figure 10.2 Limiting cases A, B and C of metal nanoparticles supported by MOFs. Reprinted with permission from ref. 28. Copyright (2010) WILEY-VCH Verlag GmbH & Co. KGaA, Weinheim.

MOFs are particularly sensitive to electron beam imaging, quite in contrast to zeolites and other purely inorganic porous matrices. Collapse of the whole framework and/or chemical degradation together with coalescence of nano-particles may occur under the high energy electron doses in TEM analysis. All information on the internal *versus* external location of guests on MOFs from electron microscopy must therefore be interpreted cautiously. Additional characterization techniques like UV-Vis, infra-red, solid state NMR and X-Ray absorption spectroscopies together with PXRD and, if applicable, chemisorption methods need to be used in order to fully characterize NP@MOF composites.

In Table 10.1 the publications reported since the seminal work of Hermes *et al.*[31] on the loading of metal and metal oxide nanoparticles on MOFs with catalytic applications are summarized. Early work focused on the use of MOF-5,[34] and eventually of HKUST-1,[35] as scaffolds, while more recent work has been extended to MIL[36,37] and ZIF[38] materials. The excellent chemical and thermal stability of these latter structures has facilitated the fair evaluation of the new composites properties. Other applications of the encapsulated NP/MOF systems apart from catalysis (and therefore not be further discussed in this chapter) deal mostly with adsorption[39] and storage,[40–42] but more recently also with optical applications.[43–50]

When it comes to catalytic systems, a critical evaluation of most presented results demonstrates that the encapsulated metal nanoparticles and metal oxides do not display an outstanding activity, while the long-term stability of the composites has hardly been addressed. The relatively low activity should not be surprising since, in most of the cases, catalytic reactions have been explored where the use of a MOF as support does not represent any advantage (*i.e.* nanoparticle size is not crucial) but might even add diffusion limitations. In contrast, the picture might change for catalytic applications where control of nanoparticle size and morphology is needed, for instance in fine chemistry applications. Cross-coupling reactions as demonstrated by Hwang *et al.*,[51] Yuan *et al.*,[52] and by Huang *et al.*[53] are clear examples of catalytic systems where careful nanoparticle size control, dispersion and framework polarity might result in catalytic systems with high activity and specificity.

Hwang *et al.*[51] prepared Pd nanoparticles (2–4 nm) inside MIL-101(Cr)[37] with amine groups anchored to the coordinatively unsaturated chromium sites of the dehydrated material. The authors prepared Pd/ED-MIL-101 and Pd/APS-MIL-101 materials (ED ethylenediamine, APS 3-aminopropyl trialk-oxysilane). No apparent loss of crystallinity was observed from the corresponding XRD, although incorporation of the metal resulted in a change of the diffraction peak intensities. TEM images revealed formation of nanoparticles in the range of 2–4 nm. The authors studied the catalytic activity of both Pd/ED-MIL-101 and Pd/APS-MIL-101 samples for the Heck C–C coupling reaction between iodobenzene and acrylic acid. Both Pd/ED-MIL-101 and Pd/APS-MIL-101 showed catalytic activities comparable to those of a commercial Pd/C catalyst under similar conditions.

Pd nanoparticles have been supported in MIL-101. Yuan *et al.* originally reported the preparation of 1 wt% well dispersed, small (1.9 nm) nanoparticles by impregnation of Pd(NO$_3$)$_2$ in DMF,[52] and later Huang *et al.*[53] used the same precursor and MOF to achieve 2.6 ± 0.5 nm particle size. The first system was tested in different Suzuki–Miyaura coupling reactions of aryl chlorides, demonstrating the applicability and high activity of the catalyst over at least 5 catalytic cycles using a broad scope of reactants and using different bases as co-catalysts. The Pd/MOF composite outperformed a commercial Pd/C catalyst for the most difficult chlorinated reactants (*i.e.* chloroanisole). The second report focuses on describing the high catalytic activity of the system for the direct C2 arylation of indoles, where solvent, base, optimal amount of catalyst and a wide scope of reactants was carried out. The system was tested to be a heterogeneous catalyst by hot filtration experiments, and it was possible to recycle it for at least five cycles.

More recently, Huang *et al.*[53] have prepared well dispersed Pd nanoparticles in the amino functionalized MIL-53(Al)[55,56] framework using a direct anionic exchange and subsequent chemical reduction with NaBH$_4$ in one of the few examples where functional groups at the organic linker are used to anchor metal precursors and nanoparticles. Li *et al.*,[54] following a similar synthesis method, obtained an excellent Pd dispersion with loadings as high as 3 wt% in MIL-101 (Cr), see Figure 10.3. However, it is important to mention that the

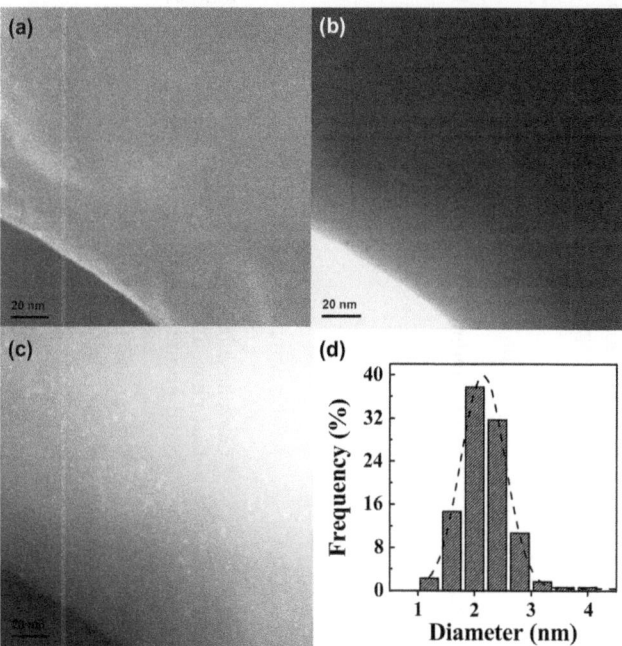

Figure 10.3 STEM images of 3%-Pd/MIL-101 catalyst: (a) SE, (b), STEM-BF, and (c) STEM-DF modes. (d) The corresponding particle size distribution histogram.
Reprinted with permission from ref. 54. Copyright (2011) American Chemical Society.

Table 10.1 Overview of reported "inorganic NP@MOF" systems used as catalysts, adapted and updated from Meilikhov et al.[28]

MOF	M precursor	Synthesis method	Reduction method	M wt%	NP size/nm	Application	Ref.
Ag@MOF							
MOF-5	AgNO$_3$	Impregnation	—	15.3	—	Propylene epoxidation	59
Ni-MOF	AgNO$_3$	Ion-exchange	Urotropine	—	1	Three component coupling	60
Ag/Au@MOFs							
ZIF-8	AgNO$_3$/HAuCl$_4$	Sequential deposition/reduction	NaBH$_4$	2/2	2–6	Reduction of 4-nitrophenol by NaBH$_4$ in H$_2$O	61
Au@MOF							
MOF-5	Me$_2$Au(acac)	Solid grinding	H$_2$	≤1	4.8	Oxidation of alcohols	32
				—	3.3±2.3	N-alkylation of primary amines	62
MIL-53(Al)				—	1.1	Oxidation of alcohols	32
				—	1.6±1.0	N-alkylation of primary amines	62
HKUST-1				—	—	Oxidation of alcohols	32
CPL-1 CPL-2				—	2.4±1.0	Oxidation of alcohols N-alkylation of primary amines	32, 62
MIL-100(Fe)	HAuCl$_4$·4H$_2$O	NP Encapsulation	K$_2$CO$_3$ HCHO	—	>50	Reduction of 4-nitrophenol to 4-aminophenol	63
ZIF-90	Au(CO)Cl	Gas phase infiltration	H$_2$	≤30	1–5	Oxidation of aldehydes	64
ZIF-8	Au(CO)Cl	Gas phase infiltration	H$_2$	≤30	1–5	Oxidation of alcohols	64
	Me$_2$Au(acac)	Solid grinding	H$_2$	≤5	3.4±1.4	CO oxidation	65
Au/MOx (M = Zn, Ti; x = 1,2)@MOF							
MOF-5	ClAuCO Et$_2$Zn/Ti(OiPr)$_4$	Gas phase infiltration	H$_2$	—	1–3	Oxidation of acohols	66
Au/Pd@MOF							
MIL-101(Cr)	HAuCl$_4$/PdCl$_2$	Colloidal impregnation	H$_2$	1	—	Aerobic oxidation of toluene	67
Cu@MOF							
MOF-5	CpCu(PMe$_3$)	Gas phase infiltration	H$_2$	13.8	3–4	MeOH synthesis	31
	CpCuL (L = PMe$_3$,CN^{t}Bu)	Gas phase infiltration	H$_2$	≤8	1–3	MeOH synthesis	68

MOF	Precursor	Method	Reducing agent			Reaction	Ref.
MIL-101(Cr) Ni@MOF	Cu(NO₃)₂	Impregnation	N₂H₄·H₂O	2	—	CO oxidation	69
MOF-5	Ni(acac)₂	Wet impregnation	H₂	7.4	2–6	Hydrogenation of crotonaldehyde	70
Zn-MOCP	Ni(acac)₂	Wet impregnation	H₂	7.5	10	Hydrogenation of crotonaldehyde	71
MesMOF-1	Ni(Cp)₂	Gas phase infiltration	H₂	≤35	—	Reduction of nitrobenzene and styrene	72
ZIF-8	Ni(Cp)₂	Chemical Vapour Deposition	H₂	≤19	2.7 ± 0.7	Hydrolysis of ammonia borane	73
Pd@MOF MOF-5	Pd(acac)₂	Wet impregnation	H₂	1	—	Hydrogenation of styrene	30
	Pd(acac)₂	Impregnation	—	1.6	—	Synthesis of H₂O₂	74
	Pd(NO₃)₂·2H₂O	Coprecipitation	—	0.5	—	Hydrogenation of ethyl cinnamate	75
	PdCl₂	Chemical deposition	N₂H₄·H₂O	3	3–6	Sonogashira coupling	76
	K₂PdCl₄	Adsorption	NaBH₄	0.5	3–12	Aminocarbonylations	77
	Pd(C₃H₅)(C₅H₅)	Chemical Vapour Deposition	H₂	0.2–4.2	2–5	Suzuki–Miyaura coupling	78
	Pd(OAc)₂	Impregnation	—	≤1.8	1.8	CO oxidation	79
NH₂-MIL-53(Al)	H₂PdCl₄	Ion exchange	NaBH₄	0.97	3.12	Suzuki–Miyaura coupling	80
				—	3.2	Heck coupling	57
	Pd(NO₃)₂	Impregnation	H₂	1–6	2–3	Hydrogenation of phenols	81
MIL-101(Cr)	Pd(acac)₂	Wet impregnation	H₂	1	1.5	Hydrogenation of styrene and acetylene/ethylene	82
	CpPd(η³-C₅H₅)	Gas phase infiltration	H₂	≤50	2.9 (RT) 1.7 (70 °C)	Reduction of aryl alkyl ketones	83
	[PdCl₄]²⁻/[PdCl₆]²⁻	Ion exchange	NaBH₄	1	2–4	Heck reaction	51
	Pd(NO₃)₂	Impregnation	H₂	1	1.9 ± 0.7	Ullmann and Suzuki–Miyaura coupling	52
	Pd(NO₃)₂	Impregnation	H₂	0.56	2.6 ± 0.5	Indole C₂-arylation	53
	Pd(acac)₂	Impregnation	H₂	1–3	2.6	Indole synthesis	54
	Pd(NO₃)₂	Impregnation	N₂H₄,H₂O	≤5	3–4	CO oxidation	69
	Pd(NO₃)₂	Wet impregnation	H₂	≤2.79	2.5	MIBK synthesis	84

Table 10.1 (*Continued*)

MOF	M precursor	Synthesis method	Reduction method	M wt%	NP size/nm	Application	Ref.
	Pd(NO$_3$)$_2$	Wet impregnation	H$_2$	1–5	0.35	Synthesis of secondary arylamines, quinolones, pyrrols, and 3-arylpyrrolidines	85
Pd/Cu@MOF							
MIL-101(Cr)	Pd(NO$_3$)$_2$/(CuNO$_3$)$_2$	Impregnation	N$_2$H$_4$ · H$_2$O	1/2	—	CO oxidation	69
Pt@MOF							
MOF-5	Pt(NH$_3$)$_4$]Cl$_2$	Wet impregnation	—	5	—	Selective oxidation of benzyl alcohol derivatives into aldehydes	86
MOF-177	Me$_3$PtCp′	Gas-phase infiltration	H$_2$	≤43	2–3	Oxidation of alcohols	87
MIL-101(Cr)	H$_2$PtCl$_6$	Double solvent	H$_2$	≤5	1.2–3	Ammonia borane hydrolysis Ammonia borane thermal dehydrogenation CO oxidation	88
	K$_2$PtCl$_4$	Wet impregnation	H$_2$	2–8	0.35	Synthesis of secondary arylamines, quinolones, pyrrols, and 3-arylpyrrolidines	85
Ru@MOF							
La(btc)	RuCl$_3$ · 3H$_2$O	Impregnation	SC CO$_2$- MeOH/H2O	0.98	2	Hydrogenation of cyclohexene and benzene	89
MOF-5	Ru(cod)(cot)	Gas-phase infiltration	H$_2$	≤30	1.5–1.7	Oxidation of alcohols	90

Abbreviations: MOF systems: HKUST-1 = Cu$_3$(btc)$_2$; Lanthanum-BTC = Ln(btc).6H$_2$O; MOF-5 = Zn$_4$O(bdc)$_3$; MOF-177 = Zn$_4$O(btb)$_2$; MOF-508 = Zn(bdc)(bipy); MIL-53(Al) = Al(OH)(bdc); NH$_2$-MIL-53(Al) = Al(OH)(bdc-NH$_2$); MIL-68(In) = In(OH)(bdc); MIL-100(Al) = Al$_3$(btc)$_2$; MIL-101(Cr) = Cr$_3$(bdc)$_3$; CPL-1 = Cu$_2$(pzdc)$_2$(pyz); CPL-2 = Cu$_2$(pzdc)$_2$(bipy); Zn-MOCP = Zn$_4$O(OH)$_2$(BDC)$_2$(H$_2$O)$_{2.7}$;ZIF-8 = Zn(MeIM)$_2$;ZIF-90 = Zn(ICA)$_2$; Linker and functional groups: bdc = 1,4-terephthalate; bdc-NH2 = 2-amino-1,4-terephthalate; bipy = 4,4′-bipyridine; btb = benzene-1,3,5-tribenzoate; btc = 1,3,5-benzene- tricarboxylate; pzdc = pyrazine-2,3-dicarboxylate; pyz = pyrazine; bpdc = 4,4′-biphenyldicarboxylate; bptc = 1,1′-biphenyl-2,2′,6,6′- tetracarboxylate; cyclam = 1,4,8,11-tetraazacyclotetradecane; MeIM = 2-methylimidazole; ICA = imidazolate-2-carboxyaldehyde; mtb = methanetetrabenzoate; ntb = 4,4′,4′′-nitrilotrisbenzoate; Cp′ = methylcyclopentadienyl; Cp = η^5-C$_5$H$_5$; cod = 1,5-cyclooctadiene; cot = 1,3,5-cycloocta-triene; acac = acetylacetonate; SAM = self-assembled monolayer; MIBK = methyl isobutyl ketone; DEF = N,N-diethylformamide; DMF = N,N-dimehtylformamide.

resulting nanoparticles, with a mean size of 3 nm, are outside of the pore channels, since the dimensions of NH_2-MIL-53(Al) pores are smaller than the nanoparticle mean diameter, avoiding diffusional problems observed before for the same MOF when using the pore's interior for catalysis.[56] Despite the location of the Pd nanoparticles, the system showed high activity and good stability in Suzuki–Miyaura cross-coupling reactions under mild conditions for a broad scope of reactants. Mix linkers have also been used to control the amount of amino groups in the MOF and therefore the loading and dispersion of Pd nanoparticles in MIL-53(Al).[57] Very recently, the same group has followed this approach on NH_2-MIL-101(Cr), and discover the high catalytic activity of Pd encapsulated in the meso-cages of NH_2-MIL-101(Cr) for the dehalogenation of aryl chlorides in water.[58] A complete overview of the reported "inorganic NP@MOF" systems can be found in Table 10.1.

10.2.2 Enzymes

The use of enzymes in industrial applications is often handicapped by their low operational stability, difficult recovery, and deactivation due to aggregation under reaction conditions. Immobilization of enzymes on solid supports can enhance enzyme stability as well as facilitate product separation and catalyst recovery.[2,91] In many cases, immobilized enzymes show higher specific activity than free enzymes, since deactivation by aggregation can be prevented.[4] Crucial aspects to consider enzyme immobilization are structural deformations and host–guest interactions, responsible for loss in specific activity and leaching, respectively.[92–94]

Several features have been delineated for an ideal host matrix: (a) large pores for enzyme access and to reduce to the minimum the structural deformation, (b) high surface area to ensure high enzyme loading, (c) large cages decorated with functional organic groups to enhance interactions reducing leaching issues, and (d) sustained framework integrity under typical reaction conditions.[96] Intensive attention has been paid to the immobilization of enzymes in porous solids like mesoporous silicas, carbons or ceramics.[97–100] In spite of the great potential of MOFs for the immobilization of enzymes, only a few works have been published in this field. Although, in principle, the microporous nature of many MOFs makes it difficult to incorporate macromolecules like enzymes, recent advances in the synthesis of robust mesoporous MOFs (such as MILs with MTN topologies[36,37] and Tb-mesoMOFs[101]) open the door for enzyme immobilization.

Lykourinou *et al.* have demonstrated that microperoxidase-11 (MP-11) can be successfully encapsulated into a mesoporous MOF (see Figure 10.4).[95] Tb-mesoMOF is formed from a solvothermal reaction between triazine-1,3,5-tribenzoic acid (H_3-TATB) and a Tb source. The three-dimensional extended network is built up by four Tb^{3+} ions connected in a trigonal-planar geometry and four of these Tb_4 units form a supertetrahedron (ST) similar to the MTN topology exhibited by MIL-100 and MIL-101 structures. Five- and six-membered ring windows are created in the framework, which, considering Van der Waals radii, have free diameters of 13.0 and 17.0 Å, respectively. Two

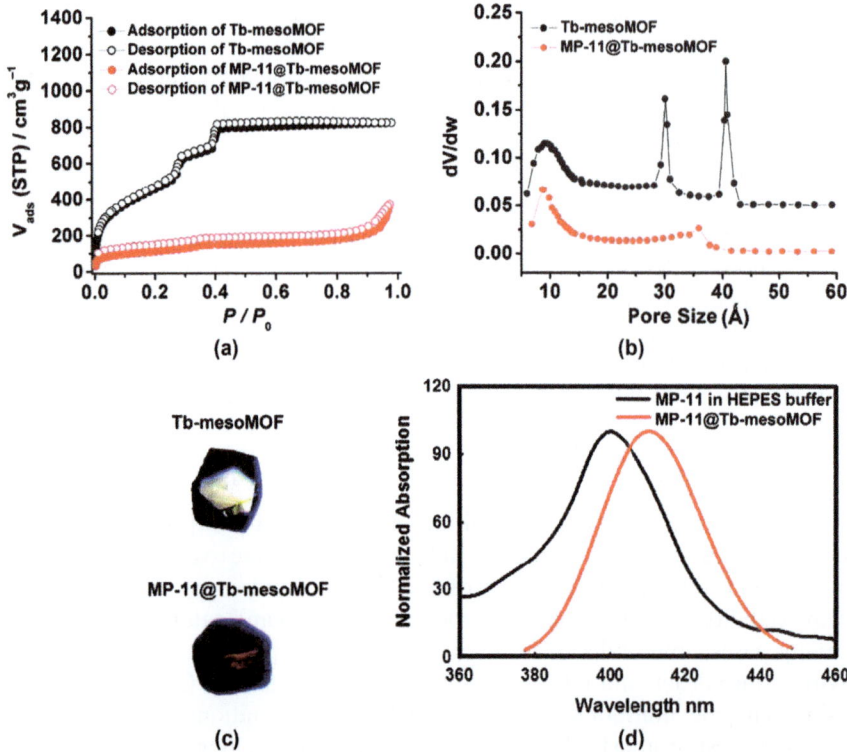

Figure 10.4   (a) N_2 sorption isotherms, and (b) pore size distributions of Tb-mesoMOF and MP-11@Tb-mesoMOF; (c) optical images of Tb-mesoMOF and MP-11@Tb-mesoMOF; (d) normalized single-crystal absorbance spectrum derived from specular reflectance for MP-11@Tb-mesoMOF (red) and solution optical spectrum for free MP-11 in buffer solution (black).
Reprinted with permission from ref. 95. Copyright (2011) American Chemical Society.

different distorted spherical cages are formed by these window openings. The smaller cage S is surrounded by 20 truncated STs and has 12 pentagonal windows. Its internal free diameter is 39.1 Å. The larger cage L, defined by 28 truncated STs and having 12 pentagonal and 4 hexagonal windows, has an internal free diameter of 47.1 Å.

The MP-11 enzyme has dimensions of $3.3 \times 1.7 \times 1.1$ nm, and therefore can easily be accommodated in the cages of Tb-mesoMOF. Using a simple impregnation procedure with MP-11 under controlled pH, loadings as high as 19 μmol MP-11 per gram of MOF are successfully incorporated after 50 h. A colour change of the crystal to dark red, and an extreme decrease in N_2 adsorption demonstrates the presence of trapped MP-11 in the framework. The shift exhibited in the Soret band indicates a strong interaction of the encapsulated enzyme with the MOF.

The composite material has been tested in the oxidation of polyols, such as catechol like compounds. Using free MP-11, a high initial reaction rate was found, but after a few reaction minutes, the activity decreases dramatically due to MP-11 aggregation. On the contrary, no decline in activity is found for MP-11@Tb-mesoMOF and M-11@MCM-41 composites, while both supports are not active for the studied reaction. Reusability tests were performed for both materials. While MP-11@MCM-41 loses 60% of its activity after the first use, M-11@Tb-mesoMOF shows stable performance during the first six cycles, with no sign of leaching or framework decomposition. However, after six cycles the catalytic activity of M-11@Tb-mesoMOF decreases more than 53%, while the frameworks remain stable. Deactivation due to leaching was observed for both MP-11 immobilized catalysts. This decrease in activity is less dramatic in the case of the MOF catalyst, ascribed to the strong interactions between the MP-11 and the hydrophobic MOF framework.

In addition, recent studies have demonstrated that large macromolecules like proteins can enter the interior cavities of MOFs. Significant conformational changes must occur during the immobilization, as suggested by fluorescence studies. Biocatalytic properties still need to be demonstrated.[102]

10.3 "Bottle around a Ship" Encapsulation Approach

10.3.1 Polyoxometalates

Polyoxometalates (POMs) present several advantages as catalysts that make them economically and environmentally attractive.[56] POMs are complex Brønsted acids that consist of metal–oxygen octahedra heteropolyanions as the basic structural units. The first characterized and the best known of these is the Keggin-type heteropolyanion, typically represented by the formula $XM_{12}O_{40}^{x-}$,[8] where X is the central atom (usually P or Si). With a very strong Brønsted acidity, approaching the superacidity region, and exhibiting fast reversible multi-electron redox transformations under rather mild conditions, POMs represent a serious alternative to other acid systems. Their acid–base and redox properties can be varied over a wide range by changing the chemical composition. This unique structure exhibits extremely high proton mobility, while heteropolyanions can stabilize cationic organic intermediates. On top of that, POMs have a good thermal stability in the solid state, better than other strong acids like ion exchange resins.[56]

Supporting POMs is crucial for catalysis applications because most bulk POMs have a low specific surface area (1–5 m^2 g^{-1}). Acidic or neutral porous solids such as SiO_2, Al_2O_3 and activated carbons have been explored as carriers.[103–107] The acidity and catalytic activity of supported POMs depend mainly on the type of carrier and the loading, *i.e.* the interaction with activated carbons is so strong that the activity of the final catalyst is much lower than that of the POM itself, while weak interactions with the POM-support lead to dramatic leaching. Encapsulation of POMs inside zeolitic cavities has been achieved by direct synthesis of the Keggin structures inside the zeolite cavities (FAU). This approach is shown to solve the problem of leaching, since the POM clusters are bigger than the windows of the zeolitic cavities, but only low loadings can be utilized (<5% wt) if diffusion limitations are to be avoided.[108]

Due to their rich structural and chemical variety,[56] POMs possess tuneable shape, size and high negative charge, and are remarkably versatile building blocks in the construction of coordination supramolecules.[109] Frequently, POMs have been shown to act as anionic templates to build three-dimensional metal organic frameworks, while the host MOF structure is not altered by this templating effect (Table 10.2).

Table 10.2 Overview of reported POMS encapsulated within MOF cages along with their catalytic applications.

Synthesis method	Encapsulated moiety	wt%	Application	Ref.
HKUST-1				
Encapsulation	$H_nXM_{12}O_{40}$ (X = Si,Ge,P,As) (M = W or Mo)	35–47	Hydrolysis of esters	119
	$H_3PW_{12}O_{40}$	~50	Hydrolysis of DMMP Esterification	120 127, 128
	$H_4SiW_{12}O_{40}$	54	Dehydration of methanol to DME	121
	$CuPW_{11}O_{39}$	~50	Oxidation of thiols to disulfides	129
MIL-101(Cr)				
Impregnation	$H_5PW_{11}XO_n$ (X = Co, Ti, Zr)	~10	Oxidation reactions	112, 114, 144
	$H_3PW_{4,12}O_{24,40}$	5–14		113,114
	$(BTBA)_4HBW_4O_{24}$	~20		145
	$H_5ZnPMo_2W_9O_{39}$	17.5		111
	$Co_4(PW_9)_2$	<50		146
Encapsulation	$H_3PW_{12}O_{40}$	≤50	Knoevenagel condensation Esterification Dehydration of methanol to DME	108
		10–30	Carbohydrates dehydration	141
		~30	Acetaldehyde-phenol condensation Benzaldehyde and methanol to dimethylacetal	143
		~30	Benzaldehyde and 2-naphtol Three-component condensation Caryophyllene epoxidation	140
		17–50	Oxidative desulfurization	147
NH₂-MIL-101(Al)				
Encapsulation (Impregnation)	$H_3PW_{12}O_{40}$ (H_2PtCl_6)	~15	Toluene hydrogenation CO oxidation PROX reaction	142

Abbreviations: MOF systems: HKUST-1 = Cu₃(btc)₂; MIL-101(Cr) = Cr₃(bdc)₃; NH₂-MIL-101(Al) = Al₃(bdc-NH₂)₃. Linker and functional groups: btc = 1,3,5-benzene- tricarboxylate; bdc = 1,4-tereph-thalate; bdc-NH₂ = 2-amino-1,4-terephthalate; BTBA = benzyltributylammonium. Reactants: DMMP = dimethyl methylphosphonate; DME = dimethyl ether.

The incorporation of POMs in MOF cavities has been widely studied. Impregnation of their salts was firstly performed by Férey and co-workers in MIL-101(Cr) as an original method to prove such large cavities.[110] A similar approach has been followed by other research groups, also using MIL-101(Cr) as host for "ship in a bottle" catalysts.[111] Specially worth a mention, Kholdeeva and co-workers have utilized impregnated POM/MIL-101(Cr) materials for oxidation reactions.[112–114] Although the direct impregnation of POM moieties into the large cavities of MIL-101 is straightforward, a number of drawbacks are to be mentioned: (a) the maximal achievable loading of POM (15 wt%) is to a certain extent low; (b) homogeneity of the sample is low; (c) dispersion of the active species is not optimal, since only the large cavities of MIL-101 are filled with POM; (d) leaching has been observed for certain reactions since immobilization is based on adsorption equilibrium.

In 2003, Yang *et al.* reported the one-pot incorporation of SiMo and PW directly during the synthesis of HKUST-1.[115] Homogeneously monodispersed POMs were encapsulated in the large (10 Å) cavities of the HKUST-1 structure. Bottle-neck connections with the medium adjacent cavities (6 Å) allow for successful POM encapsulation, avoiding leaching and strong diffusional problems. Homogeneous dispersion of POM is obtained since large and medium cavities are alternatively arranged. Detailed XRD analysis showed the formation of two pure POM anionic species rotated 90° one from the other, and a compact encapsulation of the Keggin anions into the large cativities without presence of counter-cations or water molecules. Similar systems were thoroughly characterized by Hundal *et al.*[116]

Over the last few years, POM/HKUST-1 composites have received increasing attention within the scientific community,[117,118] especially towards catalytic applications. Sun *et al.* described the incorporation of different POMs, $H_3XM_{12}O_{40}$ (X = Si, Ge, P, As; M = W, Mo, see Table 10.3), into the cavities of this so-called NENU-n series.[119]

After encapsulation, the main properties of the framework were preserved, while high POM loadings (35–45 wt%) were observed for all XM_{12} studied. These numbers exceeded by far the common loadings in traditional supported systems. Extensive structural information on the POM/MOF composite is given. Location of water and two molecules of $(CH_3)_4N^+$ cations in cavity B are essential to understand the protonation state of the POMs (in a further neutral framework). The POM/MOF conformation provides a well-dispersed composite, avoiding agglomeration and deactivation, which results in an enhancement of its catalytic properties. The acid catalytic properties of the resulting NENU composites were explored in the hydrolysis of esters in excess water, with yields reaching 65% for methyl and ethyl acetate at 333 K in

Table 10.3 Heteropolyacids encapsulated in the NENU-n series.[119]

NENU-n : $[Cu_2(BTC)_{4/3}(H_2O)_2]_6[POM] \cdot (C_4H_{12}N)_2 \cdot xH_2O$					
1	2	3	4	5	6
$[H_2SiW_{12}O_{40}]^{-2}$	$[H_2GeW_{12}O_{40}]^{-2}$	$[HPW_{12}O_{40}]^{-2}$	$[H_2SiMo_{12}O_{40}]^{-2}$	$[HPMo_{12}O_{40}]^{-2}$	$[HAsMo_{12}O_{40}]^{-2}$

aqueous solution in 5 h. No POM leaching or framework decompositions were observed, allowing recycling of the catalyst.

The same group reported that NENU-11 (encapsulated PW_{12} in HKUST-1) is an excellent candidate for eliminating nerve gas by decomposition of type-G and type-X toxic nerve agents.[120] NENU-11 showed a rapid adsorption of dimethyl methylphosphonate (DMMP) in the initial 20 min, and reaching 1.92 mmol g^{-1} within 100 minutes (15.5 DMMP molecules per formula unit). This value is much higher than previous data obtained for other MOFs, with higher surface areas attributed to the presence of POMs providing a strong interaction with DMMP through Cu–O bonds. The effect of humidity was considered, in order to investigate the DMMP decomposition under ambient conditions. The total amount adsorbed decreased as the relative humidity (RH) increased, but still the main contribution to the total amount comes from DMMP since it presents higher polarity comparing to water. The conversion of DMMP to methyl alcohol was 34% at room temperature. The conversion increased gradually with temperature, reaching the optimum of 93% at 50 °C. NENU-11a was reused for 10 cycles and maintained its structural properties.

The influence of particle size on the catalytic activity of $H_4SiW_{12}O_{40}$/NENU composites has been studied by Ya-Guang Chen et al.[121] Hydrothermal synthesis with different initial concentration of reactants and synthesis times leads to the formation of a sample range with average particle size of 23, 105, and 450 μm. The composites are provided with both Brønsted acidity from POM molecules and Lewis acidity of HKUST-1, necessary for its application in the acid catalytic dehydration of methanol to DME. NENU-1a exhibited a higher catalytic performance than the MOF support and $H_4SiW_{12}O_{40}$ supported on α-Al_2O_3. High catalytic activity was also observed in the formation of ethyl acetate, where NENU-1a exhibited comparable performance with that of $H_4SiW_{12}O_{40}$ supported on SiO_2. Internal diffusion limitations in the POM/MOF systems were avoided when the particle size was reduced to 105 μm, demonstrating that for this type of "small pore" MOF diffusion limitations might play a very important role in the catalytic performance.

Martens, Kirschhock and co-workers have reported the room temperature synthesis of similar POM/HKUST-1 composites and they have thoroughly studied the system properties, including stability[122] and POM templation effects,[123–125] and applications like thin films[126] and catalysis.[127,128] The esterification of acetic acid and 1-propanol in the absence of solvent was chosen as a model reaction to evaluate the acidity of the nano synthesized composites. The hybrid materials showed a fair catalytic performance in comparison with other acid catalysts like zeolite Y and the hydrothermally synthesized POM/HKUST-1. However, clear diffusion limitations were observed, as demonstrated by the faster kinetics shown by the smallest POM-MOF particles (50 nm).

One of the latest catalytic applications revealed for POM/HKUST-1 composites has been developed by Song et al.,[129] by encapsulating Cu containing phosphotungstic units $[Cu_2PW_{11}O_{39}]^{-5}$ in the alternative largest cavities of HKUST-1. The electrostatic interaction between POM molecules and MOF structures results in a higher reduction potential of the POM. The stability of both components is also enhanced due to synergistic effects, as reported before by Mustafa *et al.*[122] The resulting CuPOM/HKUST-1 composite is capable of catalyzing the oxidative decontamination of toxic sulfur compounds in air. H_2S is oxidatively converted into S_8 and water under ambient conditions in both liquid and gas phase. Control experiments showed the poor catalytic activity and stability of POM and MOF alone under the reaction conditions. Gas phase reactions showed the importance of CuPOM as the main active site. Due to its catalytic activity in aerobic oxidation reactions, the conversion of volatile mercaptans to less toxic and odorous disulfides was also examined. The POM/HKUST-1 was reported to produce 200 turnovers, whereas no product formation was observed without catalyst. POM and MOF alone showed no activity, while PW_{12}/HKUST-1 was also not active for the oxidation reaction. The catalyst was observed to maintain its catalytic performance for at least three cycles without significant activity loss, while characterization techniques did not show any sign of MOF or POM degradation.

In addition to the widely studied POM/HKUST-1 composites, the encapsulation of POMs in other MOF structures like MIL-101[130] and MIL-100[131,132] have been investigated.[133–138]

We reported a broad study of the catalytic performance of $HPW_{12}O_{40}$/MIL-101(Cr) composites.[139] Phosphotungstic acid (PTA or PW_{12}, $HPW_{12}O_{40}$) was encapsulated by a one-pot procedure in both middle and large cavities of MIL-101(Cr). POM and MOF survived during the harsh synthesis conditions (8 h, 493 K, pH < 1) as shown by XRD, DRIFT and N_2 adsorption. Middle-sized and large cages of MIL-101 are able to contain more than one PW_{12} per cavity (29 and 34 Å in diameter), leaving room for reactants to diffuse through. Interconnection of the cavities through 5 membered ring windows (10 Å in diameter) does not lead to diffusion of POMs, avoiding agglomeration and deactivation of these active species. High POM loadings were achieved, which could be tuned (10–50 wt% PW_{12}) by variation of the POM concentration in the synthesis mixture. We studied the catalytic application in several acid and base catalysed reactions (esterification, etherification and Knoevenagel condensation). Having fair acid catalytic properties, surprisingly the catalyst showed an outstanding activity in the Knoevenagel condensation reaction of ethyl cyano acetate with benzaldehyde.

Reusability and hot filtration experiments demonstrated that the system was stable enough, and no leaching was observed. The activity was higher than that of commercial homogeneous superbases like bicycloguanidine and of the POM used as homogeneous catalyst or impregnated in MIL-101(Cr). The reaction mechanism was elucidated by observing POM interactions with ethyl cyano acetate, while interactions with benzaldehyde, occurring with basic amino-MOFs, were absent.[56]

Along the same line, Zhang *et al.* studied the activity of the same POM/MIL-101(Cr) catalyst for the dehydration of carbohydrates in ionic liquids.[141] The best performance was found for the highest loaded catalyst (3 POM per cage), with a hydroxymethylfurfural (HMF) yield of 79% from fructose after 2.5 h. Reactions in different solvents led to the conclusion that the most feasible mechanism proceeds *via* proton exchange between the immobilized PW_{12} and the organic cations of the ionic liquid. The HMF yield was 63% using DMSO as solvent in 30 min at 130 °C, while after previous titration with NaOH the catalyst did not show any significant activity. The catalyst could be recycled under the same reaction conditions, demonstrating the true nature of POM/MOF as a solid acid catalyst for dehydration of fructose to HMF.

Recently, Hatton and coworkers have reported two additional works on PTA/MIL-101(Cr) composites and proposed several crystal structures for the composites (see Figure 10.5). The first work focuses on aldehyde–alcohol reactions like acetaldehyde–phenol (A–P) condensation and dimethylacetal formation from benzaldehyde and methanol (B–M reaction).[143] The PTA/MIL-101(Cr) composites were synthesized by one-pot encapsulation during the MOF synthesis or by impregnation of PTA after MIL-101 formation without the use of hydrofluoric acid (HF), as extensively reported in literature. When comparing particle size, one-pot samples showed 5 to 25-fold larger crystals (2–10 μm) than the impregnated or bare samples (~ 400 nm), with no significant change in the catalytic activity, suggesting that the mass transport through the MOF pores did not limit the activity. The composites show catalytic activity for both reactions, with half-lives ranging from 0.5 h (A–P) to 1.5–2 h (B–M) and turnover numbers over $14\,min^{-1}$ for A–P and over $18\,min^{-1}$ for B–M reactions, respectively. PTA/MIL-101(Cr) composites offer a synergistic effect due to the strong acidity presented by the PTA and to the mildly acidic Lewis sites of MIL-101. The catalyst activity was constant during at least four cycles of 24 h each. In a second paper, the same type of materials are further described by XRD, and tested in the Baeyer condensation of benzaldehyde and 2-naphthol, in the three-component condensation of benzaldehyde, 2-naphthol, and acetamide, and in the epoxidation of caryophyllene by hydrogen peroxide.[140] The catalysts show over 80–90% conversion of the reactants under microwave-assisted heating. The recyclability tests presented >92% maintained conversion after four consecutive reaction cycles.

Lately, a new application for POM/MOF composites has been explored: Ramos-Fernandez *et al.* reported the use of highly dispersed PW_{12} encapsulated in NH_2-MIL-101(Al) as supports for different Pt precursor species. Reduction of the composite at 473 K results in the formation of small Pt^0 clusters sitting on the heteropolyacid, as demonstrated by XPS and CO chemisorption experiments (see Figure 10.6). Further reduction at 573 K induces the formation of intermetallic Pt^0-W^{5+} species, which exhibit the best CO oxidation activity within this sample set and a higher selectivity towards CO_2 in the PROX reaction, resembling the combination of a noble metal on a reducible support.

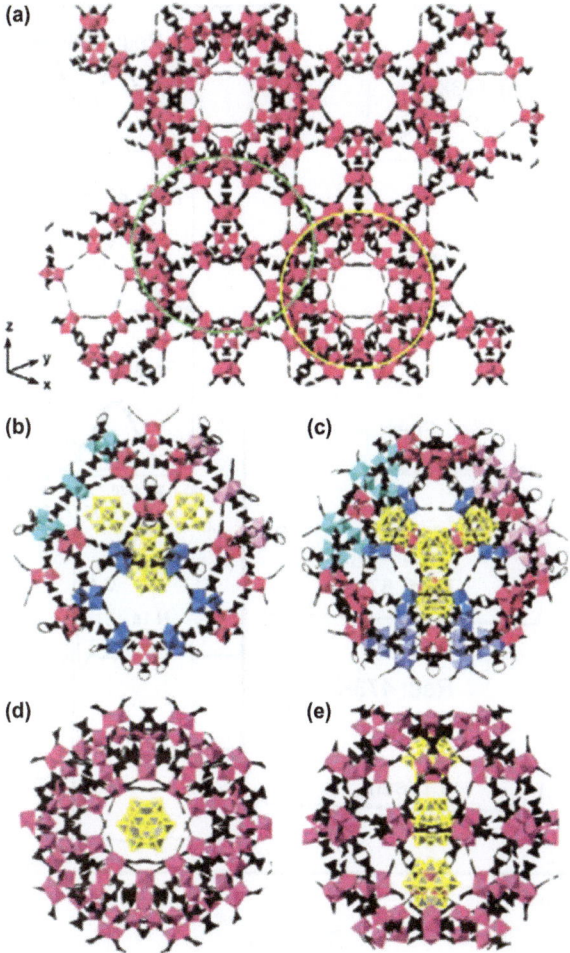

Figure 10.5 Proposed crystal structure of (a) MIL-101 and (b–e) MIL101/PTA composites. (a) MIL-101 crystal structure with the boundary of the large cage highlighted green and that of the small cage highlighted yellow. The Cr polyhedrons are coloured pink. (b, c) The PTA tetrahedron in a multicoloured large cage of MIL-101, which corresponds to the model with ordered PTA. Views from two directions are presented, b is viewed from the same direction as in a, and c is viewed facing a six-carbon ring. Each of the four six-carbon rings is coloured differently, each aligned with a PTA molecule. (d, e) A possible configuration of three PTA molecules in a small cage of MIL101 viewed (d) from the front, which is the same direction as in a, and (e) from the top. The three PTA molecules were allowed to rotate around their shared axis.
Reprinted with permission from ref. 140. Copyright (2012) American Chemical Society.

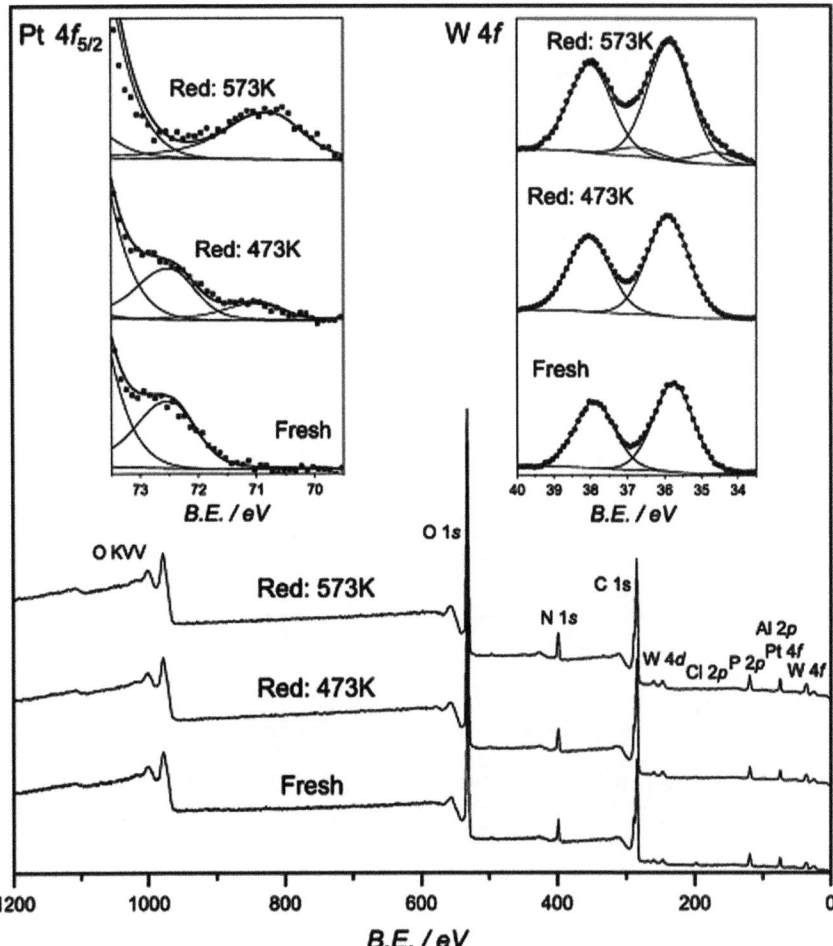

Figure 10.6 XPS survey for the sample containing Pt before and after reduction. (Left-insert) Zoom in at the Pt $4f_{5/2}$ core-level. (Right-insert) Zoom in at the W4*f* core-level.
Reprinted with permission from ref. 142. Copyright (2012) Elsevier Inc.

10.3.2 Meta Organic Macromolecules

Porphyrins and metalloporphyrins provide an extremely versatile basis for a large variety of applications. The fundamental properties of these materials come from their interaction with applied electric, magnetic or electromagnetic fields and with other chemical species. Porphyrins and metalloporphyrins have been broadly studied as field-responsive materials, particularly for potential optoelectronic applications and as chemoselective catalysts.[148] However, homogeneous metalloporphyrin-based catalysts present several drawbacks like limited lifetime activity as a result of dimerization and self-degradation. Therefore, immobilization in solid matrices is expected to overcome such downsides by enhancing stability and by protecting the active site from deactivation. Several attempts on supporting metalloporphyrins include the use of mixed oxide surfaces and porous inorganic solids like zeolites, mesoporous silicates or silica surfaces.[15,149,150] Nevertheless, aggregation, limited catalyst loading, heterogeneous distribution, and/or leaching of the metalloporphyrins are some of the current limitations.[148]

Confined spaces have been proposed as an efficient environment to protect porphyrins from aggregation, heterogeneous distribution and leaching. A suitable host matrix providing such confinement effects should offer appropriate topological characteristics: large cavity dimensions (one guest porphyrin molecule per cavity) and relatively reduced opening windows preventing catalyst leaching while preserving reactant diffusion. It goes without saying that, when following this approach, stability of the metalloporphyrins during the matrix synthesis and the framework through possible post-synthetic metalation treatments might be considered. In the last decade of the twentieth century, zeolites were used to encapsulate metalloporphyrins through direct "bottle around a ship" composite synthesis. Such structures offered many opportunities for the development of selective catalysis, illustrating the stabilizing effect of zeolites.[22,25,151–156]

In a similar approach, the encapsulation of metalloporphyrins in metal organic frameworks is expected to offer several advantages over other nano-molds, although only a few proofs of concept have been published to date (Table 10.4).[158,159]

Eddaoudi and co-workers pioneered the direct encapsulation of such moieties in MOFs.[157] The direct encapsulation of different porphyrins during the synthesis of zeolite-*like* MOF (ZMOF) with an In-based structure was investigated (see Figure 10.7). The indium source reacted with 4,5-imidazoledicarboxylic acid in DMF/acetonitrile mixture in the presence of different porphyrin concentrations. The colourless ZMOF system changed to dark red when porphyrin was present in the synthesis, suggesting the successful encapsulation.

The insoluble crystals were thoroughly washed until no residual porphyrin was detected by UV-Vis measurements in the solution. The framework unit cell was determined as $[In_{48}(HImDC)_{96}]^{48-}$, where the negative charges were balanced by cationic guest molecules. Single-crystal XRD could not verify the inclusion of porphyrin inside the cages of *rho*-ZMOF due to the porphyrin's low symmetry compared to the framework. UV-Vis characterization of the fully washed crystalline solid showed the characteristic five adsorption bands associated with the free porphyrin, assuming no metalation with In^{+3}. Diffusion measurements indicated no release of porphyrin from the framework. Dissolving the porphyrin/ZMOF composite under strongly acidic conditions allowed estimation of the porphyrin loading (2.5 wt%), which was controllable *via* variable initial concentration. Electrostatic interactions between cationic porphyrin and the anionic framework during the self-assembly of the system became a strong aspect for efficient encapsulation, since neither anionic nor neutral porphyrins were successfully incorporated in the framework. Furthermore, post-synthetic metalation was performed by contacting the

Table 10.4 Overview of reported studies of metalloporphyrins encapsulated in MOF structures.

MOF	Encapsulated molecule	wt%	Synthesis method	Application	Ref.
rho-ZMOF	M-TMPyP (M = Mn, Cu, Zn, Co)	2.5	Direct encapsulation PSM metalation	Cyclohexane oxidation	157
MOMzyme-1	M4SP (M = Fe, Mn)	33–66	Direct encapsulation of metalated porphyrin	Monooxygenation of organic subtrates	162
porh@MOM-X (X = 4–9)	M-TMPyP (M = Fe, Co, Mn, Ni, Mg, Zn)	14–88	Direct encapsulation and metalation	Olefin oxidation	164
porh@MOM-10	M-TMPyP (M = Cd, Mn, Cu)	~40	Direct encapsulation PSM metal exchange	Epoxidation of *trans*-stilbene	163
MIL-101(Cr)	MPcF$_{16}$ (M = Fe, Ru)	3.6	Wet infiltration	Aerobic tetralin oxidation	165
MIL-101(Cr)	FePcS	22	Adsorption from aqueous solution	Selective Oxidation of aromatic substrates	166

Abbreviations: MOF systems: rho-ZMOF = [In$_{48}$(HImDC)$_{96}$]$^{48-}$; MOMzyme-1 (or porph@MOM-3) = M4SP@HKUST-1;   HKUST-1 = Cu$_3$(btc)$_2$;   porph@MOM-4 = [Fe$_{12}$(btc)$_8$(S)$_{12}$]Cl$_6$ · *x*FeTMPyPCl$_5$; porph@MOM-5 = [Co$_{12}$(btc)$_8$(S)$_{12}$] · *x*CoTMPyPCl$_4$;   porph@MOM-6 = [Mn$_{12}$(btc)$_8$(S)$_{12}$] · *x*MnTMPyPCl$_5$;   porph@MOM-7 = [Ni$_{10}$(btc)$_8$(S)$_{24}$] · *x*NiTMPyP · (H$_3$O);   porph@MOM-8 = [Mg$_{10}$(btc)$_8$(S)$_{24}$].*x*MgTMPyP · (H$_3$O)$_{(4-4x)}$;   porph@MOM-9 = [Zn$_{18}$(OH)$_4$(btc)$_{12}$(S)$_{15}$] · *x*ZnTMPyP · (H$_3$O)$_{(4-4x)}$;   *x* = porphyrin loading;   S = solvent;   porph@MOM-10 = Cd$_6$(bpt)$_4$Cl$_4$(H$_2$O)$_4$ · [C$_{44}$H$_{36}$N$_8$CdCl].[H$_3$O] · S;   MIL-101(Cr) = Cr$_3$(bdc)$_3$.   Organic   linker: HImDC = 4,5-imidazoledicarboxylic   acid;   btc = 1,3,5-benzene- tricarboxylate;   bpt = biphenyl-3,4′,5-tricarboxylate; bdc = 1,4-terephthalate. Encapsulated molecules: TMPyP = *meso*-tetra(*N*-methyl-4-pyridyl)porphyrin tetratosylate;   M4SP = M(3 +)tetrakis(4-sulphonatophenyl)porphyrin;   MPcF$_{16}$ = metal   perfluorophthalocyanine; FePcS = iron tetrasulfophthalocyanine; RuBpy = tris(bipyridine)ruthenium(2 +).

free-based encapsulated porphyrin/ZMOF system with several transition metal ions like Mn, Cu, Zn or Co. Only the Mn-porphyrin containing composite was tested as catalyst for hydrocarbon oxidation. After 24 h the total yield (from cyclohexane to cyclohexanol/cyclohexanone) reached 91.5% and a TON of 23.5. Controlled reactions did not show any activity for other MOF and porphyrin combinations. Recyclability studies demonstrated maintained crystallinity, reactivity, and selectivity of the system throughout eleven cycles without observable leaching of the encapsulated Mn-porphyrin, as especially evidenced by UV-Vis spectroscopy.

After reporting these results, several publications described further investigations of different porphyrin/MOF systems.[160,161] Below, recent examples will be described.

In 2011, Larsen *et al.*, inspired by the previous results, described the encapsulation of M(3 +)tetrakis(4-sulphonatophenyl)porphyrin (M4SP, M = Fe or Mn) within the octahedral cages of HKUST-1(Cu or Zn) forming

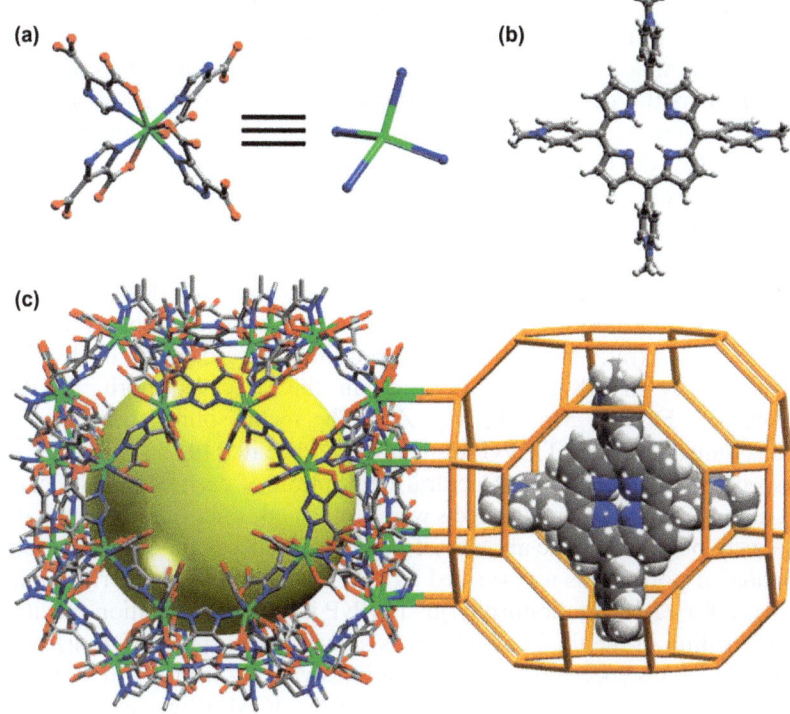

Figure 10.7 (a) Eight-coordinated InN₄ represented as a tetrahedral building unit, (b) [H₂TMPyP]$^{+4}$ porphyrin, (c) crystal structure of *rho*-ZMOF (left), hydrogen atoms omitted for clarity, and schematic representation of [H₂TMPyP]$^{+4}$ porphyrin ring enclosed in *rho*-ZMOF a-cabe (right, drawn to scale).
Reprinted with permission from ref. 157. Copyright (2008) American Chemical Society.

MOMzyme-1. The catalytic performance of the resulting solids was studied in the monooxygenation of organic substrates.[162] Crystal structures of the M4SP@HKUST-1 were determined as isostructural to HKUST-1 through single-crystal X-ray diffraction. The porphyrin ring is accessible at both sides through two square windows above and below the channel extensions. The metalloporphyrin loading was controlled by initial concentration and estimated by X-ray and spectroscopy data to range between 33% and 66%, saturating the structure at the highest loading. The presence of guests in the framework is observed by a decrease of surface area, and the corresponding optical absorption spectra Soret band slightly shifted due to interactions with the framework.

The peroxidase activity of Fe4SP@HKUST-1(Cu) was tested using 2,2′-azinobis(3-ethylbenzthiazoline)-6-sulfonate (ABTS) as a redox indicator and monitoring the rate of increase in absorbance at 660 nm after peroxide addition. The initial rate of H_2O_2 degradation is lower for the heterogeneous sample, compared with benchmark catalysts in solution (hhMb, MP-11 and Fe4SP), since samples have to diffuse within the lattice. However, conversions relative to hhMb per mol of heme are similar to MP-11 and Fe4SP in solution. Recycling test results in retention of 33% of the initial rate of H_2O_2 degradation and 66% remaining conversion of ABTS after three runs. Initial loss of activity is attributed to the degradation of porphyrinic catalysts by guest molecules within the framework that are consumed during the first run, leading to a stable conversion after the initial catalytic cycle.

After this work, two more publications from Zaworotko's group have reported templation effects of porphyrins during MOF synthesis and their catalytic application.[163,164]

Zhang *et al.* reported the templated synthesis of a series of HKUST-1 and *tbo* topology MOFs by a porphyrin guest and their performance in the oxidation of olefins.[164] *meso*-tetra(*N*-methyl-4-pyridyl)porphine tetratosylate (TMPyP) was encapsulated by in HKUST-1, resulting in the so called porh@MOM-X (X = 4–9; M = Fe, Co, Mn, Ni, Mg, Zn). The metal source, benzene 1,3,5-tricarboxylate (BTC) and TMPyP were mixed and heated to 358 K for 12, 24 or 48 h (specific procedures and structural description of each composite can be found in the article). Template effects were confirmed, since different crystalline phases were obtained in the absence of porphyrin. Successful encapsulation of the metalated porphyrins was verified by UV-Vis exhibiting the corresponding Soret band for each metalloporphyrin. TMPyP initial concentration facilitated variable loading in the final composite, as shown for porph@MOF-4(Fe) and studied by UV-Vis spectroscopy, with loadings varying from 14 to 88 wt%.

The catalytic activity was studied for porph@MOM-4(Fe) displaying 85% conversion of styrene after 10 h (TOF = 269 h^{-1}), compared with the 35% conversion obtained by the equivalent amount of Fe(III)TMPyP in solution (see Figure 10.8). Selectivity values are comparable with literature, achieving 30% styrene oxide and 57% benzaldehyde. When using larger molecules than styrene, like stilbene or triphenylethylene, a constant decrease in activity with increasing size was observed, confirming diffusional problems for larger reactants. The absence of metalloporphyrin species in solution together with similar conversion levels after several cycles proved the absence of porphyrin leaching from the framework.

The second work from Zhang *et al.* describes the formation of Porph@MOM-10 from CdTMPyP directly encapsulated in an anionic Cd(II) carboxylate framework, and its performance as heterogeneous catalyst in the epoxidation of *trans*-stilbene.[163] Direct incorporation of TMPyP in the synthesis with $CdCl_2$ and biphenyl-3,4′,5-tricarboxylate (BPT), lead to the formation of a new MOF structure, with successful encapsulation of the porphyrin Cd metalated in a cuboid box, interacting with phenyl BPT ligans, μ_2-connected chlorides, and electrostatic interactions with the framework. By immersion of the composite in $MnCl_2$ methanol solution and by UV-Vis monitoring, the transformation of CdTMPyP in MnTMPyP was observed within 1 week and MOM-10-Mn within 1 month, while direct formation of Mn-porph@MOM-10-Mn was unsuccessful. Copper incorporation in the system was less efficient and slower, 1 month was needed for 76% Cd exchange, leading to Cuporph@MOM-10-CdCu.

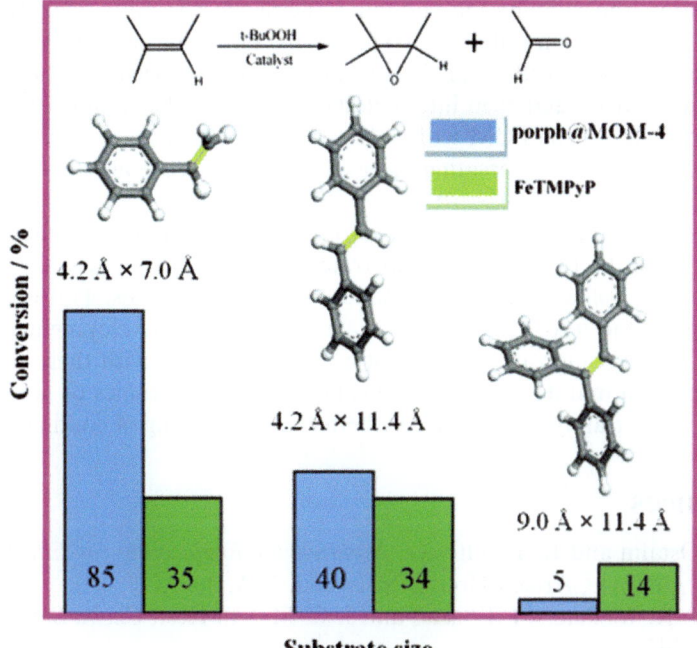

Figure 10.8 Catalytic effect of porph@MOM-4 vs FeTMPyP upon substrates of different size (styrene, trans-stilbene, and triphenylethylene), revealing size selectivity consistent with the pore size of porph@MOM-4.
Reprinted with permission from ref. 164. Copyright (2011) American Chemical Society.

Mn-porph@MOM-101-Mn exhibited 75% conversion (TON = 178 h^{-1}) in the epoxidation of *trans*-stilbene, similar to the 85% observed for the equimolar amount of Mn(III)TMPyP in solution, while Cu-porph@MOM-10-CdCu afforded a conversion of 79% (TON = 182 h^{-1}). Stilbene oxide and benzaldehyde were the major products with 60% and 20%, respectively, in both cases. Reuse of both catalysts show maintained conversions of 60% (Mn) and 70% (Cu) after six cycles of 12 h each.

10.4 Summary and Outlook

MOFs offer many advantages and opportunities for exploring the concept of a nanoreactor. The combination of an almost unlimited number of topologies with mild synthetic procedures and the absence of structure directing agents opens the door to the encapsulation of almost every molecule, macromolecule or even nanoparticle of choice, giving the required active sites confined space for performing chemical transformations. In this sense, we believe that the works published to date are only the "tip of the iceberg" and we expect a rapid development of this challenging research field in the coming decade.

In addition, the combination of different guest species within one framework will open the door to multifunctional materials. For instance, POMs have been shown to serve as excellent anchoring points for homogeneous catalysts,[102] metal nanoparticles and even intermetallic species.[167] The combination of the redox properties of the POM with stereo-selective moieties or metal nanoparticles might contribute to creating more efficient heterogeneous catalysts, as we have recently shown.[142]

Summarizing, the proofs of principle for the encapsulation of a wide range of moieties in MOFs have been demonstrated in the last few years, opening an exciting subdiscipline in the field of nano-structured materials. This is only the beginning: in the next decade we will witness a rapid development of these multifunctional composites that might eventually end up with their industrial implementation in different fields like catalysis, optoelectronics or adsorption, and that will certainly contribute to a better understanding of confined matter.

References

1. A. Ostafin and K. Landfester, *Nanoreactor Engineering for Life Sciences and Medicine*, Artech House, Norwood, MA, 2009.
2. R. J. R. W. Peters, I. Louzao and J. C. M. van Hest, *Chem. Sci.*, 2012, **3**, 335–342.
3. K. T. Kim, S. A. Meeuwissen, R. J. M. Nolte and J. C. M. van Hest, *Nanoscale*, 2010, **2**, 844–858.
4. K. Renggli, P. Baumann, K. Langowska, O. Onaca, N. Bruns and W. Meier, *Adv. Funct. Mater.*, 2011, **21**, 1241–1259.
5. T. S. Koblenz, J. Wassenaar and J. N. H. Reek, *Chem. Soc. Rev.*, 2008, **37**, 247–262.
6. Y. Zhou, W. Huang, J. Liu, X. Zhu and D. Yan, *Adv. Mater.*, 2010, **22**, 4567–4590.
7. M. Sanlés-Sobrido, M. Pérez-Lorenzo, B. Rodríguez-González, V. Salgueiriño and M. A. Correa-Duarte, *Angew. Chem., Int. Ed.*, 2012, **51**, 3877–3882.
8. K. Kita and C. Dittrich, *Exp. Opin. Drug Deliv.*, 2011, **8**, 329–342.
9. T. Ishihara and T. Mizushima, *Exp. Opin. Drug Deliv.*, 2010, **7**, 565–575.
10. R. Ravichandran, *Int. J. Green Nanotechnol.: Phys. Chem.*, 2010, **1**, 72–96.
11. D. Trimnell, B. S. Shasha, R. E. Wing and F. H. Otey, *J. Appl. Polym. Sci.*, 1982, **V 27**, 3919–3928.
12. D. Avnir, S. Braun, O. Lev and M. Ottolenghi, *Chem. Mater.*, 1994, **6**, 1605–1614.
13. L. Rashidi and K. Khosravi-Darani, *Crit. Rev. Food Sci. Nutr.*, 2011, **51**, 723–730.
14. B. F. Gibbs, S. Kermasha, I. Alli and C. N. Mulligan, *Int. J. Food Sci. Nutr.*, 1999, **50**, 213–224.
15. D. E. De Vos, B. F. Sels and P. A. Jacobs, *CATTECH*, 2002, **6**, 14–29.
16. D. Fraenkel and J. Shabtai, *J. Am. Chem. Soc.*, 1977, **99**, 7074–7076.

17. D. Fraenkel, *J. Chem. Soc., Faraday Trans. 1*, 1981, **77**, 2029–2039.
18. D. Fraenkel, *J. Chem. Soc., Faraday Trans. 1*, 1981, **77**, 2041–2052.
19. A. Markus, I. Eger, M. Mhasalkar, Z. Pelah and A. Galun, *J. Colloid Interface Sci.*, 1983, **94**, 284–285.
20. T. Bein and P. Enzel, *Angew. Chem., Int. Ed. Engl.*, 1989, **28**, 1692–1694.
21. T. Bein and P. Enzel, *Synth. Metals*, 1989, **29**, 163–168.
22. K. J. Balkus Jr, A. K. Khanmamedova, K. M. Dixon and F. Bedioui, *Appl. Catal. A*, 1996, **143**, 159–173.
23. K. J. Balkus Jr, A. A. Welch and B. E. Gnade, *Zeolites*, 1990, **10**, 722–729.
24. X. Liu, K. K. Iu and J. Kerry Thomas, *Chem. Phys. Lett.*, 1992, **195**, 163–168.
25. K. J. Balkus Jr, A. A. Welch and B. E. Gnade, *J. Inclusion Phenom. Mol. Recog. Chem.*, 1991, **10**, 141–151.
26. J. Juan-Alcaniz, J. Gascon and F. Kapteijn, *J. Mater. Chem.*, 2012, **22**, 10102–10118.
27. A. Corma, H. García and F. X. Llabrés i Xamena, *Chem. Rev.*, 2010, **110**, 4606–4655.
28. M. Meilikhov, K. Yusenko, D. Esken, S. Turner, G. Van Tendeloo and R. A. Fischer, *Eur. J. Inorg. Chem.*, 2010, 3701–3714.
29. A. Dhakshinamoorthy and H. Garcia, *Chem. Soc. Rev.*, 2012, **41**, 5262–5284.
30. M. Sabo, A. Henschel, H. Fröde, E. Klemm and S. Kaskel, *J. Mater. Chem.*, 2007, **17**, 3827–3832.
31. S. Hermes, M. K. Schröter, R. Schmid, L. Khodeir, M. Muhler, A. Tissler, R. W. Fischer and R. A. Fischer, *Angew. Chem., Int. Ed.*, 2005, **44**, 6237–6241.
32. T. Ishida, M. Nagaoka, T. Akita and M. Haruta, *Chem.– Eur. J.*, 2008, **14**, 8456–8460.
33. B. W. Jacobs, R. J. T. Houk, M. R. Anstey, S. D. House, I. M. Robertson, A. A. Talin and M. D. Allendorf, *Chem. Sci.*, 2011, **2**, 411–416.
34. H. Li, M. Eddaoudi, M. O'Keeffe and O. M. Yaghi, *Nature*, 1999, **402**, 276–279.
35. S. S. Y. Chui, S. M. F. Lo, J. P. H. Charmant, A. G. Orpen and I. D. Williams, *Science*, 1999, **283**, 1148–1150.
36. G. Férey, C. Serre, C. Mellot-Draznieks, F. Millange, S. Surblé, J. Dutour and I. Margiolaki, *Angew. Chem., Int. Ed.*, 2004, **43**, 6296–6301.
37. C. Férey, C. Mellot-Draznieks, C. Serre, F. Millange, J. Dutour, S. Surblé and I. Margiolaki, *Science*, 2005, **309**, 2040–2042.
38. K. S. Park, Z. Ni, A. P., J. Y. Choi, R. Huang, F. J. Uribe-Romo, H. K. Chae, M. O'Keeffe and O. M. Yaghi, *Proc. Natl. Acad. Sci. U. S. A.*, 2006, **103**, 10186–10191.
39. T. Uemura, Y. Kadowaki, C. R. Kim, T. Fukushima, D. Hiramatsu and S. Kitagawa, *Chem. Mater.*, 2011, **23**, 1736–1741.
40. R. K. Bhakta, J. L. Herberg, B. Jacobs, A. Highley, R. Behrens Jr, N. W. Ockwig, J. A. Greathouse and M. D. Allendorf, *J. Am. Chem. Soc.*, 2009, **131**, 13198–13199.

41. C. Zlotea, R. Campesi, F. Cuevas, E. Leroy, P. Dibandjo, C. Volkringer, T. Lolseau, G. Férey and M. Latroche, *J. Am. Chem. Soc.*, 2010, **132**, 2991–2997.

42. Y. E. Cheon and M. P. Suh, *Angew. Chem., Int. Ed.*, 2009, **48**, 2899–2903.

43. K. Sugikawa, Y. Furukawa and K. Sada, *Chem. Mater.*, 2011, **23**, 3132–3134.

44. M. Müller, X. Zhang, Y. Wang and R. A. Fischer, *Chem. Commun.*, 2009, 119–121.

45. D. Esken, H. Noei, Y. Wang, C. Wiktor, S. Turner, G. Van Tendeloo and R. A. Fischer, *J. Mater. Chem.*, 2011, **21**, 5907–5915.

46. D. Esken, S. Turner, C. Wiktor, S. B. Kalidindi, G. Van Tendeloo and R. A. Fischer, *J. Am. Chem. Soc.*, 2011, null-null.

47. Y.-Q. Lan, H.-L. Jiang, S.-L. Li and Q. Xu, *Adv. Mater.*, 2011, **23**, 5015–5020.

48. R. J. T. Houk, B. W. Jacobs, F. E. Gabaly, N. N. Chang, A. A. Talin, D. D. Graham, S. D. House, I. M. Robertson and M. D. Allendorf, *Nano Lett.*, 2009, **9**, 3413–3418.

49. M. D. Allendorf, A. Schwartzberg, V. Stavila and A. A. Talin, *Chem.–Eur. J.*, 2011, **17**, 11372–11388.

50. P. L. Feng, J. J. Perry, Iv, S. Nikodemski, B. W. Jacobs, S. T. Meek and M. D. Allendorf, *J. Am. Chem. Soc.*, 2010, **132**, 15487–15489.

51. Y. K. Hwang, D. Y. Hong, J. S. Chang, S. H. Jhung, Y. K. Seo, J. Kim, A. Vimont, M. Daturi, C. Serre and G. Ferey, *Angew. Chem., Int. Ed.*, 2008, **47**, 4144–4148.

52. B. Yuan, Y. Pan, Y. Li, B. Yin and H. Jiang, *Angew. Chem., Int. Ed.*, 2010, **49**, 4054–4058.

53. Y. Huang, Z. Lin and R. Cao, *Chem.–Eur. J.*, 2011, **17**, 12706–12712.

54. H. Li, Z. Zhu, F. Zhang, S. Xie, P. Li and X. Zhou, *ACS Catal.*, 2011, **1**, 1604–1612.

55. T. Ahnfeldt, D. Gunzelmann, T. Loiseau, D. Hirsemann, J. Senker, G. Férey and N. Stock, *Inorg. Chem.*, 2009, **48**, 3057–3064.

56. J. Gascon, U. Aktay, M. D. Hernandez-Alonso, G. P. M. van Klink and F. Kapteijn, *J. Catal.*, 2009, **261**, 75–87.

57. Y. Huang, S. Gao, T. Liu, J. Lü, X. Lin, H. Li and R. Cao, *ChemPlusChem*, 2012, **77**, 106–112.

58. Y. Huang, S. Liu, Z. Lin, W. Li, X. Li and R. Cao, *J. Catal.*, 2012.

59. U. Muller, L. Lobree, M. Hesse, O. M. Yaghi and M. Eddaoudi, WO 03/101975, 2003.

60. S. Wang, X. He, L. Song and Z. Wang, *Synlett*, 2009, 447–450.

61. H. L. Jiang, T. Akita, T. Ishida, M. Haruta and Q. Xu, *J. Am. Chem. Soc.*, 2011, **133**, 1304–1306.

62. T. Ishida, N. Kawakita, T. Akita and M. Haruta, *Gold Bull.*, 2009, **42**, 267–274.

63. F. Ke, J. Zhu, L. G. Qiu and X. Jiang, *Chem. Commun.*, 2013, **49**, 1267–1269.

64. D. Esken, S. Turner, O. I. Lebedev, G. Van Tendeloo and R. A. Fischer, *Chem. Mater.*, 2010, **22**, 6393–6401.
65. H. L. Jiang, B. Liu, T. Akita, M. Haruta, H. Sakurai and Q. Xu, *J. Am. Chem. Soc.*, 2009, **131**, 11302–11303.
66. M. Müller, S. Turner, O. I. Lebedev, Y. Wang, G. Van Tendeloo and R. A. Fischer, *Eur. J. Inorg. Chem.*, 2011, 1876–1887.
67. H. Liu, Y. Li, H. Jiang, C. Vargas and R. Luque, *Chem. Commun.*, 2012, **48**, 8431–8433.
68. M. Müller, S. Hermes, K. Kähler, M. W. E. Van Den Berg, M. Muhler and R. A. Fischer, *Chem. Mater.*, 2008, **20**, 4576–4587.
69. M. S. El-Shall, V. Abdelsayed, A. E. R. S. Khder, H. M. A. Hassan, H. M. El-Kaderi and T. E. Reich, *J. Mater. Chem.*, 2009, **19**, 7625–7631.
70. H. Zhao, H. Song and L. Chou, *Inorg. Chem. Commun.*, 2012, **15**, 261–265.
71. H. Zhao, L. Chou and H. Song, *React. Kinetics, Mech. Catal.*, 2011, 1–15.
72. Y. K. Park, S. B. Choi, H. J. Nam, D. Y. Jung, H. C. Ahn, K. Choi, H. Furukawa and J. Kim, *Chem. Commun.*, 2010, **46**, 3086–3088.
73. P.-Z. Li, K. Aranishi and Q. Xu, *Chem. Commun.*, 2012, **48**, 3173–3175.
74. U. Muller, O. Metelkina, H. Junicke, T. Butz and O. M. Yaghi, US Pat., 2004/081611, 2004.
75. S. Opelt, S. Türk, E. Dietzsch, A. Henschel, S. Kaskel and E. Klemm, *Catal. Commun.*, 2008, **9**, 1286–1290.
76. S. Gao, N. Zhao, M. Shu and S. Che, *Appl. Catal. A*, 2010, **388**, 196–201.
77. T. T. Dang, Y. Zhu, S. C. Ghosh, A. Chen, C. L. L. Chai and A. M. Seayad, *Chem. Commun.*, 2012, **48**.
78. M. Zhang, J. Guan, B. Zhang, D. Su, C. T. Williams and C. Liang, *Catal. Lett.*, 2012, **142**, 313–318.
79. W. Kleist, M. Maciejewski and A. Baiker, *Thermochim. Acta*, 2010, **499**, 71–78.
80. Y. Huang, Z. Zheng, T. Liu, J. Lü, Z. Lin, H. Li and R. Cao, *Catal. Commun.*, 2011, **14**, 27–31.
81. H. Liu, Y. Li, R. Luque and H. Jiang, *Adv. Synth. Catal.*, 2011, **353**, 3107–3113.
82. A. Henschel, K. Gedrich, R. Kraehnert and S. Kaskel, *Chem. Commun.*, 2008, 4192–4194.
83. J. Hermannsdörfer and R. Kempe, *Chem.–Eur. J.*, 2011, **17**, 8071–8077.
84. Y. Pan, B. Yuan, Y. Li and D. He, *Chem. Commun.*, 2010, **46**, 2280–2282.
85. F. G. Cirujano, A. Leyva-Pérez, A. Corma and F. X. Llabrés i Xamena, *ChemCatChem*, 2013, **5**, 538–549.
86. A. L. Tarasov, L. M. Kustov, V. I. Isaeva, A. N. Kalenchuk, I. V. Mishin, G. I. Kapustin and V. I. Bogdan, *Kinetics Catal.*, 2011, **52**, 273–276.
87. M. Müller, O. I. Lebedev and R. A. Fischer, *J. Mater. Chem.*, 2008, **18**, 5274–5281.
88. A. Aijaz, A. Karkamkar, Y. J. Choi, N. Tsumori, E. Rönnebro, T. Autrey, H. Shioyama and Q. Xu, *J. Am. Chem. Soc.*, 2012.

89. Y. Zhao, J. Zhang, J. Song, J. Li, J. Liu, T. Wu, P. Zhang and B. Han, *Green Chem.*, 2011, **13**, 2078–2082.

90. F. Schröder, D. Esken, M. Cokoja, M. W. E. Van Den Berg, O. I. Lebedev, G. Van Tendeloo, B. Walaszek, G. Buntkowsky, H. H. Limbach, B. Chaudret and R. A. Fischer, *J. Am. Chem. Soc.*, 2008, **130**, 6119–6130.

91. M. T. Reetz, *Adv. Mater.*, 1997, **9**, 943–954.

92. J. J. Davis, K. S. Coleman, B. R. Azamian, C. B. Bagshaw and M. L. H. Green, *Chem.–Eur. J.*, 2003, **9**, 3732–3739.

93. R. Plagemann, L. Jonas and U. Kragl, *Appl. Microbiol. Biotechnol.*, 2011, **90**, 313–320.

94. Y. Yu, B. Chen, W. Qi, X. Li, Y. Shin, C. Lei and J. Liu, *Microporous Mesoporous Mater.*, 2012, **153**, 166–170.

95. V. Lykourinou, Y. Chen, X.-S. Wang, L. Meng, T. Hoang, L.-J. Ming, R. L. Musselman and S. Ma, *J. Am. Chem. Soc.*, 2011, **133**, 10382–10385.

96. S. Hudson, J. Cooney and E. Magner, *Angew. Chem., Int. Ed.*, 2008, **47**, 8582–8594.

97. M. Hartmann, *Chem. Mater.*, 2005, **17**, 4577–4593.

98. K. M. de Lathouder, D. Lozano-Castelló, A. Linares-Solano, S. A. Wallin, F. Kapteijn and J. A. Moulijn, *Microporous Mesoporous Mater.*, 2007, **99**, 216–223.

99. S. Mitchell and J. Pérez-Ramírez, *Catal. Today*, 2011, **168**, 28–37.

100. J. F. Díaz and K. J. Balkus Jr, *J. Mol. Catal., B*, 1996, **2**, 115–126.

101. Y. K. Park, B. C. Sang, H. Kim, K. Kim, B. H. Won, K. Choi, J. S. Choi, W. S. Ahn, N. Won, S. Kim, H. J. Dong, S. H. Choi, G. H. Kim, S. S. Cha, H. J. Young, K. Y. Jin and J. Kim, *Angew. Chem., Int. Ed.*, 2007, **46**, 8230–8233.

102. Y. Chen, V. Lykourinou, C. Vetromile, T. Hoang, L.-J. Ming, R. W. Larsen and S. Ma, *J. Am. Chem. Soc.*, 2011, **133**, 16322.

103. S. S. Wu, J. Wang, W. H. Zhang and X. Q. Ren, *Catal. Lett.*, 2008, **125**, 308–314.

104. H. Atia, U. Armbruster and A. Martin, *J. Catal.*, 2008, **258**, 71–82.

105. M. A. Schwegler, P. Vinke, M. van der Eijk and H. van Bekkum, *Appl. Catal. A.*, 1992, **80**, 41–57.

106. J. Haber, K. Pamin, L. Matachowski and D. Mucha, *Appl. Catal. A*, 2003, **256**, 141–152.

107. I. V. Kozhevnikov, A. Sinnema, R. J. J. Jansen, K. Pamin and H. Bekkum, *Catal. Lett.*, 1994, **30**, 241–252.

108. P. Ferreira, I. M. Fonseca, A. M. Ramos, J. Vital and J. E. Castanheiro, *Catal. Commun.*, 2009, **10**, 481–484.

109. X. Y. Zhao, D. D. Liang, S. X. Liu, C. Y. Sun, R. G. Cao, C. Y. Gao, Y. H. Ren and Z. M. Su, *Inorg. Chem.*, 2008, **47**, 7133–7138.

110. G. Ferey, C. Mellot-Draznieks, C. Serre, F. Millange, J. Dutour, S. Surble and I. Margiolaki, *Science*, 2005, **309**, 2040–2042.

111. Z. Saedi, S. Tangestaninejad, M. Moghadam, V. Mirkhani and I. Mohammadpoor-Baltork, *J. Coord. Chem.*, 2012, **65**, 463–473.

112. N. V. Maksimchuk, M. N. Timofeeva, M. S. Melgunov, A. N. Shmakov, Y. A. Chesalov, D. N. Dybtsev, V. P. Fedin and O. A. Kholdeeva, *J. Catal.*, 2008, **257**, 315–323.

113. N. V. Maksimchuk, K. A. Kovalenko, S. S. Arzumanov, Y. A. Chesalov, M. S. Melgunov, A. G. Stepanov, V. P. Fedin and O. A. Kholdeeva, *Inorg. Chem.*, 2010, **49**, 2920–2930.

114. N. V. Maksimchuk, O. A. Kholdeeva, K. A. Kovalenko and V. P. Fedin, *Israel J. Chem.*, 2011, **51**, 281–289.

115. L. Yang, H. Naruke and T. Yamase, *Inorg. Chem. Commun.*, 2003, **6**, 1020–1024.

116. G. Hundal, Y. K. Hwang and J.-S. Chang, *Polyhedron*, 2009, **28**, 2450–2458.

117. F. Ma, S. Liu, D. Liang, G. Ren, C. Zhang, F. Wei and Z. Su, *Eur. J. Inorg. Chem.*, 2010, **2010**, 3756–3761.

118. F.-J. Ma, S.-X. Liu, D.-D. Liang, G.-J. Ren, F. Wei, Y.-G. Chen and Z.-M. Su, *J. Solid State Chem.*, 2011, **184**, 3034–3039.

119. C. Y. Sun, S. X. Liu, D. D. Liang, K. Z. Shao, Y. H. Ren and Z. M. Su, *J. Am. Chem. Soc.*, 2009, **131**, 1883–1888.

120. F. J. Ma, S. X. Liu, C. Y. Sun, D. D. Liang, G. J. Ren, F. Wei, Y. G. Chen and Z. M. Su, *J. Am. Chem. Soc.*, 2011, **133**, 4178–4181.

121. D. D. Liang, S. X. Liu, F. J. Ma, F. Wei and Y. G. Chen, *Adv. Synth. Catal.*, 2011, **353**, 733–742.

122. D. Mustafa, E. Breynaert, S. R. Bajpe, J. A. Martens and C. E. A. Kirschhock, *Chem. Commun.*, 2011, **47**, 8037–8039.

123. S. R. Bajpe, C. E. A. Kirschhock, A. Aerts, E. Breynaert, G. Absillis, T. N. Parac-Vogt, L. Giebeler and J. A. Martens, *Chem.–Eur. J.*, 2010, **16**, 3926–3932.

124. S. R. Bajpe, E. Breynaert, D. Mustafa, M. Jobbágy, A. Maes, J. A. Martens and C. E. A. Kirschhock, *J. Mater. Chem.*, 2011, **21**, 9768–9771.

125. L. H. Wee, C. Wiktor, S. Turner, W. Vanderlinden, N. Janssens, S. R. Bajpe, K. Houthoofd, G. Van Tendeloo, S. De Feyter, C. E. A. Kirschhock and J. A. Martens, *J. Am. Chem. Soc.*, 2012, **134**, 10911–10919.

126. S. Kayaert, S. Bajpe, K. Masschaele, E. Breynaert, C. E. A. Kirschhock and J. A. Martens, *Thin Solid Films*, 2011, **519**, 5437–5440.

127. L. H. Wee, S. R. Bajpe, N. Janssens, I. Hermans, K. Houthoofd, C. E. A. Kirschhock and J. A. Martens, *Chem. Commun.*, 2010, **46**, 8186–8188.

128. L. H. Wee, N. Janssens, S. R. Bajpe, C. E. A. Kirschhock and J. A. Martens, *Catal. Today*, 2011, **171**, 275–280.

129. J. Song, Z. Luo, D. K. Britt, H. Furukawa, O. M. Yaghi, K. I. Hardcastle and C. L. Hill, *J. Am. Chem. Soc.*, 2011, **133**, 16839–16846.

130. J. Juan-Alcañiz, M. Goesten, A. Martinez-Joaristi, E. Stavitski, A. V. Petukhov, J. Gascon and F. Kapteijn, *Chem. Commun.*, 2011, **47**, 8578–8580.

131. J. Juan-Alcaniz, M. G. Goesten, E. V. Ramos-Fernandez, J. Gascon and F. Kapteijn, *New J. Chem.*, 2012.

132. R. Canioni, C. Roch-Marchal, F. Sacheresse, P. Horcajada, C. Serre, M. Hardi-Dan, G. Farey, J. M. Grenache, F. Lefebvre, J. S. Chang, Y. K. Hwang, O. Lebedev, S. Turner and G. Van Tendeloo, *J. Mater. Chem.*, 2011, **21**, 1226–1233.

133. X. Kuang, X. Wu, R. Yu, J. P. Donahue, J. Huang and C.-Z. Lu, *Nature Chem.*, 2010, **2**, 461–465.

134. D.-d. Wang, J. Peng, P.-p. Zhang, X. Wang, M. Zhu, M.-g. Liu, C.-l. Meng and K. Alimaje, *Inorg. Chem. Commun.*, 2011, **14**, 1911–1914.

135. C. Zou, Z. Zhang, X. Xu, Q. Gong, J. Li and C. D. Wu, *J. Am. Chem. Soc.*, 2012, **134**, 87–90.

136. Y. Yang, S. Liu, C. Li, S. Li, G. Ren, F. Wei and Q. Tang, *Inorg. Chem. Commun.*, 2012.

137. J. Q. Sha, J. W. Sun, C. Wang, G. M. Li, P. F. Yan, M. T. Li and M. Y. Liu, *CrystEngComm*, 2012, **14**, 5053–5064.

138. X. Kuang, X.-Y. Wu, J. Zhang and C.-Z. Lu, *Chem. Commun.*, 2011, **47**, 4150–4152.

139. J. Juan-Alcañiz, E. V. Ramos-Fernandez, U. Lafont, J. Gascon and F. Kapteijn, *J. Catal.*, 2010, **269**, 229–241.

140. L. Bromberg, Y. Diao, H. Wu, S. A. Speakman and T. A. Hatton, *Chem. Mater.*, 2012, **24**, 1664–1675.

141. Y. Zhang, V. Degirmenci, C. Li and E. J. M. Hensen, *ChemSusChem*, 2010, **4**, 59–64.

142. E. V. Ramos-Fernandez, C. Pieters, B. Van Der Linden, J. Juan-Alcañiz, P. Serra-Crespo, M. W. G. M. Verhoeven, H. Niemantsverdriet, J. Gascon and F. Kapteijn, *J. Catal.*, 2012, **289**, 42–52.

143. L. Bromberg and T. A. Hatton, *ACS Appl. Mater. Interfaces*, 2011, **3**, 4756–4764.

144. O. A. Kholdeeva, N. V. Maksimchuk and G. M. Maksimov, *Catal. Today*, 2013, **157**, 107–113.

145. I. C. M. S. Santos, S. S. Balula, M. M. Q. Simões, L. Cunha-Silva, M. G. P. M. S. Neves, B. de Castro, A. M. V. Cavaleiro and J. A. S. Cavaleiro, *Catal. Today*, 2013, **203**, 87–94.

146. S. S. Balula, C. M. Granadeiro, A. D. S. Barbosa, I. C. M. S. Santos and L. Cunha-Silva, *Catal. Today*, 2013, **203**, 95–102.

147. X. Hu, Y. Lu, F. Dai, C. Liu and Y. Liu, *Microporous Mesoporous Mater.*, 2013, **170**, 36–44.

148. K. S. Suslick, N. A. Rakow, M. E. Kosal and J. H. Chou, *J. Porphyrins Phthalocyanines*, 2000, **4**, 407–413.

149. D. E. De Vos, M. Dams, B. F. Sels and P. A. Jacobs, *Chem. Rev.*, 2002, **102**, 3615–3640.

150. R. F. Parton, I. F. J. Vankelecom, M. J. A. Casselman, C. P. Bezoukhanova, J. B. Uytterhoeven and P. A. Jacobs, *Nature*, 1994, **370**, 541–544.

151. S. R. Batten, B. F. Hoskins and R. Robson, *J. Am. Chem. Soc.*, 1995, **117**, 5385–5386.

152. V. Faraon and R. M. Ion, *Optoelectron. Adv. Mater., Rapid Commun.*, 2010, **4**, 1135–1140.

153. M. Radha Kishan, V. Radha Rani, P. Sita Devi, S. J. Kulkarni and K. V. Raghavan, *J. Mol. Catal., A*, 2007, **269**, 30–34.

154. M. Moghadam, S. Tangestaninejad, V. Mirkhani, I. Mohammadpoor-Baltork and M. Moosavifar, *J. Mol. Catal., A*, 2009, **302**, 68–75.

155. F. C. Skrobot, A. A. Valente, G. Neves, I. Rosa, J. Rocha and J. A. S. Cavaleiro, *J. Mol. Catal., A*, 2003, **201**, 211–222.

156. F. C. Skrobot, I. L. V. Rosa, A. P. A. Marques, P. R. Martins, J. Rocha, A. A. Valente and Y. Iamamoto, *J. Mol. Catal., A*, 2005, **237**, 86–92.

157. M. H. Alkordi, Y. Liu, R. W. Larsen, J. F. Eubank and M. Eddaoudi, *J. Am. Chem. Soc.*, 2008, **130**, 12639–12641.

158. C. Zou and C.-D. Wu, *Dalton Trans.*, 2012, **41**, 3879–3888.

159. K. Ono, M. Yoshizawa, T. Kato, K. Watanabe and M. Fujita, *Angew. Chem., Int. Ed.*, 2007, **46**, 1803–1806.

160. R. W. Larsen and L. Wojtas, *J. Phys. Chem. A*, 2012, **116**(30), 7830–7835.

161. R. Sen, S. Koner, A. Bhattacharjee, J. Kusz, Y. Miyashita and K. I. Okamoto, *Dalton Trans.*, 2011, **40**, 6952–6960.

162. R. W. Larsen, L. Wojtas, J. Perman, R. L. Musselman, M. J. Zaworotko and C. M. Vetromile, *J. Am. Chem. Soc.*, 2011, **133**, 10356–10359.

163. Z. Zhang, L. Zhang, L. Wojtas, P. Nugent, M. Eddaoudi and M. J. Zaworotko, *J. Am. Chem. Soc.*, 2011, **134**, 924–927.

164. Z. Zhang, L. Zhang, L. Wojtas, M. Eddaoudi and M. J. Zaworotko, *J. Am. Chem. Soc.*, 2011, **134**, 928–933.

165. E. Kockrick, T. Lescouet, E. V. Kudrik, A. B. Sorokin and D. Farrusseng, *Chem. Commun.*, 2011, **47**, 1562–1564.

166. O. V. Zalomaeva, K. A. Kovalenko, Y. A. Chesalov, M. S. Mel'gunov, V. I. Zaikovskii, V. V. Kaichev, A. B. Sorokin, O. A. Kholdeeva and V. P. Fedin, *Dalton Trans.*, 2011, **40**(7), 1441–1444.

167. N. Janssens, L. H. Wee, S. Bajpe, E. Breynaert, C. E. A. Kirschhock and J. A. Martens, *Chem. Sci.*, 2012, **3**, 1847–1850.

CHAPTER 11

Asymmetric Catalysis with Chiral Metal Organic Frameworks

JOSEPH M. FALKOWSKI, SOPHIE LIU AND WENBIN LIN*

Department of Chemistry, CB#3290, University of North Carolina, Chapel Hill, NC 27599, USA
*Email: wlin@email.unc.edu

11.1 Introduction

11.1.1 Why Asymmetric Catalysts, And Why Metal Organic Frameworks?

The importance of chiral molecules in biological systems cannot be overstated. The various amino acids and sugars that form the building blocks of life are nearly exclusively L-amino acids and D-sugars.[1–3] This reality has important implications in medical and pharmaceutical applications. New drugs, for example, must be not only chemically pure but enantiopure as well. Contaminants of an undesirable enantiomer in a drug can often have undesirable and sometimes lethal repercussions for a patient. While the resolution of chiral molecules is a well-established method for producing enantiopure molecules, the intrinsically low atom efficiency of chiral resolutions makes it desirable to synthesize chiral molecules directly from achiral substrates using asymmetric catalysts. Many examples of asymmetric catalysts used for the synthesis of chiral molecules appear in the literature, but these catalysts often

RSC Catalysis Series No. 12
Metal Organic Frameworks as Heterogeneous Catalysts
Edited by Francesc X. Llabrés i Xamena and Jorge Gascon
© The Royal Society of Chemistry 2013
Published by the Royal Society of Chemistry, www.rsc.org

rely on elaborate and synthetically challenging chiral ligands.[4–11] To make these homogeneous systems commercially viable, it is often desirable to form their heterogenized analogs. Heterogenized catalysts are often easier to recycle and reuse and can be removed without costly chromatography or distillation procedures. As the most common porous heterogeneous catalyst system in use, zeolites have found a wide range of applications as strong acid catalysts in petroleum refining. The harsh synthetic methods used to synthesize zeolites, however, would destroy any features of chirality and preclude them from being a viable option for a solid-state asymmetric catalyst.

Like zeolites, metal organic frameworks (MOFs) have been demonstrated in the last decade to be competent catalysts for asymmetric transformations. This ability, which is not available to zeolite materials, is due to the relatively mild conditions under which MOFs are synthesized. Solvothermal and diffusion crystallizations used for MOF synthesis are much milder methods than the calcination process to remove templates from zeolites, allowing for the incorporation of delicate chiral moieties into a MOF without racemization or decomposition of the chiral sites.

In addition to their mild synthetic routes, the tunability and versatility of MOFs enable a wide catalytic scope. As we have seen thus far, MOF catalysis can occur either at the secondary building unit (SBU) or at rationally designed sites present on the linking ligand itself. Furthermore, as we shall see in some cases, the same material can utilize both types of catalytic sites cooperatively in order to perform multi-step transformations. While the metal of the SBUs can be modulated to effect different degrees of reactivity, the scope of SBU-catalyzed reactions remains limited to Lewis acid-type reactivity. The stereoselectivity that SBU-based catalysis exhibits is also often limited due to the active sites' distance from the chiral centers. By incorporating the catalytically active site and the chiral centers directly into the linking ligand, more diverse and selective catalysis can be conducted. Using catalytic linking ligands also allows for the opportunity to heterogenize well-studied homogeneous catalytic systems by incorporating them into MOFs. MOFs can thus serve as a platform bridging heterogeneous and homogeneous asymmetric catalysis and combine the benefits of both: the system exhibits the facile recyclability of heterogeneous systems while still maintaining the reproducibility and single-site nature of homogeneous systems.

11.1.2 Sources of Chirality

Asymmetric catalysis cannot be accomplished without a source of chirality. Before discussing how asymmetric catalysis is conducted with MOFs, it is best to understand the methods by which chiral MOFs can be synthesized. Of all the methods, three general schemes predominate: seeding, templating, and the use of enantiopure linking ligands (Scheme 11.1).[12]

One route for synthesizing chiral MOFs is to grow them from achiral starting materials, which can be achieved if the framework material crystallizes in a chiral space group. A well-studied example of this chiral self-resolution

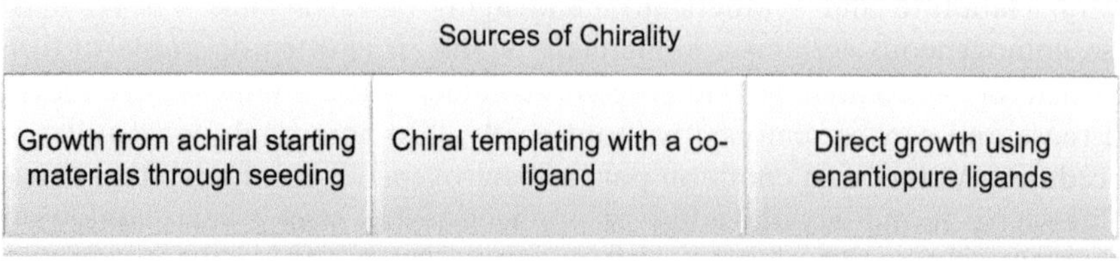

Sources of Chirality		
Growth from achiral starting materials through seeding	Chiral templating with a co-ligand	Direct growth using enantiopure ligands

Scheme 11.1 The various methods of introducing chirality into a metal organic framework.

phenomenon was discovered by Ezuhara.[13] In this example, [5-(9-anthracenyl)] pyrimidine was grown with cadmium nitrate tetrahydrate to form enantiopure crystals. It was discovered that if crystals of a particular handedness of the material were used to seed the crystal growth, a bulk material with the corresponding handedness would result. However, relying on framework materials to crystallize into chiral space groups as a source of chirality has some drawbacks. Not all systems will undergo this chiral self-resolution, and even if single crystals of a given framework do grow into chiral space groups, the bulk material in the absence of any outside influence will be racemic, consisting of an equal distribution of both enantiomers.

While the possibility of achieving a homochiral MOF without the use of chiral ligands is enticing, it relies on the chance crystallization of a ligand into a chiral space group. A more direct method involves the use of a chiral co-ligand to direct the handedness of the crystal growth. Rosseinsky demonstrated the use of 1,2-propanediol to direct the handedness of helices formed from nickel nitrate and benzenetricarboxylic acid.[14] In this example, the chiral co-ligand, while present in the growth solutions, did not constitute any part of the desolvated structure. Other examples of this method of introducing chirality into a MOF include the zinc(L-lac) material reported by Kim *et al.*; this material was grown with L-lactic acid as the co-ligand, which in this case remained part of the structure after crystal growth.[15] As we will see later in this chapter, the use of chiral templating has also been used in the synthesis of catalytically active framework materials.

A third option, and often the most effective, is to directly grow a MOF from a chiral ligand. The vast majority of the chiral MOFs presented in the literature are synthesized with commercially available chiral molecules, such as amino acids, hydroxyl carboxylic acids, nucleic acids, and biotin.[16–19] A variant of this approach is to incorporate these chiral molecules into a MOF *via* post-synthetic modification (PSM) after the formation of the framework structure; chiral MOFs produced in this manner have been demonstrated to catalyze reactions with high enantioselectivities.[20] The disadvantage of using these abundant natural sources of chirality is the lack of catalytic sites that can be exploited inside of the framework channels. For this reason, a great deal of research has been conducted on the modification of privileged catalytic systems such as BINOL and salen systems. Homogeneous catalysts based on these ligand systems are well studied, and the scope of their reactivity is broad due to the wide range of metals that can be used with these ligands. These aspects make them ideal candidates for their immobilization in MOFs.

11.1.3 Benefits of MOFs in Asymmetric Catalysis

MOFs, like other heterogenized catalysts, can be easily removed by simple filtration methods and then recycled. The ease of catalyst removal from a reaction medium also helps reduce heavy metal contamination in the product. This advantage of heterogeneous catalysts over their homogeneous counterparts is of particular importance in the pharmaceutical industry, where trace heavy metal contamination is often not acceptable. The ease of recycling also allows for MOFs that utilize more expensive or synthetically challenging chiral ligands in their synthesis; because the catalyst is retrievable after the reaction, it becomes economically viable to utilize more efficient yet costly chiral ligands on large scales.

MOFs can also offer enhanced activity over other heterogenized or homogeneous systems. Because of the high degree of order that the active catalyst sites exhibit, it is possible to limit decomposition pathways within framework materials. This increased activity is exemplified in the manganese-salen framework reported by Lin *et al.*[21] In this example, the bimolecular decomposition pathway that leads to catalyst deactivation is suppressed by restraining the motion of the catalytic moieties to the walls of the framework channels. Beyond enhancing the activity of the catalyst, the selectivity of MOF-based systems can be enhanced *via* confinement effects and restrained motion of the chiral ligand.[20,22] Several reports by the groups of Kim and Duan demonstrated increases in the selectivity of a catalyzed reaction as a result of immobilizing the catalyst inside the framework.

Beyond the benefits in selectivity and stability afforded by MOF catalysts, the crystalline nature of these materials also allows for precise structure–function relationships to be elucidated. By utilizing single-crystal X-ray diffraction, the structure of the active catalytic sites in a MOF and the effect of the environment around the catalytic center can be determined. Lin *et al.* were able to use the structure of an isolated titanium BINOLate-functionalized framework to rationalize the decreased selectivities that were observed in the catalytic alkylation of aldehydes.[23] In this case, an intermolecular cross-linking was responsible for the change in the active site, which, in turn, reduced the chiral induction imparted on the catalytic active site (Figure 11.1).

11.2 First Examples of Asymmetric Catalysis

In 2000, Kim and coworkers were the first to publish the asymmetric synthesis of a chiral product using a homochiral MOF as the catalyst. In this work, a chiral framework was synthesized from a tartaric acid derivative and zinc nitrate hexahydrate (Figure 11.2).[24] The resulting 2-dimensional structure contained trinuclear zinc cluster SBUs, which were connected by six ligands

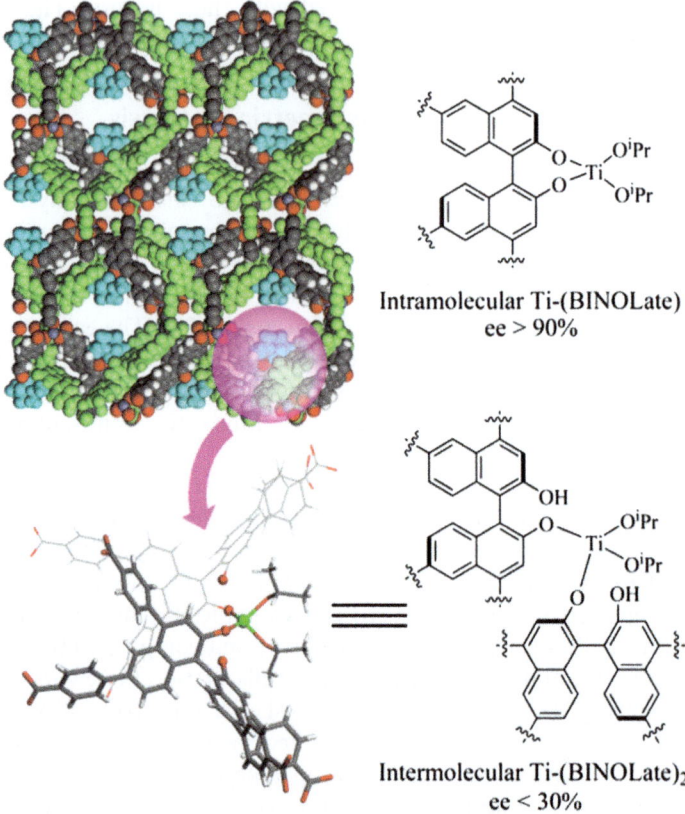

Figure 11.1   Inter- *vs.* intra-molecular coordination of titanium as revealed by X-ray diffraction. The crystallinity of MOFs offers unique opportunities to elucidate structure–function relationships.

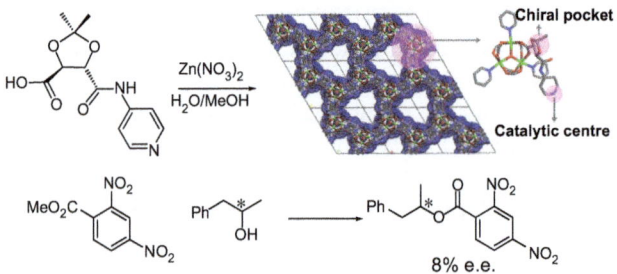

Figure 11.2   The synthesis of a face-on hexagonal two-dimensional framework and the kinetic resolution of 1-phenyl-2-propanol as reported by Kim and coworkers.

bound to the SBUs through their carboxylate moieties. Three of these ligands were coordinated to adjacent SBUs through their pyridine functionalities, resulting in a face-on hexagonal 2-dimensional framework.

The remaining three ligands do not participate in the network structure; instead, they have their pyridine groups dangling in the open channels. Kim *et al.* believed that these accessible basic sites could be utilized for catalysis and screened the material for activity in the asymmetric transesterification reaction of ethyl-2,4-dintrobenzoate with racemic 1-phenylproanol. They observed that the resulting transesterified product contained a slight enantiomeric excess (e.e.) of 8%.

At around the same time, Lin and coworkers synthesized a chiral porous framework from lanthanum group metals and a phosphate-substituted diethoxy-BINOL ligand.[25] The phosphates coordinated to the lanthanum ion SBUs to form a 2-dimensional lamellar framework (Figure 11.3).

This material was screened for activity in the cyanosilylation of aldehydes; nearly racemic products were obtained with all of the substrates that were screened. When the material was screened in the ring opening of meso cyclic anhydrides, small enantioselectivies were observed (5% e.e.). The disappointing selectivities exhibited by these two examples reveal the inefficiencies of utilizing remote induction of chirality onto a Lewis acidic SBU to produce an enantioselective catalyst.

Realizing the ineffectiveness of the Lewis acid catalysis with respect to enantioselectivity, Lin and coworkers developed a rationally designed catalyst in which the chiral and catalytically active centers were spatially adjacent, increasing the degree of chiral induction that can be exerted in an asymmetric reaction.[26,27] In these materials, zirconium phosphonate coordination polymers

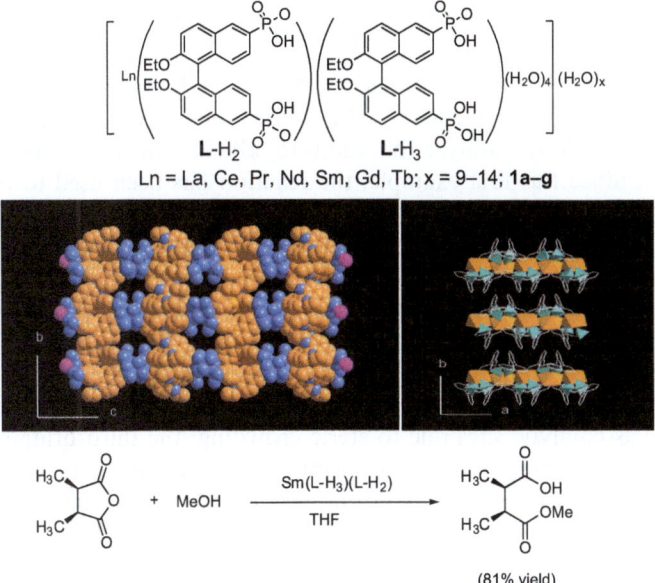

Figure 11.3 A 2-dimensional, lamellar, chiral framework and its activity in the asymmetric ring opening of meso-epoxides.

were prepared from a ruthenium-BINAP complex and zirconium tetrachloride. The resulting amorphous material was then screened in the asymmetric hydrogenation of α,β-unsaturated ketones. This resulted in high yields and high enantioselectivities of up to 95% with quantitative conversion. This material represented the first time that a rationally designed coordination polymer catalyst was utilized for asymmetric reactions. Since this investigation, a series of advances have progressed the field of heterogeneous asymmetric catalysis with MOFs, with the scope of their reactivity broadening with the progression of time.

11.3   Reactivity Scope

11.3.1   Additions to Carbonyl Groups

The carbonyl functionality offers a wide range of potential reactivity. Because of its role as an electrophile and as a potential nucleophile, it can be harnessed in a wide range of asymmetric reactions. Additionally, their activation by Lewis acid and Lewis base catalysis makes them ideal substrates for MOF-based catalysts.

11.3.1.1   Diethylzinc Additions

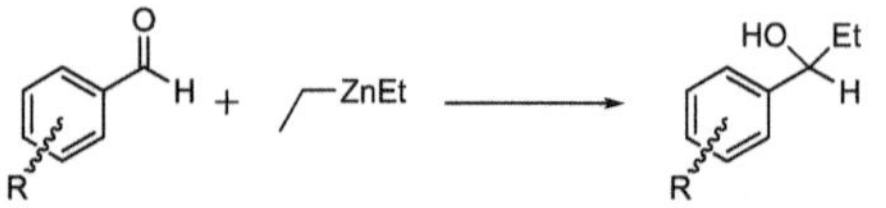

Lin and coworkers first demonstrated the MOF-catalyzed diethylzinc addition to aromatic aldehydes in 2004.[28] The amorphous zirconium phosphonate materials were able to catalyze the addition, albeit with low selectivities and conversions. Since this first example, the reaction has been used to probe the reactivity and utility of new BINOL-based MOF catalysts. Following up on this work, Lin and coworkers synthesized a crystalline material from a 4,4′-dipyridine-substituted BINOL linker and cadmium nitrate in 2005.[29] The 3-dimensional material consisted of Cd(μ-Cl)₂ chains bridged by pairs of interlocking, hydrogen-bonded ligands, forming a 2-dimensional network. A third ligand then bridged these 2-dimensional networks into a 3-dimensional structure. While the hydroxyl groups of the tightly bound ligand pairs were inaccessible as catalytic sites due to steric crowding, the third bridging ligand possessed chelating sites that were accessible to the open channels (Figure 11.4).

After activating the material with titanium tetra(isopropoxide), the efficacy of the material toward the asymmetric addition of diethylzinc to aromatic aldehydes was determined. Using various substituted aromatic aldehydes, the reactions exhibited high yields and enantioselectivities comparable to those of the homogeneous catalyst. It was also observed that the activity of the catalyst was inversely related to the size of the substrate. This observation provided

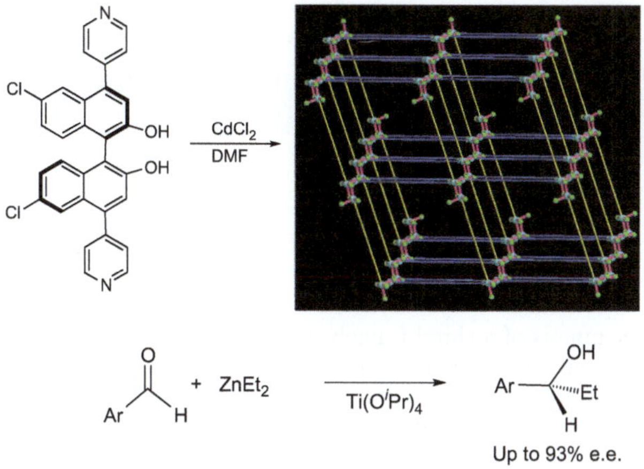

Figure 11.4 The synthesis of a 3-dimensional chiral framework from a 4,4'-pyridyl substituted BINOL ligand. The framework consists of a 2-dimentional network consisting of zig-zag chains of Cd(μ-Cl)$_2$ SBUs (red chains) connected by hydrogen-bonded ligand pairs (blue rods). The 2-dimensional layers are connected by a third ligand (yellow rods) to form the 3-dimensional framework.

evidence that the reaction was indeed proceeding at the catalytic sites in the open channels and not at sites located on the surface of the material.

11.3.1.2 Cyanosilylation Reactions

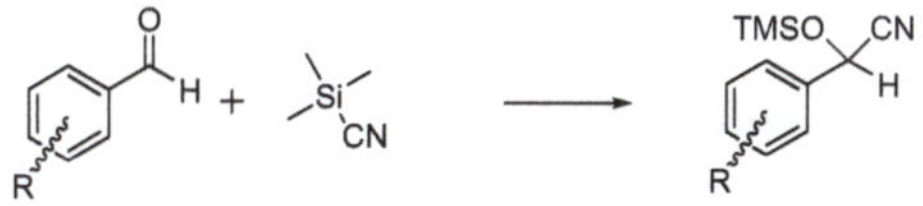

Several groups have utilized the cyanosilylation of aldehydes as a model reaction for MOF catalysts. Duan and coworkers demonstrated a unique system in which a chiral framework was synthesized from achiral starting materials.[22] This system was constructed through the use of the chiral templating agents L- and D-pyrrolidine-2-yl-imidizole (PYI). When methyl-enediisophthalic acid was reacted with cerium nitrate in the presence of this templating agent, an enantiopure framework resulted; the handedness could be controlled by changing the handedness of the templating agent (Figure 11.5).

Using 2 mol% loading, this MOF was able to catalyze the asymmetric addition of trimethylsilylcyanide to aromatic aldehydes with >98% enantios-electivity. This is the only instance of chiral induction in an enantioselective MOF-catalyzed reaction that results purely from the chirality of the framework itself; additionally, this result represents a rare instance of a highly selective catalyst that utilizes the SBUs as the catalytic active sites.

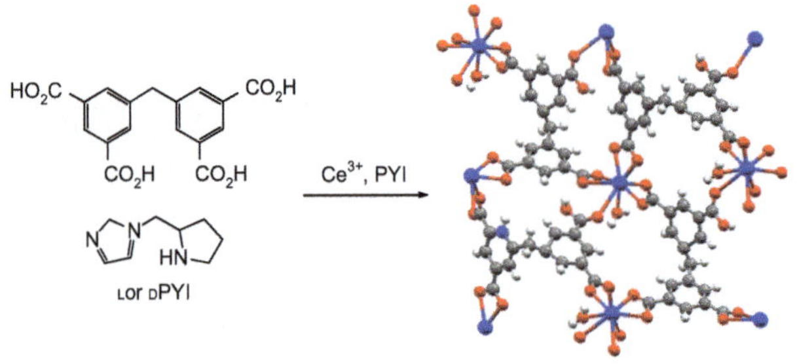

Figure 11.5 Synthesis of a chiral framework from achiral starting materials utilizing
L- and D-pyrrolidine-2-yl-imidizole (PYI) as chiral templating agents.
The resulting enantiopure MOF crystals are active catalysts for the
cyanosilylation of aldehydes.

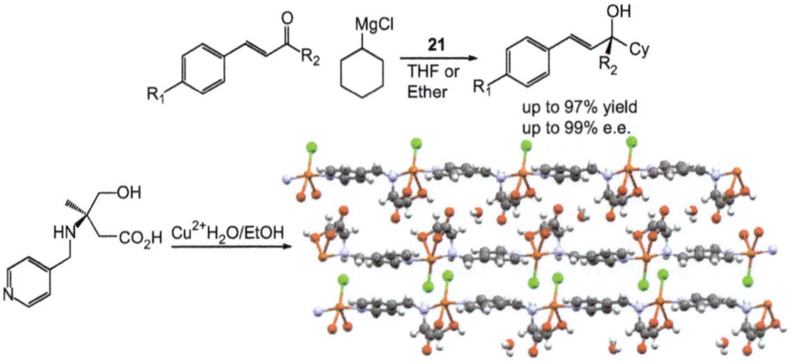

Figure 11.6 The 1-dimensional coordination polymer reported by Wu *et al.*
constructed of modified serine ligands and copper SBUs (orange).

11.3.1.3 Grignard Additions

Grignard reagents are potent and versatile nucleophiles that can be conjugated
with aldehydes and ketones to make a host of secondary and tertiary alcohols.
Wu *et al.* reported a 2-dimensional lamellar framework constructed from a
modified serine derivative (Figure 11.6).[30] By using the amine sites, this lamellar
framework could selectively catalyze the addition of cyclohexylmagnesium
chloride to only one face of a prochiral ketone, forming the tertiary alcohol
product in up to 98% enantiomeric excess. In these reactions, the observed
selectivity was greater than that of the corresponding reaction catalyzed by the
free ligand. This increase in activity was the result of a second chiral center
generated from the complexation of the amine functionality with the copper
SBU. Experiments with this system showed that due to the relatively small size
of the pores, the catalysis that was observed was mainly performed by the active
sites on the surface of the material rather than active sites in the MOF channels.
This material was also screened for its activity in the Biginelli reaction because
of the presence of Cu(II) centers; however, despite excellent conversion, no
enantioselectivity was observed.

11.3.1.4 Aldol Reactions

Aldol reactions are a powerful way to utilize carbonyl compounds as both an electrophile and nucleophile in organic transformations. Several MOFs have been reported to catalyze asymmetric aldol condensation reactions. In 2009, Kim and coworkers reported a post-synthetically modified MIL-101 framework in which a pyridine-derived L-proline co-ligand was coordinated to the unsaturated chromium sites in the MIL-101 SBUs.[20,31] Using the pendant proline groups, the modified framework was able to catalyze the aldol reaction between aromatic aldehydes and either acetone or cyclopentanone. The aldol products were obtained at up to 91% yields and up to 76% e.e., respectively. The unmodified material demonstrated limited reactivity (what little reactivity that was observed was attributed to the Lewis acid activation by the chromium SBUs), while the free ligand was able to catalyze the reaction albeit with lower enantioselectivities (29% e.e.). The researchers attributed this e.e. enhancement in the MOF system to the restricted movement of the substrates in the microporous channels of the MIL-101 framework as well as the presence of multiple chiral induction sites in the MOF channels.

Using a different approach, Duan and coworkers obtained a chiral framework from 1,3,5-tris(4-carboxyphenyl)benzene, cadmium nitrate, and an enantiopure co-ligand, L- or D-pyrrolidine-2-yl-imidizole (Figure 11.7).[22]

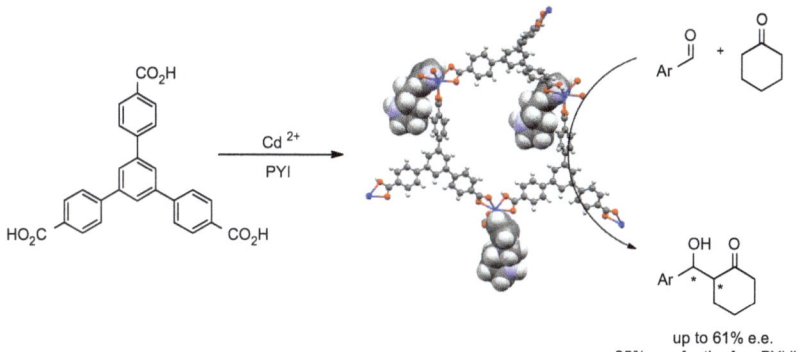

up to 61% e.e.
25% e.e. for the free PYI ligand

Figure 11.7 Synthesis of a chiral 2-dimensional framework from 1,3,5-tris(4-carbo-xyphenyl)benzene in the presence of the co-ligand, PYI. The chiral PYI ligands (drawn as space filling atoms) coordinate to the cadmium SBUs (blue) and occupy the space between 2-D layers. The MOF-supported PYI ligands are able to catalyze aldol reactions with much higher selectivities than the free PYI ligand.

The tetrahedral cadmium SBUs were bridged by three benzene tricarboxylate ligands and the nitrogen atom of an imidazole ring. The Cd^{2+} ions and the three benzene tricarboxylate ligands formed a honeycomb 2-dimensional framework. The imidizole ligands, located above and below the 2-D network, contained active N–H pyrrolidine groups that were exposed to the open channels and were shown to be active in aldol reactions between aromatic aldehydes and cyclohexanone. While the yields of the MOF-catalyzed reactions approached those of the homogeneous system, the enantioselectivity afforded by the MOF catalyst far exceeded that of the free ligand alone. Duan attributed the enhanced chiral induction to the restriction of substrate movement inside the channels of the framework.

Telfer *et al.*, using a new ligand derived from 4,4′-biphenyldicarboxcylic acid and a pendant proline group, was able to synthesize a new MOF with a chiral precatalyst incorporated into the linking ligands.[32] In this example, the pendant proline group was protected as a tertbutylamide (BOC). This protecting group served to prevent the basic –NH group from coordinating and becoming deactivated during the crystal growth. The reaction of the ligand with zinc nitrate yielded a cubic MOF in which the pendant proline group resided in the open channels. Upon heating the framework, the BOC group thermally decomposed, yielding the free, catalytically active amine with minimal racemization of the chiral pendant group (Figure 11.8).

Using the free proline-decorated MOF, the organocatalytic aldol reactions of *para*-nitrobenzaldehyde with different linear and cyclic ketones were found to give quantitative conversions; however, only very modest selectivities of 29% were observed for the reaction of the aromatic aldehyde with acetone. In this case, the larger channel sizes and limited rigidity of the proline group allow for limited differentiation of the ri and si face of the ketone, resulting in lowered selectivity.

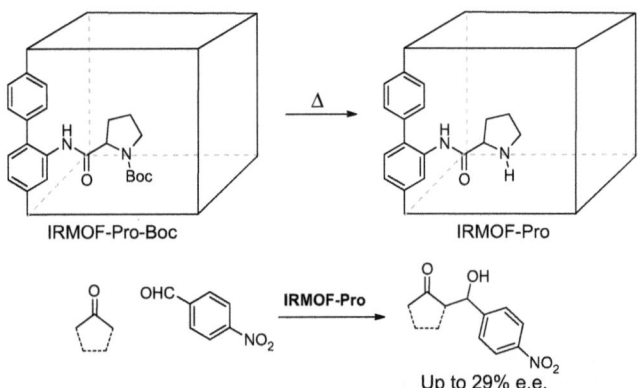

Figure 11.8 Synthesis of a cubic framework and the thermal deprotection of the catalytic sites.

11.3.2 Atom Transfer Reactions

11.3.2.1 Epoxidation Reactions

$$X = (CH_2)_{0 \to 1}, O$$

Like carbonyl groups, alkenes offer a wide scope of reactivity from which new functionalities can be constructed. Specifically, the oxidation of alkenes by transition metal catalysts offers a method of forming useful synthons. Hupp and coworkers reported the first MOF-based catalyst for the asymmetric epoxidation of alkenes.[33] In this system, a pyridine-derived salen catalyst was metalated with manganese and then subsequently oxidized to form the manganese(III) salen chloride complex. Treating this ligand with zinc nitrate in the presence of 4,4′-bipyridine led to a layered structure composed of sheets of salen ligands connected by dinuclear zinc paddlewheel SBUs. These 2-dimensional layers were connected by the 4,4′-bipyridine co-ligands to form a 3-dimensional structure (Figure 11.9).

Using the manganese MOF, chromene derivatives were epoxidized in high yields (71%) and high selectivities (82%) with the mild oxidant 2-(*tert*-butylsulfonyl)iodosylbenzene. Recycling the MOF showed only a mild loss of activity without a decrease in selectivity. Leaching experiments showed that between 4 and 7% of the initial manganese leached out of the sample during the reaction either as the free molecular species or as particles too small to be trapped by filtration. Heterogeneity tests, however, showed that this small amount of leaching did not contribute to the overall production of epoxide product. While the MOF catalyst was unable to match the selectivity of the homogeneous catalyst, turnover numbers of >3000 were obtained for the MOF *versus* ~1000 for the free catalyst under identical conditions. The increase in the turnover number exemplifies the advantage of a highly ordered MOF

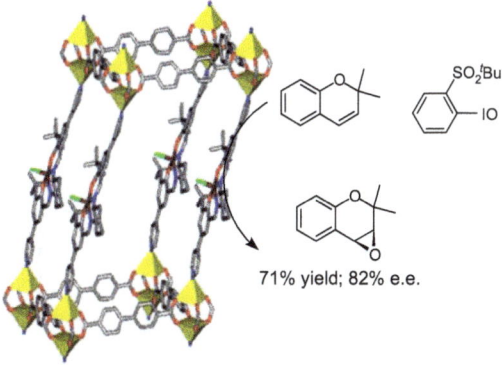

71% yield; 82% e.e.

Figure 11.9 Epoxidation of 2,2,-dimethyl-2H-chromene using a pillared MOF framework reported by Hupp and coworkers.

catalyst; by restricting the motion of catalytic units *via* incorporation into the framework, intermolecular decomposition pathways are eliminated, thus increasing the lifetime of the catalyst.

Using a similar reaction scheme, Lin and coworkers developed an isoreticular series of frameworks constructed from carboxylate-derived manganese-salen ligands in 2010.[21] Reacting the ligand with zinc nitrate resulted in a 3-dimensional cubic network with tetranuclear zinc cluster SBUs (Figure 11.10).

By varying the ligand length as well as the catenation of the formed networks, the researchers were able to modulate the size of the channels, which was found to affect the rate of substrate diffusion during catalysis. The rate of substrate diffusion, in turn, affected the rate of the epoxidation and could be measured quantitatively (Figure 11.11).

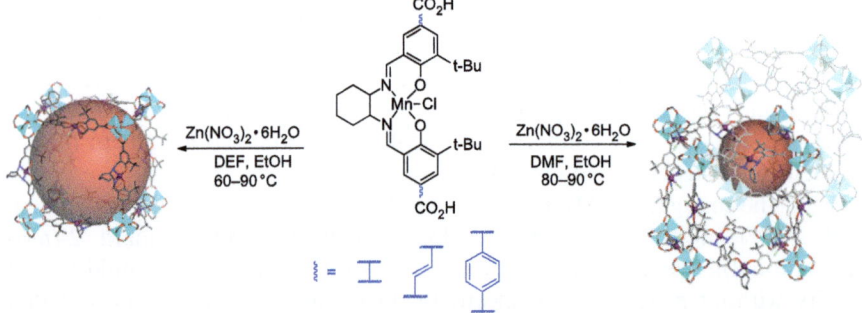

Figure 11.10   Synthesis of a series of chiral MOFs with varying channel sizes controlled by framework catenation and ligand length.

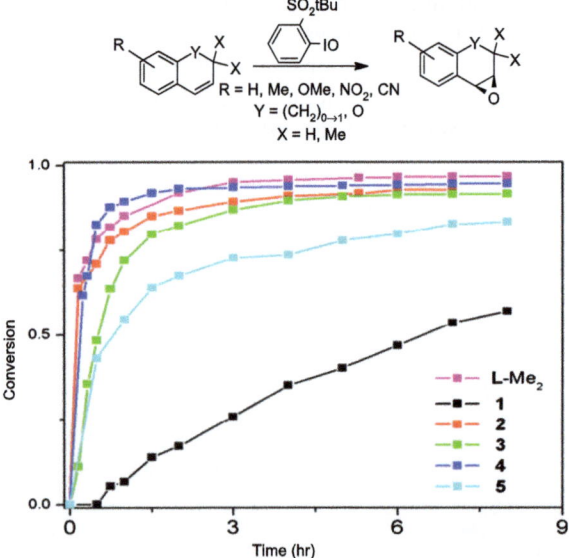

Figure 11.11   Rate of substrate formation when using an isoreticular series of Mn(salen) frameworks of varying channel dimensions.

The MOFs were found to be active epoxidation catalysts with a broad scope of alkene substrates. Screening the reaction of 2,2-dimethyl-2H-chromene with the isoreticular MOF series revealed the dependence of the epoxidation rates on the channel dimensions of the framework. It was observed that as the channel size increased, the reaction rate with the MOF approached and even surpassed the rate with a homogeneous control catalyst. Furthermore, for several chromene derivatives, MOFs gave higher yields of epoxidation products than the homogeneous control catalyst did. Finally, the isoreticular MOF system also exhibited higher levels of stereoselectivity than other previous salen-based MOF catalysts.

11.3.2.2 Carbene Transfer Reactions

In addition to the oxygen atom transfer to alkenes, metallosalen complexes have also been used in the transfer of other functional groups, specifically carbene moieties to yield cyclopropane fragments.[33] Inspired by work of Nguyen *et al.* on the homogeneous ruthenium-salen catalyzed carbene transfer reactions, Lin *et al.* synthesized a dicarboxylate-derived ruthenium-metallated salen ligand. The oxidized ruthenium-salen ligand was then used to synthesize a cubic 3-dimensional framework with tetranuclear zinc cluster SBUs isostructural to the previously reported manganese-salen-based materials.[34] Treating the ruthenium(III) MOF with lithium triethylborohydride yielded the active ruthenium(II) MOF, which could then be utilized in catalyzing asymmetric carbene transfer reactions (Figure 11.12).

This reduction resulted in no loss of crystallinity and could be monitored spectroscopically. The reduction of the MOF was demonstrated to be completely reversible upon exposure in air, which converted the reduced MOF back to the

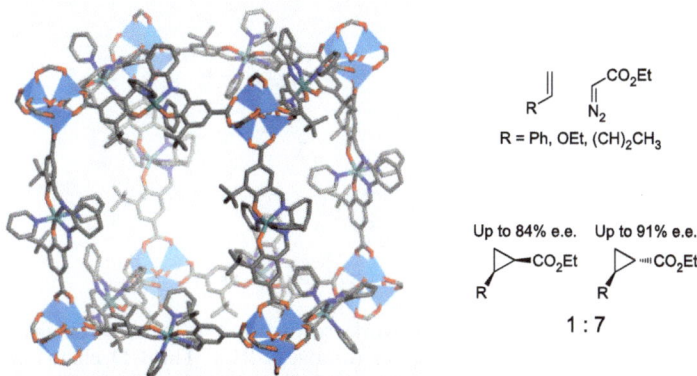

Figure 11.12 MOF-immobilized, redox-actuated Ru(salen) carbene transfer catalyst.

original material as determined by UV-Vis spectroscopy. Using ethyl-2-diazoacetate (EDA) as the carbene precursor, the reduced ruthenium MOF was able to catalyze the transfer of the ethyl ester moiety to various alkenes with relatively high yield (54%) and excellent enantioselectivity (up to 92%).

11.3.3 Asymmetric Ring Opening Reactions

11.3.3.1 Ring Opening of Epoxides

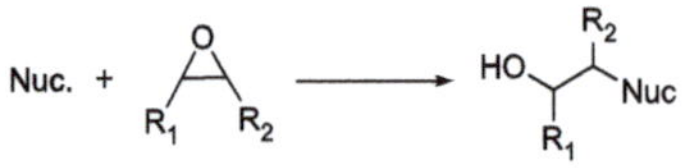

Chiral epoxides are often desirable functionalities due to the ease of their transformations into other optically active functional groups. We have already reviewed one method of synthesizing optically active epoxides directly through the epoxidation of double bonds. Kinetic resolution is another popular method of producing chiral epoxides. In this method, a difference in the rate of reaction of one enantiomer of the epoxide substrate results in an enantiomeric excess in the products. Often the resolution is carried out with water as the nucleophile, resulting in the corresponding diol, which is discarded as a byproduct, leaving the desired resolved epoxide. In some cases, the nucleophile can be chosen so that both the starting material and product are desirable compounds. The ring-opening of epoxides has been explored with MOFs using both Lewis-acidic as well as Brønsted-acidic sites.

Using the chiral ligand L-aspartate (L-asp), Rosseinsky and coworkers synthesized a 2-dimensional framework by reacting the chiral ligand with copper chloride and 4,4′-bipyridine (Figure 11.13).[35] The framework consisted of 1-dimensional chains of L-asp ligands and copper SBUs bridged by the bipyridine ligands. Upon treatment of this material with anhydrous HCl, a protonated carboxylate formed, which could then be characterized by IR

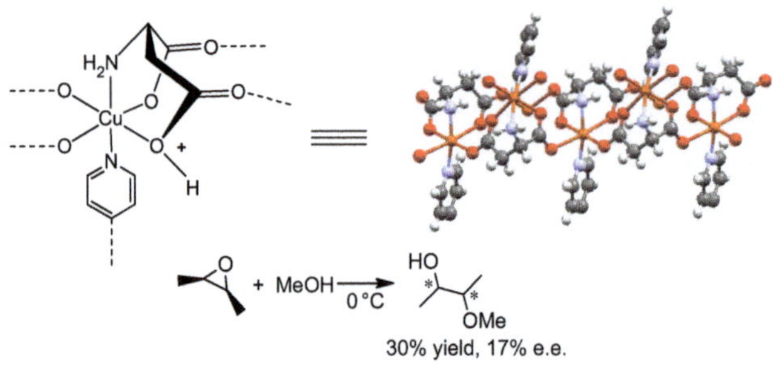

30% yield, 17% e.e.

Figure 11.13 2-Dimenstional framework constructed from 1-D chains of L-asp ligands bound to copper (orange) SBUs. The 1-D chains coordinate with 4,4′-bipyridine to form a 2-dimensional network. The protonated carboxylic acid can act as a Brønsted acid catalyst in the asymmetric ring opening of propylene oxide.

spectroscopy. Using this site as a Brønsted acid, the activity of this material in the asymmetric ring-opening of propylene oxide (PO) was determined. Using methanol as the nucleophile in the ring-opening of PO, the corresponding methoxy alcohols were synthesized in up to 17% e.e. and 65% yield. While the activity and selectivity of this catalyst were modest compared to other enantioselective reactions catalyzed by MOFs, this example presents a rare case of a Brønsted acid's use as a catalytic moiety in a MOF. Typically, the growth conditions would deprotonate any acidic groups that are present on the ligand during crystal growth. The inherent instability of MOFs at low pH typically precludes the option of post-synthetically modifying a MOF under acidic conditions.

Using a different approach, the Tanaka group utilized the Lewis-acidic character of the copper paddlewheel SBUs of a MOF to catalyze the kinetic resolution of styrene oxide.[36,37] Using a BINOL-based MOF grown with copper nitrate, a 2-dimensional structure was obtained in which the BINOL units were connected by copper paddlewheels (Figure 11.14).

In this structure, the axially bound solvent molecules on the copper padd-lewheels were removed under vacuum to yield the active catalytic species. Using methanol as a nucleophile, the methanolytic kinetic resolution of epoxides was performed. In this system, the enantiomeric excess of the starting material and the products varied based on the temperature. When run at room temperature, the e.e. of the isolated ring-opened product was 81%, while residual starting material was obtained in only 5% e.e. These two compounds were isolated at 5% and 83% yields, respectively. When the reaction was carried out at higher temperatures (60°C), the starting material could be recovered in up to 29% yield with an e.e. of up to 98%. This same system could also catalyze the asymmetric ring opening of epoxides using aniline as the nucleophile. This

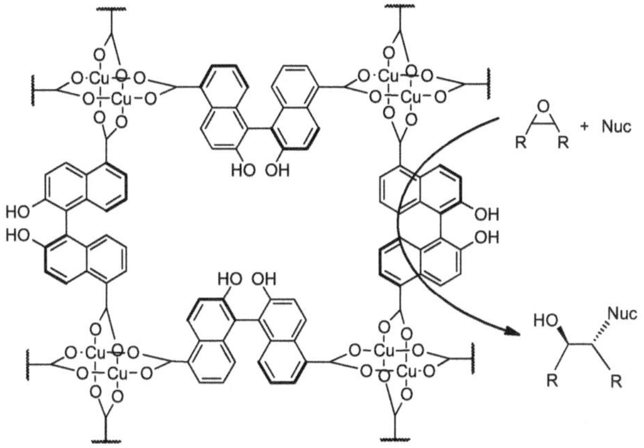

Figure 11.14 A 2-dimensional MOF reported by Tanaka and coworkers utilized coordinately unsaturated copper paddlewheel SBUs to catalyze the ring opening of epoxides with methanol and aniline.

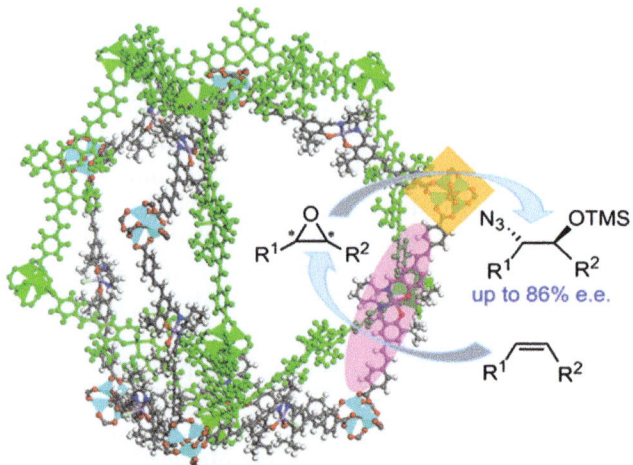

Figure 11.15   Doubly interpenetrated salen-based MOF with **icy** topology. Utilizing
both salen and zinc centers as active catalysts, Lin and coworkers were
able to demonstrate sequential catalysis using a MOF.

reaction proceeded with higher selectivity, with up to 51% e.e. and 50% yield of
the ring-opened product.

Thus far, the production of optically active molecules through the reaction of
racemic epoxides has depended on the differences of the rate of reaction
between the enantiomers of the starting material. This approach, however, has
the disadvantage of limiting the yield of any single enantiomer to 50%. An
alternative route explored by Lin and coworkers involves the synthesis of a
chiral epoxide directly from achiral starting materials and then subsequently
performing a ring-opening reaction utilizing the same MOF material as a
catalyst for both reactions.[38] By reacting a diacid-substituted salen ligand with
zinc nitrate in dibutylformamide and ethanol, a framework with the **icy**
topology was formed (Figure 11.15).

The resulting framework contains two catalytically active centers: the Mn-
salen site for asymmetric epoxidation and the tetranuclear Zn SBUs as Lewis
acidic sites for epoxide ring-opening reactions with trimethylsilyl azide. After
the epoxidation and subsequent ring opening, the ring-opened product could be
obtained in yields of up to 60% with enantioselectivities of up to 81%. The
ring-opening step was highly regioselective, with only one pair of enantiomers
being obtained from a possibility of four.

11.4   Framework Structure-Dependent Stereocontrol

Due to the crystalline nature of MOFs, the relationship between the structure
and the observed catalytic performance can be readily elucidated *via* X-ray
diffraction studies. Utilizing a previously reported BINOL-based ligand, the
Lin group was able to grow two networks of different topologies using

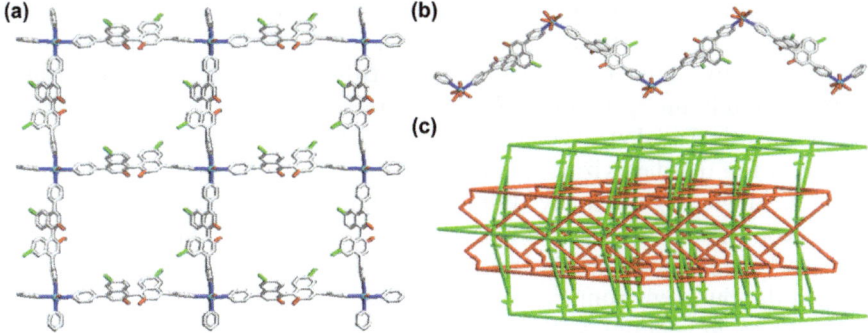

Figure 11.16 (a) The formation of 2-dimensional nets from the cadmium SBUs and the 4,4′-substituted ligands. (b) The formation of the 1-dimensional zigzag chains. (c) 3-dimensional framework constructed of bridged 2-D nets and 1-D chains.

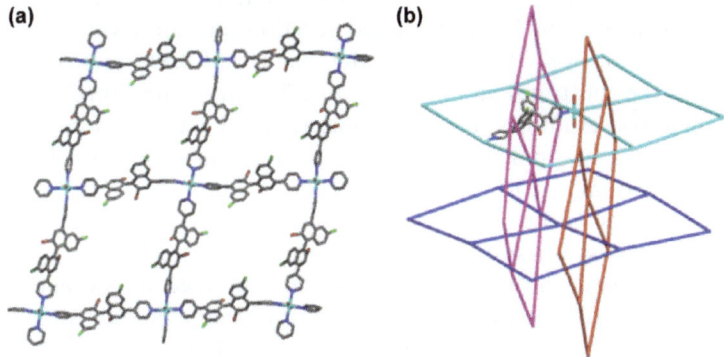

Figure 11.17 (a) 2-Dimensional network formed from cadmium SBUs. (b) Interpenetration of two orthogonal 2-dimensional networks.

cadmium as the connecting metal.[39] Through alterations in the crystal growth conditions, two materials were formed. The first material contained two unique frameworks that consisted of 2-dimensional networks (Figure 11.16a) stacked with 1-dimensional zigzag chains occupying the space between the 2-dimensional layers. The zigzag chains and 2-D layers were linked through bridging nitrate groups. The second material was formed from a pair of orthogonal interlocking 2-dimensional networks (Figure 11.17). Treatment of the former MOF with Ti(^{i}OPr)$_4$ led to a competent catalyst for the addition of diethylzinc to aromatic aldehydes with high conversion and good enantioselectivities (up to >99% yield and 90% e.e.). Treatment of the latter MOF with Ti(^{i}OPr)$_4$ under the same conditions did not afford an active catalyst for diethylzinc addition reactions. It was rationalized, based on the X-ray structures, that steric crowding around the dihydroxyl groups prevented the coordination of Ti(^{i}OPr)$_4$ to the BINOL ligand in the latter MOF.

In another study, Lin *et al.* were able to elucidate the enantioselectivity differences among MOF catalysts based on the formation of typically unobservable active catalytic species. Using a BINOL-derived tetracarboxylate ligand, a doubly interpenetrated Zn MOF framework was obtained.[23] Upon treatment with Ti(^{i}OPr)$_4$, the resulting catalyst gave much lower enantioselectivities for the diethylzinc addition products than the other MOF-based catalysts. Single crystal X-ray diffraction studies showed that the titanium centers coordinated to the BINOL moieties of two different folds of the framework. This intermolecular *versus* intramolecular coordination drastically changed the chiral environment around the active center.

Differences in enantioselectivities of MOF catalysts as a result of different open channel sizes were also observed by Lin *et al.*[40] Using a series of 4,4´,6,6´ substituted BINOL ligands, a isoreticular series of MOFs were synthesized with channels ranging from 1 nm to 3 nm. After activating these materials with Ti(^{i}OPr)$_4$ they were screened in the asymmetric diethylzinc addition to aromatic aldehydes. It was observed that the enantioselectivity of the reaction scaled with the open channel sizes of the MOF. This relationship was attributed to a competition between the MOF-catalyzed reaction and an asymmetric background reaction which predominated as the channels of the framework became too small for the substrates and products to efficiently diffuse through.

While these two examples show how unique chiral active centers can result due to the network topology of the MOF, a recent report by Lin *et al.* demonstrated the enzyme-like remote chiral induction of the MOF framework on the reactive sites. Using phosphoric acid substituted ligands, chiral Brønsted acid catalysts were synthesized with copper nitrate (Figure 11.18).[41]

These MOF-based chiral acid catalysts were screened in the asymmetric Friedel–Crafts reaction between indole and various sulfonamides. An inversion of selectivity was observed when using the MOF derived from the 3,3´,6,6´-substituted BINOL ligand when compared to the homogeneous control. Through the use of molecular mechanics/quantum mechanics calculations, this

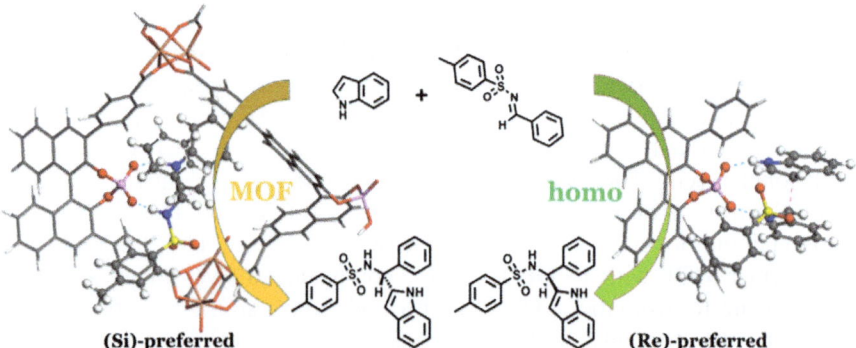

Figure 11.18   Schematic showing the pocket environment of chiral MOF cavity leads to the opposite chirality in asymmetric catalysis using the Brønsted acidic MOF built from a 3,3´,6,6´-substituted BINOL phosphoric acid ligand and the copper paddlewheel SBU.

inversion of selectivity was attributed to chiral induction from the walls of the MOF channels. To further substantiate this claim, a MOF based on the 4,4′,6,6′-substituted BINOL phosphoric acid was screened in the Friedel-Crafts reaction. Due to the lack of steric bulk in the 3,3′ positions, the 4,4′,6,6′-based material did not show inversion of chirality in the Friedel–Crafts products.

11.5 Summary

The body of work to date has seen MOF-based asymmetric catalysts evolve from simple Lewis acid catalysts to complex and robust systems that exhibit enhanced activity, unique selectivity, enzyme-like behavior, and a high degree of recyclability. In the future, advancements in structural motifs may allow for the synthesis of more robust and versatile asymmetric MOF catalysts that can tolerate harsher reaction conditions and catalyze a broader reaction scope. As advancements are made on this front, asymmetric MOF catalysts will have the potential to be applied in larger scale organic transformations.

Acknowledgements

The authors wish to thank National Science Foundation (CHE-1111490) for financial support. J.M.F. is supported by a DOE Office of Science graduate fellowship under the DOE contract number DE-AC05-06OR23100. S.L. was supported by a UNC William W. and Ida W. Taylor Fellowship.

References

1. R. Noyori, *Angew. Chem., Int. Ed.*, 2002, **41**, 2008.
2. W. S. Knowles, *Adv. Synth. Catal.*, 2003, **345**, 3.
3. V. Farina, J. T. Reeves, C. H. Senanayake and J. J. Song, *Chem. Rev.*, 2006, **106**, 2734.
4. Y. G. Zhou, W. Tang, W. B. Wang, W. Li and X. Zhang, *J. Am. Chem. Soc.*, 2002, **124**, 4952.
5. T. Ireland, K. Tappe, G. Grossheimann and P. Knochel, *Chem.–Eur. J.*, 2008, **14**, 3509.
6. T. Ireland, K. Tappe, G. Grossheimann and P. Knochel, *Chem.–Eur. J.*, 2002, **8**, 843.
7. N. W. Boaz, S. D. Debenham, E. B. Mackenzie and S. E. Large, *Org. Lett.*, 2002, **4**, 2421.
8. J. M. Thomas, T. Maschmeyer, B. F. G. Johnson and D. S. Shephard, *J. Mol. Catal. A: Chem.*, 1999, **141**, 139–144.
9. D. Sinou, *Adv. Synth. Catal.*, 2002, **344**, 221.
10. C. E. Song and S. G. Lee, *Chem. Rev.*, 2002, **102**, 3495.
11. P. McMorn and G. J. Hutchings, *Chem. Soc. Rev.*, 2004, **33**, 108.
12. W. Lin, *MRS Bull.*, 2007, **32**, 544.
13. T. Ezuhara, K. Endo and Y. Aoyama, *J. Am. Chem. Soc.*, 1999, **121**, 3279–3283.

14. C. J. P. Kepert, T. J. Prior and M. J. Rosseinsky, *J. Am. Chem. Soc.*, 2000, **122**, 5158–5168.

15. D. N. Dybtsev, A. L. Nuzhdin, H. Chun, K. P. Bryliakov, E. P. Talsi, V. Fedin and K. Kim, *Angew. Chem., Int. Ed.*, 2006, **45**, 916.

16. J. D. Ranford, J. J. Vittal, D. Wu and X. Yang, *Angew. Chem., Int. Ed.*, 1999, **38**, 3498.

17. B. F. Abrahams, M. Moylan, S. D. Orchard and R. Robson, *Angew. Chem., Int. Ed.*, 2003, **42**, 1848.

18. W. S. Sheldrick, *Acta Crystallogr.*, 1981, **B37**, 1820.

19. K. Aoki and W. Saenger, *J. Inorg. Biochem.*, 1983, **19**, 269.

20. M. Banerjee, S. Das, M. Yoon, H. J. Choi, H. H. Hyun, S. M. Park, G. Seo and K. Kim, *J. Am. Chem. Soc.*, 2009, **131**, 7524–7525.

21. F. Song, C. Wang, J. M. Falkowski, L. Ma and W. Lin, *J. Am. Chem. Soc.*, 2010, **132**, 15390–15398.

22. D. Dang, P. Wu, C. He, Z. Xie and C. Duan, *J. Am. Chem. Soc.*, 2010, **132**, 14321–14323.

23. L. Ma, C.-D. Wu, M. M. Wanderley and W. Lin, *Angew. Chem., Int. Ed.*, 2010, **49**, 8244–8248.

24. J. S. Seo, D. Whang, H. Lee, S. I. Jun, J. Oh, Y. L. Jeon and K. Kim, *Nature*, 2000, **404**, 982–986.

25. O. R. Evans, H. L. Ngo and W. Lin, *J. Am. Chem. Soc.*, 2001, **123**, 10395–10396.

26. A. Hu, H. L. Ngo and W. Lin, *Angew. Chem., Int. Ed.*, 2003, **42**, 6000.

27. A. Hu, H. L. Ngo and W. Lin, *J. Am. Chem. Soc.*, 2003, **125**, 11490.

28. H. L. Ngo, A. Hu and W. Lin, *J. Mol. Catal. A: Chem.*, 2004, **215**, 177–186.

29. C.-D. Wu, A. Hu, L. Zhang and W. Lin, *J. Am. Chem. Soc.*, 2005, **127**, 8940–8941.

30. M. Wang, M.-H. Xie, C.-D. Wu and Y.-G. Wang, *Chem. Commun.*, 2009, 2396–2398.

31. G. Férey, C. Mellot-Draznieks, C. Serre, F. Millange, J. Dutour and S. Surble, *Science*, 2005, **309**, 2040.

32. D. J. Lun, G. I. N. Waterhouse and S. G. Telfer, *J. Am. Chem. Soc.*, 2011, **133**, 5806–5809.

33. S.-H. Cho, B. Ma, S. T. Nguyen, J. T. Hupp and T. E. Albrecht-Schmitt, *Chem. Commun.*, 2006, 2563–2565.

34. J. M. Falkowski, C. Wang, S. Liu and W. Lin, *Angew. Chem., Int. Ed.*, 2011, **50**, 8674–8678.

35. M. J. Ingleson, J. P. Barrio, J. Bacsa, C. Dickinson, H. Park and M. J. Rosseinsky, *Chem. Commun.*, 2008, 1287–1289.

36. K. Tanaka and K.-i. Otani, *New. J. Chem.*, 2010, **34**, 2389–2391.

37. K. Tanaka, S. Oda and M. Shiro, *Chem. Commun.*, 2008, 820–822.

38. F. Song, C. Wang and W. Lin, *Chem. Commun.*, 2011, **47**, 8256–8258.

39. C.-D. Wu and W. Lin, *Angew. Chem.*, 2007, **119**, 1093–1096.

40. L. Ma, J. Falkowski, C. Abney and W. Lin, *Nat. Chem.*, 2010, **2**, 838–846.

41. M. Zheng, Y. Liu, C. Wang, S. Liu and W. Lin, *Chem. Sci.*, 2012, **3**, 2623.

CHAPTER 12

Photocatalysis by MOFs

HERMENEGILDO GARCÍA*[a,b] AND BELÉN FERRER[b]

[a] Instituto de Tecnología Química UPV-CSIC, Universidad Politécnica de Valencia, Consejo Superior de Investigaciones Científicas, Spain; [b] Departamento de Química, Universidad Politécnica de Valencia, Spain
*Email: hgarcia@qim.upv.es

12.1 Photochemistry in Heterogeneous Media. Rationalization of the Use of MOFs

In the early days of organic photochemistry most of the studies were carried out in homogeneous phase looking for media as isotropic as possible. These studies in gas and liquid phases have served as the foundation of photochemistry as a science and to determine the photoreactivity of different functional groups.[1,2]

Typically, due to the high energy of electronically excited states, there is a wide range of physical and chemical pathways that can occur to various extents simultaneously upon light excitation of organic molecules. Thus, in addition to radiationless deactivation of excited states, a molecule can deactivate by emitting a photon of longer wavelength than the one used for excitation (fluorescence and phosphorescence) or can undergo bond cleavage and rearrangement triggering chemical transformations.[1,2] It is also very common that the selectivity for a single product in photochemical reactions is low and several compounds are present in the final reaction mixture. For this reason, even though light absorption is a universal phenomenon that occurs in almost every organic molecule, homogenous photochemistry has not been used as a favorite synthetic methodology in contrast to thermal activation.[3]

RSC Catalysis Series No. 12
Metal Organic Frameworks as Heterogeneous Catalysts
Edited by Francesc X. Llabrés i Xamena and Jorge Gascon
© The Royal Society of Chemistry 2013
Published by the Royal Society of Chemistry, www.rsc.org

With the aim of gaining control on the photochemical reactivity of molecules it soon became evident that confinement and anisotropy are very powerful tools. For this reason a natural evolution of photochemistry was to move from homogeneous to heterogeneous systems in which parameters such as conformational mobility, diffusion of reactive species, photosensitization and other parameters not encountered in solution or gas phase can determine the selective formation of a single product among all the possible products formed in homogeneous phase.[4,5]

In fact, irradiation of compounds on a solid surface restricts the conformational freedom of the substrate by "freezing" the molecule with surface–molecule interactions such as hydrogen bonds, dipole forces or simple Van der Waals interactions. If the surface defines a cavity then the intermediates generated inside the cavity tend to react isolated from the rest of other molecules ("confinement effect") and the result can be the prevalent formation of rearrangement products arising from geminated or "in-cage" intermediates.

For photochemistry in heterogeneous media,[5] porous solids have been found among the favorite materials used as media to perform the photochemical reactions of incorporated guests (Figure 12.1). In this context, besides layered aluminosilicates (such as montmorillonite) and double metal hydroxides (hydrotalcites), zeolites and related aluminosilicates have been hard steel, being widely used. Due to the chemical composition of zeolites (aluminosilicates)[6,7] light of wavelengths above 250 nm can be transmitted through the solid. Even though the physical appearance of powdered zeolites are as white opaque solids in which light can not be transmitted through millimetric lengths layers due to optical phenomena like scattering and reflection, at the submillimetric scale light can penetrate inside the crystals. It has been estimated by simple optical measurements that for white zeolite powders the penetration of the light is about 10 microns, which is more than sufficient to produce photochemical excitation of included guest molecules occupying the internal pores of the crystals that are about 1 nm size.[8]

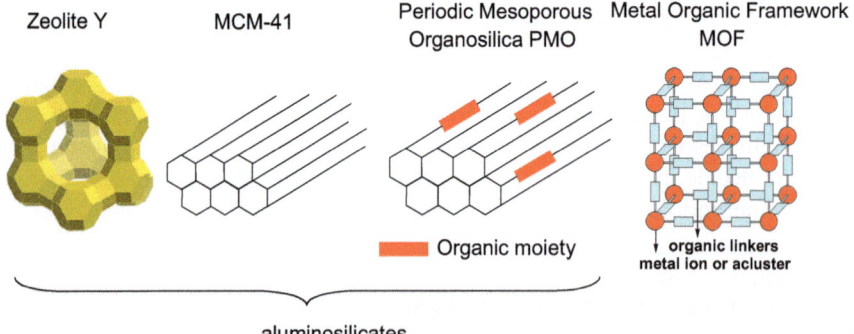

Figure 12.1   3D structure of four different porous solids.

Besides the appropriated chemical composition, zeolites offer, as a solid matrix for heterogeneous photochemistry, many advantages including easy characterization, high crystallinity and most importantly a large internal pore volume with a strictly regular pore dimension. This internal pore volume can be accessed from the exterior, allowing mass transfer and the incorporation inside the crystals of photoactive guest molecules that can undergo photo-excitation inside the zeolite framework. In fact, intra-zeolite photochemistry has been an active field of research since the early eighties and the outcome of this intensive research has been to provide remarkable examples of control of the photochemical behavior of incorporated molecules with respect to their behavior in solution.[8] Photo-stability and enhancement of the lifetime of excited states and transient species are among the most generally observed phenomena in this type of materials.[9]

A logical extension of intra-zeolite photochemistry was to employ as host mesoporous aluminosilicates with MCM-41 or SBA-15 structures comparable to zeolites. Structured mesoporous silica offer larger pore size, overcoming the important limitation on the dimensions of the molecules that can be included in conventional zeolites.[10,11]

Subsequently, at the beginning of the 2000s, reports on the preparation of hybrid periodic mesoporous organosilica (PMOs)[12,13] offered new opportunities in the field of heterogeneous photochemistry. The most important one was that, in contrast to the purely inorganic materials and particularly aluminosilicates, which are photochemically inactive, the presence of organic components in PMOs allowed photochemistry of the matrix to be performed. Thus, we were among the first to report that viologen containing PMOs undergo photochromism and photoinduced electron transfer, with remarkable differences in terms of lifetime with respect to other situations in which this electron acceptor unit was not covalently grafted to the walls and the pores were completely packed with a surfactant acting as spectator molecule.[14-16] Also, we have shown that the pore size of the PMO can be in a certain range modulated by performing *trans-cis* isomerization of C=C of diaryl benzenes forming part of the PMO structure.[17]

The transition from amorphous silica to zeolites and then to periodic mesoporous silica and PMOs illustrates very clearly the continuous interest of heterogeneous photochemistry to exploit the potential of any newly available porous material. In this regard, not surprisingly, the synthesis of metal organic frameworks[18,19] offer new opportunities because in a simple view they can host any photochemical probe, like zeolites or mesoporous inorganic solids, but also they can allow photochemistry to be performed on the solid itself if the organic linkers exhibit some interesting photoactivity.[20,21]

12.2 The Use of Solids as Photocatalysts

In the previous section we have commented mostly on the use of a porous matrix as a provider of a rigid reaction cavity in which a photochemical reaction of a photoactive guest can be performed. In the previous section, the solid plays mainly a "passive" role while the actor of the process is confined in the interior of the solid matrix. As commented above, the main reason is the photochemical inertness of aluminosilicates and actually the chemical composition of quartz and glasses, commonly used as photochemical cells and for optics, is not very different from those solids. In contrast to this situation, there are some other solids that can absorb photons because they present an absorption band in optical spectroscopy above to 200 nm. In these cases, the most general phenomenon taking place is charge separation with the generation of electrons and holes. One typical case of this type of solids are transition metal oxides and chalcogenides. In many of these cases, light excitation leads to an electron transfer from the negative non metallic element (oxide or sulfide) to the transition metal cation with the generation of a positive electron hole in the non metallic element and an electron trapped with different energies into the transition metal cation. Very frequently, for highly crystalline solids, the atomic orbitals overlap in such a way that the density of states of the solid has a valence band with not much conductivity and the empty bands where an electron can move more or less freely throughout the particle (conduction band). The term semiconductor applies for these solids in which before irradiation they act as insulator but in the charge separated state the electron can diffuse through the conduction band and then in this state behaves as a conductive material. Scheme 12.1 presents the most common phenomena occurring upon absorption of a photon of energy higher than the valence conduction band gap in a semiconductor.

When the charge migrates to the external surface of the particle, electrons and holes can react with organic substrates and the same particle can act simultaneously as an oxidizing (the positive hole) and reducing (electrons in the conduction band) agent, with different redox potential depending on the position of the conduction and valence bands. Typically for reactions carried out in the atmosphere and considering the concentration and the electron acceptor ability of oxygen, irradiation of a semiconductor in the open air leads to the generation of superoxide (O_2^-) and species derived from it that are noted as reactive oxygen species (ROS). Also very common, for those irradiations carried out in water, this molecule can act as hole quencher leading to a proton and the formation of highly aggressive hydroxyl radicals, that are also

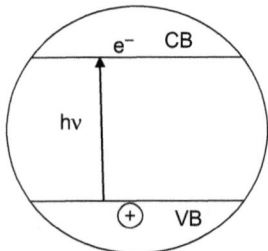

Scheme 12.1   Phenomena occurring in a semiconductor upon absorption of a photon of energy higher than the valence conduction band gap.

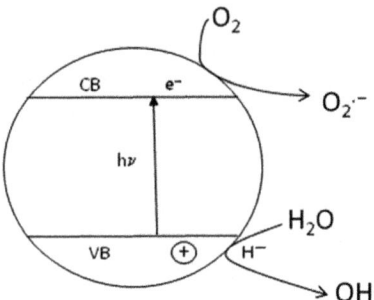

Scheme 12.2   Schematic representation of a semiconductor acting simultaneously as oxidizing (the positive hole) and reducing (electrons in the conduction band) agent.

considered as one particular case of ROS. Scheme 12.2 shows some of the general processes that can take place by reaction of electrons in the conduction band with oxygen and positive holes in the valence band with water.

Alternatively, depending on the substrate concentration and its relative redox potentials with respect to oxygen and water, it can compete with these species giving raise initially to the generation of radical ions (anions or cations). This transient species can undergo a variety of chemical processes that can be very useful to promote their degradation.

One typical problem in semiconductors, is their photo-stability under operation conditions. Very frequently, illumination of the semiconductor in aqueous media leads to its corrosion, with the migration of metal ions to the solution that eventually can lead to the complete dissolution of the solid. Transition metal sulfides are typically considered as highly unstable narrow band-gap semiconductors and their irradiation typically produces the dissolution of the solid. However, other metal oxides such as iron oxides with various stoichiometries, silver oxide and copper oxides are prone to photo-corrosion. Due to its stability, its wide availability and lack of toxicity, TiO_2 is the most widely used semiconductor that enjoys a remarkable chemical and photochemical stability.[22] In fact, dissolution of TiO_2 is very difficult and requires concentrated acids at high temperatures (nitric acid in autoclave above 150 °C) to take place. However, in spite of these advantages, TiO_2 has two general limitations. One of this is its lack of visible light photo-response, since the onset of the absorption band for the anatase crystal form is 380 nm. Another problem is its efficiency for some reactions that is limited by the lifetime of charge separation, electron and hole mobility and the limited surface area.[22] For many applications and for catalysis the preferred TiO_2 are nano-particles of small dimensions predominantly in the anatase phase with some schottky barrier to retard charge separation. In this context, commercial P25 is TiO_2 from Degussa with an average particle size of 25 nm, specific surface area of about $50 \, m^2 \, g^{-1}$ and with a composition of about 80% anatase, 20% rutile. The need for more efficient photocatalysts has led to different approaches to alter the performance of TiO_2. In this regard, it is worth commenting that small clusters and particles of TiO$_2$ have been included inside the micro/mesopores of structured aluminosilicates such as zeolites and MCM-41. Even isolated Ti atoms grafted on walls of aluminosilicates have been found to exhibit photo-catalytic activity. However, even though the proof of principle of the photo-activity of this type of solids have been established, the main problem of the strategy of encapsulation/isolation of Ti atoms in highly porous host matrices arises from the operation of quantum size effects that, as a consequence of the confinement in a very small number of atoms, leads to higher band gap energies with the onset of a blue shift of the absorption band. In other words, excitation of these systems requires UV light of short wavelength and therefore these materials can not be excited with visible or solar light.

One way to circumvent this problem consists of doping the small clusters with nitrogen or the formation of charge transfer complexes between the TiO$_2$ clusters and polyhydroxy aromatic compounds.

One alternative to the use of TiO$_2$ as a semiconductor would be to exploit MOFs as semiconductors.[20] As it will be shown later, the fact that small metal clusters are in intimate connection with organic linkers can be a powerful strategy to effect sensitization of the metal clusters by the organic linker through a phenomenon termed the antenna effect. In this regard, the porosity of the MOFs will ensure a large surface area and accessibility of the substrates. Thus, provided that photo-stability is granted, the use of linkers presenting visible absorption bands can serve to photosensitize the small metallic clusters leading to efficient charge separation. This topic will constitute a key section of the present chapter.

12.3 Structure of MOFs

Although there are several possible terms to denote the type of materials that are the focus of the present chapter, in the present case we consider metal organic frameworks as crystalline porous solids whose structure is constituted by a metal ion or a cluster of metal ions connected through coulombic inter-actions with rigid bi- or poly-podal organic linkers.[18,19] According to this definition, besides crystallinity, the key features of the materials under discussion are open porosity and relatively strong linker interactions. Thus, for the present case we can not consider those materials that, even though they have metal ion coordination bonds, do not have porosity or the porosity is only present when certain solvents cause the swelling of the material.[23] Also excluded from the present chapter is the photochemical behavior of those solids in which hydrogen bonds are the structural motif maintaining the framework of the material.

Due to the directionality of the metal–ligand interaction and the low weight of the typical linkers, MOFs are currently up the top of the list of the most porous materials, exhibiting very large surface area and very high internal pore volume. Zeolites have typical specific surface area values of 400 m^2 g^{-1} or below and pore volumes of 0.2 cm^3 g^{-1} or smaller. Periodic mesoporous aluminosilicates such as MCM-41 or SBA-15 can have a BET area over $1000\,m^2\,g^{-1}$ and pore volume of $0.5\,cm^3\,g^{-1}$. In the case of the MOFs, there have been reported structures with specific surface area above $5000\,m^2\,g^{-1}$ and pore volumes above $1\,cm^3\,g^{-1}$. Thus, MOFs constitute the materials with lowest framework density meaning that the weight per nm^3 is the lowest ever reported and the space is filled with the lowest possible mass.[24]

Typical organic linkers are aromatic polycarboxylates such as terephthalate (BDC) and 1,3,5-benzenetricarboxylate (BTC). Other examples that can be more relevant in the present chapter are those aromatic carboxylates having condensed rings such as those derived from naphthalene, binaphtyl, anthracene and pyrenyl among others (see Scheme 12.3).

Besides aromatic carboxylates, other linkers can act as ligands through other group such as nitrogen atoms forming part of basic heterocylces and phosphonate groups. The series of ZIF materials are based on imidazole and benzeneimidazolates that mimic the angles present in zeolites.[25,26]

Concerning the nature of the metal ion that can be in MOFs, almost all transitions metals as well as rare earth metals can be used to obtain MOFs. The nature of the transition metal and its coordination with oxygen determines that in some cases the node is constituted by a single metal ion. This is the case of $Al_2(BDC)_3$.[27] In other cases, such as $Cu_3(BTC)_2$, the node is constituted by a pair of Cu^{2+} ions linked to three common carboxylate groups forming a paddlewheel motive.[28] In the case of MOF-5, the node is formed by a central oxygen atom coordinated to four Zn^{2+} ions and these Zn_4O clusters then coordinate with the carboxylate groups.[29] The diversity of the metal clusters is very large, but what is interesting for the present chapter is that when the number and dimensions of the nodes increases they can reach a point in which

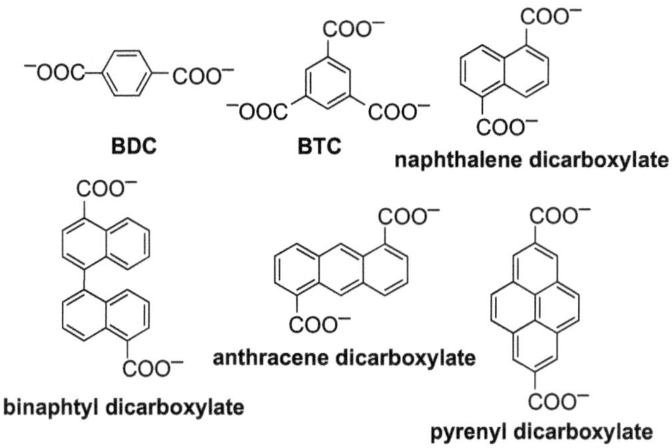

Scheme 12.3 Chemical structure of some typical aromatic polycarboxylate linkers in MOFs.

the oxometallic cluster can be consider as a semiconducting quantum dot. For instance, MOF UiO-66 has a node of six Zr atoms connected with oxygen forming an octahedron that is bonded to the terephthalate linkers.[30] One important example of photoactive metal clusters is that present in the MIL125 MOF.[31] This material contains a cyclic octamer of edge and corner sharing TiO_5 (octahedral) joined by BDC ligands. Considering that TiO_2 in the anatase form is the most important photocatalyst, the potential of this MOF is evident, provided that there is an efficient photoinduced charge separation between the linker and the $Ti_8O_8(OH)_4$ units. The metal to ligand interaction develops in the tri directional space, defining empty spaces, denoted as cages, that can be accessed through smaller windows or channels.

Due to the large surface area and pore volume, one of the main applications of MOFs is as absorbents for gas separation and also as the stationary phase in liquid chromatography. MOFs also hold promise for hydrogen and CO_2 absorption.[32–34]

There are certain structures, such as $Cu_3(BTC)_2$ (BTC- 1,3,5-bennzene-tricarboxylate), Fe(BTC) or MIL-100(Cr), in which the metal nodes have coordination positions compromised with the structure and that are occupied by solvent molecules or by mono-nodal ligands such as F^-. The ligands at these positions can be exchanged without damaging the crystal structure of the MOF, and then these materials hold promise as heterogeneous catalysts, provided that the reaction conditions including solvent, temperature and reagents are compatible with the structural stability of the MOF. In fact, considering the similarities of MOFs with other porous materials and particularly zeolites widely used as solid catalysts, the use of MOFs in heterogeneous catalysis is a fast growing area.[35]

Related to the applicability of MOFs in catalysis, one further step is to check the photocatalytic activity of these materials. In order to develop efficient photocatalysts one general tendency has been to decrease the particle size of the semiconductor and in this regard the limit will be to have a material with isolated clusters of semiconducting quantum dots. In this sense, some of the metal clusters acting as structural nodes in MOF can be considered as the limit of this approach since the number of metal atoms has been decreased to the minimum possible number and the linker ensures that no agglomeration of these quantum dots can take place. Also from this perspective, the linker could play a role as antenna, not only isolating the semiconducting quantum dots, but also absorbing light of the appropriate wavelength and transferring this energy to the non absorbing quantum dot. In this sense, provided that the linker has adequate absorption bands and the metallic clusters can behave as quantum dots, the structure of the MOF will be perfectly suited as photocatalyst. In fact, a photocatalyst has to combine some of the features of a catalyst, such as high surface area, single site composition and robustness, with other specific properties necessary for an adequate interaction with the light, such as an adequate absorption spectrum, efficient photophysical processes and long lifetime of excited states. In the next section, we will comment on the possibilities that MOF composition and structure offer in the development of photoactive materials, and then we will comment on some of the specific examples, mostly from our group, that have been found in the literature reporting photo-responsive MOFs.

12.4 Photochemical Activity of MOFs

To develop photoactive materials based on MOFs there are at least three different strategies, as is illustrated in Figure 12.2. One of the simplest possibilities will be to use MOF as a porous matrix to encapsulate a chromophore that will be responsible for the photochemical properties. In this strategy, the MOF is playing a passive role and can contribute to the control and alteration of the photochemical activity of the encapsulated chromophore in different ways by defining a reaction cavity in which the reaction occurs. Thus, in this approach the rigid structure of the MOF will act primarily by isolating the chromophore, but it can also immobilize and promote in-cage and intramolecular events.

It is expected that this strategy, still to be developed and exploited for MOFs, will be similar in the effects that can be observed as those already reported for zeolites and other microporous and mesoporous solids.

A second approach will be to synthesize MOFs with photo-responsive organic ligands or to modify these linkers after the synthesis to convert them into photo-responsive units.[36–39] Considering that most organic groups have a rich photochemistry, this approach can be used to develop a whole range of photo-responsive solids with application as sensors in photochromics, photoluminescence and others based on the intrinsic activity of organic compounds.

A third possibility will be to exploit the semiconducting quantum dots properties derived from the presence of metallic clusters.[20] In this approach, light absorption on the organic linker should lead to charge separation, with the creation of a hole in the electron rich linker and an electron in the metal cluster. This state of charge separation is similar to that occurring in most of the semiconductors and can serve to convert light into chemical energy. The charge separated state occurring in one particle on the millisecond time scale can serve to promote oxidation of electron rich substrates (by the positive electron hole) or chemical reduction (by the electron in the metallic cluster).[40–43]

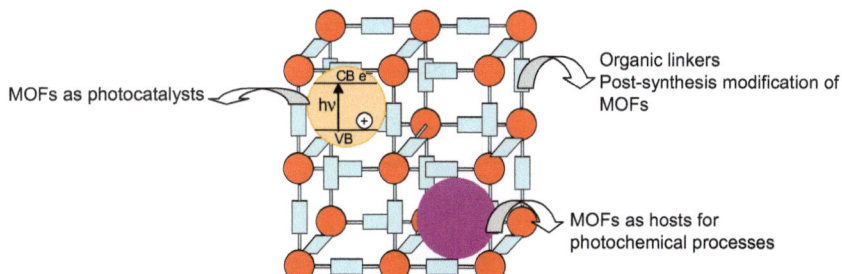

Figure 12.2   Different strategies for developing photoactive materials based on MOFs.

12.5   MOFs as Photocatalysts

Soon after Zechina's groups reported the photoluminescence of MOF-5,[44] suggesting that this emission arises from the interaction of terephthalate with the Zn_4O clusters, we reported the first example showing that MOF-5 can behave as a photocatalyst.[45] A laser flash photolysis study of this solid allowed the detection of a transient spectrum that was attributed to the charge separated state arising from the electron transfer from the terephthalate unit to the Zn_4O cluster. The broad absorption band from 500 to 840 nm was assigned to delocalized electrons existing on the microsecond time scale, and most probably, occupying conduction bands. These photogenerated electrons are able to reduce electron acceptors, such as methylviologen. On the other hand, the photogenerated positive holes are able to oxidize *N,N,N,N*-tetramethyl-*p*-phenylenediamine to its corresponding radical cation. The actual conduction band energy value was estimated to be 0.2 V *versus* NHE with a band gap of 3.4 eV. The charge-separation state, with electrons in the conduction band and holes in the valence band, makes MOF-5 behave as a semiconductor that exhibits photocatalytic activity for the degradation of phenol in aqueous solutions comparable in efficiency to that of a P-25 titanium dioxide standard. Scheme 12.4 summarizes the behavior of MOF-5 as semiconductor.

As a photocatalyst, a second and more relevant conclusion is that MOF-5 exhibits reverse shape-selectivity in which large phenolic molecules (*i.e.*, molecules with a high steric hindrance with respect to the pore openings of MOF-5) that cannot access the interior of the micropores are degraded significantly faster than those others that can enter into the pores.[46]

Typically efficient semiconductors do not emit. The reason for this is that the electronically excited state is in reality better considered as a charge separation state. In other words, when an organic molecule absorbs a photon, an electron jump from the HOMO to a LUMO orbital occurs giving a singlet or triplet excited state. However, in a material like a semiconductor, the electron jump occurs from the valence band (commonly negative oxygen in metal oxides) to a conduction band that is delocalized throughout the particle, and therefore instead of electronic excitation, the photon absorption is better described as

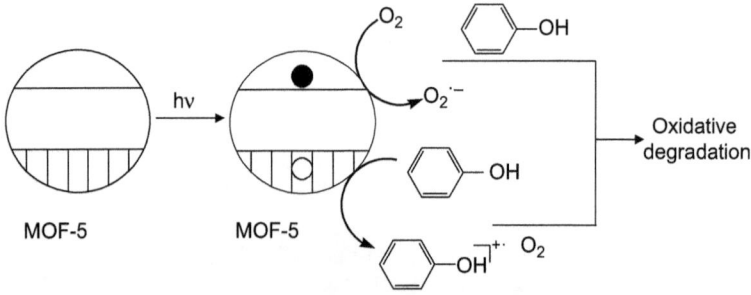

Scheme 12.4   A mechanistic proposal for the photodegradation of phenol using MOF-5 as semiconductor.[45]

charge separation. Photoluminescence would require that electrons and holes recombine and the energy liberated in the recombination is dissipated in an emissive way. For this reason, if the semiconductor is in operation or in the presence of oxygen (electron quencher) or water (hole quencher), recombination is not possible and emission does not occur. Nevertheless, when the semiconductor is not in the presence of electron and hole quenchers and is not under operation, the process of charge separation can lead to the observation of photoluminescence, albeit in low efficiency, arising from charge recombination. Scheme 12.5 illustrates the origin of photoluminescence in semiconductors and the need for an absence of electron or hole quenchers.

In this context, the pioneering observation of photoluminescence in MOF-5 is interesting.[44]

In a comprehensive study, Majima and co-workers determine the energy barrier of trapped electrons in MOF-5 and compare these energies with that of a ZnO semiconductor.[47] By monitoring the photoluminescence intensity as a function of the temperature in the range of 110 to 302 K, it was determined that in ZnO electrons are trapped from 50 to 190 meV below the conduction band edge of this non porous semiconductor. The trapping sites in ZnO were proposed to be structural defects consisting of oxygen vacancies. Similar studies of the influence of the temperature on the photoluminescence intensity of MOF-5 show that the intensity of the photoluminescence first decreases in the range of temperatures from 302 to 170 K and then increases in the 170–110 K interval. This abnormal behavior was interpreted by assuming an activation energy for the thermal non-radiative decay of 120 meV as well as a second energy barrier of 5 meV. The higher activation energy of the non-radiative decay indicates that electrons in the state of charge separation in MOF-5 are more deeply trapped than in ZnO. However, the origin of the

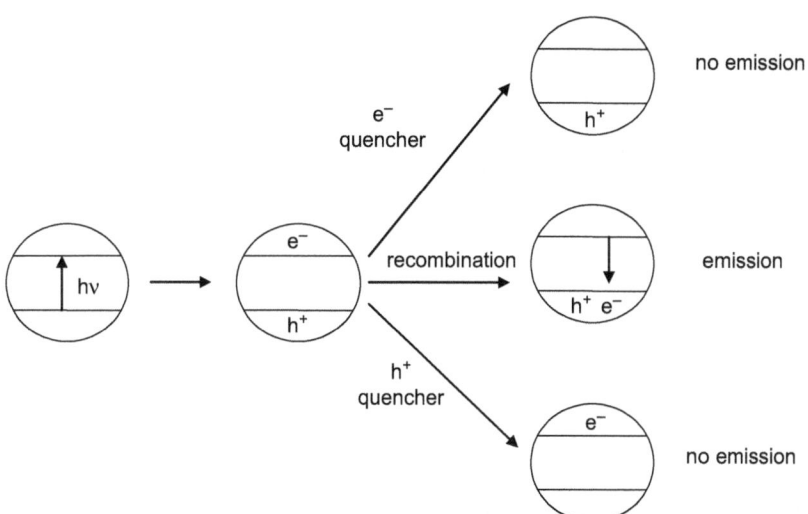

Scheme 12.5   Illustration of the origin of photoluminescence in semiconductors.

second energy barrier is still unclear and it is probably a manifestation of different types of sites.

In conclusion, the comparative behavior of photoluminescence as a function of the temperature in ZnO and MOF-5 indicates that the reorganization energy of the charge separation state in MOF-5 originates from larger structural changes than in ZnO, which was considered as a model of this MOF.

Photoluminescence also provides indirect evidence of the preferential absorption of some quenchers in MOF-5, as consequence of the microporosity and absorption sites in this MOF compared to ZnO.[47] Thus, systematic quenching studies of the photoluminescence of MOF-5 by a series of electron donors with different redox potentials (see Figure 12.3) have shown that in general most of the quenchers exhibit a similar relative quenching constant for ZnO and MOF-5.

However, there were three quenchers that exhibited much higher quenching constants for photoluminescence quenching in MOF-5 than in ZnO. This contrasting behavior was attributed to the adsorption of these quenchers inside MOF-5 pores resulting in a pre-concentration of the quencher and a strong binding of these molecules with MOF-5 sites, these processes leading to a much more efficient quenching than would be expected based on their redox potentials.

The observation of "enhanced quenching" of some electron donors due to porosity of MOF-5, as well as the previously commented different photo-catalytic degradation of phenols depending on their sites and their ability to be protected inside the pores, offers the opportunity of shape selectivity effects in photocatalysis as a unique opportunity in MOFs compare to non porous, dense

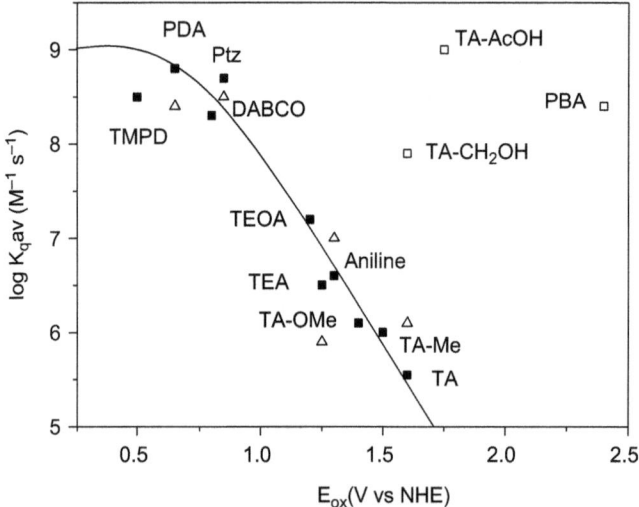

Figure 12.3   Relationship between the redox potentials, E_{ox}, of a series of electron donors and log k_q the quenching rate constants of the photoluminescence of MOF-5 (squares) and ZnO (triangles).[47]

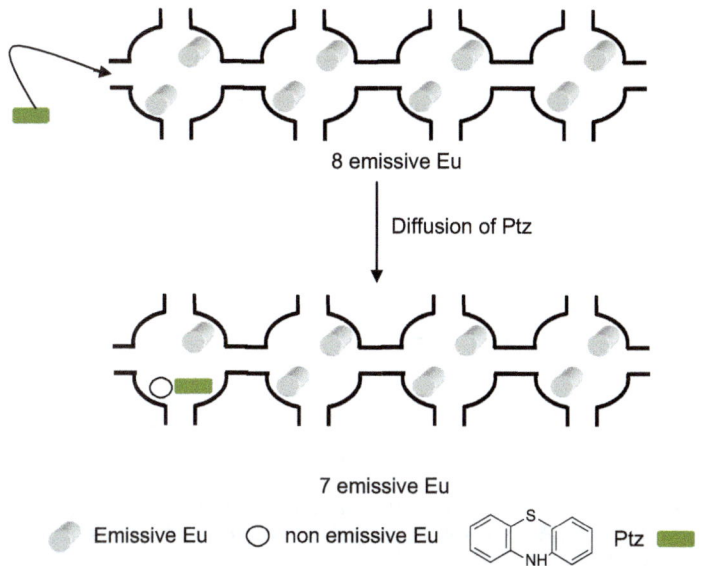

8 emissive Eu

Diffusion of Ptz

7 emissive Eu

Emissive Eu ○ non emissive Eu Ptz

Scheme 12.6 Pictorial illustration of how Confocal Fluorescence Microscopy can be used to monitor the diffusion of phenothazine (Ptz) inside Eu-MOF by monitoring the switching off of the emission.[48]

semiconductors. Although examples exploiting this shape selectivity in photocatalysis are still very scarce and clearly this is an under exploited field, Majima and coworkers using Eu-MOFs demonstrated the probing of the principle using spatially resolved confocal fluorescence microscopy.[48] These authors observed that large molecules quench the photoluminescence of Eu-MOF crystals more slowly and in an inhomogeneous manner from the exterior to the interior of the crystal, compared to small molecules that can diffuse faster and also quench Eu photoluminescence homogeneously (see Scheme 12.6). According to this shape selectivity, those molecules that are small enough to penetrate inside the micro-pores should experience a different environment and photocatalytic activity than those others that are too large to enter into the pores and, therefore, can only probe the external surface.

12.6 Photocatalytic Water Splitting

One hot topic in photocatalysis, related to alternative energy resources and the conversion of solar light into fuels, is photocatalytic water splitting to generate hydrogen. Due to the limited water stability of many of the reported MOFs, one prerequisite for this application is the stability of the MOF crystal structure to water. In this regard, one of the most stable MOFs is the Zr-BDC that is commonly known as UiO-66. This MOF can resist heating in water at 100 °C for 1 hour without collapsing. We have observed that UiO-66 exhibits photocatalytic activity for hydrogen generation from water using methanol as

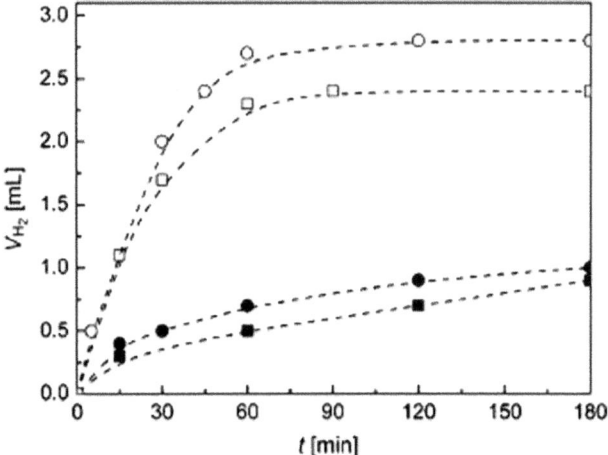

Figure 12.4   Temporal evolution of the photocatalytic hydrogen generation using UiO-66 (■), UiO-66/Pt (i□), UiO-66(NH$_{i2}$) (●) and UiO-66(NH$_2$)/Pt (○).[49]

sacrificial electron donor upon irradiation at wavelengths longer than 300 nm.[49] Furthermore, as expected in view of the catalytic activity for hydrogen evolution, the presence of Pt in the system significantly increases the photo-catalytic hydrogen production. Figure 12.4 shows the temporal evolution of the photocatalytic hydrogen generation for UiO-66 without and with Pt nanoparticles.

In addition to the presence of Pt nanoparticles, and to illustrate the versa-tility of MOFs as photocatalysts, a similar UiO-66 solid was prepared by using aminoterephthalate as linker instead of the parent terephthalate that is the organic component in conventional UiO-66.[49] The presence of the amino group introduces a new absorption band in the UV-visible region (band from 300 to 410 nm) that plays a favorable role in absorbing visible light. As consequence of the presence of this amino group, the efficiency of the material for hydrogen generation increases (see Figure 12.4). The apparent quantum yield for hydrogen generation using 370 nm monochromatic light in a 3 to 1 water/methanol mixture for UiO-66(NH$_2$) was 3.5%. Also in this case, laser flash photolysis has detected the generation of a long lived charge separated state.

12.7   Potential Applications of MOFs as Semiconductors

A photocatalyst converts the energy of light into charge separation and, as we have already exemplified in the cases of phenol degradation, charge separation can serve to promote chemical reactions initiated by oxidation, reduction or by the effect of reactive oxygen species generated in the medium by the trapping of electrons by oxygen or holes by water.

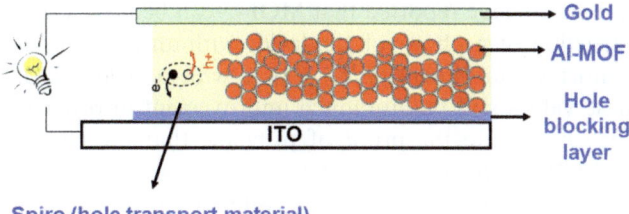

Figure 12.5 Schematic composition and operation of a photovoltaic device. Reprinted with permission from ref. 50. Copyright (2011) American Chemical Society.

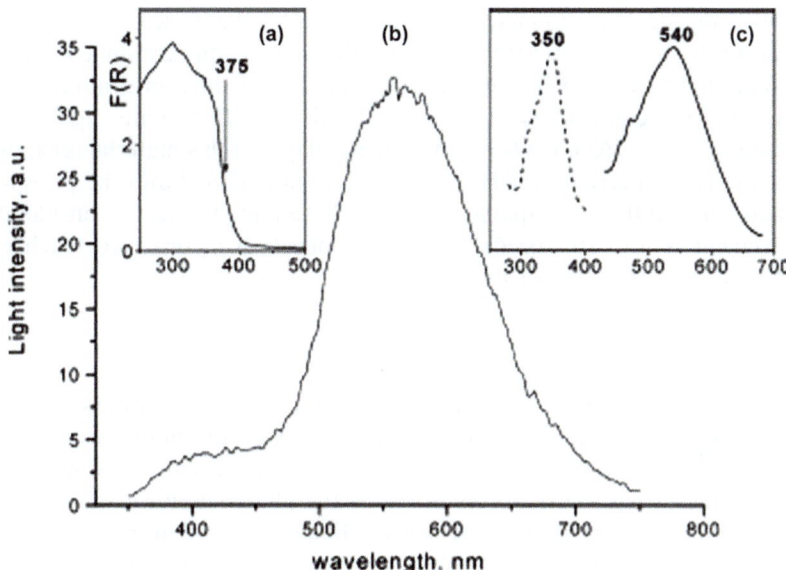

Figure 12.6 (a) Diffuse reflectance UV-vis spectrum of MOF-5. (b) Phosphor emission spectrum of MOF-5, obtained with a square alternating current voltage of 60 V and a frequency of 180 Hz. (c) Photoluminescence excitation (---, monitored at 540 nm) and emission (—, exciting at 350 nm) spectra of MOF-5.
Reprinted with permission from ref. 46. Copyright (2007) American Chemical Society.

In addition to these photocatalytic reactions, other typical applications of semiconductors are in photovoltaic cells and as phosphors. In photovoltaic cells, light is converted into electrical energy and the recombination of electrons and holes is carried out by a series of individual steps in which electrons in the conduction band and high energy state are transported through an external circuit and, then, later by means of an electrolyte to the valence band. Figure 12.5 presents a simplified scheme of the operation of a photovoltaic cell.

In this context, we have reported that MOF-5 can be used as semiconductor for the construction of a photovoltaic device with an open circuit voltage of 0.33 V and short circuit intensity of 0.7 μA with a field factor of 44%.[46] Although these values still require much improvement in order to become of some use, they constitute the proof of principle that the charge separation occurring in MOFs can be transported by electrodes through an external circuit. To illustrate the large number of possibilities in which this photovoltaic activity can be improved, we also reported that inclusion of 1,4-dimethoxy-benzene as hole quencher inside $Al_2(BDC)_3$ enhances the photovoltaic response of this MOF.[50] What is needed in this area are linkers that act as light harvesters to collect some light, metallic clusters that are efficient in charge separation, fast charge mobility and a high density of charge carriers in order to increase the quantum efficiency of the devices, as well as higher robustness of the crystalline structure to increase the durability of the devices.

Concerning the use of MOFs as phosphors, this application consists of promoting light emission by submitting a thin film of the semiconductor to a voltage. In this context, we have reported that MOF-5 emits light corresponding to $\lambda_{max} = 565$ nm when a micrometric layer of this material is exposed to an alternating current (180 Hz) of 60 V.[46] Figure 12.6 shows the phosphor emission of MOF-5 compared to its photoluminescence. Considerable improvement is also still needed for this application in order to reach efficiencies closer to the state of the art.

12.8   Conclusions

The previous sections have exemplified that MOFs can act as a passive matrix to accommodate in the internal voids photoactive chromophores that can exhibit a distinctive response due to the properties of the surrounding media, but more interesting are the cases in which the photoactivity of the material derives from the interaction of the organic linker with the inorganic clusters. Comparison with analogous systems in solution show that the crystal lattice of MOF, in which the linker and the organic cluster have an intimate contact and exhibit a single site arrangement, originates a photochemical response that is not found for the homogeneous solution. All these current examples illustrate the possibility to obtain interesting photoresponses that can open applications beyond absorption and catalysis. Among these applications we have commented the use of MOFs as photocatalysts for environmental remediation and for the generation of solid fuels, particularly hydrogen, as well as applications in optoelectronics, including solar cells and phosphors. All these applications derive from the consideration of MOFs as semiconductors.

However, the area of photocatalysis by MOFs is still in its infancy, since it has been developed mainly for MOFs that have been synthesized for purposes other than to achieve optimal photochemical response. In this regard, and considering the flexibility that MOFs offer in the design and preparation of new materials, it can be easily anticipated that "on purpose" MOFs can be obtained to enhance the photoactivity of these materials by exploiting the large surface area, the composition and the possibility to control the morphology of the sample, either as nanoparticles, large crystals or thin films. In particular, organic linkers that can act as antenna to absorb radiation in the visible region should be prepared. For solar light applications, it would be very interesting to design panchromatic dyes absorbing in the largest possible region of the visible spectrum. Also, MOFs containing metal ions or clusters having d^0 or d^{10} electronic configuration seem more appropriate for their use as photocatalysts.

The photochemical response must always be accompanied by crystal stability from the point of view of mechanical, thermal and chemical stability. But photostability, particularly in the case that dyes are used as linkers, and the resistance to undergo photobleaching, should also be considered.

Since a large variety of rigid organic linkers and transition metals can be used for the preparation of novel MOFs that could even accommodate quantum dots in the internal cavities, it can be easily anticipated that in contrast to the case of porous aluminosilicates MOFs will have a bright future as semiconductors, photocatalysts and photoresponsive materials in general.

References

1. M. Montalti, A. Credi, L. Prodi and M. T. Gandolfi, *Handbook of Photochemistry*, CRC Press Taylor & Francis Group, 2006.
2. J. C. Scaiano, *Handbook of Organic Photochemistry*, CRC Press 1989.
3. L. S. Kaanumalle, A. Natarajan and V. Ramamurthy, *Molecular and Supramolecular Photochemistry* 2005, 12(Synthetic Organic Photochemistry), 553.
4. C. Bhone, R. W. Redmond and J. C. Scaiano, *Photochemistry in Organized and Constrained Media*, John Wiley & Sons Inc, 1991.
5. K. Kalyansundaram, *Photochemistry in Microheterogeneous Systems*, Plenum Press, 1987.
6. R. M. Barrer, *Zeolites and clay minerals as sorbents and molecular sieves*, Academic Press, 1978.
7. H. v. Bekkum, E. M. Flanigen and J. C. Jansen, *Introduction to Zeolite Science and Practice*, Elsevier, 1991.
8. J. C. Scaiano and H. Garcia, *Acc. Chem. Res.*, 1999, **32**, 783.
9. H. Garcia and H. D. Roth, *Chem. Rev.*, 2002, **102**, 3947.
10. J. S. Beck, J. C. Vartuli, W. J. Roth, M. E. Leonowicz, C. T. Kresge, K. D. Schmitt, C. T.-W. Chu, D. H. Olson, E. W. Sheppard, S. B. McCullen, J. B. Higgins and J. L. Schlenker, *J. Am. Chem. Soc.*, 1992, **114**, 10834.
11. C. T. Kresge, M. E. Leonowicz, W. J. Roth, J. C. Vartuli and J. S. Beck, *Nature*, 1992, **359**, 710.
12. T. Asefa, M. J. MacLachlan, N. Coombs and G. A. Ozin, *Nature*, 1999, **402**, 867.
13. S. Inagaki, S. Guan, Y. Fukushima, T. Oshuna and O. Terasaki, *J. Am. Chem. Soc.*, 1999, **121**, 9611.

14. M. Álvaro, B. Ferrer, V. Fornes and H. Garcia, *Chem. Phys. Chem.*, 2003, **4**, 612.
15. M. Álvaro, B. Ferrer, V. Fornes and H. García, *Chem. Commun.*, 2001, 2546.
16. M. Álvaro, B. Ferrer, H. García, S. Hashimoto, M. Hiratsuka, T. Asahi and H. Masuhara, *Chem. Phys. Chem.*, 2004, **5**.
17. M. Álvaro, B. Ferrer, V. Fornes and H. García, *Chem. Commun.*, 2002, 2012.
18. J. L. C. Rowsell and O. M. Yaghi, *Microporous and Mesoporous Materials*, 2004, **73**, 3.
19. M. O'Keeffe, *Chem. Soc. Rev.*, 2009, **38**, 1215–1217.
20. C. G. Silva, A. Corma and H. Garcia, *J. Mater. Chem.*, 2010, **20**, 3141.
21. D. Zhao, D. J. Timmons, D. Yuan and H. C. Zhou, *Acc. Chem. Res.*, 2011, **ACS ASAP**.
22. A. Fujishima, X. Zhang and D. A. Trykc, *Surf. Sci. Rep.*, 2008, **63**, 515.
23. P. Haider, J. D. Grunwaldt and A. Baiker, *Catal. Today*, 2009, **141**, 349.
24. M. Eddaoudi, J. Kim, N. Rosi, D. Vodak, J. Wachter, M. O'Keeffe and O. M. Yaghi, *Science*, 2002, **295**, 469.
25. M. P. Suh, Y. E. Cheon and E. Y. Lee, *Coord. Chem. Rev.*, 2008, **252**, 1007.
26. S. Natarajan and P. Mahata, *Chem. Soc. Rev.*, 2009, **38**, 2304.
27. T. Loiseau, C. Serre, C. Huguenard, G. Fink, F. Taulelle, M. Henry, T. Bataille and G. Ferey, *Chem.-Eur. J.*, 2004, **10**, 1373.
28. I. A. Baburin, S. Leoni and G. Seifert, *J. Phys. Chem. B*, 2008, **112**, 9437.
29. O. M. Yaghi, M. O'Keeffe, N. W. Ockwig, H. K. Chae, M. Eddaoudi and J. Kim, *Nature*, 2003, **423**, 705.
30. J. H. Cavka, S. Jakobsen, U. Olsbye, N. Guillou, C. Lamberti, S. Bordiga and K. P. Lillerud, *J. Am. Chem. Soc.*, 2008, **130**, 13850.
31. M. Dan-Hardi, C. Serre, T. Frot, L. Rozes, G. Maurin, C. Sanchez and G. Ferey, *J. Am. Chem. Soc.*, 2009, **131**, 10857.
32. S. Barman, H. Furukawa, O. Blacque, K. Venkatesan, O. M. Yaghi and H. Berke, *Chem. Commun.*, 2010, **46**, 7981.
33. J. R. Li, R. J. Kuppler and H. C. Zhou, *Chem. Soc. Rev.*, 2009, **38**, 1477.
34. J. R. Long and O. M. Yaghi, *Chem. Soc. Rev.*, 2009, **38**, 1213.
35. A. Corma, H. Garcia and F. X. Llabrés i Xamena, *Chem. Rev.*, 2010, **110**, 4606.
36. A. Michaelides, S. Skoulika and M. G. Siskos, *Chem. Commun.*, 2011, **47**, 7140.
37. A. Modrow, D. Zargarani, R. Herges and N. Stock, *Dalton Trans.*, 2011, **40**, 4217.
38. K. K. Tanabe, C. A. Allen and S. M. Cohen, *Angew. Chem., Int. Ed.*, 2010, **49**, 9730.
39. X. Y. Wang, Z. M. Wang and S. Gao, *Chem. Commun.*, 2007, **11**, 1127.
40. J. Gascon, M. D. Hernandez-Alonso, A. R. Almeida, G. P. M. van Klink, F. Kapteijn and G. Mul, *Chem Sus Chem*, 2008, **1**, 981.
41. H. Khajavi, J. Gascon, J. M. Schins, L. D. A. Siebbeles and F. Kapteijn, *J. Phys. Chem. C*, 2011, **115**, 12487.

42. C. Wang, Z. Xie, K. E. deKrafft and W. Lin, *J. Am. Chem. Soc.*, 2011, **133**, 1.

43. M. H. Xie, X. L. Yang, C. Zou and C. D. Wu, *Inorg. Chem.*, 2011, **50**, 5318.

44. S. Bordiga, C. Lamberti, G. Ricchiardi, L. Regli, F. Bonino, A. Damin, K. P. Lillerud, M. Bjorgen and A. Zecchina, *Chem. Commun.*, 2004, 2300.

45. M. Alvaro, E. Carbonell, B. Ferrer, F. X. Labrés i. Xamena and H. Garcia, *Chem.–Eur. J.*, 2007, **13**, 5106.

46. F. X. Labrés i. Xamena, A. Corma and H. Garcia, *J. Phys. Chem. C*, 2007, **111**, 80.

47. T. Tachikawa, J. R. Choi, M. Fujitsuka and T. Majima, *J. Phys. Chem. C*, 2008, **112**, 14090.

48. J. R. Choi, T. Tachikawa, M. Fujitsuka and T. Majima, *Langmuir*, 2010, **26**, 10437.

49. C. G. Silva, I. Luz, F. X. Labrés i. Xamena, A. Corma and H. Garcia, *Chem. Eur. J.*, 2010, **16**, 11133.

50. H. A. Lopez, A. Dhakshinamoorthy, B. Ferrer, P. Atienzar, M. Alvaro and H. Garcia, *J. Phys. Chem. C*, 2011, **115**, 22200.

CHAPTER 13

Catalysis by Covalent Organic Frameworks (COFs)

MARCUS ROSE AND REGINA PALKOVITS*

Lehrstuhl für Nanostrukturierte Katalysatoren, Institut für Technische und Makromolekulare Chemie, RWTH Aachen University, Worringerweg 1, 52074 Aachen, Germany
*Email: palkovits@itmc.rwth-aachen.de

13.1 Introduction to COFs

Covalent organic frameworks are a class of porous materials similar to MOFs. Instead of coordinative bonds between the organic linkers and the inorganic knots, the organic linkers in COFs are connected exclusively by covalent bonds between each other or *via* covalent bonded heteroatoms or through inorganic clusters as connecting units. The term "covalent organic frameworks" was introduced by Omar Yaghi's group in 2005 with their first publication on COFs derived by a condensation reaction of multifunctional organic boronic acids.[1] Since then, a substantial research effort involving several groups has led to a variety of similar materials, which can be derived by applying different synthetic reactions originating from organic as well as polymeric chemistry.[2–6] Many of those materials have been summarized in classes such as CTFs, CMPs, PAFs, EOFs, PIMs and HCPs, which are explained in more detail in this chapter. Although these materials are all based on different reaction concepts compared to the "classical" COFs, all these materials are covalent organic frameworks in a broad sense and thus are treated as such.

Due to the exclusively covalent bonded framework structures, COFs in general, with a few exceptions, exhibit a significantly higher hydrolytic stability

RSC Catalysis Series No. 12
Metal Organic Frameworks as Heterogeneous Catalysts
Edited by Francesc X. Llabrés i Xamena and Jorge Gascon
© The Royal Society of Chemistry 2013
Published by the Royal Society of Chemistry, www.rsc.org

Table 13.1 Classification of porous organic framework materials.

	Modular Organic Frameworks		*Classical Polymeric Materials*	
Concept	modulare concept of MOFs		crosslinked polymers	rigid monomers
Structure	ordered	unordered	unordered	
Porosity	exactly definable	controllable	poorly controllable	
Examples	COFs, CTFs	CMPs, EOFs, PAFs	HCPs	PIMs

than MOFs. Thus, they prove suitable as catalysts and catalyst supports even for reactions in aqueous media in which many MOFs are readily decomposed. Their high stability in combination with the organic scaffold is ideally suited to post-synthetic functionalization, such as nitration, amination or sulfonation, for the introduction of additional functional groups. Similar to MOF structures, COFs provide an open porosity with a high accessible surface area. Also, the modular concept applies, by which the pore size and the network morphology can be tailored by varying the size, geometry and connectivity of the linkers. Within the enormous variety of such materials, significant differences in terms of crystallinity and porosity are observed depending on the applied synthetic routes. Despite huge amounts of analytical data, the influence of specific synthesis parameters on the resulting material's properties such as the degree of crosslinking is not yet clearly understood.

COFs can be produced by classical organic synthesis methods utilizing multifunctional organic linker molecules and applying the modular concepts known from MOF synthesis. Alternatively, from an approach derived from classical polymer chemistry, and utilized for decades as *Davankov* resins, porous polymers can be obtained by chemical crosslinking of monomers as in the case of hyper-crosslinked polymers (HCPs). All framework materials that can be obtained by these methods provide an open porosity, usually of the size of micropores as well as small mesopores. The material's porosity is mainly influenced by the degree of crosslinking and thus by the long-range order/crystallinity of the networks. As a result, organic framework materials can be classified as shown in Table 13.1. Additionally, an overview of the most important materials and synthesis routes is given in this chapter, as a brief introduction to this topic. For further details, some extensive reviews on these materials have been published in recent years.[2–6]

13.1.1 Ordered COFs

Ordered covalent organic frameworks, in which the modular concept of MOFs applies, can be obtained by thermodynamically controlled synthesis methods. Thus, the network formation reaction has to be reversible under the applied reaction conditions, which facilitates the reorganization of the framework structure, with the goal of achieving a higher long-range order and, in an ideal case, measureable crystallinity.

In the case of the COFs invented by the group of Omar Yaghi, the multi-functional aromatic linkers are covalently connected *via* heteroatoms or clusters (Scheme 13.1a–d). These connectors are typically generated during the self-condensation reaction of three boronic acid groups to six-membered boroxine rings (B_3O_3). Alternatively, connectors can be produced by the este-rification of one boronic acid group with a vicinal hydroxyl functionalized aromatic system yielding five-membered C_2O_2B-rings.[1,3,7] Additionally, a few COF materials have been reported based on a borosilica connector (COF-202)[8] as well as imine linkages (COF-300).[9] COFs are typically synthesized by solvothermal processes using inert solvents. More efficient syntheses in terms of a shorter reaction time and lower temperatures have been demonstrated utilizing microwave radiation instead of conventional heating, *e.g.*, the synthesis of COF-5 conventionally at 120 °C for 72 h,[1] which can be carried out at 100 °C in only 20 min.[10] Due to the reversible condensation reactions in network formation, the reorganization of the linkers is facilitated. Thus, high degrees of crosslinking are achieved, resulting in crystalline materials with high specific surface areas of up to 4000 $m^2 g^{-1}$ as for COF-103.[7] On the contrary, the reversibility of condensation reactions, especially when boronic acids are involved, poses a disadvantage in terms of hydrolytic stability. Most of these COFs are easily decomposed upon contact with water and are thus not well suited for many applications except under inert or at least water-free conditions.

Overcoming this drawback, another class of partially ordered organic framework materials was reported in 2008 by the group of Arne Thomas – the covalent triazine-based frameworks (CTFs).[11] These compounds are synthesized from multifunctional nitriles *via* an ionothermal synthesis route in molten zinc chloride as reaction medium (Scheme 13.1e).[12,13] The powdery monomer is mixed with Lewis acidic zinc chloride and the solid mixture is heated in quartz glass ampules. Crosslinking of the monomers occurs *via* the cyclotrimerization reaction of the nitrile groups under the formation of triazine rings (C_3N_3). At temperatures of 400–600 °C this reaction is reversible enabling a partial crystallization and the formation of ordered domains. Since no water is formed as a by-product as in the condensation reaction of COFs, the CTFs are insensitive to water. In fact, these materials show very high chemical as well as thermal stability due to the exclusively aromatic linkers and connecting units which are formed above 400 °C. Additionally, they are not just stable in the presence of water, but they are also stable under highly acidic and harsh conditions, as shown by Palkovits *et al.*[14] In these investigations, CTFs were used as catalyst supports for the direct oxidation of methane to methanol, which was carried out in oleum as reaction medium at 215 °C. Beside these outstanding properties, CTFs provide high specific surface areas of up to 2475 $m^2 g^{-1}$.[15] However, it was shown that higher synthesis temperatures as well as excess amounts of $ZnCl_2$ result in higher specific surface areas. In this case, amorphous materials were obtained and a lower amount of nitrogen was identified in the framework structure as would be expected from the ideal composition. For example, in the case of CTF-1 an increase of the specific

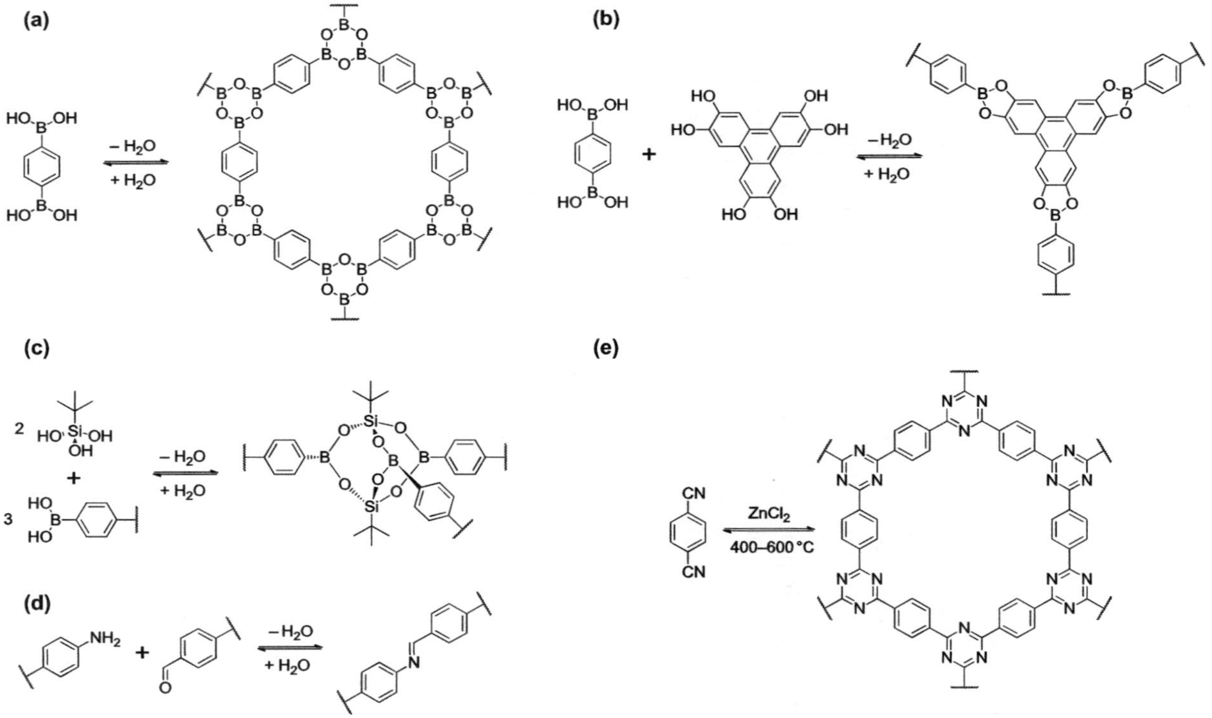

Scheme 13.1 Formation of COFs based on boronic acid linkers by self-condensation (a),[1] esterification (b),[1] condensation to borosilica connectors (c),[8] as well as an imine linkage (d).[9] CTF materials are derived by the cyclotrimerization of multifunctional nitrile monomers (e).[11]

surface area from 791 for the crystalline material to $1123 \, \text{m}^2 \, \text{g}^{-1}$ for the amorphous derivative has been observed.[11] The lower nitrogen content is explained by the partial decomposition of the nitrile groups at elevated temperatures, which results in a widened pore system that can be selectively tailored by variation of the above mentioned parameters.[15,16] In analogy to the COFs, the CTF synthesis can also be carried out more rapidly by using microwave heating, *e.g.* reducing the synthesis time of CTF-1 from several days to below one hour.[17]

13.1.2   Unordered COFs

Beside the above presented COFs and CTFs, which possess at least partially ordered structures, a large number of unordered porous organic materials have been reported in recent years. Despite no long-range order in these frameworks, the modular concept for tuning the pore size and related properties still applies and facilitates a certain control over the pore size and geometry.

One of the most important classes of these materials is designated as conjugated microporous polymers (CMPs). They were first reported in 2007 by the group of Andrew Cooper.[18] Most CMPs are poly(arylene ethynylenes) produced by the crosslinking of multifunctional aromatic halogenide and ethynyl monomers *via* the Pd-catalyzed Sonogashira-Hagihara cross-coupling (Scheme 13.2a). Since the catalyzed crosslinking reaction involving a C–C bond formation is not thermodynamically controlled and thus not reversible under the reaction conditions, CMPs are amorphous materials. Nevertheless, control of the pore size distribution, specific surface area and pore volume is still possible.[19] However, the increase of the size of a linker does not necessarily result in an increase of the specific surface area, as in crystalline materials.

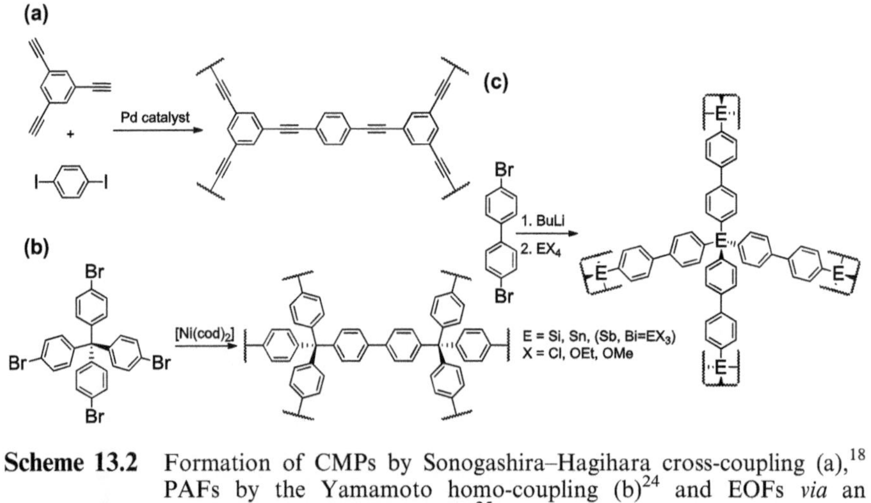

Scheme 13.2   Formation of CMPs by Sonogashira–Hagihara cross-coupling (a),[18] PAFs by the Yamamoto homo-coupling (b)[24] and EOFs *via* an organolithium intermediate (c).[25]

In fact, large linkers with a greater flexibility result in more efficiently packed amorphous frameworks and decreased specific surface areas and pore volumes. This effect can be diminished by using linkers with more functional groups enabling a higher crosslinking degree, *e.g.* the combination of trigonal with tetrahedral aromatic linkers, which results in the highest specific surface areas (up to $1200 \, m^2 \, g^{-1}$) for materials produced by this coupling reaction.[20,21]

Beside the Sonogashira–Hagihara reaction, other cross-coupling methods have been applied. For example, the related Pd-catalyzed Suzuki–Miyaura coupling of aryl-halogenides with aryl-boronic acids resulted in organic framework materials with specific surface areas of up to $1380 \, m^2 \, g^{-1}$.[22,23] Exceptional results have been obtained by the production of porous aromatic frameworks (PAFs) by the Yamamoto reaction, a Ni-catalyzed version of the Ullmann homo-coupling applied to tetrahedral aryl halides (Scheme 13.2b).[24] The resulting PAF-1 shows very few and broad peaks in a powder X-ray diffraction pattern indicating only rudimentary long-range order of the framework-building units. Nevertheless, this material exhibits, with $5640 \, m^2 \, g^{-1}$, one of the highest specific BET surface areas ever reported.

The crosslinking of exclusively organic linkers by covalent C–C bonds is not the only option to yield porous organic framework materials. In element organic frameworks (EOFs) the network formation occurs by covalent bonds between carbon atoms of the organic linkers and heteroatoms *via* an organolithium intermediate (Scheme 13.2c). Multifunctional aryl halogenides are reacted with butyl lithium to obtain strong nucleophilic linkers. By addition of electrophilic precursors of various elements, the network formation occurs and the highly crosslinked networks precipitate immediately. Due to the kinetically controlled reaction pathway, the resulting framework materials are amorphous and show no long-range order. Nevertheless, specific surface areas of up to $1040 \, m^2 \, g^{-1}$ have been obtained for EOF-2 using silicon atoms as connectors.[25] Expanding this concept to other elements, the formation of porous framework materials containing either tin, antimony or bismuth has been shown.[26] Since these materials provide good hydrolytic and thermal stability, applications as catalysts are of interest, due to the highly disperse metal centers.

13.1.3 Unordered Porous Polymers

The aforementioned classes of porous organic framework materials are based on the modular concept and the use of defined linkers and connection geometries comparable to MOFs. Alternatively, mainly microporous polymers with high specific surface areas can be produced by crosslinking methods known from classical polymeric chemistry. Hyper-crosslinked polymers (HCPs) have been known for decades and are produced commercially as ion exchange resins or ad-/absorbents, *e.g. Davankov* resins.[27] They have again attracted increasing attention within the field of COFs.[28] Classically, polymers such as polystyrene can be swollen in certain solvents. In the swollen state the chains can be chemically crosslinked, *e.g.* by Friedel Crafts alkylation reactions, using internal (*e.g.* chloromethylene groups on the polymer chain) or external electrophiles (*e.g.* dimethoxymethane as additional crosslinking substrate). Upon drying of the polymer networks, the chains are not able to pack efficiently due to the crosslinking groups, and thus, the remaining voids are accessible as pore space (Scheme 13.3a). The synthesis of HCPs is not restricted to pre-formed polymers. Also monomers with the respective functional groups such as chloromethylene groups can be used for crosslinking from the respective monomer solution (Scheme 13.3b). The precipitating porous and amorphous network materials show a high crosslinking degree. Thus, they are not soluble and provide an open porosity as well as a high chemical and thermal stability. By variation of the monomers and the synthetic conditions to tune the degree of crosslinking, micropores as well as small mesopores can be obtained and tuned within certain boundaries. Within these materials specific surface areas of up to $2000 \, m^2 \, g^{-1}$ can be achieved.[29,30]

A significantly different concept for the production of porous organic materials was realized for polymers of intrinsic microporosity (PIMs), which were developed and reported for the first time in 2002 by the groups of Peter Budd and Neil McKeown.[31–36] For such materials, crosslinking is not necessary to obtain accessible pore volume. In fact, PIMs consist of linear polymeric chains that are not able to pack efficiently since they consist of sterically demanding, tilted and highly rigid monomers. The applied polymerization reaction is the two-fold condensation of bifunctional alcohols with fluoro- or chloro-functionalized monomers under the elimination of the respective hydrogen halides yielding dioxane rings as connecting species (Scheme 13.3c).

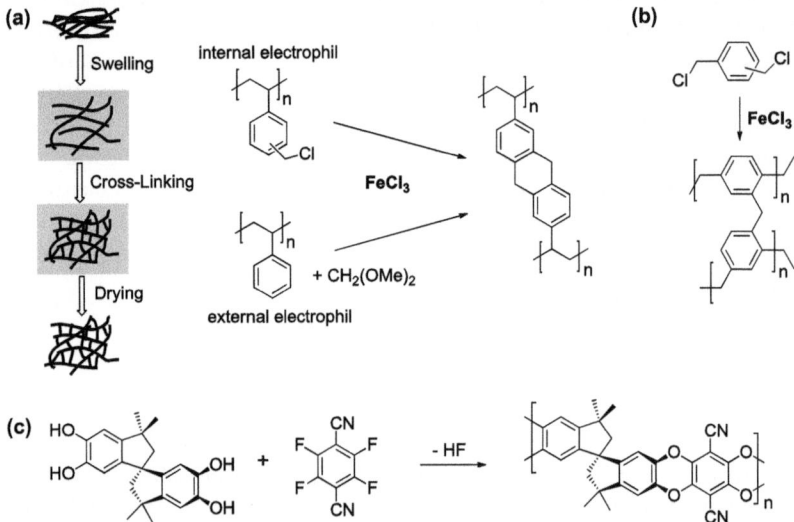

Scheme 13.3   Formation of HCPs from pre-formed polymers *via* internal or external electrophiles (a)[28] and from multifunctional monomers (b)[29,30] by Friedel–Crafts alkylation in comparison to the formation of PIMs from rigid monomers by a condensation reaction under the formation of dioxane rings as connecting species (c).[31–36]

The highest specific surface areas of up to $1760\,m^2\,g^{-1}$ were obtained with a triptycene-based PIM.[37] Instead of relatively simple aromatic monomers as in most COFs, this class of materials is mainly dominated by more complex monomers, *e.g.* based on derivatives of spirobisindane,[39–40] binaphthyl,[33] porphyrin[31,41,42] or phthalocyanine.[32,41,42] In particular, the latter two are of great interest for the coordinative immobilization of single metal atoms for catalytic purposes since they allow a high dispersion of such species with excellent accessibility of the active sites due to the open pore structure. The unique property of PIMs is the combination of microporosity with a soluble network structure due to the missing crosslinks. Since the porosity is preserved after dissolving and drying, the PIMs enable various possibilities for further processing. As an application of great importance in various fields, the formation of microporous membranes has been investigated by a slow evaporation of the solvent of a PIM solution.[38] Processing catalytic active and porous PIMs by this method could feasibly enable the production of reactive membranes as high performance catalysts.

13.2 Catalysis by Metal-containing COFs

Due to their chemically well-defined and high inner surface areas in combination with a high chemical and thermal stability due to covalently connected frameworks, COFs are ideally suited as versatile catalysts or catalyst supports. Recent developments in this field have been summarized in extensive and thorough reviews.[43,44] In the case of the immobilization of catalytic active metal species, COFs can be used as conventional supports for metal nanoparticles or clusters. Alternatively, well-defined functional groups on the inner surface allow for the immobilization of molecular metal species. By this approach, the advantages of molecular catalyst species applied in homogeneous catalysis are combined with the advantages of solid catalysts, such as a simplified separation and recycling.

13.2.1 COFs with Metal Nanoparticles and Clusters

COFs can be applied as porous support materials for the immobilization of metal nanoparticles or defined metal clusters. They pose certain advantages over conventional porous supports, such as activated carbons (ACs) or ceramics. In comparison to the often heterogeneously functionalized inner surfaces of ACs, COFs provide a well-known and much better tunable surface chemistry due to their modular construction from organic monomers. Ceramic supports such as metal oxides or nitrides often show a much lower accessible porosity than COFs. Additionally, it is much more difficult to equip ceramic supports with defined functional groups (except by certain grafting methods), while in COFs they can be easily incorporated in the framework structures using mild pre- or post-synthetic methods according to their organic chemistry.

Simply utilizing the porous structure of hyper-crosslinked polystyrene (HCPS), platinum nanoclusters were incorporated by impregnation with organic solutions of hexachloroplatinic acid.[45] Drying under mild conditions resulted in a mixture of $Pt^0/Pt^{II}/Pt^{IV}$ species. This catalyst was tested in the oxidation of L-sorbose with molecular oxygen to 2-keto-L-gulonic acid (Scheme 13.4a). It was shown that the activity of the catalyst could be increased by using a HCPS with a hierarchical micro-/macroporous instead of an exclusively microporous support structure. This effect can be attributed to hindered transport processes of the relatively large substrate and product molecules within pores of 2 nm diameter that are additionally filled with Pt-clusters of approximately 1.3 nm, emphasizing the importance of tailored porosity in heterogeneous catalysis.

Similar catalysts were tested in the catalytic wet air oxidation of phenol (Scheme 13.4b) proving the high stability of these materials under relatively harsh conditions under which most MOFs would be hydrolyzed very quickly.[46] It was also shown, that these Pt/HCPS catalysts exceed commercial Pt/Al_2O_3 or Pt/AC catalysts with regard to productivity and selectivity.

If COFs are used as supports that contain heteroatoms in their frameworks, an additional stabilizing effect on immobilized metal nanoparticles occurs due to the Lewis basic properties. CTFs containing a high content of aromatic nitrogen atoms have been used as a support for palladium nanoparticles.[47] These materials were obtained by a sol immobilization technique in which a solution of pre-formed Pd nanoparticles with a diameter of approximately 3 nm is used for impregnation. This catalyst was tested in the oxidation of glycerol with molecular oxygen (Scheme 13.4c). The catalytic performance was directly compared to such Pd nanoparticles supported on AC as a nitrogen-free support. Pd/CTF showed a slightly higher initial rate as well as a slightly increased selectivity for glyceric acid than Pd/AC. In terms of stability, Pd/CTF resulted in nearly full conversion of glycerol after 3 h with a decreasing activity

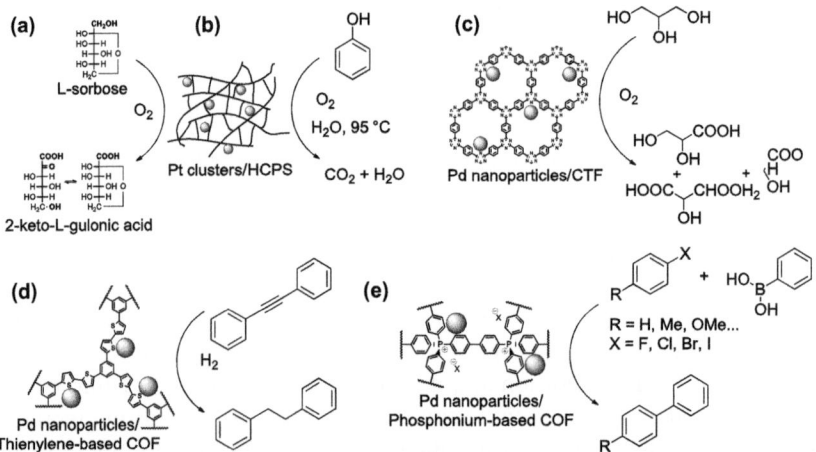

Scheme 13.4   Noble metal nanoparticles and clusters in different COFs enable catalytic oxidation (a–c),[45–47] hydrogenation (d)[48] as well as C–C coupling reactions (e).[50]

beginning with the 4th cycle, while Pd/AC did not exceed a conversion of 30% after 1 h and longer in the first cycle. The quick deactivation of Pd/AC was shown by TEM investigation to be a result of immediate particle aggregation. In contrast, aggregation was delayed for Pd/CTF due to the stabilizing effect of the coordinating nitrogen atoms of the supporting framework material. The deactivation after three cycles is additionally attributed to typically catalyst poisoning effects.

COFs containing Pd nanoparticles have also been prepared by several different routes. For example, microporous poly(thienylene arylene) networks were synthesized by a simple impregnation method with a $PdCl_2$ solution.[48] Monodisperse Pd^0 nanoparticles with an average diameter of 1.5 nm, which are stabilized by coordination to sulfur atoms in the framework, have been obtained after reduction. These catalysts have been successfully applied in the hydrogenation reaction of diphenylacetylene to 1,2-diphenylethane, showing full conversion after 2 h and 100% selectivity to the desired product (Scheme 13.4d). An alternative approach is to obtain highly dispersed Pd nanoparticles within the pores of a CMP framework material using super-critical CO_2 as an impregnation solvent for Pd^{II} precursors.[49] Due to the negligible surface tension of the supercritical fluid all pores are penetrated and the precursor is distributed very homogeneously within the porous framework. Upon thermal decomposition of the precursor, highly monodisperse Pd^0 nanoparticles of 1–3 nm in diameter are obtained. Recently, ion exchange was also used to obtain well dispersed Pd^0 nanoparticles.[50] The support material was a COF consisting of aromatic linkers connected *via* quaternary phosphonium groups. Due to the charged connectors, the halogenide counter-ions within the pores could be exchanged using H_2PdCl_4 in solution to obtain a uniform distribution of Pd. The Pd species was then chemically reduced using $NaBH_4$ to obtain Pd^0 nanoparticles. This catalyst has been applied in the Suzuki–Miyaura cross-coupling reaction applied to various substrates (Scheme 13.4e). In all experiments, conversions over 95% could be observed. Due to its high activity, this catalyst can even be applied in the Suzuki–Miyaura coupling reaction if fluorobenzene with a very low reactivity is involved as a reactant. Even in this case a yield of 87% was observed.

13.2.2 COFs with Molecular Metal Species

Beside the application of COFs as support materials for metal nanoparticles and their supporting effects due to defined functionalities on the inner surface, these materials have attracted much greater attention with regard to the possibilities of immobilizing defined molecular metal species. In this approach, the modular tunable functional groups enable the heterogenization of homogeneous catalyst species either by coordinative or even covalent bonds.

Various examples have been reported utilizing porphyrin as well as phthalocyanine linkers in catalytic active PIM as well as CMP materials.[42,51–54] These nitrogen containing macrocyles facilitate the coordination of single metal atoms in chemically highly defined vicinities. In their molecular form, these ligands are densely packed, providing a very low specific surface area, and most of the active metal centers are not accessible. Thus, within porous organic frameworks the accessibility of the active sites is significantly increased by preventing co-facial packing and thereby enabling their utilization as solid molecular catalysts. Three different synthetic strategies can be employed. One option is the application of the respective pre-formed metal-complexes as linkers in the network formation as in the case of a Co-phthalocyanine-PIM[42] and a Fe-porphyrin-CMP.[52] In the case of a slightly different Co-phthalo-cyanine-PIM, the metallation is carried out during the network formation, since the crosslinking reaction in this case is the formation of the phthalo-cyanine macrocycle containing the metal atom.[42] In the third case, the metal-free framework is synthesized first, with a subsequent, post-synthetic metal-lation step. This method allows for a higher flexibility in the choice of the metals within identical framework structures, as shown for imine-based porphyrin COFs containing iron or manganese, respectively.[54]

The improved catalytic performance of these porphyrin- and phthalocyanine-based materials in comparison to their molecular analogues has been proven in various test reactions. The Co-phthalocyanine-PIM exhibited a 20 times increased initial activity in the decomposition of hydrogen peroxide due to the improved accessibility of the metal centers.[42] Other successfully tested reactions are the oxidation of cyclohexene, cyclohexane and hydroquinone.[42,54] Environmentally of interest is the oxidation of sulfides in aqueous media. Due to its high stability, the Co-phthalocyanine-PIM was shown to efficiently catalyze the oxidation of sulfide ions to elemental sulfur in water (Scheme 13.5a).[51] Also the Fe-porphyrin-CMP showed very high activity and good selectivity in the oxidation of organic sulfides to the respective sulfones (Scheme 13.5b).[52] Additionally, the self-decomposition of the catalyst by intermolecular reactions of the porphyrin rings is restricted by the spatial separation of the linkers in the framework. Porphyrin-based COFs including iron as well as manganese have been tested in epoxidation reactions that are of

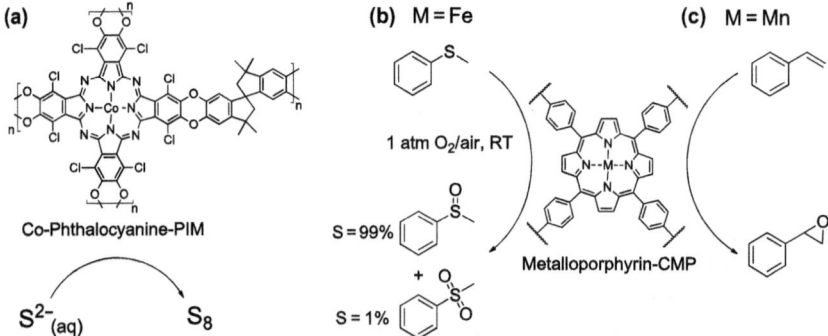

Scheme 13.5 Metallophthalocyanines and -porphyrins have been applied in various catalyzed reaction, such as the oxidation of sulfide ions in water (a),[51] the oxidation of organic sulfides to sulfones (b)[52] as well as the epox-idation of styrene (c).[53]

significant industrial interest. Fe-porphyrin-CMP showed a high activity, selectivity towards the epoxide and stability using diphenylethene as substrate with molecular oxygen as the oxidizing agent.[53] The advantage of hetero-geneous porphyrin frameworks over their molecular derivatives has also been shown for the epoxidation of styrene catalyzed by a Mn-porphyrin-framework (Scheme 13.5c).[54] It showed TONs >2000 while the molecular Mn-porphyrin species did not exceed 780.

COFs synthesized from pre-formed metal-containing linkers pose the advantage of exactly defined metal species within the resulting frameworks. For example, diethynyl bipyridyl linkers with coordinated Ir and Ru species have been incorporated into porous framework structures as co-monomers by applying the Co-catalyzed cyclotrimerization reaction of ethynyl groups for the network formation.[55] Besides a high porosity as well as a good thermal and chemical stability, these materials proved useful as photocatalysts. Predominantly higher yields compared to their homogeneous analogues have been observed for various test reactions, such as the light-driven aza-Henry reaction, the α-arylation of bromomalonate and the oxyamination of aldehydes (Scheme 13.6a–c). Due to their high stability even under catalytic conditions, the catalysts can be recycled several times. Also no metal leaching occurred during these model reactions. Other metal-containing bipyridyl linkers have

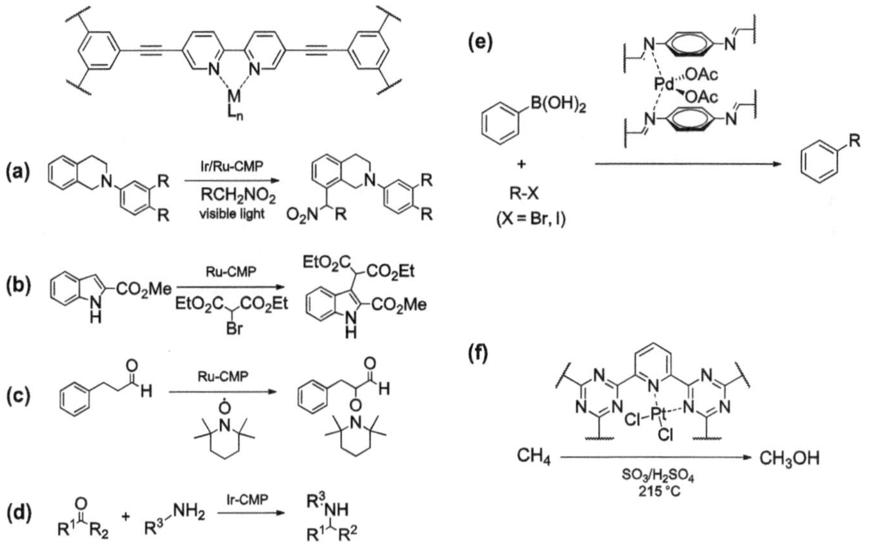

Scheme 13.6 Nitrogen-containing COFs are ideally suited as supports for coordinated molecular metal species. E.g. bipyridyl-based CMPs have been tested in various reactions (a–d).[55,56] A molecular Pd-species coordinated in between two sheets of a 2D imine-based COF was applied as catalyst in the Suzuki–Miyaura coupling (e).[57] Pt/CTF as solid analogue to the molecular Periana catalyst was used under very harsh conditions in the direct oxidation of methane to methanol, proving its high stability (f).[14]

been used for the formation of metallated CMPs.[56] These materials containing Ir, Rh and Re species have been investigated in the reductive amination of various ketones (Scheme 13.6d). Interestingly, the Ir catalyst with the lowest specific surface area showed the highest catalytic activity. This discrepancy is not yet fully understood.

Nitrogen-containing frameworks for the coordination of metal species are the most investigated materials among COFs. An imine-based COF with immobilized Pd^{II} species has been successfully demonstrated as a solid catalyst in the Suzuki–Miyaura coupling (Scheme 13.6e).[57] The catalyst was stable over at least four cycles under relatively harsh conditions (base-containing solution at 150 °C). It showed only negligible metal leaching but high product yields for various substrates. The incorporation of the $Pd(OAc)_2$ was proposed to occur in between the semi-crystalline layered structure by two-fold coordination of the Pd species by imine-nitrogen atoms from two opposing layers.

In general, the characterization of the coordinated metal species in the often amorphous frameworks poses quite a challenge. Often TEM is used to confirm the absence of metal nanoparticles. In combination with electron microscopy, EDX element mapping also provides valuable insights in the dispersion of the metal in the framework materials. X-Ray photoelectron spectroscopy (XPS) is often used to determine the oxidation state of the metal species. Together with more specialized X-ray absorption spectroscopy methods such as EXAFS and XANES based on synchrotron radiation, the chemical neighbourhood of coordinated metals can be determined accurately even in solid materials. More convenient methods such as NMR spectroscopy of dissolved metal-ligand complex species have been used, *e.g.* for binding studies of Pd in bipyridyl moieties of a PIM network, proving the suitability of the modular concept for framework immobilized molecular metal species.[41]

An outstanding example of the advantages of COFs, such as their high thermal and chemical stability and the suitability for catalyst immobilization, was shown for a platinum containing CTF material.[14,58] Due to its high nitrogen content, a high concentration of bipyridyl moieties are available in which $PtCl_2$ has been coordinated by wet impregnation methods. With several of the aforementioned methods, such as TEM, EDX mapping and XPS analysis, the molecular coordination of the metal species within the porous framework was proven. A comparable molecular analogue is the platinum bipyrimidine complex, also known as *Periana* catalyst. As a catalytic test reaction in which the Periana catalyst has been applied successfully, the one-pot oxidation of methane to methanol with oxygen has been chosen (Scheme 13.6f). This "dream reaction" is of great importance with regard to the utilization of natural gas for chemical energy storage and its efficient transport in form of a liquid compound. Unfortunately, this reaction requires the presence of highly concentrated sulfuric acid with free sulfur trioxide (oleum) as oxidant. It is involved in the reaction mechanism by the formation of methyl bisulfate as an intermediate. The formed sulfur dioxide is finally re-oxidized by molecular oxygen. However, oleum is used as the reaction medium in which the molecular Periana catalyst provides sufficient stability. The heterogenization of this catalyst is not possible using most of the conventional catalyst supports. Also, MOFs and presumably most COFs would be immediately oxidized due to their organic nature. So far, the only exception is the Pt/CTF that has been successfully tested in this reaction. Similar results compared to the homogeneous Periana catalyst could be achieved with TONs of 200–300 and selectivity to methanol of above 75%. This exceptional stability and catalytic performance under the very harsh conditions (oleum as a reaction medium at 215 °C) is maintained over at least six reaction cycles rendering this material a suitable catalyst for small-scale methane activation processes.

Beside nitrogen-containing metal coordination sites, other Lewis basic functional groups are also of interest for post-synthetic metallation. For example, hyper-crosslinked polymers with high specific surface areas of up to $1800 \, m^2 \, g^{-1}$ containing dibenzofuran as well as dibenzothiophene as co-monomers, which pose great potential as catalyst supports due to their respective oxygen and sulfur functionalities.[59] Other examples are COFs that contain catechol derivatives as functional groups. It was shown that the catechol moieties could be nearly quantitatively loaded with various molecular metal species, such as magnesium, copper and manganese, in a chemically highly defined environment.[60] A microporous polyisocyanurate has been derived from bifunctional isocyanate linkers by a cyclotrimerization reaction catalyzed by a *N*-heterocyclic carbene.[61] The linkers contain methoxy groups in close proximity to the carbonyl groups of the isocyanurate connectors. This structural feature was shown to efficiently coordinate $FeCl_2$ with a high stability in a catalytic model reaction. As such, the oxidation of benzyl alcohol to benzaldehyde by hydrogen peroxide in aqueous solution has been investigated (Scheme 13.7a). The catalytic performance of the Fe/COF with a TON of 100 and 98% selectivity towards benzaldehyde exceeds that of homogeneously applied $FeCl_2$ with a TON of 29 and a low selectivity of 51%, as well as a nanoparticulate Fe_2O_3 exhibiting a TON of 12–32 at selectivities of 35–97%. The high stability of the novel catalyst was emphasized by recycling experiments exhibiting no loss in activity and selectivity for at least eight cycles.

A first example of asymmetric catalysis by metal-containing COFs was recently reported. Tetraalkynyl binaphthyl linkers of various length have been polymerized by the cobalt-catalyzed cyclotrimerization reaction of ethynylene functional groups.[62] By post-synthetic metallation with $Ti(O^iPr)_4$ *via* covalent bonds at the 1,1'-hydroxyl groups, chiral catalysts have been derived. As a catalytic model reaction, the asymmetric addition of diethylzinc to aldehydes was chosen (Scheme 13.7b). The catalysts exhibit enantioselectivities of 55–81% (ee), which are lower than for molecular catalysts or comparable MOF catalysts. It was suggested that different active sites are present within the frameworks due to inhomogeneities within these amorphous structures. Nevertheless, these catalysts could be successfully recycled up to ten times without loss of activity or selectivity.

A significantly different concept of metal-containing covalent organic frameworks was realized within the EOF materials. In such examples, different metals such as tin, antimony and bismuth have been used as connectors for

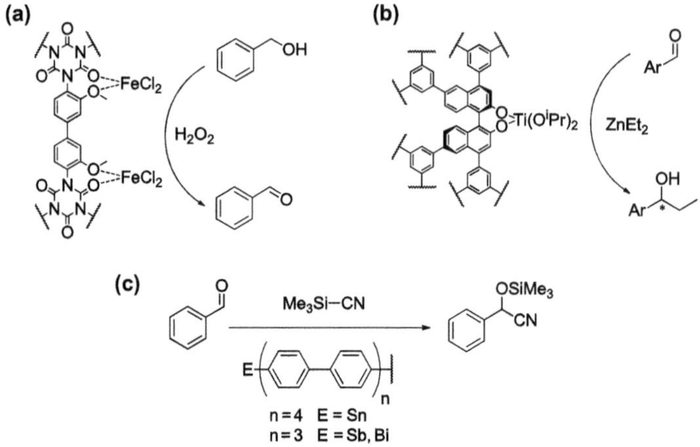

Scheme 13.7 Oxygen-containing COFs were used as support materials for the coordinative (a)[61] or covalent (b)[62] immobilization of metal species as viable solid catalyst even in asymmetric catalysis. The intrinsic catalytic activity of EOFs that contain covalently bonded metal atoms as connectors was proven by the cyanosilylation of benzaldehyde as a model reaction (c).[26]

organic aromatic linkers exclusively by covalent bonds.[26] Due to the synthesis route *via* a highly reactive organolithium intermediate, framework defects such as hydroxyl groups on the connector atoms have been identified. Nevertheless, these resulting materials provide sufficient stability to be handled in air and even in aqueous environments. Since metals are used as connectors, no post-synthetic metallation step is necessary. Their intrinsic catalytic activity has been investigated in the cyanosilylation of benzaldehyde (Scheme 13.7c). They show full conversion after 15 to 25 h, with the highest activity displayed by the Bi-EOF. Filtration tests after a short time and only partial conversion showed no further reaction that might occur due to leaching effects. Together with the successful recycling tests, the high stability of these materials under the applied conditions was proven. In particular, the tin-containing EOF might be of great interest for industrial applications within oleochemical processes. Currently, molecular tin species with a high toxicity, dependent on the functional groups, are often used as catalysts. Since great efforts are put into Sn separation after the reaction, a solid Sn-containing catalyst would pose a valuable alternative in the case of a comparable catalytic performance.

13.3 Metal-free COFs in Organocatalysis

Beside the immobilization of molecular metal species within porous organic frameworks for catalysis, the application of metal-free organocatalysts due to their intrinsic catalytic properties is possible. Polymeric supports are known for a long time to be easily functionalized by classical electrophilic substitution reactions. For example, solid acidic materials are obtained by sulfonation.

In this regard, various ion exchange resins are used commercially in several applications.[63] Functionalized COFs can be produced by utilizing pre-synthetic functionalized linkers, as shown for several COFs and CMPs.[64,65] Alternatively, COFs that provide a sufficient stability also allow for various post-functionalization reactions, as shown for HCPs by post-synthetic bromination and nitration.[66] Beside such groups for a further surface functionalization, acid/basic COF catalysts are also conceivable. Many of the COFs discussed in the previous chapters that provide heteroatom functionalities within their framework structures are viable for Lewis-base catalytic applications. Selected examples of organocatalytic applications within COF materials are presented in the following section.

A HCP with hierarchical micro- and macroporosity has been produced by high internal phase emulsion polymerization (polyHIPE) of chloromethylene styrene and divinylbenzene as monomers yielding a macroporous monolithic polymer.[67] Subsequently, the polymer was swollen and crosslinked by partial use of the chloromethylene groups as internal electrophiles to obtain a microporous network. The residual chloromethylene functionalities have been used for post-synthetic covalent anchoring of 4-(N-methylamino)pyridine (MAP). The resulting highly porous solid base material has been investigated in the acylation of tertiary methylcyclohexanol with acetic anhydride (Scheme 13.8a). It showed a high catalytic activity comparable to the molecular (dimethylamino)pyridine. Furthermore, the advantage of the high surface area due to the crosslinking over a non-crosslinked material was shown by comparison of catalyst performances. While the microporous HCP showed full conversion, the material with a significantly lower surface area achieved only 65%. Since these materials appear in monolithic form and the macropore system enables a fast mass transport, they are of great interest as solid catalysts in fixed-bed continuous flow reactors.

The immobilization of conventional homogeneous organocatalysts can also be carried out by incorporating the respective molecular species as framework linkers. This concept was realized within a CMP material that included a Tröger's base derivative as a building block.[68] This material was tested in the catalytic addition of diethylzinc to 4-chlorobenzaldehyde (Scheme 13.8b). Compared to molecular derivatives of the catalyst exhibiting product yields of 65 and 74%, the CMP showed slightly decreased yields of 50–60%. Nevertheless, due to its heterogeneity the catalyst could be easily separated from the reaction mixture and successfully re-used in at least three cycles without loss of activity.

A similar approach was used for the immobilization of molecular N-heterocyclic carbenes (NHC) that are of great importance in the field of organocatalysis.[69] Bifunctional aryl halogenide-substituted imidazolium salts are incorporated within porous organic frameworks by cross-coupling with tetrahedral aromatic linkers. Results were reported simultaneously by two different groups applying the Suzuki–Miyaura coupling (NHC-EOF)[70] as well as the Sonogashira–Hagihara coupling (NHC-CMP)[71] for the network formation. The catalytic performance of the NHC-EOFs was tested in the

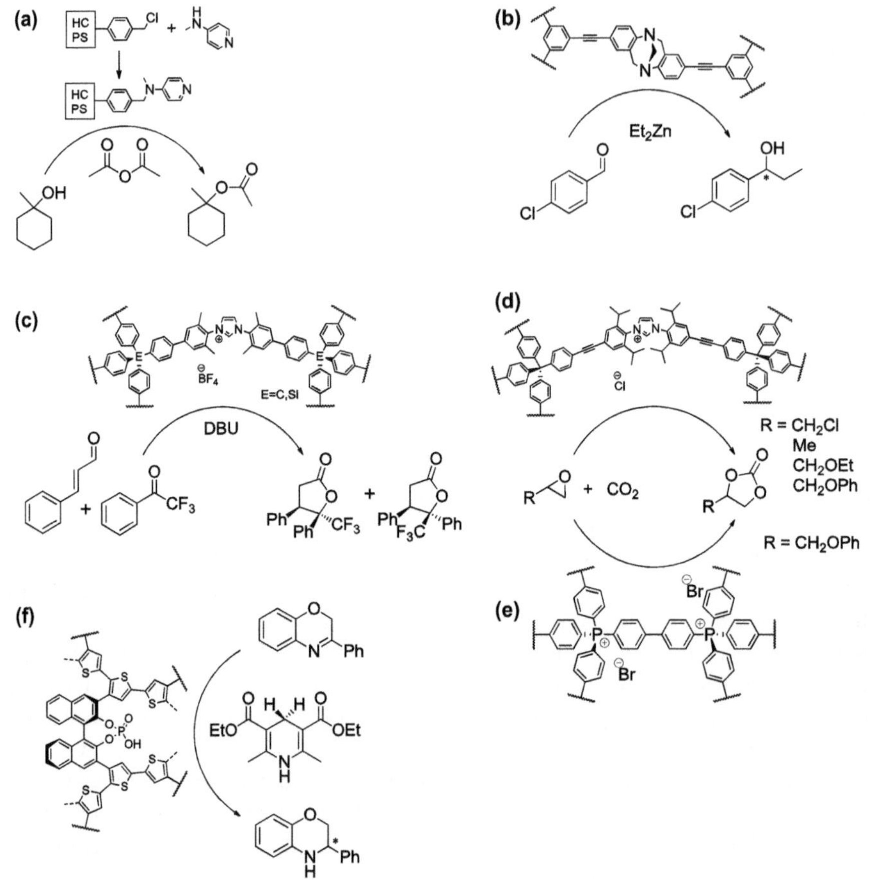

Scheme 13.8 Due to their well-defined framework structure, COFs are ideally suited as solid, metal-free organocatalysts. A HCPS was post-synthetically functionalized with (methylamino)pyridine as a Lewis base (a).[67] Other well-known organocatalysts, such as Tröger's base (b)[68] and imidazolium salts as precursors for *N*-heterocyclic carbenes (c,d),[70,71] can be inserted as organic linkers within the framework structure. A NHC-CMP (d),[71] as well as a phosphonium-based COF (e),[50] were successfully applied as catalysts in the activation of epoxides and the subsequent conversion with CO_2 to cyclic carbonates. A phosphoric acid-containing binaphthyl-based COF showed catalytic activity in an asymmetric transfer hydrogenation reaction (f).[72]

conjugated umpolung of an α,β-unsaturated cinnamaldehyde with trifluoroacetophenone (Scheme 13.8c).[70] The ionic imidazolium groups are converted to the NHC *in situ* prior to the reaction by deprotonation with mild bases. The NHC-EOFs exhibit a catalytic performance comparable to their homogeneous analogues in terms of product yields as well as stereoselectivity. Also, heterogeneity of the reaction was proven by filtration tests. No leaching of the catalytic active compound has been observed. The catalyst could also be recycled up to four times with a minor loss in activity but constant stereoselectivity.

In contrast, the NHC-CMP was tested in a model reaction of high industrial importance.[71] NHC is known to activate epoxides for the conversion with CO_2 to obtain cyclic carbonates (Scheme 13.8d). The solid catalyst showed high catalytic activity in this reaction. Also, the recyclability was proven. Interestingly, an advantage of modular framework materials was observed. Since the pore size can be tuned by variation of the linkers and connection geometries, size selectivity has been obtained allowing or restricting the access of substrate molecules to the pore systems as well as to the catalytic active centers. Thus, NHC-CMP showed a decrease of catalytic activity with an increase of the substrate size from epichlorohydrin to glycidyl phenyl ether, while the activity of non-porous solid catalysts was similar for these substrates.

The same reaction starting from the glycidyl phenyl ether has been catalyzed by the aforementioned phosphonium-based COF (Scheme 13.8e).[50] The results are not directly comparable since different reaction conditions have been applied, such as higher catalyst loadings of the phosphonium COF (1–4% instead of 0.065%) as well as longer reaction times (20–70 h instead of 10h). On the other hand, lower reaction temperatures have been investigated (90–140 °C instead of 150 °C), and instead of 10 bar CO_2 pressure the reaction with the phosphonium COF has been carried out at 1 atm under constant CO_2 flow. Thus, yields of the cyclic carbonate of up to 98% have been reported. After at least five cycles, the product yield is still higher than 90% demonstrating the stability and recyclability of the phosphonium COF.

In the case of the metal-free COFs with intrinsic activity, only one example has been reported thus far dealing with the application as solid asymmetric organocatalyst. Binaphthyl phosphoric acid (BNPPA) has been incorporated as a building block within a porous organic framework material. The chiral unit was two-fold functionalized with thiophene groups that are necessary for an oxidative polymerization reaction catalyzed by iron(III)chloride.[72] The resulting BNPPA-COF provided a BET specific surface area of merely $88 \, m^2 \, g^{-1}$, while the phosphoric acid-free polymer exhibited $560 \, m^2 \, g^{-1}$. This significant reduction in porosity is a result of the additional weight due to the functionalization as well as partial pore blocking. Nevertheless, interesting results have been obtained in asymmetric catalysis since amorphous materials of this kind provide a moderate flexibility within their frameworks. Thus, depending on the solvent, they are able to swell and provide access to the pore system within the widened framework. However, the chosen catalytic reaction is the enantioselective transfer hydrogenation of dihydro-2*H*-benzoxazine (Scheme 13.8f). In general, the enantioselective transfer hydrogenation of imines and unsaturated nitrogen heterocycles is of great importance since it enables the production of chiral amines. The aforementioned reaction was chosen as model system since molecular analogues of the catalyst species, *i.e.*, phenyl-substituted BNPPA, gave low selectivities of only 6% ee.[73] By an increase of the size of the aromatic substituent, the selectivity could also be significantly increased. The same effect was observed for the BNPPA-COF as catalyst, in which the binaphthyl unit is part of a highly crosslinked framework that results in a comparable effect as for the larger substituents. However, the solid catalyst exhibited nearly full conversion in combination with enantioselectivities of approximately 50% ee, which could be maintained for at least four cycles. This positive effect of the framework structure on the enantioselectivity was explained by a higher steric hindrance resulting in the formation of the preferred enantiomer.

13.4   Conclusion

In this chapter we have given an overview of the developments in an emerging field of novel porous materials. In recent years, a huge variety of COFs have been developed rapidly, displaying versatile and useful properties. These materials combine the advantages of both the modular concept and the tunable properties of MOFs with the high stability due to covalent bonds of classical porous polymers. Thus, most COFs are viable catalyst supports as well as catalysts with intrinsic activity. COFs can also be applied as supports for catalytically active metal nanoparticles or clusters comparable to conventional porous supports. In the case of heteroatom containing materials, an additional stabilizing effect for the nanoparticles is observed by coordination to the framework support preventing or at least delaying the particle aggregation. Much higher potential is provided by the ability of chemically well-defined heteroatom moieties for the coordination of molecular metal species. In this regard, conventional molecular organometallic catalysts can be easily tailored into their solid analogues providing the corresponding advantages, such as a simplified separation and recycling. The third option is the utilization of intrinsic catalytic properties of the metal-free organic frameworks in organocatalysis. In particular, the heterogenization of known organocatalytic reactions as well as the development of new high-performance catalysts enables the development of novel advanced processes. Thus, there is also great potential for an increased use of organocatalysis within industrial chemical processes, since solid catalysts enable a facilitated upscaling of chemical processes.

Many proof-of-principle studies have been reported up to now describing materials of mainly scientific interest. A lot of work is still necessary for a comprehensive characterization of COFs and a more detailed understanding of the respective structure–property–activity relations within these chemically well-defined materials. Future work in this field should focus on the identification of the specific strength of such materials in catalysis. The development of novel catalysts and processes based on the great potential of COFs should thus be oriented on certain challenges in heterogeneous catalysis in which conventional catalysts fail.

References

1. A. P. Côté, A. I. Benin, N. W. Ockwig, M. O'Keeffe, A. J. Matzger and O. M. Yaghi, *Science*, 2005, **310**, 1166.

2. A. I. Cooper, *Adv. Mater.*, 2009, **21**, 1291.
3. M. Mastalerz, *Angew. Chem., Int. Ed.*, 2008, **47**, 445.
4. N. B. McKeown and P. M. Budd, *Macromolecules*, 2010, **43**, 5163.
5. C. Weder, *Angew. Chem., Int. Ed.*, 2008, **47**, 448.
6. A. Thomas, *Angew. Chem., Int. Ed.*, 2010, **49**, 8328.
7. H. M. El-Kaderi, J. R. Hunt, J. L. Mendoza-Cortés, A. P. Côté, R. E. Taylor, M. O'Keeffe and O. M. Yaghi, *Science*, 2007, **316**, 268.
8. J. R. Hunt, C. J. Doonan, J. D. LeVangie, A. P. Côté and O. M. Yaghi, *J. Am. Chem. Soc.*, 2008, **130**, 11872.
9. F. J. Uribe-Romo, J. R. Hunt, H. Furakawa, C. Klöck, M. O'Keeffe and O. M. Yaghi, *J. Am. Chem. Soc.*, 2009, **131**, 4570.
10. N. L. Campbell, R. Clowes, L. K. Ritchie and A. I. Cooper, *Chem. Mater.*, 2009, **21**, 204.
11. P. Kuhn, M. Antonietti and A. Thomas, *Angew. Chem., Int. Ed.*, 2008, **47**, 3450.
12. P. Kuhn, A. Forget, D. Su, A. Thomas and M. Antonietti, *J. Am. Chem. Soc.*, 2008, **130**, 13333.
13. P. Kuhn, A. Thomas and M. Antonietti, *Macromolecules*, 2009, **42**, 319.
14. R. Palkovits, M. Antonietti, P. Kuhn, A. Thomas and F. Schüth, *Angew. Chem. Int. Ed.*, 2009, **48**, 6909.
15. P. Kuhn, A. Forget, J. Hartmann, A. Thomas and M. Antonietti, *Adv. Mater.*, 2009, **21**, 897.
16. M. J. Bojdys, J. Jeromenok, A. Thomas and M. Antonietti, *Adv. Mater.*, 2010, **22**, 2202.
17. W. Zhang, C. Li, Y.-P. Yuan, L.-G. Qiu, A.-J. Xie, Y.-H. Shen and J.-F. Zhu, *J. Mater. Chem.*, 2010, **20**, 6413.
18. J.-X. Jiang, F. Su, A. Trewin, C. D. Wood, N. L. Campbell, H. Niu, C. Dickinson, A. Y. Ganin, M. J. Rosseinsky, Y. Z. Khimyak and A. I. Cooper, *Angew. Chem. Int. Ed.*, 2007, **46**, 8574.
19. J.-X. Jiang, F. Su, A. Trewin, C. D. Wood, H. Niu, J. T. A. Jones, Y. Z. Khimyak and A. I. Cooper, *J. Am. Chem. Soc.*, 2008, **130**, 7710.
20. J.-X. Jiang, A. Trewin, F. Su, C. D. Wood, H. Niu, J. T. A. Jones, Y. Z. Khimyak and A. I. Cooper, *Macromolecules*, 2009, **42**, 2658.
21. E. Stöckel, X. Wu, A. Trewin, C. D. Wood, R. Clowes, N. L. Campbell, J. T. A. Jones, Y. Z. Khimyak, D. J. Adams and A. I. Cooper, *Chem. Commun.*, 2009, 212.
22. M. Rose, N. Klein, W. Böhlmann, B. Böhringer, S. Fichtner and S. Kaskel, *Soft Matter*, 2010, **6**, 3918.
23. J. Weber and A. Thomas, *J. Am. Chem. Soc.*, 2008, **130**, 6334.
24. T. Ben, H. Ren, S. Ma, D. Cao, J. Lan, X. Jing, W. Wang, J. Xu, F. Deng, J. M. Simmons, S. Qiu and G. Zhu, *Angew. Chem., Int. Ed.*, 2009, **48**, 9457.
25. M. Rose, W. Böhlmann, M. Sabo and S. Kaskel, *Chem. Commun.*, 2008, 2462.
26. J. Fritsch, M. Rose, P. Wollmann, W. Böhlmann and S. Kaskel, *Materials*, 2010, **3**, 2447.
27. S. V. Rogozhin and M. P. Tsyurupa, *USSR Pat.*, UdSSR 299165, 1969.

28. M. P. Tsyurupa and V. A. Davankov, *React. Funct. Polym.*, 2006, **66**, 768.
29. C. D. Wood, B. Tan, A. Trewin, H. Niu, D. Bradshaw, M. J. Rosseinsky, Y. Z. Khimyak, N. L. Campbell, R. Kirk, E. Stöckel and A. I. Cooper, *Chem. Mater.*, 2007, **19**, 2034.
30. C. D. Wood, B. Tan, A. Trewin, F. Su, M. J. Rosseinsky, D. Bradshaw, Y. Sun, L. Zhou and A. I. Cooper, *Adv. Mater.*, 2008, **20**, 1916.
31. N. B. McKeown, S. Hanif, K. J. Msayib, C. E. Tattershall and P. M. Budd, *Chem. Commun.*, 2002, 2782.
32. N. B. McKeown, S. M. Makhseed and P. M. Budd, *Chem. Commun.*, 2002, 2780.
33. P. M. Budd, B. S. Ghanem, S. M. Makhseed, N. B. McKeown, K. J. Msayib and C. E. Tattershall, *Chem. Commun.*, 2004, 230.
34. P. M. Budd, N. B. McKeown and D. Fritsch, *J. Mater. Chem.*, 2005, **15**, 1977.
35. N. B. McKeown and P. M. Budd, *Chem. Soc. Rev.*, 2006, **35**, 675.
36. N. B. McKeown, P. M. Budd, K. J. Msayib, B. S. Ghanem, H. J. Kingston, C. E. Tattershall, S. M. Makhseed, K. J. Reynolds and D. Fritsch, *Chem.–Eur. J.*, 2005, **11**, 2610.
37. B. S. Ghanem, M. Hashem, K. D. M. Harris, K. J. Msayib, M. Xu, P. M. Budd, N. Chaukura, D. Book, S. Tedds, A. Walton and N. B. McKeown, *Macromolecules*, 2010, **43**, 5287.
38. P. M. Budd, E. S. Elabas, B. S. Ghanem, S. M. Makhseed, N. B. McKeown, K. J. Msayib, C. E. Tattershall and D. Wang, *Adv. Mater.*, 2004, **16**, 456.
39. M. Carte, K. J. Msayib, P. M. Budd and N. B. McKeown, *Org. Lett.*, 2008, **10**, 2641.
40. B. S. Ghanem, N. B. McKeown, P. M. Budd and D. Fritsch, *Macromolecules*, 2008, **41**, 1640.
41. P. M. Budd, B. S. Ghanem, K. J. Msayib, N. B. McKeown and C. E. Tattershall, *J. Mater. Chem.*, 2003, **13**, 2721.
42. H. J. Mackintosh, P. M. Budd and N. B. McKeown, *J. Mater. Chem.*, 2008, **18**, 573.
43. R. Dawson, A. I. Cooper and D. J. Adams, *Prog. Polym. Sci.*, 2012, **37**, 530.
44. P. Kaur, J. T. Hupp and S. T. Nguyen, *ACS Catal.*, 2011, **1**, 819.
45. L. M. Bronstein, G. Goerigk, M. Kostylev, M. Pink, I. A. Khotina, P. M. Valetsky, V. G. Matveeva, E. M. Sulman, M. G. Sulman, A. V. Bykov, N. V. Lakina and R. J. Spontak, *J. Phys. Chem. B*, 2004, **108**, 18234.
46. E. M. Sulman, V. G. Matveeva, V. Y. Doluda, A. I. Sidorov, N. V. Lakina, A. V. Bykov, M. G. Sulman, P. M. Valetsky, L. M. Kustov, O. P. Tkachenko, B. D. Stein and L. M. Bronstein, *Appl. Catal., B*, 2010, **94**, 200.
47. C. E. Chan-Thaw, A. Villa, P. Katekomol, D. Su, A. Thomas and L. Prati, *Nano Lett.*, 2010, **10**, 537.
48. J. Schmidt, J. Weber, J. D. Epping, M. Antonietti and A. Thomas, *Adv. Mater.*, 2009, **21**, 702.

49. T. Hasell, C. D. Wood, R. Clowes, J. T. A. Jones, Y. Z. Khimyak, D. J. Adams and A. I. Cooper, *Chem. Mater.*, 2010, **22**, 557.
50. Q. Zhang, S. Zhang and S. Li, *Macromolecules*, 2012, **45**, 2981.
51. S. Makhseed, F. Al-Kharafi, J. Samuel and B. Ateya, *Catal. Commun.*, 2009, **10**, 1284.
52. L. Chen, Y. Yang and D. Jiang, *J. Am. Chem. Soc.*, 2010, **132**, 9138.
53. L. Chen, Y. Yang, Z. Guo and D. Jiang, *Adv. Mater.*, 2011, **23**, 3149.
54. A. M. Shultz, O. K. Farha, J. T. Hupp and S. T. Nguyen, *Chem. Sci.*, 2011, **2**, 686.
55. Z. Xie, C. Wang, K. E. deKrafft and W. Lin, *J. Am. Chem. Soc.*, 2011, **133**, 2056.
56. J.-X. Jiang, C. Wang, A. Laybourn, T. Hasell, R. Clowes, Y. Z. Khimyak, J. Xiao, S. J. Higgins, D. J. Adams and A. I. Cooper, *Angew. Chem., Int. Ed.*, 2011, **50**, 1072.
57. S.-Y. Ding, J. Gao, W. Qiong, Y. Zhang, W.-G. Song, C.-Y. Su and W. Wang, *J. Am. Chem. Soc.*, 2011, **133**, 19816.
58. F. Schüth, R. Palkovits, C. Baltes, M. Antonietti and A. Thomas, WO2011/009429A1, 2011.
59. M. G. Schwab, A. Lennert, J. Pahnke, G. Jonschker, M. Koch, I. Senkovska, M. Rehahn and S. Kaskel, *J. Mater. Chem.*, 2011, **21**, 2131.
60. M. H. Weston, O. K. Farha, B. G. Hauser, J. T. Hupp and S. T. Nguyen, *Chem. Mater.*, 2012, **24**, 1292.
61. Y. Zhang, S. N. Riduan and J. Y. Ying, *Chem.–Eur. J.*, 2009, **15**, 1077.
62. L. Ma, M. M. Wanderley and W. Lin, *ACS Catal.*, 2011, **1**, 691.
63. B. C. Gates, in *Handbook of Heterogeneous Catalysis*, ed. G. Ertl, H. Knözinger, F. Schüth and J. Weitkamp, Wiley-VCH, Weinheim, 2008, pp. 278.
64. R. Dawson, A. Laybourn, R. Clowes, Y. Z. Khimyak, D. J. Adams and A. I. Cooper, *Macromolecules*, 2009, **42**, 8809.
65. D. N. Bunck and W. R. Dichtel, *Angew. Chem., Int. Ed.*, 2012, **51**, 1885.
66. J. Germain, J. M. J. Fréchet and F. Svec, *Polym. Mater. Sci. Eng.*, 2007, **97**, 272.
67. I. Pulko, J. Wall, P. Krajnc and N. R. Cameron, *Chem.–Eur. J.*, 2010, **16**, 2350.
68. X. Du, Y. Sun, B. Tan, Q. Teng, X. Yao, C. Su and W. Wang, *Chem. Commun.*, 2010, **46**, 970.
69. D. Enders, O. Niemeier and A. Henseler, *Chem. Rev.*, 2007, **107**, 5606.
70. M. Rose, A. Notzon, M. Heitbaum, G. Nickerl, S. Paasch, E. Brunner, F. Glorius and S. Kaskel, *Chem. Commun.*, 2011, **47**, 4814.
71. H. C. Cho, H. S. Lee, J. Chun, S. M. Lee, H. J. Kim and S. U. Son, *Chem. Commun.*, 2011, **47**, 917.
72. C. Bleschke, J. Schmidt, D. S. Kundu, S. Blechert and A. Thomas, *Adv. Synth. Catal.*, 2011, **353**, 3101.
73. M. Rueping, A. P. Antonchick and T. Theissmann, *Angew. Chem., Int. Ed.*, 2006, **45**, 6751.

CHAPTER 14

Towards Future MOF Catalytic Applications

FRANCESC X. LLABRÉS I XAMENA*[a] AND JORGE GASCON[b]

[a] Instituto de Tecnología Química UPV-CSIC, Universidad Politécnica de Valencia, Consejo Superior de Investigaciones Científicas, Avda. de los Naranjos, s/n, 46022, Valencia, Spain; [b] Catalysis Engnieering, Technical University of Delft, Julianalaan 136, 2628 BL Delft, The Netherlands
*Email: fllabres@itq.upv.es

14.1 Introduction

Throughout the previous pages of this book, we have described the synthesis, characterization and catalytic properties of MOFs. In this last chapter, we would like to highlight some additional aspects that in our opinion are also relevant for the future application of MOFs in catalytic processes at the industrial scale. On one hand, we will comment on the potential of (multi-functional) MOF catalysts for one-pot tandem reactions or multicomponent coupling reactions as a means of process intensification to improve the economy of the system. On the other hand, we will review the different approaches followed for the shaping of MOF catalysts, one of the critical steps before industrial implementation.

RSC Catalysis Series No. 12
Metal Organic Frameworks as Heterogeneous Catalysts
Edited by Francesc X. Llabrés i Xamena and Jorge Gascon
© The Royal Society of Chemistry 2013
Published by the Royal Society of Chemistry, www.rsc.org

14.2 Process Intensification with (Multi-functional) MOFs: One-pot Multi-component Couplings and Tandem Reactions

As the field of MOF catalysis matures, it becomes increasingly clear that MOFs will hardly replace other more conventional catalysts (such as mineral acids and bases, metal salts and complexes, or zeolites) for the synthesis of bulk chemicals, especially in processes that do not require highly specific catalysts. These catalysts are usually cheaper and/or more stable than MOFs and will certainly be the choice of industry. The circumstantial evidence of this is that currently there is not yet a single industrial application of MOFs in catalysis. In our opinion, if MOFs are to have a future in catalysis, we should direct our efforts towards reactions where the superior tunability of MOFs can be exploited, without being hampered by stability issues, and where other catalysts may find severe limitations. For instance, introduction of chiral centers is relatively easy in MOFs, either during synthesis or *a posteriori* by post-synthesis modification. Conversely, attempts to prepare chiral zeolites had only limited success, with a few chiral zeolites being known so far (SU-32,[1] ITQ-37[2] or goosecreekite[3]). Thus, MOFs could become preferred over zeolites and other catalysts for the synthesis of (chiral) drugs and other high added value fine chemicals. The syntheses of these complex substances will usually demand highly efficient and (enantio)selective catalysts, especially when polyfunctional substrates or chiral centers are involved. Meanwhile, the use of MOFs in one-pot multi-component coupling reactions (MCRs) and sequential (tandem) reactions allows process simplification, avoiding costly time- and energy-consuming isolation and purification of intermediate products. The high added value of fine chemical products on the one hand and the design of new more economic processes on the other hand could largely compensate for the possible higher costs of MOFs as compared to other alternative catalysts. In our opinion, the successful application of MOFs in catalysis should go in one of these directions: (i) asymmetric catalysis; (ii) one-pot multi-component coupling reactions; (iii) multifunctional MOFs for one-pot tandem reactions; or (iv) a combination of them. In this book, each individual chapter has been dedicated to different approaches to building catalytic functionalities in MOFs, including chiral moieties for asymmetric catalysis by MOFs (*viz.*, Chapter 11). We will now discuss the combination of several functionalities in one single scaffold, along with our ideas that MOFs should follow in order to become effective industrial catalysts.

14.2.1 MOFs as Multifunctional Heterogeneous Catalysts

In one-pot tandem processes, the consecutive steps occur under the same reaction conditions or, when this is not possible, they are carried out in two or more stages under different reaction conditions, with the correct addition sequence of reactants. Heterogeneous catalysts are promising candidates to perform multi-step transformations, especially when different and incompatible active sites (for instance acid–base) are required in each step. In these cases, it is not possible to use a combination of homogeneous catalysts. In contrast, solid catalysts allow the generation of robust site-isolated and well-defined multisite catalysts in which the active sites will never cancel each other and in which they can act either in a cooperative way or in different steps in a given tandem process.

The high tunability of MOFs and the possibility to introduce well defined catalytic active sites at the metallic nodes, at the organic linkers or inside the pore system using very different strategies (see Chapter 7), allow the preparation of multi-functional MOFs. In this way, a material containing simultaneously various types of functionalities (acid–base, metal–acid, metal–base or metal–metal) can be readily prepared, in which the catalytic sites are used cooperatively to perform multi-step transformations in one-pot.

Although reports on multifunctional MOF catalysts are still scarce and very recent, more are yet to come in the near future. In the following we will review some of the most representative reports appeared so far.

14.2.1.1 Bifunctional Acid–Metal Systems

In 2010, Pan *et al.*[4] reported on the use of MIL-101(Cr^{3+}) containing palladium nanoparticles (NPs) as a bifunctional Lewis acid/hydrogenation catalyst, which allowed the one-pot synthesis of methyl isobutyl ketone from acetone and H_2, through a condensation, dehydration and hydrogenation one-pot process (see Scheme 14.1). The Lewis acid properties of this composite material come from the coordinatively unsaturated Cr^{3+} centers of the MOF, while the Pd NPs conferred hydrogenation potential. When the parent MIL-101 MOF was used as catalyst, 60.1% acetone conversion with 74.9% selectivity to product **2** (45% yield) was obtained, due to the Cr^{3+} acid sites of the MOF. The addition of Pd significantly promoted the hydrogenation of **2**, and thus enhanced the conversion of acetone and selectivity to the desired product **3**.

The authors concluded that the formation of the desired product **3** needs the cooperation between acidic and metallic sites, while the production of isopropanol, coming from the reduction of the carbonyl group of acetone, only requires metallic sites. Thus, as the concentration of Pd increased, the palladium nanoparticles are deposited on the external surface, while the acidic sites (Cr^{3+}) which are necessary for the condensation and dehydration are mainly located inside the pores. As a result, the direct hydrogenation of acetone to isopropanol is favored while the condensation is significantly suppressed for

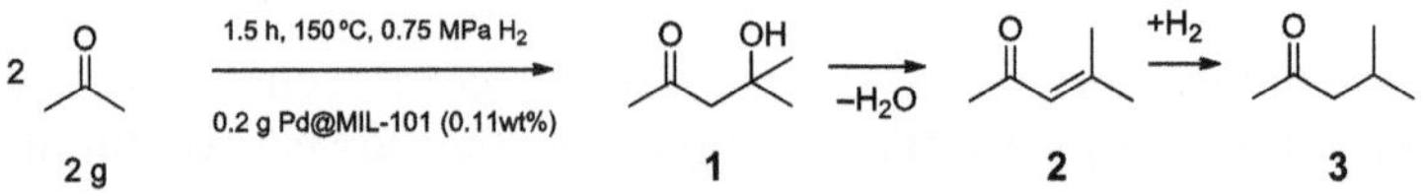

Scheme 14.1

the catalyst with high Pd loading, leading to reduced acetone conversion and selectivity to **3**.

The above example clearly illustrates that if we want to design an efficient multifunctional catalyst, it is not enough to introduce all the required types of functionalities in a solid, but a delicate interplay must exist between them, concerning concentration, location and distance.

We have recently used a similar Pd@MIL-101 system as bifunctional catalyst for the sequential conversion of citronellal into isopulegol and hydrogenation to menthol, according to Scheme 14.2.

We first studied the Cr^{3+} catalyzed carbonyl-ene reaction. We found that when the parent MIL-101(Cr^{3+}) was used as catalyst, citronellal was quantitatively converted in 18 h at 80 °C, with full selectivity to isopulegol and with a disatereoselectivity of 74% to the most industrially relevant isopulegol isomer. When Pd@MIL-101 was used as catalyst, we noticed an increase in the reaction rate (full conversion after 12 h) without affecting the selectivity or diastereoselectivity. These results were in line with previous reports on Ir-containing H-beta zeolite catalysts,[5] although were far from the performance described for Sn-beta.[6] In a second step, isopulegol was quantitatively hydrogenated over Pd@MIL-101 after an additional 6 h, producing menthol with an overall 86% selectivity and 81% diastereoselectivity.

We later extended the preparation of bifunctional acid–hydrogenation systems based on MIL-101(Cr^{3+}) to materials containing Pd or Pt,[7] either in the form of encapsulated metal nanoparticles (Pd@MIL-101 and Pt@MIL-101) or in the form of isolated transition metal Schiff base complexes (MIL-101-SI-Pd and MIL-101-SI-Pt) prepared according to Scheme 14.3.

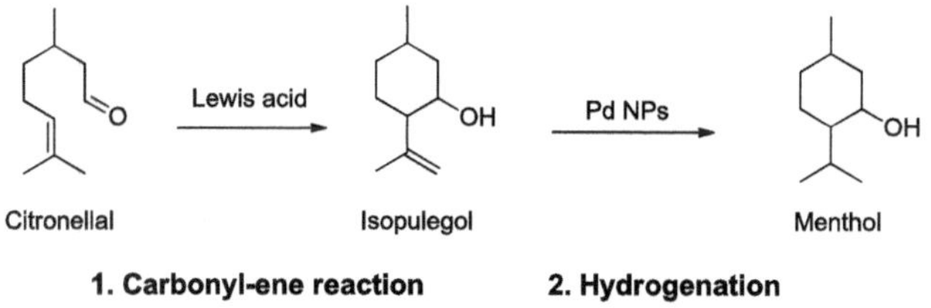

1. Carbonyl-ene reaction   **2. Hydrogenation**

Scheme 14.2

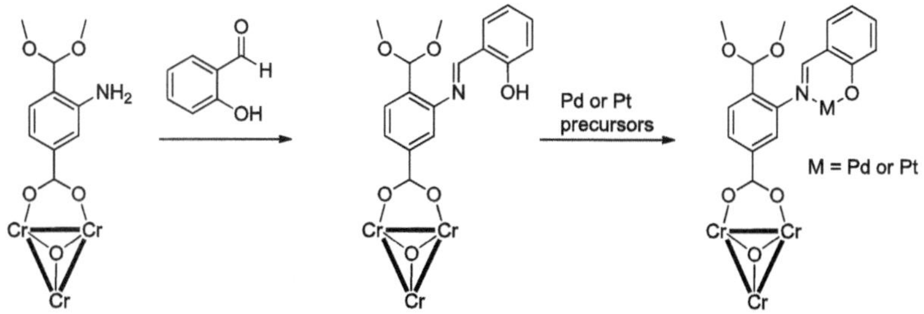

Scheme 14.3

The Pd- and Pt- containing MOFs were used as bifunctional catalysts for the one-pot synthesis of secondary arylamines through hydrogenation of nitroarene compounds followed by reductive amination of aldehydes and ketones, according to the general Scheme 14.4.

The method and the performance of the catalysts was demonstrated for various carbonyl compounds (benzaldehyde, acetophenone and cyclohexanone).[7] The tandem process was also applied to the synthesis of N-containing heterocycles of interest, such as (tetrahydro)quinolines (Scheme 14.5a), while the coupling of nitroarene reduction with Paal–Knorr condensation or Michael addition of

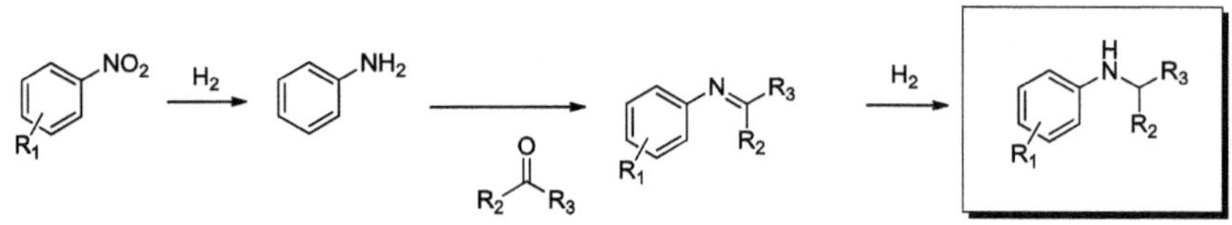

Scheme 14.4

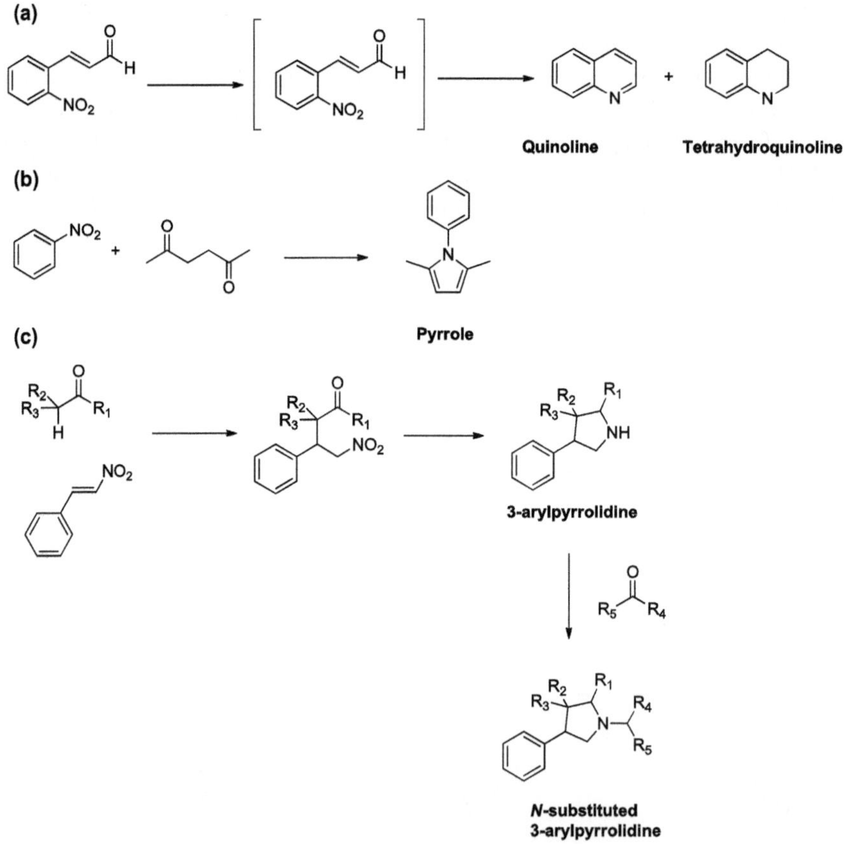

Scheme 14.5

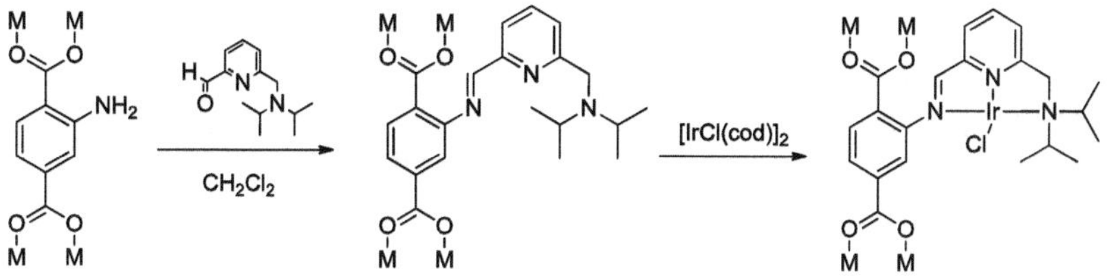

Scheme 14.6

suitable substrates lead to the formation of pyrrole (Scheme 14.5b) or 3-arylpyrrolidine and *N*-substituted 3-arylpyrrolidine (Scheme 14.5c), respectively.

In the above reactions catalyzed by bifunctional MOFs, we demonstrated the beneficial interplay between Lewis acid sites of the MOF and the hydrogenation properties of the metal species. As a result, the MOFs surpassed the performance of commercially available Pd/C, Pt/C, Pd/Al$_2$O$_3$, and Pt/Al$_2$O$_3$ catalysts under the same conditions. These catalysts, having only marginal acidity of the respective supports (carbon or Al$_2$O$_3$) cannot perform as well as the MIL-101 materials, which features highly active Cr^{3+} Lewis acid sites. This was especially clear for the synthesis of pyrrole and pyrrolidines, for which the limitation of the commercial catalysts was evident.

Nitroarene reduction followed by reductive amination using an Ir-containing bifunctional MOF was also reported very recently by Pintado-Sierra.[8] These authors used amino-containing UiO-66-NH$_2$ (or IRMOF-3) MOFs as supports, and introduced the iridium imino-pincer complex shown in Scheme 14.6 by post-synthesis modification as hydrogenation function. The scope of the reaction with this catalyst was demonstrated for various benzaldehydes and nitrobenzenes.

14.2.1.2 Bifunctional Acid–Base Systems

De Vos and co-workers were among the first to describe a bifunctional acid–base MOF catalyst.[9] The authors studied the use of the zirconium terephthalate UiO-66-NH$_2$ and they reached the conclusion that controlled thermal treatment under vacuum produced the reversible dehydroxylation of the [Zr$_6$O$_4$(OH)$_4$]$^{12+}$ clusters to [Zr$_6$O$_6$]$^{12+}$ at temperatures between 373 and 523 K, thus leaving coordinatively unsaturated positions in the triangular faces of the cluster (shared by 3 Zr centers) in close proximity to the amino groups of the organic linkers. Moreover, of the twelve linkers surrounding each Zr cluster in the ideal structure, approximately three were found to be systematically missing for the real materials. Such linker deficiency allows coordinatively unsaturated sites on Zr to be identified as the active sites. According to the authors, the resulting dehydroxylated material acted as a bifunctional acid base catalyst, as demonstrated for the cross aldol condensation reaction between benzaldehyde and heptanal.

Kim and co-workers have reported on the use of Al^{3+}-MIL-101-NH$_2$ as a bifunctional Lewis acid (coordinatively unsaturated Al^{3+} sites) and Brönsted base (–NH$_2$ groups) catalyst for the tandem Meinwald rearrangement of

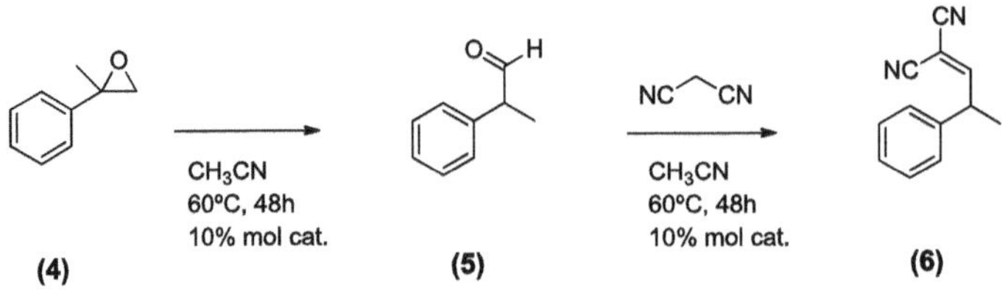

Scheme 14.7

epoxides and Knoevenagel condensation of the resulting aldehyde with activated methylene groups,[10] according to Scheme 14.7.

Reaction of the epoxide (4) and malononitrile in the presence of 10 mol% Al^{3+}-MIL-101-NH_2 resulted in the tandem epoxide ring opening to (5), followed by condensation and subsequent dehydration to (6) with and overall 70% yield. Although this reaction certainly demonstrated the bifunctional character of the Al^{3+}-MIL-101-NH_2, the scope of the reaction was somewhat limited: Meinwald rearrangement failed when aliphatic epoxides, styrene oxide or *trans*-stylbene oxide where used as substrates. Meanwhile, only a highly activated methylene compound, malononitrile, was tested for the Knoevenagel condensation.

A similar acid–base tandem process was reported by Zhou and co-workers consisting of a sequential deacetalization followed by Knoevenagel condensation catalyzed by PCN-124.[11] This MOF contains weakly acidic coordinatively unsaturated Cu^{2+} and basic pyridine and amide groups provided by the ligand 5,5′-((Pyridine-3,5-dicarbonyl)bis(azanediyl))diisophthalate (7).

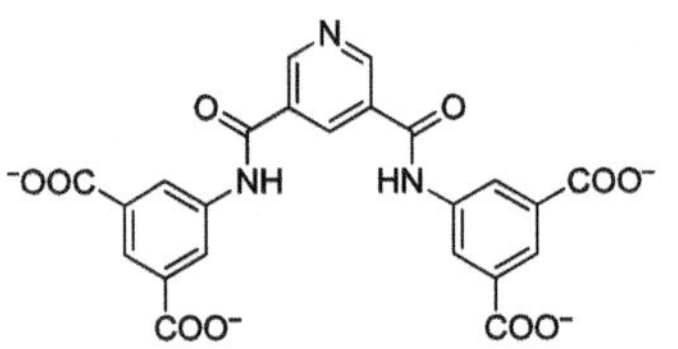

(7)

The first step of the tandem reaction studied consisted of the acid-catalyzed deacetalization of dimethoxymethylbenzene to give benzaldehyde. The second step produced benzylidene malononitrile through the Knoevenagel reaction between benzaldehyde and malononitrile. Control experiments proved that both open Cu^{2+} sites and amide groups are essential for the tandem reaction, and they work cooperatively.

We have recently drawn the attention to an intriguing point.[12] Sometimes we might be dealing with a bifunctional catalyst without being aware of it. We demonstrated that this could be the case of the zinc aminoterephthalate IRMOF-3. If we look into the "*ideal*" structure of the MOF, we only identify the amino groups of the linkers as the catalytic function, since the Zn^{2+} ions are completely blocked by the ligands and in principle are not accessible to catalysis. However, we found that in the "*real*" MOF, and depending critically on the synthesis conditions, a non negligible concentration of framework defects can be present, along with zinc oxide and hydroxide nanoparticles that may be formed during the MOF synthesis. These defect species, which in principle should not be present in the MOF, can contribute with Lewis acidity and, together with the amino groups of the linkers, lead to an unexpected bifunctional acid–base catalyst. Although the introduction of a (defective) second catalytic functionality to the material may seem advantageous for certain reactions, as we showed for the Knoevenagel condensation reaction, we must take this into account when dealing with other reactions where the Lewis sites can activate unwanted side reactions. In this latter case, preparation procedures and manipulation of the MOFs have to be carefully considered to generate or to avoid these second types of sites.

14.2.1.3 Bifunctional Metal–Metal Systems

Arnanz et al.[13] have used the coordinatively unsaturated Cu^{2+} sites of CuBTC as anchoring points for introducing either 4-aminopyridine or a pyridine terminated Schiff base complex, followed by palladium coordination, as shown in Scheme 14.8.

The resulting bimetallic Cu-Pd MOF was found to catalyze the tandem Sonogashira/click reaction starting from 2-iodobenzyl bromide, sodium azide and different alkynes, leading to the one-pot synthesis of a series of triazolo[5,1-a]isoindoles (Scheme 14.9).

Lin and co-workers[14] reported on the preparation of a bifunctional MOF containing tetranuclear Zn clusters as Lewis acid centers and a chiral Mn-salen metalloligand. With this material, the authors studied the formation of a chiral epoxide directly from achiral substrates catalyzed by the Mn-salen complex, followed by the acid catalyzed ring-opening with trimethylsilyl azide (see Scheme 14.10). Yields of the final ring-openend product of up to 60% with enantioselectivities of up to 81% were obtained. Moreover, the ring-opening

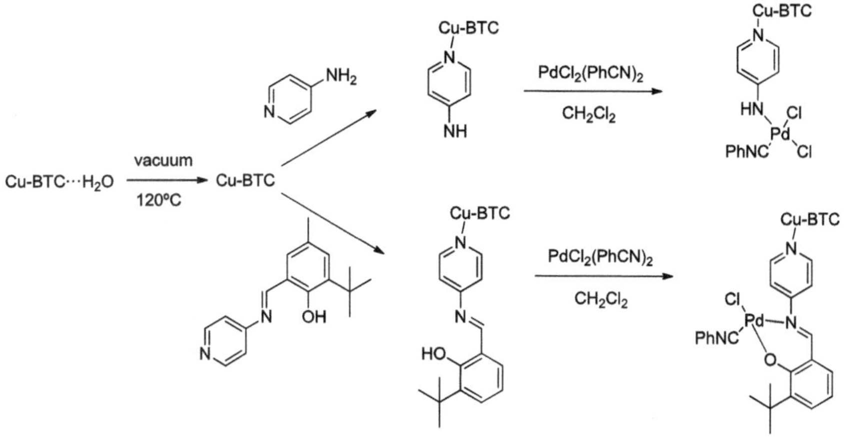

Scheme 14.8

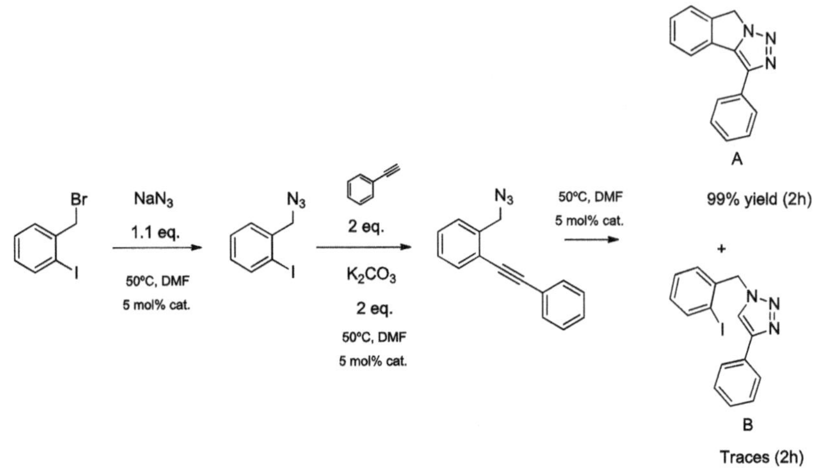

Scheme 14.9

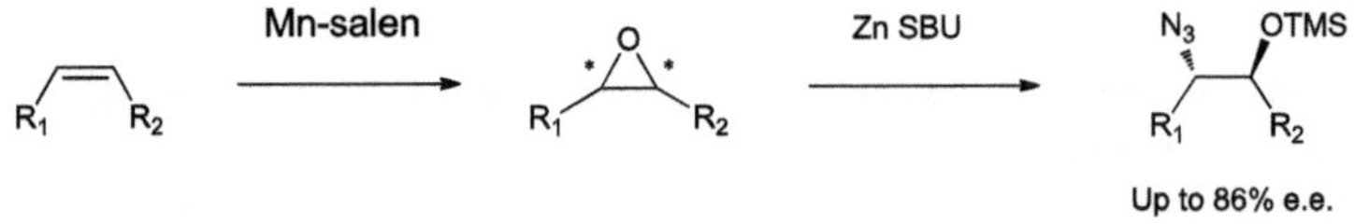

Scheme 14.10

step was highly regioselective, with only one pair of enantiomers being formed of the four possible.

14.2.2 MOFs for Multicomponent Coupling Reactions

Multicomponent coupling reactions (MCRs) are also reactions occurring in one reaction vessel and involve more than two starting reagents that form a single product which contains the essential parts of the starting materials. MCR procedures are powerful tools in modern drug discovery processes, providing an important source of molecular diversity by systematically using variants of each of the components involved in the reaction. Moreover, the simple experimental procedures and its one-pot character, make MCRs highly suitable for automated and high throughput generation of organic compounds.

MOFs could become very interesting catalysts for MCRs due to the possibility to add shape and/or transition state selectivity and the fine tuning of the electronic properties and (chiral) environments of the metal sites. This provides unprecedented privileged systems in which the flexibility of design of metal coordination complexes is combined with the advantages of heterogeneous catalysts. In this section we will revise some of the successful examples on the use of MOFs as catalysts for MCRs described so far.

The three-component coupling of amines, alkynes and aldehydes (the so called A[3] reaction) leads to propargylamines, according to Scheme 14.11. By

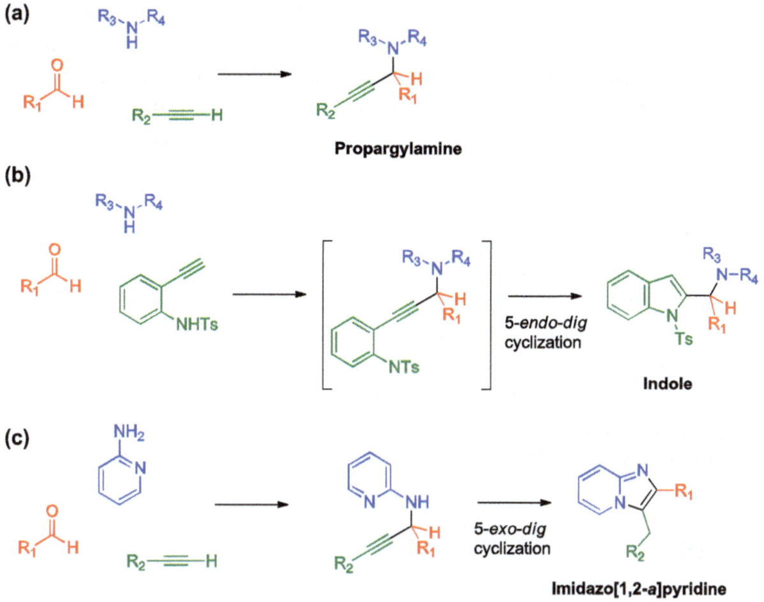

Scheme 14.11

properly selecting the substrates of this reaction, these propargylamines can be further converted into other interesting fine chemicals, such as indoles (Scheme 14.11b) and imidazo[1,2-*a*]pyridines (Scheme 14.11c). We have reported that this MCR is successfully catalyzed by various MOFs. First, we demonstrated the use of a MOF containing a Au^{3+}-Schiff base complex for the synthesis of indoles through an A^3 reaction.[15] We found that the Schiff base complex was very effective in stabilizing cationic gold species, avoiding their spontaneous reduction to metallic Au^0 and the corresponding loss of activity. Therefore, the Au-MOF was found to be very effective for the A^3 reaction.

We later extended our study to various Cu^{2+}-containing MOFs for the synthesis of propargylamines, indoles and imidazopyridines.[16] We found particularly good results for the synthesis of imidazopyridines with the lamellar [Cu(BDC)•DMF] MOF, surpassing the results obtained with other homogenous catalysts while allowing the reusability of the catalyst. Furthermore, when working with homogeneous copper catalysts, it is usually necessary to work under an inert atmosphere to avoid the Glaser coupling reaction between two terminal alkynes, which is a side reaction that can compete with the formation of the propargylamine. However, we found that the Cu-MOFs studied were not able to catalyze the Glaser coupling, and therefore the use of an inert atmosphere was not required. This is a further advantage of the Cu-MOFs with respect to homogenous catalysts.

Later, other studies on MOF catalyzed A^3 coupling reactions were also reported by Jayaramulu *et al.*,[17], Yang *et al.*[18] and Liu *et al.*[19]

Bromberg *et al.* have recently reported on the use of MIL-101 containing encapsulated phosphotungstic acid (PTA) as catalyst under microwave irradiation for various three-component coupling reactions leading to bioactive drug intermediates: synthesis of dibenzoxanthene by condensation of benzaldehyde and two molecules of 2-naphtol (Scheme 14.12a); and synthesis of 1-amidoalkyl-2-naphthol by coupling of benzaldehyde, 2-naphthol and acetamide (Scheme 14.12b).[20]

Li *et al.* have reported the use of two isostructural Zn^{2+} and Cd^{2+} MOFs prepared with the ligands 1,2-bis(4-pyridyl)ethylene and 1,3-benzenedisulfonic acid, for the synthesis of dihydropyrimidinone derivatives through the Biginelli three-component coupling reaction (Scheme 14.13).[21]

One year later, the same group also reported the use of these two MOFs as catalysts for the three component coupling reaction of aldehydes, malononitrile, and thiophenols leading to the one-pot synthesis of 2-amino-3,5-dicarbonitrile-6-thio-pyridines (Scheme 14.14).[22]

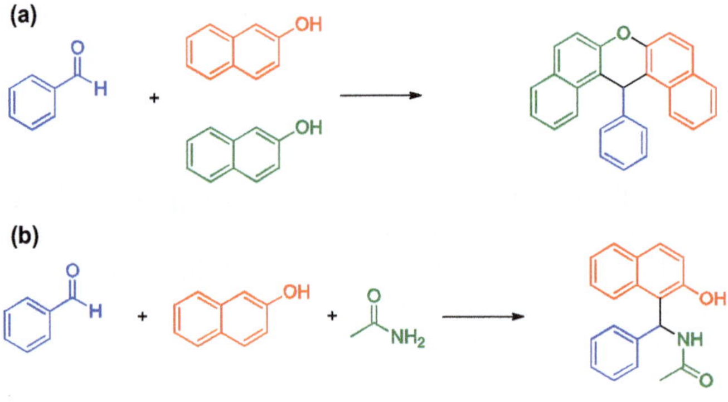

Scheme 14.12

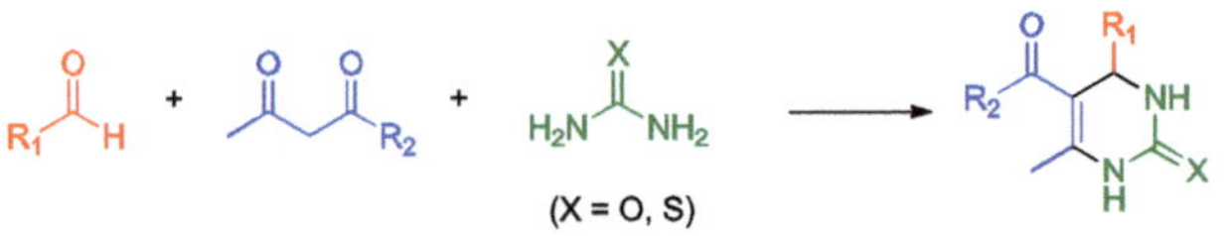

Scheme 14.13

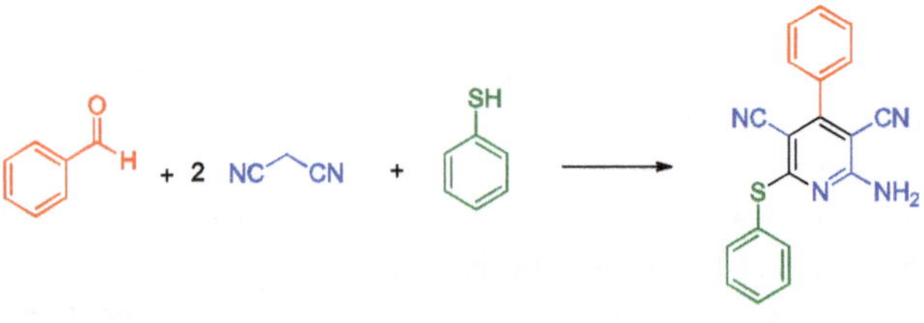

Scheme 14.14

Although this is only the beginning, we are witnessing an increasing interest in developing new catalytic applications of MOFs, focused on the synthesis of high added value products and designing new, more efficient, one-pot synthesis procedures. In this sense, it is evident that the use of MOFs as multifunctional catalysts and in MCRs will certainly trigger thrilling research.

14.3 From the Lab Scale to Application: MOF Formulation as Key Step

Most industrial heterogeneous catalysts are used in the form of tablets, rings, spheres, extrudates or limps that are prepared under relatively high pressures.[23] On the one hand, in order to secure a high activity per unit volume and to avoid diffusion limitations, catalysts need to be prepared in a very fine state of subdivision: crystallites from 5 to 500 Å diameter are common in industrial catalysts. On the other hand, these crystals must be brought into contact with the reagent gas or liquid and then separated from the product while producing a minimum pressure drop in fixed bed reactors and avoiding catalyst attrition in slurry operation, as clearly exemplified for IRMOF-3 in Figure 14.1. Because of these reasons, crystallites are usually aggregated into bigger particles (50 μm to 10 mm) that must be porous to allow diffusional traffic of reagents and products.[24]

As one would expect, after formulation of a catalyst particle, the size distribution of the voids or pores will be dependent on the processes of preparation and will critically affect catalyst performance. Moreover, some intrinsic catalyst porosity might be compromised during the formulation procedure. Roughly, three sorts of pore may exist: (i) those corresponding to the crystal lattice (that survived the formulation), (ii) those formed by the spaces between aggregated particles, which usually range in the 20–500 Å and (iii) those formed by compaction of a powder made from aggregates of these precipitates in, for instance, a tabletting machine. The spaces left between the crushed together aggregates usually falls in the range 500–2000 Å. The porosity of the final catalyst is of the highest importance, since diffusion limitations

Figure 14.1 SEM micrographs of a IRMOF-3 crystal before (left) and after (right) slurry catalytic operation under stirring.

might influence selectivity, especially in those cases where the targeted product is intermediate in a series of consecutive reactions.

Another important factor in shaping is the stability or hardness of a shaped body. Normally the stability correlates with the pressure used to from the shaped body. The hardness of the body is closely associated with its stability. On one hand a stable shaped body is desired, on the other hand the necessary compression to obtain the shaped body will decrease the active surface area.

From the above paragraphs, it is easy to realize that formulation is one of the critical steps before industrial implementation of a catalyst can be made possible. Surprisingly, outside of the vast patent literature there is little publicly available information on the formulation of industrial heterogeneous catalysts. Moreover, in spite of the importance of this step, in most cases industrial catalysts are shaped following trial and error procedures based on experience rather than on understanding.

When it comes to metal organic frameworks, the compaction of powders into the desired shape by pressing might result in a considerable decrease in surface area due to destruction of the crystalline structure. At this point, the mechanical behavior of the MOF, the way a given framework behaves under pressure, becomes very important. Surprisingly, despite the apparent importance of the research into the mechanical behavior of MOFs, only a small number of studies on the mechanical properties of MOFs have been reported to-date. Chapman *et al.* described the change in lattice parameters of CuBTC upon pressurization by different guest molecules.[25] Later, Spencer *et al.*, studied the phase transition of a zinc imidazolate in the range of 0.5–0.8 GPa.[26] The compressibility and bulk moduli of single crystals and polycrystalline MOF powders have been mainly studied by high-pressure X-ray crystallography in a diamond anvil cell.[27,28] Amorphization in most studied MOFs has been observed at pressures as low as 0.34 GPa for ZIF-8[6] and 3.5 MPa for MOF-5.[29] The reversible pressure-induced amorphization of ZIF-4 was reported to occur at 6.5 GPa irrespective of pore occupancy and takes place *via* a high-pressure phase (formed at 3.7 GPa). On the other hand, based on the pressure-induced changes in lattice volume obtained from high pressure X-ray diffraction, the bulk moduli (K_0) of MOFs were determined from the Birch–Murnaghan equation of state,[30] with values varying from 3 to 30 GPa, mostly depending on the density of the studied MOF. More recently, Gascon *et al.*[31] reported that in the case of flexible MOFs, the compressibility is much higher, with NH_2-MIL-53(In) amorphisizing only at pressures higher than 20 GPa. Figure 14.2 summarizes the bulk modulus *versus* physical density maps for metal organic frameworks plotted alongside major classes of materials (adapted from ref. 32). As can be seen in the figure, most MOFs are softer than zeolites and very likely to collapse under typical pressures used during pelletization. This may be explained by the high pore volume resulting in a high fragility of the organic framework structure in comparison to other crystalline nano-structured materials like zeolites.

As discussed above for industrial catalysts in general, in the case of MOFs it is also the case that outside patent literature[33–35] there is little publicly available

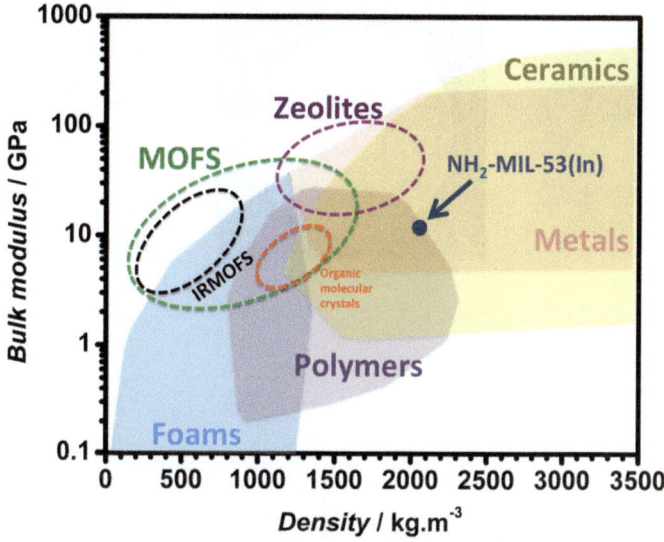

Figure 14.2 Bulk modulus *versus* physical density map for NH$_2$-MIL-53(In), plotted alongside major classes of materials (adapted from ref. 31 and 32).

information on formulation, with BASF being several steps ahead in this field of research. Already in 2006 BASF scientists claimed that *"shaping of metal–organic framework powders into industrially widespread geometries of mm-sized tablets, extrudates, honeycombs, etc. can be performed on MOFs as well without any major obstacle"*.[36] In US patent 2009/0155588,[11] the inventors claim that *"it is found that even though the surface area of a shaped body with a certain weight is lower than that of a respective amount of powder the situation is completely converted in a situation where surface areas are compared at a predetermined volume. The predominating effect of destroying 3-dimensional structures during the conversion of the powder into a shaped body is surprisingly "delayed" in such a way that firstly the ratio surface area per volume of the shaped body containing a MOF to that of the respective powder increases before the 3-D destroying effect resulting from the shape forming step prevails and the abovementioned ratio decreases in the expected manner. As a consequence, shaped bodies can be prepared according to the present invention by a process comprising the step of converting a MOF containing powder into the shaped body, wherein the ratio of the surface area per volume of the shaped body to the surface area per volume of the powder is greater than 1.6"*. CuBTC bodies prepared according to this patent were later tested by Rodrigues and co-workers (see Figure 14.3)[37] in the separation of propane/propylene mixtures. Results demonstrate that even in case of such a fragile MOF, pellets with similar adsorptive behavior as the parent material can be manufactured. Whether this method can be extended to other MOF topologies with different mechanical properties is still to be unraveled.

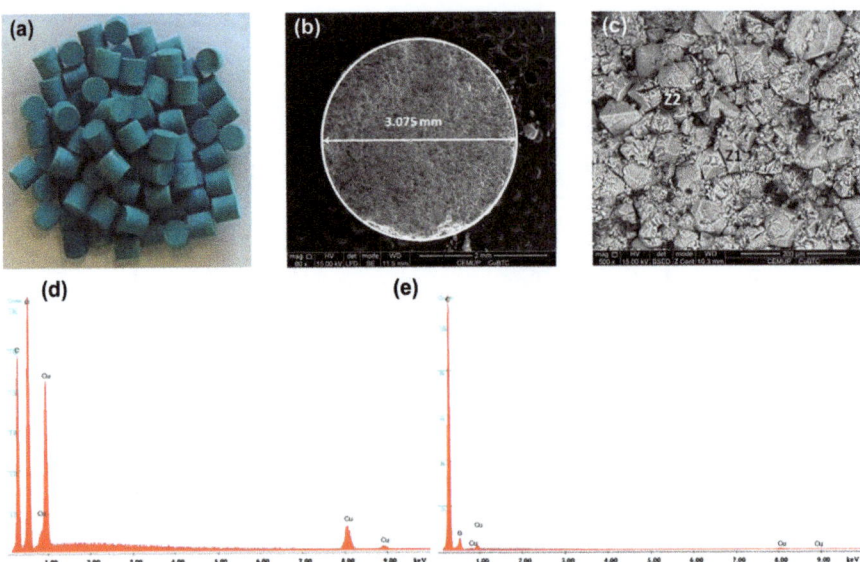

Figure 14.3   (a) Photograph of hydrated Cu-BTC tablets; (b) SEM micrograph of a
                   pellet at 60× magnifications; (c) SEM micrograph of the sample at 500×
                   magnifications; EDS analysis of (d) Cu-BTC crystal (Z1 in c) and (e)
                   binder (Z2 in c).
                   Reproduced from ref. 37 with permission from Elsevier.

A very elegant way to avoid serendipity and possible framework collapse
during catalyst formulation is the use of structured reactors or supports: a
structured reactor contains an structured internal which can be made out of
ceramics, metals or carbon, placed inside a reactor. It can be considered as an
intensified form of a randomly packed bed reactor. A monolith is an example of
a structured reactor; in fact, the borderline between catalyst and reactor vanishes
for this application. The advantage of a structured reactor is that it may be
designed in full detail up to the local surroundings of the catalyst, allowing
ultimate precision. Structured reactors show flexibility with respect to different
length scales, *i.e.*, diffusion lengths, voidage *etc.* (Figure 14.4). They effectively
allow the decoupling of intrinsic reaction kinetics, transport phenomena and
hydrodynamics. Structured internals such as monoliths can be coated with
catalysts, supports or their precursors. Immobilization of MOFs into supports
can offer several advantages over the use of self-supported pellets, especially when
one considers that the most promising MOF catalytic applications involve
reactions in the liquid phase and slurry operation, where catalyst pellets are not
only more likely to collapse but also more likely to provoke diffusion limitations.
In contrast, when using a coated catalyst, due to the short diffusion distances, the
catalytically active material can be utilized more efficiently.[38-41] Once more, the
moderate thermal stability of MOFs does not allow for common immobilization
techniques applied in structured reactors, namely washcoating followed by thermal
compaction. Therefore, alternative preparation routes are needed.

Figure 14.4 A MIL-101(Cr) coated monolith.
Figures adapted from ref. 42.

The immobilization of MOFs into different structured supports, mostly for application in gas separation is a relatively mature field of research.[43] After some early works on MOF dense coatings[44,45] on porous substrates and layer by layer deposition of MOFs on non porous supports,[46,47] the first MOF membranes displaying separation properties different from Knudsen diffusion control were reported by the groups of Tsapatsis[48] and Caro.[49,50] At the same time, the first self supported MOF membrane was reported by Guo and coworkers.[51] In the latter, a HKUST-1 membrane was produced by hydro-thermally oxidizing part of a copper wire mesh, creating a local supersaturation that promoted the formation of the MOF material. This filled up the space in between the mesh wires, resulting in a kind of reinforced self-supported membrane. This work, together with the electrochemical synthesis of MOF coatings reported by Ameloot *et al.*,[52] clearly exemplify some of the unique possibilities that MOFs offer for the facile formation of dense coatings. Other synthetic methods utilized vary from direct hydrothermal synthesis[53] to the use of secondary growth in combination with microwave heating[50] and the use of covalent linkers like 3-aminopropyltriethoxysilane to favor crystal attachment to a support.[54] At the same time, some groups have made use of the intrinsic properties of MOFs to promote the growth of dense coatings: Hu *et al.* used the porous support as the inorganic source reacting with the organic precursor to grow a seeding layer used afterwards in the synthesis of MIL-53(Al) membranes.[55] Along the same line, Jeong *et al.* developed a surface modifi-cation involving the hot treatment of the support with the methyl imidazole linker used for the synthesis of ZIF-8 membranes.[56]

In spite of the already vast MOF membrane literature, to the best of our knowledge only a few works report on the use of immobilized MOFs in catalytic applications.[42,57,58] In pioneering work, Ramos-Fernandez *et al.*[42] reported the synthesis, *via* secondary growth, of MIL-101(Cr) monoliths. The resulting structured catalysts where applied in a stirring monolithic reactor configuration in the oxidation of tetraline. The catalysts could be recovered and re-used for tens of times, experiencing neither catalyst attrition nor leaching. In the same spirit, Aguado and co-workers have reported the immobilization of the so called SIM-1 (a porous zinc carbonylimidazolate) by direct hydrothermal treatment in both monolithic supports and Al_2O_3 beads and the application of the resulting structured support in ketone transfer hydrogenations and Knoevenagel condensations.[57,58] Even when using the Al_2O_3 beads under vigorous magnetic stirring (400 rpm), no notable weathering was observed and once the beads could easily be separated from the reaction mixture by removing the solution.

Summarizing, MOF shaping is of the upmost importance for their application in catalysis. As shown above, neither pelletization nor immobilization into structured supports seem to be insurmountable barriers to overcome. However, much more work is needed in this area in order to answer already long-standing questions such as *"how many times can a MOF catalyst be recycled?"*.

References

1. L. Tang, L. Shi, C. Bonneau, J. Sun, H. Yue, A. Ojuva, B.-L. Lee, M. Kritikos, R. G. Bell, Z. Bacsik, J. Mink and X. Zou, *Nature Mater.*, 2008, **7**, 381–385.
2. J. L. Sun, C. Bonneau, A. Cantin, A. Corma, M. J. Diaz-Cabanas, M. Moliner, D. L. Zhang, M. R. Li and X. D. Zou, *Nature*, 2009, **458**, 1154–U1190.
3. R. C. Rouse and D. R. Peacor, *Am. Mineral.*, 1986, **71**, 1494–1501.
4. Y. Pan, B. Yuan, Y. Li and D. He, *Chem. Commun.*, 2010, **46**, 2280–2282.
5. F. Neatu, S. Coman, V. Parvulescu, G. Poncelet, D. E. De Vos and P. A. Jacobs, *Top. Catal.*, 2009, **52**, 1292–1300.
6. A. Corma and M. Renz, *Chem. Commun.*, 2004, 550–551.
7. F. G. Cirujano, A. Leyva-Pérez and A. Corma, and F. X. Llabrés i Xamena, *ChemCatChem*, 2013, **5**, 538–549.
8. M. Pintado-Sierra, A. Rasero-Almansa, A. Corma, M. Iglesias and F. Sanchez, *J. Catal.*, 2013, **299**, 137–145.
9. F. Vermoortele, R. Ameloot, A. Vimont, C. Serre and D. E. De Vos, *Chem. Commun.*, 2011, **47**, 1521–1523.
10. R. Srirambalaji, S. Hong, R. Natarajan, M. Yoon, R. Hota, Y. Kim, Y. H. Ko and K. Kim, *Chem. Commun.*, 2012, **48**, 11650–11652.
11. J. Park, J.-R. Li, Y.-P. Chen, J. Yu, A. A. Yakovenko, Z. U. Wang, L.-B. Sun, P. B. Balbuena and H.-C. Zhou, *Chem. Commun.*, 2012, **48**, 9995–9997.
12. F. X. Llabrés i Xamena, F. G. Cirujano and A. Corma, *Microporous Mesoporous Mater.*, 2012, **157**, 112–117.
13. A. Arnanz, M. Pintado-Sierra, A. Corma, M. Iglesias and F. Sanchez, *Adv. Synth. Catal.*, 2012, **254**, 1347–1355.
14. F. Song, C. Wang and W. Lin, *Chem. Commun.*, 2011, **47**, 8256–8258.

15. X. Zhang, F. X. Llabrés i Xamena and A. Corma, *J. Catal.*, 2009, **265**, 155–160.
16. I. Luz, F. X. Llabrés i Xamena and A. Corma, *J. Catal.*, 2012, **285**, 285–291.
17. K. Jayaramulu, K. K. R. Datta, M. V. Suresh, G. Kumari, R. Datta, C. Narayana, M. Eswaramoorthy and T. K. Maji, *ChemPlusChem*, 2012, **77**, 743–747.
18. J. Yang, P. Li and L. Wang, *Catal. Commun.*, 2012, **27**, 58–62.
19. L. Liu, X. Zhang, G. Jinsen and C. Xu, *Green Chem.*, 2012, **14**, 1710–1720.
20. L. Bromberg, Y. Diao, H. Wu, S. A. Speakman and T. A. Hatton, *Chem. Mater.*, 2012, **24**, 1664–1675.
21. P. Li, S. Regati, R. J. Butcher, H. D. Arman, Z. Chen, S. Xiang, B. Chen and C.-G. Zhao, *Tetrahedron Lett.*, 2011, **52**, 6220–6222.
22. M. Trimmaiah, P. Li, S. Regati, B. Chen and J. C.-G. Zhao, *Tetrahedron Lett.*, 2012, **53**, 4870–4872.
23. J. A. Moulijn, M. Makkee and A. van Diepen, *John Wiley & Sons, Chichester,* England, 2001.
24. S. P. S. Andrew, *Chem. Eng. Sci.*, 1981, **36**, 1431–1445.
25. K. W. Chapman, G. J. Halder and P. J. Chupas, *J. Am. Chem. Soc.*, 2008, **130**, 10524–10526.
26. E. C. Spencer, R. J. Angel, N. L. Ross, B. E. Hanson and J. A. K. Howard, *J. Am. Chem. Soc.*, 2009, **131**, 4022–4026.
27. S. A. Moggach, T. D. Bennett and A. K. Cheetham, *Angew. Chem., Int. Ed.*, 2009, **48**, 7087–7089.
28. K. W. Chapman, G. J. Halder and P. J. Chupas, *J. Am. Chem. Soc.*, 2009, **131**, 17546–17547.
29. Y. H. Hu and L. Zhang, *Phys. Rev. B*, 2010, **81**, 174103.
30. F. Birch, *J. Geophys. Res.*, 1952, **57**, 227–286.
31. P. Serra-Crespo, E. Stavitski, F. Kapteijn and J. Gascon, *RSC Adv.*, 2012, **2**, 5051–5053.
32. J. C. Tan and A. K. Cheetham, *Chem. Soc. Rev.*, 2011, **40**.
33. M. Hesse, U. Mueller and O. Yaghi, US Patent US2009155588-A1, 2009.
34. M. Hesse, U. Muller and O. M. Yaghi, US Patent US2006099398-A1, 2006.
35. J. S. Chang, Y. K. Hwang, Y. K. Seo, J. W. Yoon, Y. Ji Woong, J. Jong San, S. You Kyoung, H. Young Kyu, J. S. Jang, K. H. Young, S. J. Jong, K. S. You and W. Y. Ji, World Patent WO2009128636-A2, 2009.
36. U. Mueller, M. Schubert, F. Teich, H. Puetter, K. Schierle-Arndt and J. Pastre, *J. Mater. Chem.*, 2006, **16**, 626–636.
37. M. G. Plaza, A. F. P. Ferreira, J. C. Santos, A. M. Ribeiro, U. Müller, N. Trukhan, J. M. Loureiro and A. E. Rodrigues, *Microporous Mesoporous Mater.*, 2012, **157**, 101–111.
38. R. K. Edvinsson Albers, M. J. J. Houterman, T. Vergunst, E. Grolman and J. A. Moulijn, *AIChE J.*, 1998, **44**, 2459–2464.
39. I. Hoek, T. A. Nijhuis, A. I. Stankiewicz and J. A. Moulijn, *Chem. Eng. Sci.*, 2004, **59**, 4975–4981.

40. F. Kapteijn, J. J. Heiszwolf, T. A. Nijhuis and J. A. Moulijn, *CATTECH*, 1999, **3**, 24–41.
41. T. A. Nijhuis, A. E. W. Beers, T. Vergunst, I. Hoek, F. Kapteijn and J. A. Moulijn, *Catal. Rev. Sci. Eng.*, 2001, **43**, 345–380.
42. E. V. Ramos-Fernandez, M. Garcia-Domingos, J. Juan-Alcañiz, J. Gascon and F. Kapteijn, *Appl. Catal. A*, 2011, **391**, 261–267.
43. J. Gascon, F. Kapteijn, B. Zornoza, V. Sebastián, C. Casado and J. Coronas, *Chem. Mater.*, 2012, **24**, 2829–2844.
44. M. Arnold, P. Kortunov, D. J. Jones, Y. Nedellec, J. Karger and J. Caro, *Eur. J. Inorg. Chem.*, 2007, 60–64.
45. J. Gascon, S. Aguado and F. Kapteijn, *Microporous Mesoporous Mater.*, 2008, **113**, 132–138.
46. S. Hermes, F. Schroder, R. Chelmowski, C. Woll and R. A. Fischer, *J. Am. Chem. Soc.*, 2005, **127**, 13744–13745.
47. S. Hermes, D. Zacher, A. Baunemann, C. Woll and R. A. Fischer, *Chem. Mater.*, 2007, **19**, 2168–2173.
48. R. Ranjan and M. Tsapatsis, *Chem. Mater.*, 2009.
49. H. Bux, F. Liang, Y. Li, J. Cravillon, M. Wiebcke and J. Caro, *J. Am. Chem. Soc.*, 2009, **131**, 16000–16001.
50. Y.-S. Li, F.-Y. Liang, H. Bux, A. Feldhoff, W.-S. Yang and J. Caro, *Angew. Chem., Int. Ed.*, 2010, **49**, 548–551.
51. H. Guo, G. Zhu, I. J. Hewitt and S. Qiu, *J. Am. Chem. Soc.*, 2009, **131**, 1646–1647.
52. R. Ameloot, E. Gobechiya, H. Uji-i, J. A. Martens, J. Hofkens, L. Alaerts, B. F. Sels and D. E. D. Vos, *Adv. Mater.*, 2010, **22**, 2685–2688.
53. S. Aguado, J. Gascon, D. Farrusseng, J. C. Jansen and F. Kapteijn, *Microporous Mesoporous Mater.*, 2011, **146**, 69–75.
54. A. Huang, H. Bux, F. Steinbach and J. Caro, *Angew. Chem., Int. Ed.*, 2010, **49**, 4958–4961.
55. Y. Hu, X. Dong, J. Nan, W. Jin, X. Ren, N. Xu and Y. M. Lee, *Chem. Commun.*, 2011, **47**, 737–739.
56. M. C. McCarthy, V. Varela-Guerrero, G. V. Barnett and H.-K. Jeong, *Langmuir*, 2010, **26**, 14636–14641.
57. S. Aguado, J. Canivet and D. Farrusseng, *Chem. Commun.*, 2010, **46**, 7999–8001.
58. S. Aguado, J. Canivet and D. Farrusseng, *J. Mater. Chem.*, 2011, **21**, 7582–7588.

Subject Index

catalysis, 25
    by ancillary ligands, 303–304
    by metal-containing
      COFs, 391–398
    at organic ligands, 251–258,
      294–304
      binapthyl-based linkers,
        301–302
      catalysis by ancillary
        ligands, 303–304
      metalloporphyrin-based
        linkers, 298–300
      organocatalysis from
        linkers, 302–303
      salen-based linkers,
        294–298
catalytic testing, 279–280
catenation, 293
cavitation leads, 12
C–C multiple bonds, 46–49
cethyltrimethylammonium bromide
    (CTAB), 23
chiral molecule, 344
CI doubles (CID), 217
CI singles (CIS), 217
Claisen rearrangement, 51
CO adsorption, 100–103
complete active space (CAS),
    216
complete active space SCF
    (CASSCF), 216–217
computational modeling of
    MOF, 230–232
concurrent tandem catalysis (CTC),
    290, 302
configurational interaction (CI),
    216–217
confocal fluorescence
    microscopy, 377
conjugated microporous polymer
    (CMP), 388
Co-phthalocyanine-PIM, 394
correlation energy, 216
coupled cluster (CC)
    doubles (CCD), 218
    method, 217–218

    singles, doubles and triples
      (CCSDT), 218
    singles and doubles (CCSD),
      218
covalent organic framework
    (COF), 4, 384
    with metal nanoparticles and
      clusters, 391–393
    with molecular metal
      species, 393–398
    ordered, 385–388
    in organocatalysis
      (metal-free), 398–402
    unordered, 388–389
covalent post-synthetic
    modification, 32, 36–54
    alcohols transformation, 46
    amines transformation, 37–44
    azides transformation, 44–45
    bridging ligands, reduction and
      oxidation, 53–54
    C–C multiple bonds
      transformation, 46–49
    deprotection reactions, 52–53
    reactions at heteroatoms, 51–52
    transformations at carbon,
      49–51
covalent triazine-based framework
    (CTF), 386
CPO-27-Ni, 95–96, 117–119, 183–187
critical size, 10
cross-coupling reaction, 316
crystal growth, 23–24
crystallization
    aspects in synthesis of solid
      compounds, 11
    mechanisms and methods,
      10–13
    and MOF synthesis, 20–22
    principles, 9–10
    process, 11
crystallographic unit cell, 155
Cu-based MOF STAM-1
    synthesis, 17
cyanosilylation reactions, 351–352
cyclobutane rings, 48